BEHAVIOURAL PHARMACOLOGY OF 5-HT

Edited by

Paul Bevan
Duphar, B.V., The Netherlands

Alexander R. Cools
Katholieke Universiteit, The Netherlands

Trevor Archer
University of Gothenburg, Sweden

LONDON AND NEW YORK

First published 1989 by
Lawrence Erlbaum Associates, Inc.

2 Park Square, Milton Park, Abingdon, Oxon OX14 4RN
711 Third Avenue, New York, NY 10017, USA

*Routledge is an imprint of the Taylor & Francis Group,
an informa business*

First issued in paperback 2016

Transferred to Digital Printing 2009 by Psychology Press

Copyright © 1989 Taylor & Francis

All rights reserved. No part of this book may be reprinted or reproduced or utilised in any form or by any electronic, mechanical, or other means, now known or hereafter invented, including photocopying and recording, or in any information storage or retrieval system, without permission in writing from the publishers.

Notice:
Product or corporate names may be trademarks or registered trademarks, and are used only for identification and explanation without intent to infringe.

Library of Congress Cataloging in Publication Data

Behavioral Pharmacology of 5-HT / edited by Paul Bevan, Alexander R. Cools, Trevor Archer.
      p.   cm.
  Includes bibliographies and index.
  ISBN 0-8058-0135-9
  1. Serotonin--Psychotropic effects. I. Bevan, Paul L. T.
II. Cools, Alexander Rudolph, 1941- . III. Archer, Trevor.
  [DNLM: 1. Behavior--drug effects. 2. Serotonin--pharmacology.
QV 126 B419]
RM666.S46B44 1989
615'.78--dc20
DNLM/DLC
for Library of Congress   89-1439
                                                       CIP

Publisher's Note
The publisher has gone to great lengths to ensure the quality of this reprint but points out that some imperfections in the original may be apparent.

ISBN 978-0-8058-0135-4 (hbk)
ISBN 978-1-138-96452-5 (pbk)

TABLE OF CONTENTS

SECTION	I	Introduction A.R. Cools	1
	1	Behavioural pharmacology of 5–HT: an introduction A.R. Green	3
	2	Is rotational behaviour in the rat a $5-HT_{1A}$ or $5-HT_{1B}$ mediated response? T.P. Blackburn, D.A. Martin, P. Slater	21
	3	Serotonin–mediated behaviour following central or peripheral administration of the selective NK–3 tachykinin agonist senktide A.J. Stoessl, C.T. Dourish, S.D. Iversen	27
SECTION	II	5–HT and sexual behaviour P. Bevan	33
	4	New aspects on the serotonergic modulation of male rat sexual behaviour S. Ahlenius, K. Larsson	35
	5	Differential roles of 5–HT receptor subtypes in female sexual behaviour S.D. Mendelson, B.B. Gorzalka	55
	6	Hypothalamic sites of action of the dual effect of 5–HT on female sexual behaviour in the rat M.D. James, S.M. Lane, D.R. Hole, C.A. Wilson	73
	7	$5-HT_{1A}$ and $5-HT_{1B}$ agonists produce opposite effects on the ejaculatory response induced by amphetamine in rats L. Rényi & T. Lewander	79
	8	Functional interplay of serotonin (5–HT)–receptor subtypes H.H.G. Berendsen, R.J.M. Smets, C.L.E. Broekkamp	83
SECTION	III	5–HT & aggression P. Bevan	87
	9	Modulatory action of serotonin in aggressive behaviour B. Olivier, J. Mos, M. Tulp, J. Schipper and P. Bevan	89
	10	Brain 5–HT system and inhibition of aggressive behaviour K.A. Miczek, P. Donat	117

	11	Differential effects of benzodiazepines and 5–HT$_{1A}$ agonists on defensive patterns in wild rattus D.C. Blanchard, K. Hori, R.J. Rodgers, C.A. Hendrie, R.J. Blanchard	145
	12	Competition for sucrose–pellets in triads of male, Wistar rats, the effects of eight 5–HT–agonists C. Gentsch, M. Lichtsteiner, H. Feer	149
SECTION IV		Depression and psychosis A.R. Cools	153
	13	Towards a theory of serotonergic dysfunction in depression P. Willner	157
	14	5–HT receptor subtypes in depression J.F.W. Deakin	179
	15	Modulation of the 5–HT system and antipsychotic activity J.A.M. van der Heyden	205
	16	Mechanism of action of 8–OH–DPAT on a rat model for human depression G.A. Kennet, G. Curzon	225
	17	Reversal of helpless behaviour in rats by serotonin uptake inhibitors P. Martin, A.M. Laporte, P. Soubrie, S. El Mestikawy, M. Hamon	231
	18	Ritanserin (R 55 667), an original thymosthenic T.F. Meert, C.J.E. Niemegeers, Y.G. Gelders, P.A.J. Janssen	235
	19	The abilities of 5–HT$_3$ antagonists to inhibit the hyperreactivity caused by dopamine infusion into the nucleus accumbens of the rat N.M. Barnes, B. Costall, A.M. Domeney, M.E. Kelly, R.J. Naylor, M.B. Tyers	239
	20	5–HT$_2$ antagonists increase tactile startle habituation in an animal model of a habituation deficit in schizophrenia. M.A. Geyer	243
SECTION V		Feeding behaviour A.R. Cools	247
	21	Pharmacological evidence for the involvement of serotonergic mechanisms in human feeding E. Goodall, T. Silverstone	249

22 Serotonin and feeding
    R. Samanin   259

23 Opposite effects of 5-HT$_{1A}$ and 5-HT$_{1B/1C}$
    agonists on food intake
    P.H. Hutson, G.A. Kennett, T.P. Donohoe,
    C.T. Dourish, G. Curzon   283

24 Effects of selective serotonergic agonists on
    palatability-induced ingestion
    J.D. Leander   287

25 8-OH-DPAT reliably increases ingestion of solid
    but not liquid diets
    A.M.J. Montgomery, P. Willner, R. Muscat   291

26 Pharmacological manipulation of 5-HT: Effect on
    intake of diets supplemented with either sweet or
    bland carbohydrates
    C.L. Lawton, J.E. Blundell   295

SECTION VI 5-HT, pain and anxiety
    T. Archer   299

27 The behavioural pharmacology of serotonin in pain
    processes
    O-G. Berge, C. Post and T. Archer   301

28 The relationship between various animal models of
    anxiety, fear-related psychiatric symptoms and
    response to serotonergic drugs
    C.L. Broekkamp, F. Jenck   321

29 5-HT$_{1A}$ receptors: a bridge between anxiety and
    depression?
    P. Soubrie   337

30 A review of the evidence supporting the anxiolytic
    potential of 5-HT$_3$ receptor antagonists
    M.B. Tyers   353

31 Ultrasonic vocalizations by rat pups as an animal
    model for anxiolytic activity: effects of
    serotonergic drugs
    J. Mos, B. Olivier   361

32 Effects of the 5-HT$_{1A}$ agonist ipsapirone on the
    behavioural, endocrine and neurochemical responses
    to conditioned fear
    S.A. Lorens, H. Mitsushio, L.D. Van de Kar   367

33 Buspirone, ipsapirone and 5-HT$_{1A}$-receptor
    agonists show an anxiogenic-like profile in the
    elevated plus-maze
    P. Moser   371

34 Antianxiety effect of various putative 5–HT$_1$ receptor agonists on the conditioned defensive burying paradigm
A. Fernandez–Guasti, E. Hong ... 377

35 The anxiolytic activities of the 5–HT$_3$ receptor antagonists GR38032F, ICS 205–930 and BRL 43694
B. Costall, A.M. Domeney, P.A. Gerrard, B.J. Jones, M.E. Kelly, N.R. Oakley, M.B. Tyers ... 383

36 Behavioural effects of 5–HT$_3$ antagonists in animal models for aggression, anxiety and psychosis
J. Mos, J.A.M. van der Heyden, B. Olivier ... 389

SECTION VII 5–HT: Drug discrimination and aversion
T.U.C. Järbe and T. Archer ... 397

37 Receptor mechanisms of the discriminative stimulus properties of putative serotonin agonists
W. Koek, F.C. Colpaert ... 407

38 Serotonin and aversion.
F.G. Graeff ... 425

39 Stimulus properties of arylpiperazine second generation anxiolytics
R.A. Glennon, R. Young, M.E. Pierson ... 445

40 Behavioural, neurochemical and discriminative stimulus effects of spiroxatrine, a putative 5–HT$_{1A}$ receptor antagonist
J.E. Barrett, S.N. Olmstead, Z. Lei, C. Harrod, S.M. Hoffmann, K. Glover, M.A. Nader, B.A. Weissman ... 449

41 Rapid training of the stimulus properties of selective 5–hydroxytryptamine$_{1A}$ agonists
I. Lucki, J.A. South ... 453

SECTION VIII Behavioural pharmacology of 5–HT: Possible progress in the laterality of concepts
P. Bevan, S. Lorens and T. Archer ... 459

Author Index   475
Subject Index   503

PREFACE

This volume consists of 16 full-length chapters devoted to selected topics on the behavioural pharmacology of 5-HT. To date we know of no single book aimed solely at an analysis of the state of knowledge on the functional aspects of this neurotransmitter system in the central nervous system. The chapters presented, whether main or auxilliary, discuss a wide range of behavioural characteristics that are modulated by 5-HT neurotransmission. Obviously, the reader will note that perhaps not all aspects of 5-HT and behaviour are covered, or that some are dealt with more superficially than others. In spite of the constraints inherent in an enterprise of this undertaking, the book opens exciting new approaches to the interdisciplinary study of behaviour and pharmacology with special reference to ethology, endocrinology, neuroanatomy and comparative aspects of drug action. It will be noted that much of the book is devoted to new developments in therapeutic drugs of the future. Thus, in this context it will be suitable to consider the various formal topics into which the volume is divided: Anxiety, depression, feeding, sexual behaviour, aversions, pain and drug discrimination. The auxilliary chapters adjoining each of the main chapters provide new and innovative methods or results that offer exciting and up-to-date insights into the actions of 5-HT.

The book is based largely on the proceedings of the symposium "Behavioural Pharmacology of 5-HT" held in Amsterdam in October 1987. The meeting was sponsored by several pharmaceutical companies including Astra, Beecham, Duphar B.V., Duphar Nederland B.V., Organon International B.V., Organon Nederland B.V., Sandoz Nederland, Servier, Campden Instruments Ltd., London.

P. Bevan
A. Cools
T. Archer

INTRODUCTION

A.R.Cools
Psychoneuropharmacological Research–Unit, Dept. Pharmacology, University of Nijmegen, P.O.Box 9101, 6500 HB Nijmegen, The Netherlands

This book contains a selected series of papers presenting a broad view of our current understanding of the behavioural pharmacology of 5-HT (serotonin). The vast accummulation of knowledge about serotonergic receptors and about their differential role in the control of various bodily processes and behaviour has made serotonin the current focus of interest in many concepts about aetiology, pathophysiology and pharmacotherapy of bodily, affective and mental diseases in man. The recent developments in this field are so encouraging that there is an urgent need to depict the actual state of the art. The present book is constructed to provide an up-to-date overview of these new data.

It was the discovery of the vasotonic action of a chemically unknown agent in the serum that has given birth to serotonin in the fifties. During the following thirty years it became evident that serotonin has a widespread distribution throughout the body and the brain. Knowledge about its function in controlling bodily processes and behaviour is rapidly growing. Studies analysing serotonin binding sites are immensely successful. Recent binding studies have led to the classification of more than five distinct serotonin binding sites. The success of these in vitro studies is so overwhelming that – for many scientists – the aim of the research is identified with the search for agents that interact with one or more subclasses of serotonergic binding sites. At this point it is necessary to incorporate knowledge of analogous developments in adjacent fields such as that emerging from the research on dopamine. A few years ago, the number of distinct dopamine binding sites was steadily increasing. Today, we know – and classical practice has proven it – that only few of the newly discovered agents that selectively interact with one or another subtype of dopamine binding sites are clinically more effective than those interacting with the classic ones.

Thus, clinical practice has challenged the concept of nature that would explain away the complex and reduce it to the simplicity of the smallest possible components. This insight underlines the notion that in vivo studies are crucial. In other words, behaviour pharmacology is an essential prerequisite for understanding the pharmacology of serotonin.

One of the major challenging features of the research on serotonin arises as soon as one focuses the attention on the basic properties of organs and brain structures in which serotonin is released. The following might illustrate this. Serotonin is known to be released in serotonergic synapses within the nucleus accumbens, viz. a mesolimbic brain structure that plays an essential role in the transformation of emotion into motion. This brain structure contains target sites for atypical and classic neuroleptics as well as target sites for atypical and tricyclic antidepressants. Many fascinating questions emerge: Are the serotonergic binding sites found in the nucleus accumbens indeed serotonergic receptors?

Are these sites/receptors similar or dissimilar to those found in the other terminal regions of serotonergic fibres such as the hippocampus, neostriatum and hypothalamus? Does the nucleus accumbens contain both presynaptic and postsynaptic serotonin receptors? Are these receptors similar or dissimilar to those located on the serotonergic cell bodies in the median and/or dorsal raphe nuclei? Does the nucleus accumbens contain target sites for serotonergic agents which are effective in animal models for depression, anxiety, psychosis, etc? Does serotonin in the nucleus accumbens play a role in the pathophysiology of bodily, affective and mental diseases in man? Given the mutual interplay between serotonergic fibres and those containing dopamine or noradrenaline one can even wonder whether patients suffering from psychomotor diseases which are currently treated with dopaminergic and/or noradrenergic drugs will also benefit from serotonergic agents. These and related questions form our present-day challenges.

The development of research in this area requires the interaction and exchange of knowledge between basic scientists and clinical researchers from many disciplines. For that reason, a symposium was held in Amsterdam, in November 1987, where invited guests, both basic scientists and clinical researchers, discussed in depth the behavioural pharmacology of serotonin. The programme was constructed to present an up-to-date overview of our current understanding of the function of serotonin in sexual behaviour, aggressive behaviour, feeding behaviour, drug discrimination, aversion, anxiety and psychosis. It was, to the knowledge of the organizers, the first major occasion when neurobiologists, neuropharmacologists, neurochemists, neurophysiologists, neuro-ethologists, neurologists, psychiatrists, psychologists and other researchers gathered together for three days of intensive meetings. The present book is based on this symposium. It is the intention of the editors that the contents constitute the most recent developments in this exciting frontier area of serotonin and behaviour.

BEHAVIOURAL PHARMACOLOGY OF 5-HT: AN INTRODUCTION

A. Richard Green
Astra Neuroscience Research Unit, 1 Wakefield Street London WC1N 1PJ, United Kingdom

Introduction

The term 'behavioural pharmacology' encompasses two rather different approaches used in neuropharmacology. The first approach is the use of a behavioural change as a marker of 5-hydroxytryptamine (5-HT) function. Use of this approach allows studies to be made of the ways that psychoactive compounds alter a particular response and could thus be altering the function of the neurotransmitter more generally. In this regard therefore the behaviour is essentially a bioassay system and the fact that a drug used for a particular clinical condition alters the response should never be taken to imply that the behaviour is a model of the clinical condition. However the results can be used to suggest that the drug might be producing a similar change in 5-HT function when given clinically and thus at least allow speculation that this change is involved in the therapeutic mechanism of action. The second approach involves study of physiological and psychological responses of experimental animals and the assessment, by the use of specific drugs, of the involvement of a specific neurotransmitter of neurotransmitter receptor sub-types in these responses.

Both approaches are vital in increasing our understanding of neurotransmitter function in the brain and the way that drugs can be used to treat pathological conditions, particularly since neurochemistry is, in general, unable to demonstrate whether changes in neurotransmitter biochemistry have functional importance.

The recent and continuing explosion of neurochemical information on the existence of 5-HT receptor sub-types has resulted in a marked increase in research activity on 5-HT behavioural pharmacology. This book reports on much of this work and this chapter is designed to introduce some of the approaches (predominantly the use of behavioural pharmacology as a bioassay system) and hopefully point out some of the problems.

5-HT receptor sub-types

Until the 1970's neurotransmitter receptor sub-types had been defined soley on the basis of their functional differences in response to the action of drugs upon them. Examples include the nicotine and muscarinic acetylcholine receptors, α- and β- (and of course subsequently β_1- and β_2-) adrenoceptors and the peripheral 5-HT D and M receptors. The advent of ligand-receptor binding in the 1970's changed this. Receptors started to be defined purely on the basis of differences in ligand binding characteristics. This proved to be dangerous since information was lacking on whether the "receptors" so characterised were functional receptors and receptor subtypes or merely "acceptors". That is, the data did not tell us whether the subtypes defined by ligand binding had different functional characteristics or were merely sites to which the ligands bound because of slightly different physico-chemical characteristics and were therefore not really reflections of receptor heterogeneity.

Worse still the possibility always existed that these "receptors" were sites with no physiological function as receptors at all. This is not to say that neuropharmacology was not advanced enormously by the use of ligand binding, merely that the enthusiasm for designating a binding site as a receptor was sometimes misplaced. A major example was the initial classification of at least 4 different dopamine receptors in the brain which has been followed by the certain identification of, at least currently, no more than 2 functional sub-types.

The suggestion that 5-HT had receptor sub-types was first made by Gaddum and Picarelli in 1957 using peripheral tissues. However in the brain initial suggestions on receptor heterogeneity were not made until the late 1970's (Peroutka & Snyder 1979; Leysen et al., 1978). Both groups used ligand binding studies but Peroutka and Snyder soon suggested that the two receptors ($5-HT_1$ and $5-HT_2$) could be functionally differentiated using behavioural pharmacology (Peroutka et al., 1981 and see later). At the same time Pedigo et al. (1981) published ligand binding data to suggest that the $5-HT_1$ receptor could also be sub-classified into $5-HT_{1A}$ and $5-HT_{1B}$. Subsequent ligand binding studies have now suggested further sub-types $5-HT_{1C}$ (Pazos et al., 1984) and $5-HT_{1D}$ (Heuring & Peroutka, 1987).

In the periphery there is reasonable pharmacological evidence to suggest that the old 5-HT-D receptor is the $5-HT_2$ receptor while the 5-HT-M receptor is distinct again and this has now been called the $5-HT_3$ receptor (see Bradley et al., 1986). Very recent evidence has indicated that the $5-HT_3$ receptor probably also exists in the brain (Kilpatrick et al., 1987).

In order to state unequivocally that receptor sub-types exist functionally we require both agonists and antagonists and it is here that we run into problems. $5-HT_{1A}$ and $5-HT_{1B}$ agonists exist (respectively 8-hydroxy-2(di-n-propylamino) tetralin and RU 24969). However the best $5-HT_1$ antagonists are the β-adrenoceptor antagonists, particularly (-)-pindolol and whilst these are selective for $5-HT_1$ versus $5-HT_2$ they are of course also potent β-adrenoceptor antagonists as well. To my knowledge there are currently no highly selective and specific $5-HT_{1C}$ and $5-HT_{1D}$ agonists or antagonists.

Specific $5-HT_2$ antagonists do exist (ketanserin, pirenpirone and ritanserin for example) but while they show high selectivity for $5-HT_2$ versus $5-HT_1$ some also have antagonist actions at other neurotransmitter receptors; pirenperone is also an excellent dopamine antagonist (Green et al., 1983). Ritanserin however, appears to be both selective and specific (Goodwin & Green, 1985). Many hallucinogenic indolamines are selective $5-HT_2$ agonists (Glennon, 1985). Both agonists and antagonists for $5-HT_3$ receptors exist and will also be discussed in later chapters.

Current studies can thus call on a variety of drugs to study 5-HT receptors sub-types but investigators should continue to exercise caution in interpretation in the absence of selective and specific agonists and antagonists at each site.

Functional models to 5-HT$_{1A}$ receptors

It has been known for many years that increasing the 5-HT concentration in the brain (for example by giving a monoamine oxidase inhibitor and L-tryptophan) results in rats displaying a behavioural syndrome consisting of reciprocal fore-paw padding, head weaving, hind limb abduction, locomotor activity, tremor and a variety of other more subtle behavioural changes (see Grahame-Smith, 1971a for a full description and the reviews of Green & Grahame-Smith, 1976a and Jacobs, 1976).

Various other compounds known to be 5-HT agonists also induce the syndrome including 5-methoxy-N,N-dimethyltryptamine (Grahame-Smith, 1971b), 5-methoxytryptamine (Green et al., 1975), tryptamine (Foldes & Costa, 1975; Marsden & Curzon, 1978), LSD (Trulson et al., 1976) and quipazine (Green et al., 1976). The 5-HT releasing drugs fenfluramine and p-chloramphetamine (Trulson & Jacobs, 1976; Green & Kelly, 1976) also produce the behavioural changes. The behaviour is initiated in the hind brain and spinal cord (Jacobs & Klemfuss, 1975).

In 1982 Hjorth and colleagues reported that 8-OH-DPAT would produce the syndrome and the subsequent report that this compound is a selective agonist at 5-HT$_{1A}$ receptor (Middlemiss & Fozard, 1983; Tricklebank et al., 1984) suggested that the behavioural syndrome was a 5-HT$_{1A}$ mediated response.

This syndrome had in fact already been used extensively as a model of central 5-HT function using precursor loading and less selective agonists (see Green & Grahame-Smith, 1976a; Jacobs, 1976).

Some of the early studies had used automated activity meters to measure the syndrome (see Green and Grahame-Smith, 1976a) while Jacobs (1976) had suggested scoring the behavioural changes on the brain of their presence or absence. Deakin and Green (1978) used a 3 point scoring system based on the severity of individual behaviours (0=absent, 1=just present; 2=definite; 3=severe) and the advantage of this approach was indicated by a study of antagonists by Green et al. (1981) when it was found that many of the behavioural changes could be inhibited but that when this was accompanied by an increase in locomotor activity (as for example after administration of methysergide) the resulting automated counts could be deceptive; that is they did not reflect the decrease in 5-HT-mediated behaviours (see Figure 1.1). This study also demonstrated that antagonists would sometimes inhibit the behaviours initiated by administration of one agonist but not another (see Table 1.1). This indicates that care should always be exercised in accepting or rejecting 5-HT as a mediator of behaviour of other responses unless several agonists or antagonists have been examined.

The use of 8-OH-DPAT to induce the behaviours has resulted in a new sophistication in approach insofar as the behaviour can now be used as an index of 5-HT$_{1A}$ mediated responses. The view that the behaviour is 5-HT$_{1A}$ receptor mediated is strengthened by the observations that it is blocked by the 5-HT$_{1A}$ antagonist (-)-propranolol but not the 5-HT$_2$ antagonist ritanserin (Goodwin & Green, 1985).

If the behaviour is induced by non-selective agonists then it is probable that

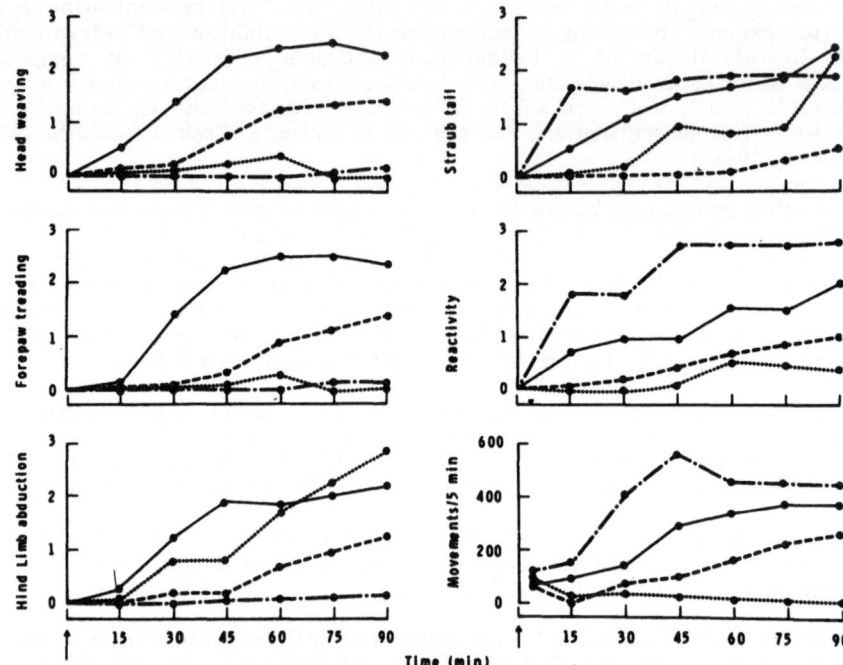

Fig. 1.1 Effects of methysergide, methiothepin and propranolol on behaviour elicited by tranylcypromine and L-tryptophan in rats: (-o-) saline pretreated; (-o-o-o-) methysergide (10 mg/kg); (.....o.....) methiothepin (5 mg/kg); (---o---) (-)-propranolol (20 mg/kg). Tryptophan (100 mg/kg) was injected 30 min after tranylcypromine (120 mg/kg) in each group). Animals were grouped in threes for activity recording. Individual animals were rated at the same time as activity recording. Behavioural ratings show means of 9 or more saline pretreated animals and 6 or more in each of the drug pretreated groups. Differences between saline and drug pretreated animals are significant ($p<0.05$ or better) for at least 3 times points on all measures except Straub tail. Reproduced from Green et al. (1981) with permission of Macmillan Press.

5-HT$_2$ receptors may also play a role. Evidence for this view comes from the observation that when the behaviour is induced by quipazine both (-)-propranolol and ritanserin can antagonise the response (Goodwin & Green, 1985).

As a model of 5-HT$_{1A}$ function relatively few studies have yet been performed. However, in an investigation on neurotransmitter interactions Tricklebank et al. (1984) revealed the involvement of catecholamines in the response in intact animals.

There have also been studies on the pharmacology of the response which is a necessary prerequisite for use of the behaviour in a functional model. The ED$_{50}$, that is the dose required to produce about half-maximal severity of the behaviour, is 2 mg/kg^{-1} i.p. and depletion of cerebral 5-HT stores by either lesioning with 5,7-dihydroxytryptamine (5,7-DHT) or the synthesis inhibitor parachlorophenylalanine (PCPA) did not alter the behavioural response, suggesting that the receptors were probably located post-synaptically (Goodwin et al., 1987a) as has been suggested by others both on the basis of behavioural data (Hjorth et al., 1982; Tricklebank et al., 1984; Goodwin et al., 1986a) and ligand binding studies (Hall et al., 1985). This view on the location of the 5-HT$_{1A}$ receptor was also supported by the failure of quipazine, a drug shown to be an antagonist at presynaptic 5-HT receptors (Moret, 1985) to decrease the behavioural response (Goodwin et al., 1987a).

The behavioural model has been used to investigate the effect of antidepressant treatments on postsynaptic 5-HT$_{1A}$ receptor function in rats (Goodwin et al., 1987b). Basically repeated treatment with antidepressant drugs (Fig. 1.2) or electroconvulsive shock (ECS) decreased the behavioural response suggesting that these treatments attenuated the function of the receptor. The anxiolytic drug flurazepam was without effect. In contrast administration of lithium, a drug used in the prophylactic treatment of mania resulted in a marked increase in the severity of the behavioural syndrome (Goodwin et al., 1986b).

In addition to the behavioural syndrome administration to rats of 8-OH-DPAT also produces hypothermia (Goodwin & Green, 1985; Hjorth, 1985). Because this response is complicated by the appearance of the behavioural syndrome it has not been exploited as a model of 5-HT$_{1A}$ function, in contrast to the hypothermic response in mice, a species that does not display 8-OH-DPAT induced behavioural changes (see later). It is however worth mentioning that the hypothermic response in rats is antagonised by propranolol, but not ritanserin (Goodwin & Green, 1985) and in general displays a rather similar pharmacological profile to the response seen in mice (see review of Green & Goodwin, 1988).

The pharmacology of the hypothermic response in mice which follows 8-OH-DPAT injection has been studied in some detail since it is not complicated by behavioural changes. Like the response in the rat the response appears to result from 8-OH-DPAT acting presynaptically since it is attenuated by a 5,7-DHT lesion or PCPA pretreatment. The response is also blocked by quipazine although surprisingly (-)-propranolol is without effect (Goodwin et al., 1985a).

Drugs which are antagonists at α- or β-adrenoceptors, the benzodiazepine and 5-HT$_2$ receptors were without effect. Again ligand binding data does

Table 1.1 Summary of the effects of putative 5-HT antagonists on the putative 5-HT agonists studied

5-HT or 5-MeODMT	H.W.	F.T.	HLA	R
Cyproheptadine	−	−	−	−
Cinanserin	−	−	−	−
Mianserin	−	−	−	−
Methysergide	↓	↓	↓	↑
Methergoline	↓	↓	↓	↑
Propranolol	↓	↓	↓	↓
Methiothepin	↓	↓	↓	↓
Haloperidol	↓	↓	↓	↓

Quipazine	H.W.	F.T.	HLA	R
Cyproheptadine	↓	↓	↓	↑
Cinanserin	↓	↓	↓	−
Mianserin	↓	↓	−	↓
Methysergide	↓	↓	−	↑
Methergoline	↓	↓	↓	↑
Propranolol	↓	↓	−	↓
Methiothepin	↓	↓	−	↓
Haloperidol	↓	↓	−	↓

H.W., Headweaving; F.T., forepaw treading; HLA, hind limb abduction; R, reactivity.
Direction of arrow indicates increase or decrease of behaviour; − indicates no effect.

suggest that 5-HT_{1A} receptors can be presynaptic as well as postsynaptic (Verge et al., 1985; Hall et al., 1985).

In functional studies it was observed that antidepressant drugs and ECS markedly attenuated the response, the effect of ECS lasting nearly a month after the last ECS administration (Goodwin et al., 1985b). Lithium also attenuated the response (Goodwin et al. 1986c). In this species ipsapirone had an antagonist effect at 5-HT_{1A} receptors, although in rats data suggested it to be a partial agonist (Goodwin et al., 1986a). When 8-OH-DPAT was given daily for 14 days the hypothermic response was attenuated. The 5-HT_{1A} receptor therefore appears to be desensitised by repeated administration of an agonist drug (DeSouza et al., 1986).

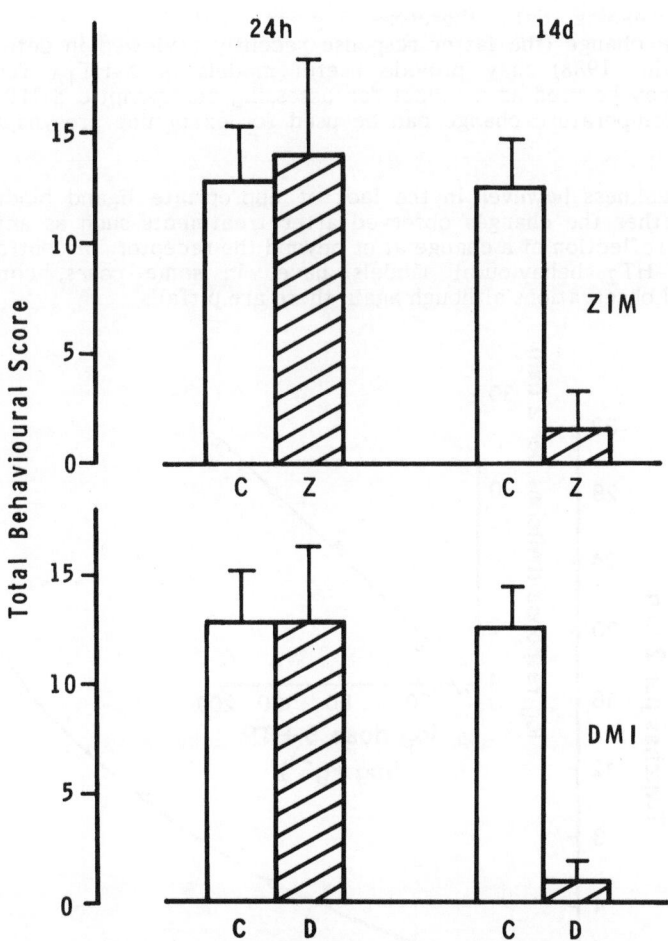

Fig. 1.2 The behavioural response to 8-OH-DPAT (0.75 mh/kh SC) 24h after a single injection or following repeated daily injection (IP for 14 days) of zimeldine (Z) or desipramine (D) at a dose of 20 mg/kg. Total scores (as mean ± SD) shown by total sum for forepaw treading, head weaving, flattening of body posture and hind-limb abduction at 10, 20 and 30 min after injection of 8-OH-DPAT. Responses after 14d are attenuated in Z- and D-treated groups compared with saline injected 0.1 ml/100 g) controls (C). P<0.025. Reproduced from Goodwin et al. (1987b) with permission of Springer-Verlag.

All the foregoing data therefore suggest that both the behaviour and temperature change (the latter response recently reviewed in detail by Green and Goodwin, 1988) may provide useful models of 5-HT_{1A} function. The behaviour may be used as a model for assessing postsynaptic 5-HT_{1A} function while the temperature change can be used for examining presynaptic 5-HT_{1A} function.

A major weakness however in the lack of appropriate ligand binding data to clarify whether the changes observed after treatments such as antidepressant drugs are a reflection of a change at or beyond the receptor. In contrast the data for the 5-HT_2 behavioural models have, in some cases, complimentary biochemical observations although again there are pitfalls.

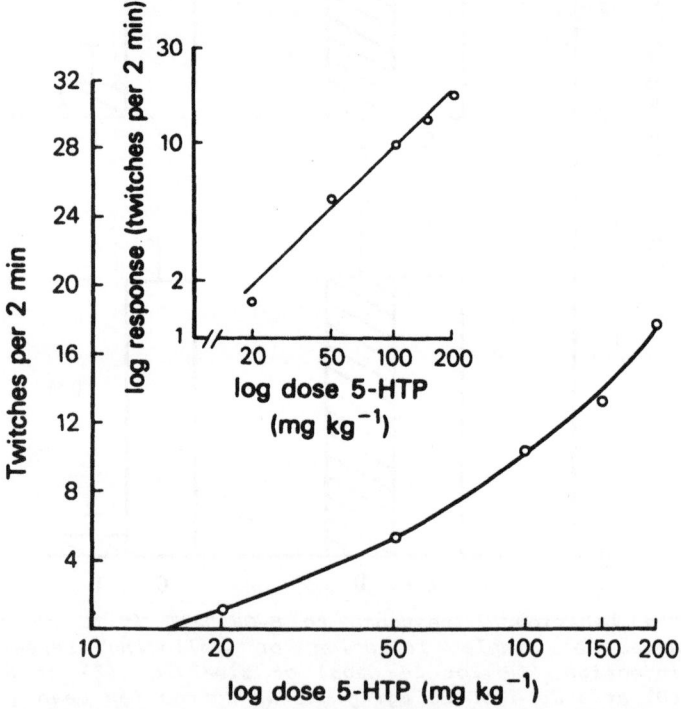

Fig. 1.3 *Effects of dose of 5-hydroxytryptophan (5-HTP) on the behavioural response to carbidopa and 5-HTP. Mice were given carbidopa (25 mg/kg^{-1}) followed by various doses of 5-HTP. Main graph shows the log dose of 5-HTP versus total number of head twitches in 2 min, 30 min after the 5-HTP. Small graph shows the log dose-log response curve. Reproduced from Green et al. (1983) with permission of Macmillan Press.*

Table 1.2 Effect of antidepressant treatments on head twitch behaviour
and 5-HT$_2$ receptor characteristics

Treatment	Behaviour (Head twitching/2min)	5-HT$_2$ receptor binding B_{max}	K_D
Control	7.2 (8)	200 ± 20 (4)	1.42 ± 0.20 (4)
Zimeldine	2.4 (8)**	151 ± 27 (4)**	1.36 ± 0.24 (4)
Desipramine	4.3 (8)**	145 ± 10 (4)**	1.50 ± 0.17 (4)
Control	5.9 (8)	207 ± 28 (3)	1.57 ± 0.23 (3)
ECS	14.2 (8)**	248 ± 25 (7)*	1.48 ± 0.92 (7)

Mice were injected with zimeldine (20 mg/kg^{-1}) or desipramine (27 mg/kg^{-1}) orally daily for 14 days or ECS 5 times over 10 days during halothane anaesthesia. Controls were respectively injected with saline or given halothane anaesthesia. Results are shown as mean head twitch response B_{max} in fmol/mg^{-1} protein, K_D in nM with number of observations in brackets.

Different from control, *p<0.05; **p<0.01.

Models of 5-HT$_2$ receptor function

The first report that increasing brain 5-HT content in mice resulted in a head-twitch response was by Corne et al. (1963). These authors showed that the response was dose dependent with 5-HTP. When a peripheral decarboxylase inhibitor such as carbidopa is given then the dose of 5-HTP administered can be reduced. A dose response curve in mice given carbidopa before 5-HTP is shown in Figure 1.3.

The response can also be induced by agonists such as quipazine (Malick et al., 1977) LSD (Corne & Pickering, 1967) and 5-MeODMT (Friedman & Dallob, 1979). A very similar behaviour, called the 'wet dog shake' can be elicited in rats by the same range of drugs (Bedard & Pycock, 1977; Matthews & Smith, 1980).

The first indication that the head twitch response was 5-HT$_2$ receptor mediated came with the report of Peroutka et al. (1981) which presented data showing a close relationship between the potency of drugs to inhibit the response and their ability to bind to the 5-HT$_2$ receptor. Subsequently the availability of drugs with high selectivity for the 5-HT$_2$ receptor has allowed investigation of the receptor controlling the head twitch response. Both pirenperone (Green et al., 1983) and ritanserin (Goodwin & Green, 1985) are potent inhibitors of head twitch behaviour in mice (ritanserin, ED$_{50}$ = 35 μg/kg) while (−)-propranolol at high dose had no effect on the response (Fig. 1.4). Ketanserin and pirenperone are also potent inhibitors of wet dog shake behaviour in rats (Yap & Taylor, 1983; Colpaert & Janssen, 1983).

A good relationship has been shown between the increase in number of 5-HT$_2$ binding sites in the frontal cortex after a 5,7-DHT lesion and the increase in head twitch behaviour (Heal et al., 1985). However caution should be exercised in such studies since the behaviour is almost certainly initiated not in the cortex

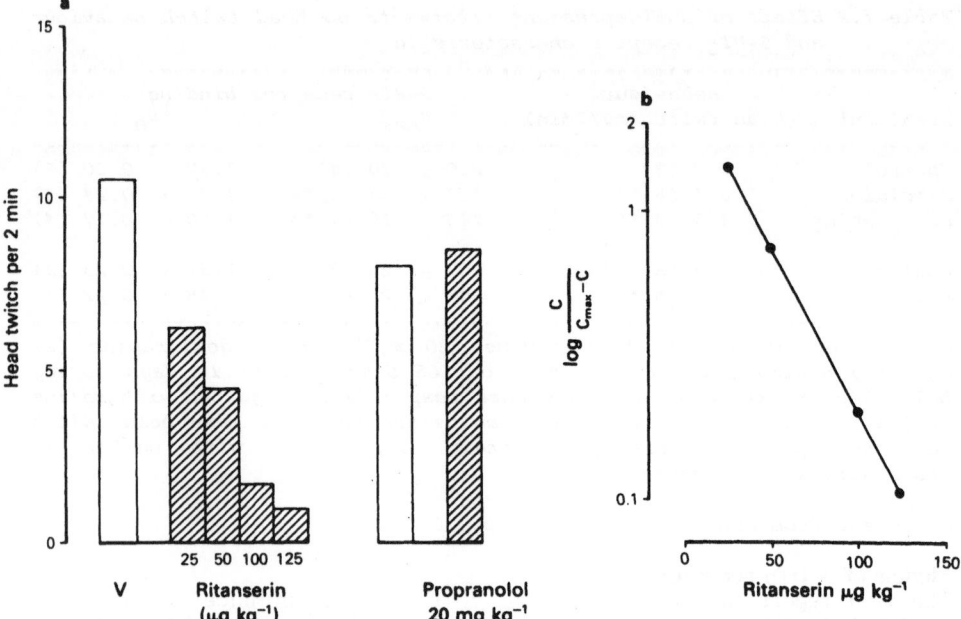

Fig. 1.4 Effect of ritanserin and propranolol on 5-hydroxytryptophan (5-HTP)-induced head twitch behaviour in mice. (a) Ritanserin was given at the doses shown 60 min before carbidopa (25 mg/kg^{-1}). 5-HTP (100 mg/kg^{-1}) was given after a further 15 min and the head twitches measured over a 2 min period after a further 15 min. Results show mean of 6 animals in each group. (-)-Propranolol (20 mg/kg^{-1}) was given 15 min before the carbidopa and the experiment repeated as above.
(b) The ED_{50} ritanserin calculated from Figure 1.1 was 35 mg/kg^{-1}; C = head-twitches in drug treated mice; C_{max} = head twitches in the vehicle treated mice. Ritanserin-treated mice significantly different from vehicle-injected (V): $P<0.01$ or better at all doses examined. Propranolol-treated mice not significantly different from saline-injected controls. Reproduced from Goodwin & Green (1985) with permission of Macmillan Press.

but in the hind brain and spinal cord (Corne et al., 1963). However the lesioning technique produced similar depletion of 5-HT in both the cortex and hind brain so the 5-HT$_2$ receptor number probably increased comparably in both regions. In rats a serotonergic lesion has been found to increase the wet dog shake behaviour in several studies (Bednarcyk & Vetulani 1978; Barbeau & Bedard, 1981).

Many studies have been performed on the involvement of other neurotransmitters in the head twitch response and wet dog shake behaviour. A review of this area is beyond the scope of this chapter but the reader is refered to Green & Heal (1985) for further information.

Peroutka & Snyder (1980) were the first to show that most antidepressant drugs decreased the density of $5-HT_2$ receptors in the cortex.

In a study of the functional consequences of this change Goodwin et al. (1984) demonstrated that a variety of antidepressant drugs decreased both the head twitch response and the density of $5-HT_2$ receptors in the cortex following repeated drug administration and while the animals were still being treated (Table 1.2). ECS in contrast increased both parameters (Table 1.2).

On withdrawal however (2 days) the mice given desipramine continued to show a decrease in the density of $5-HT_2$ receptors in the cortex but a marked enhancement in head twitch behaviour. This presumably results from the withdrawal of the desipramine and the attentent changes in the function of other neurotransmitters involved in the expression of the head twitch response. It does however emphasize the problems in equating behaviour and biochemistry.

Another example of the divergence between biochemical and functional measures came in a study of the consequences of longer term 8-OH-DPAT administration.

As might be expected from a drug which decreases 5-HT synthesis and thus presumably function longer term (14 days) 8-OH-DPAT administration to mice resulted in the appearance of an enhanced head twitch response (De Souza et al., 1986, Figure 1.5). However the $5-HT_2$ receptor density in the cortex was decreased (Table 1.3). Furthermore repeated treatment with the partial $5-HT_{1A}$ agonist ipsapirone resulted in a decrease in head twitch behaviour (Figure 1.5) but no change in cortical binding parameters (Table 1.3). It is hard to explain these binding and behavioural data other than to suggest that they reflect regional variation in the changes occuring after injection of 8-OH-DPAT.

Other functional tests for other 5-HT receptor subtypes

At present there are no other defined, reliable and simple tests for $5-HT_{1B}$ function. The $5-HT_{1B}$ agonist RU 24969 produces marked locomotor activity (Green et al., 1984). However the failure of known 5-HT antagonists to markedly influence this behaviour means that caution must be exercised in ascribing the behaviour to a change in 5-HT function.

The pharmacology of $5-HT_1$ mediated feeding behaviour has been extensively investigated but since this is discussed elsewhere in this book it will not be reviewed here.

A rather different approach to the examination of $5-HT_1$ function in vivo is the use of drug discrimination studies. Basically the animals are trained to produce a particular response under one set of conditions (eg following administration of a drug) and a different response under another set of conditions (eg a non-drug state). Glennon (1985) has reviewed the use of this approach to

Table 1.3 Effect of chronic ipsapirone and 8-OH-DPAT administration on 5-HT_2 receptor binding characteristics in frontal cortex of mice

Injected	B_{max}	K_D	n
Saline	426 ± 58	0.83 ± 0.17	3
Ipsapirone	445 ± 47	0.92 ± 0.39	3
8-OH-DPAT	280 ± 34*	0.70 ± 0.08	3

Results reported as mean ± s.d. Saline and both drugs given once daily, ipsapirone at 10 mg/kg^{-1} and 8-OH-DPAT at 5 mg/kg^{-1}. B_{max} expressed in fmol-mg^{-1} protein and K_D in nM.

*Different from control value $p<0.01$.

examine whether animals recognise various serotonergic drugs as being "the same" or not. What this means is that one can deduce whether a series of compounds are likely to be stimulating the same receptor or not. Combined with either receptor affinity studies (Glennon used a fundic strip preparation) or ligand binding techniques this approach provides powerful in vivo information in the way that drugs are being recognised by specific receptors in the brain.

Conclusions
This introduction has briefly reviewed some of the approaches used in behavioural pharmacology. There is now substantial evidence for the involvement of 5-HT in sexual behaviour, aggression, temperature control, feeding, learning, pain, anxiety, depression and psychosis. Much of this book reviews this evidence and I have therefore not covered these aspects here but rather concentrated on the use of functional tests. It should be emphasized that alone functional test provide limited information and that they are at their most powerful when combined with other approaches such as ligand binding and other neurochemical techniques. Nevertheless there are two points that should be made. Firstly, ligand binding can never tell an experimenter whether the acceptor site being studied is a functional receptor, functional studies are required. Ligand binding cannot even tell the experimenter whether a compound is an agonist or an antagonist.

Again functional studies are required. Behavioural models therefore come into their own in allowing one to assess whether changes in receptor number have any functional consequence and also whether changes in synthesis or release of 5-HT or alterations in the function of other neurotransmitters induce, in either the short or long term changes in the function of 5-HT.

Secondly, behavioural models can also be used as a preliminary and often fairly rapid method of examining the action of compounds on 5-HT function before the

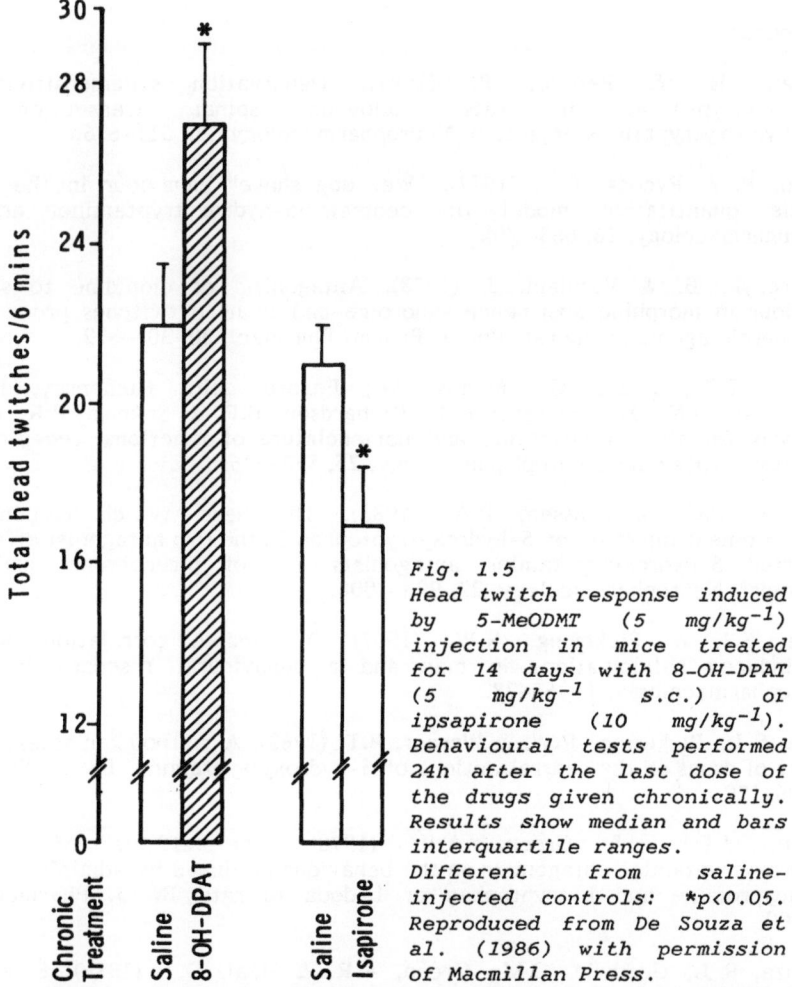

Fig. 1.5
Head twitch response induced by 5-MeODMT (5 mg/kg^{-1}) injection in mice treated for 14 days with 8-OH-DPAT (5 mg/kg^{-1} s.c.) or ipsapirone (10 mg/kg^{-1}). Behavioural tests performed 24h after the last dose of the drugs given chronically. Results show median and bars interquartile ranges.
Different from saline-injected controls: *$p<0.05$. Reproduced from De Souza et al. (1986) with permission of Macmillan Press.

use of neurochemical techniques. Thus we reported that propranolol could act as a 5-HT antagonist in the brain (Green & Grahame-Smith, 1976) before it was shown that the β-adrenoceptor antagonists bound to 5-HT receptors (Middlemiss et al., 1977) and, of course, several years before it was finally shown to have a high affinity for 5-HT$_{1A}$ sites (Middlemiss & Fozard, 1983).

Nevertheless behavioural models should always be used with the knowledge that they can give misleading information. For example responses can alter because of changes in neurotransmitter systems other than the one under investigation. Pharmacokinetic parameters such as drug metabolism and brain penetration can also play a major role in the way that a drug acts or fails to act in a behavioural model.

References

Barbeau, H. & Bedard, P. (1981). Denervation supersensitivity to 5-hydroxytryptophan in rats following spinal transection and 5,7-dihydroxytryptamine injection. Neuropharmacology, 20, 611-616.

Bedard, P. & Pycock, C.J. (1977). "Wet dog shake" behaviour in the rat: a possible quantitative model of central 5-hydroxytryptamine activity. Neuropharmacology, 16, 663-670.

Bednarczyk, B. & Vetulani, J. (1978). Antagonism of clonidine to shaking behaviour in morphine abstinence syndrome and to head twitches produced by serotonergic agents in the rat. Pol. J. Pharm. Pharmac., 30, 307-322.

Bradley, P.B., Engel, G., Fenuik, W., Fozard, J.R., Humphrey, P.P.A., Middlemiss, D.N., Mylecharane, E.J., Richardson, B.P. & Saxena, P.R. (1986). Proposals for the classification and nomenclature of functional receptors for 5-hydroxytryptamine. Neuropharmacology, 25, 563-576.

Colpaert, F.C. & Janssen, P.A. (1983). The head twitch response to intraperitoneal injection of 5-hydroxytryptophan in the rat: antagonist effects of purported 5-hydroxytryptamine antagonists and of pirenperone, and LSD antagonist. Neuropharmacology, 22, 993-1000.

Corne, S.J. & Pickering, R.W. (1967). A possible correlation between drug-induced hallucinations in man and a behavioural response in mice. Psychopharmacologia, 11, 65-78.

Corne, S.J., Pickering, R.W. & Warner, B.T. (1963). A method for assessing the effects of drugs on the central actions of 5-hydroxytryptamine. Br. J. Pharmac., 20, 106-120.

Deakin, J.F.W. & Green, A.R. (1978). The effects of putative 5-hydroxytryptamine antagonists on the behaviour produced by administration of tranylcypromine and L-tryptophan or L-dopa to rats. Br. J. Pharmac., 64, 201-209.

DeSouza, R.J., Goodwin, G.M., Green, A.R. & Heal, D.J. (1986). Effects of chronic treatment with 5-HT_1 agonist (8-OH-DPAT and RU 24969) and antagonist (ipsapirone) drugs on the behavioural responses of mice to 5-HT_1 and 5-HT_2 agonists, Br. J.Pharmac., 89, 377-384.

Foldes, A. & Costa, E. (1975). Relationship of brain monoamines and locomotor activity in rats. Biochem. Pharmac., 24, 1617-1621.

Friedman, E. & Dallob, A. (1979), Enhanced serotonin receptor activity after chronic treatment with imipramine or amitryptyline. Commun. Psychopharmac., 3, 89-92.

Gaddum, J.H. & Picarelli, Z.P. (1957). Two kinds of tryptamine receptor. Br. J. Pharmac., 12, 323-328.

Glennon, R.A. (1985). Involvement of serotonin in the action of hallucinergic drugs. In: Neuropharmacology of serotonin. (Ed. Green, A.R.) pp 253-280. Oxford University Press, Oxford.

Goodwin, G.M. & Green, A.R. (1985). A behavioural and biochemical study in mice and rats of putative selective agonists and antagonists for $5-HT_2$ receptors. Br. J. Pharmac., **84**, 734-753.

Goodwin, G.M., DeSouza, R.J. & Green, A.R. (1985a). The pharmacology of the hypothermic response in mice to 8-hydroxy-2-(di-n-propylamino) tetralin (8-OH-DPAT), a model for presynaptic $5-HT_1$ function. Neuropharmacology, **24**, 1187-1194.

Goodwin, G.M., DeSouza, R.J. & Green, A.R. (1985b). Presynaptic serotonin receptor-mediated response in mice attenuated by antidepressant drugs and electroconvulsive shock, Nature, **317**, 531-533.

Goodwin, G.M., DeSouza, R.J. & Green, A.R. (1986a). The effects of a $5-HT_1$ receptor ligand ipsapirone (TVX Q 7821) on 5-HT synthesis in mice and rats. Psychopharmacology, **89**, 382-387.

Goodwin, G.M., DeSouza, R.J., Green, A.R. & Wood, A.J. (1986b). The enhancement by lithium of the $5-HT_{1A}$ mediated serotonin syndrome produced by 8-OH-DPAT in the rat: evidence for a post synaptic mechanism. Psychopharmacology, **90**, 488-493.

Goodwin, G.M., DeSouza, R.J., Wood, A.J. & Green, A.R. (1986c). Lithium decreases $5-HT_{1A}$ and $5-HT_2$ receptor and α_2-adrenoceptor-mediated function in mice. Psychopharmacology, **90**, 482-487.

Goodwin, G.M., DeSouza, R.J., Green, A.R. & Heal, D.J. (1987a). The pharmacology of the behavioural and hypothermic responses of rats to 8-hydroxy-2-(di-n-propylamino) tetralin (8-OH-DPAT). Psychopharmacology, **91**, 506-511.

Goodwin, G.M., DeSouza, R.J. & Green, A.R. (1987b). Attenuation by electroconvulsive shock and antidepressant drugs of the $5-HT_{1A}$ receptor mediated hypothermia and serotonin syndrome produced by 8-OH-DPAT. Psychopharmacology, **91**, 500-505.

Goodwin, G.M., Green, A.R. & Johnson, P. (1984). $5-HT_2$ receptor characteristics in frontal cortex and $5-HT_2$ receptor mediated head-twitch behaviour following antidepressant treatment to mice. Br. J. Pharmac., **83**, 235-242.

Grahame-Smith, D.G. (1971a). Studies in vivo on the relationship between brain tryptophan, brain 5-HT synthesis and hyperactivity in rats treated with a monoamine oxidase inhibitor and L-tryptophan. J. Neurochem., **18**, 1053-1066.

Grahame-Smith, D.G. (1971b). Inhibitory effect of chlorpromazine on the syndrome of hyperactivity produced by L-tryptophan or

5-methoxy-N,N-dimethyltryptamine in rats treated with a monoamine oxidase inhibitor. Br. J. Pharmac., 43, 856-864.

Green, A.R. & Goodwin, G.M. (1987). The pharmacology of the hypothermic response of rodents to 8-OH-DPAT administration and the effects of psychotropic drug administration on this response. In: Brain 5-HT$_{1A}$ receptors: behavioural and neurochemical pharmacology. (Eds. Dourish, C.T., Ahlenius, S.R., Hutson, P.H.) pp 161-176, Ellis Horwood, Chichester.

Green, A.R. & Grahame-Smith, D.G. (1976a). The effects of drugs on the processes regulating the functional activity of brain 5-hydroxytryptamine. Nature, 260, 487-491.

Green, A.R. & Grahame-Smith, D.G. (1976b). (-)-Propranolol inhibits the behavioural responses of rats to increased 5-hydroxytryptamine in the central nervous system. Nature, 262, 594-496.

Green, A.R. & Heal, D.J. (1985). The effects of drugs on serotonin-mediated behavioural models. In: Neuropharmacology of serotonin. (Ed. Green, A.R.) pp. 326-365. Oxford University Press.

Green, A.R. & Kelly, P.H. (1976). Evidence concerning the involvement of 5-hydroxytryptamine in locomotor activity produced by amphetamine or tranylcypromine plus L-dopa. Br. J. Pharmac., 57, 141-147.

Green, A.R., Guy, A.P. & Gardner, C.R. (1984). The behavioural effects of RU 24969, a 5-HT$_1$ receptor agonist in rodents and the effect on the behaviour of various antidepressant treatments. Neuropharmacology, 23, 655-661.

Green, A.R., Hall, J.E. & Rees, A.R. (1981). A behavioural and biochemical study in rats of 5-hydroxytryptamine agonists and antagonists with observations on structure-activity requirements for the agonists. Br. J. Pharmac., 73, 703-720.

Green, A.R., Heal, D.J., Johnson, P., Laurence, B.E. & Nimgoankar, V.L. (1983). Antidepressant treatments: effects in rodents on dose response curves of 5-hydroxytryptamine and dopamine mediated behaviour and 5-HT$_2$ receptor number in frontal cortex. Br. J. Pharmac., 80, 377-385.

Green, A.R., Huges, J.P. & Tordorff, A.F.C.(1975). The concentration of 5-hydroxytryptamine in rat brain and its effects on behaviour following its peripheral injection. Neuropharmacology, 14, 601-606.

Green, A.R., O'Shaugnessy, K., Hammond, M., Schachter, M. & Grahame-Smith, D.G. (1983) Inhibition of 5-hydroxytryptamine mediated behaviours by the putative 5-HT$_2$ agonists pirenperone. Neuropharmacology, 22, 574-578.

Green, A.R., Youdim, M.B.H. & Grahame-Smith, D.G. (1976). Quipazine: its effects on rat brain 5-hydroxytryptamine metabolism, monoamine oxidase activity and behaviour. Neuropharmacology, 15, 173-179.

Hall, M.D., El Mestikawy, S., Emerit, M.B., Pichat, L., Hamon, M. & Gozlan, H.

(1985) [^{3}H]8-hydroxy-2-(di-n-propylamino) tetralin binding to pre- and postsynaptic 5-hydroxytryptamine sites in various regions of the rat brain. J. Neurochem., 44, 1685-1696.

Heal, D.J., Philpot, J., Molyneux, S.G. & Metz. A. (1985). Intracerebroventricular administration of 5,7-dihydroxytryptamine to mice increase both head-twitch response and the number of cortical 5-HT$_2$ receptors. Neuropharmacology, 24, 1201-1205.

Heuring, R.E. & Peroutka, S.J. (1987). Characterisation of a novel ^{3}H-5-hydroxytryptamine binding site subtype in bovine brain membranes. J. Neurosci, 7, 894-903.

Hjorth, S. (1985). Hypothermia in the rat induced by the potent serotonergic agent 8-OH-DPAT. J.Neurol. Transm., 61, 131-135.

Hjorth, S., Carlsson, A., Lindberg, P., Sanchez, D., Wikstrom, A., Arvidsson, L.-E., Hacksell, U. & Nilsson, J.L.G. (1982). 8-hydroxy-2-(di-n-propylamino) tetralin, 8-OH-DPAT, a potent and selective simplified ergot congener with central 5-HT-receptor stimulating activity. J. Neural. Trans., 55, 169-188.

Jacobs, B.L. (1976). An animal behaviour model for studying central serotonergic synapses. Life Sci., 19, 777-786.

Jacobs, B.L. & Klemfuss, H. (1975). Brainstem and spinal cord mediation of serotonergic behavioural syndrome. Brain Res., 100, 450-457.

Kilpatrick, G.J., Jones, B.J. & Tyers, M.B. (1988). Direct labelling of 5-HT$_3$ receptors in rat brain using [^{3}H]-GR 65630: characterisation and distribution. Br. J. Pharmac. (In press)

Leysen, J.E., Niemegeers, C.J.E., Tollanaere, J.P. & Laduron, P.M. (1978). Serotonergic components of neuroleptic receptors. Nature, 272, 168-170.

Malick, J.B., Dosen E. & Barnett, A. (1977). Quipazine-induced head twitch in mice. Pharmac. Biochem. Behav., 6, 325-329.

Marsden, C.A. & Curzon, G. (1979). The contribution of tryptamine to behavioural effects of L-tryptophan in tranylcypromine-treated rats. Psychopharmacology, 57, 71-76.

Matthews, W.D. & Smith, C.D. (1980). Pharmacological profile of a model for central serotonin receptor activation. Life Sci., 26, 1397-1403.

Moret, C. (1985). In: Neuropharmacology of serotonin. (Ed. Green, A.R.) pp 21-49, Oxford University Press, Oxford.

Middlemiss, D.N. & Fozard, J.R. (1983). 8-hydroxy-2-(di-n-propylamino) tetralin discriminates between subtypes of 5-HT$_1$ recognition site. Eur. J. Pharmac., 90, 151-153.

Middlemiss, D.N., Blakeborough, L. & Leather, S.R. (1977). Direct evidence for an interaction of adrenergic blockers with the 5-HT receptor. Nature, **267**, 289-290.

Pazos, A., Hoyer, D. & Palacios, J.M. (1984). The binding of serotonergic ligands to the porcine choroid plexus: characterisation of a new type of serotonin recognition site. Eur. J. Pharmac., **106**, 539-546.

Pedigo, N.W., Yamamura, H.I. & Nelson, D.L. (1981). Discrimination of multiple [^{3}H] 5-hydroxytryptamine binding sites by the neuroleptic spiperone in rat brain. J. Neurochem., **36**, 220-226.

Peroutka, S.J. & Snyder, S.H. (1979). Multiple serotonin receptors: Differential binding of [^{3}H] lysergic acid diethylamide and [^{3}H] spiroperidol. Mol. Pharmac., **16**, 687-699.

Peroutka, S.J. & Snyder, S.H. (1980). Long term antidepressant treatment decrease spiroperidol-labelled serotonin receptor binding. Science, **210**, 88-90.

Peroutka, S.J., Lebovitz, R.M. & Snyder, S.H. (1981). Two distinct central serotonin receptors with different physiological functions. Science **212**, 827-829.

Tricklebank, M.D., Forler, C. & Fozard, J.R. (1984). The involvement of subtypes of the 5-HT$_1$ receptor and the catecholaminergic system in the behavioural response to 8-hydroxy-2-(di-n-propylamino) tetralin in the rat. Eur. J. Pharmac., **106**, 271-282.

Trulson, M.E. & Jacobs, B.L. (1976). Behavioural evidence for the rapid release of CNS serotonin by PCA and fenfluramine. Eur. J. Pharmac., **36**, 149-154.

Trulson, M.E., Ross, C.A. & Jacobs, B.L. (1976). Behavioural evidence for the stimulation of CNS serotonin receptor by high doses of LSD. Psychopharmac. Commun., **2**, 149-164.

Verge, D., Daval, G., Patey, A.., Gozlan, H., El Mestikawy, S. & Hamon. M. (1985). Presynaptic 5-HT autoreceptors on serotonergic cell bodies and/or dendrites but not terminals are of the 5-HT$_{1A}$ subtype. Eur. J. Pharmac., **113**, 463-464.

Yap, C.Y. & Taylor, D.A. (1983). Involvement of 5-HT$_2$ receptors in the wet dog snake behaviour induced by 5-hydroxytryptophan in the rat. Neuropharmacology, **22**, 801-804.

IS ROTATIONAL BEHAVIOUR IN THE RAT A $5-HT_{1A}$ OR $5-HT_{1B}$ MEDIATED RESPONSE?

T.P.Blackburn, D.A.Martin, P.Slater.
ICI Pharmaceuticals Division, Bioscience Dept. II, Alderley Park, Macclesfield, Cheshire and Dept. of Physiological Sciences, University of Manchester, United Kingdom.

Introduction
It is generally accepted that two major subtypes of the serotonin (5-HT) receptor exist in the central nervous system. The differentiation of 5-HT receptors into $5-HT_1$ and $5-HT_2$ subtypes was based on the radioligand binding studies of Peroutka & Snyder (1979). Although the $5-HT_2$ receptor has been characterised only relatively recently (Leysen et al., 1981), progress on the $5-HT_1$ receptor has been more rapid. At least three subdivisions of this receptor, $5-HT_{1A}$, $5-HT_{1B}$ and $5-HT_{1C}$, have been proposed on the basis of binding studies (Pedigo et al., 1981; Schnellmann et al., 1984; Pazos & Palacios, 1985; Pazos et al., 1985a,b) and autoradiographic investigations (Deshmukh et al., 1983; Marcinkiewicz et al., 1984). In terms of receptor function, there are few, if any, well defined tests for $5-HT_1$ receptors, in contrast to $5-HT_2$ receptors for which in vivo tests have been reported (Arnt et al., 1984).

It is known that 5-HT agonists induce rotational behaviour in rats with asymmetric lesions of the 5-HT neurones in the dorsal raphe nucleus (DRN, Slater, 1980). In a recent study, we concluded that the rotational response was mediated by $5-HT_1$ receptors since circling was induced by agonists with a high affinity for this receptor, whereas selective $5-HT_2$ antagonists (ketanserin, pirenperone) failed to inhibit the behavioural response (Blackburn et al., 1984b). More recently, we have shown that 8-OH-DPAT and RU 24969, $5-HT_{1A}$ and mixed $5-HT_{1A}$ and $5-HT_{1B}$ agonists respectively, had significant affinity for $5-HT_1$ cortical binding sites labelled with [^{3}H]5-HT (pIC_{50} 7.0 and 6.9 respectively), whereas they had little or no affinity for $5-HT_2$ sites labelled with [^{3}H]spiperone (pIC_{50} <5.0 for both agonists).

However, both 8-OH-DPAT and RU 24969 were similarly active in the raphe lesioned turning model. These data point to the importance of the $5-HT_1$ receptor subtypes in the circling model. The present study was undertaken to further define the subclass of $5-HT_1$ receptor involved.

Methods
5-HT neurones in the DRN were lesioned as described previously (Blackburn et al., 1980). Adult male rats of the Alderley Park strain (180-220 g) were anaesthetised with halothane (3% v/v/ in oxygen) and were secured in a stereotaxic frame. An asymmetric lesion of the DRN (co-ordinates: A = 0.2, L = 0.3, H = 5.0 mm from dura, according to the atlas of Konig & Klippel, 1963) was made by injection of 5,7-dihydroxytryptamine (5,7-DHT). 5,7-DHT (16 μg) was dissolved in 0.2% ascorbic acid solution and was injected (2 μl) 1 h after an i.p. injection of pargiline (50 mg/kg).

Four days after DRN lesioning, each rat was assessed for turning behaviour. 8-OH-DPAT (synthesised by Dr.M.T.Cox) was dissolved in saline solution and administered s.c. (2 mg/kg). Rats were placed in automated rotometers and full turns were recorded for 2 h (Blackburn et al., 1980).

Rats were killed by decapitation 10 days after 8-OH-DPAT administration, the brains were removed and frozen in isopentane at -35°C. Brains were sectioned (20 μm) on a cryostat and the sections were collected on slides. Glass-mounted sections were incubated in Tris HCl buffer (170 mM, pH 7.6) for 30 min. This was followed by incubation for 1 h in buffer containing 4 mM $CaCl_2$, 0.01% ascorbic acid and 2 nM propyl-[^{3}H]-8-OH-DPAT at room temperature. For non-specific binding, 1 μM of 5-HT was added to the incubation medium. The sections were washed twice (2 min) in fresh buffer, dipped briefly in cold distilled water and dried in a stream of air. For autoradiography, labelled brain sections were apposed to tritium-sensitive film in X-ray cassettes for 4 weeks. Other sections were scraped from the slides into vials for liquid scintillation counting. Sections containing the DRN were stained for histological verification of the injection site.

Results
The administration of 8-OH-DPAT to DRN lesioned rats caused contraversive circling behaviour. [^{3}H]8-OH-DPAT binding studies were performed on rats that made at least 200 complete contraversive turns in the 2 h recording period. The specific (displaceable) binding of [^{3}H]8-OH-DPAT to rat forebrain sections averaged 95% of the total binding (n = 4). Thus nearly all of the autoradiographic image was attributable to specific labelling with [^{3}H]8-OH-DPAT. Autoradiographs prepared from sections cut from the brains of circling rats showed wide variations in the distribution of [^{3}H]8-OH-DPAT binding.

There was a high density of binding sites in the hippocampus and septum and moderate levels of binding in many areas, including the dorsal raphe. However, most importantly in relation to circling behaviour, there was a complete absence of [^{3}H]8-OH-DPAT binding (1A sites) in all of the basal ganglia nuclei. Thus, in the rats that circled when given 8-OH-DPAT, there was no binding in the substantia nigra, caudate-putamen or globus pallidus on either side of the brain (Fig. 2.1).

Discussion
Treatment of rats with 5,7-DHT increased the binding of [^{3}H]5-HT in the substantia nigra (Blackburn et al., 1984a; Weissman et al., 1986). This apparent supersensitivity of 5-HT receptors may be responsible for the circling behaviour produced by 5-HT agonists in rats with unilateral 5-HT lesions. It is assumed that the circling produced by 8-OH-DPAT in the present study was caused by activation of $5-HT_{1A}$ receptors since 8-OH-DPAT is regarded as a selective $5-HT_{1A}$ agonist.

However, in the past, several compounds which act as agonists or antagonists at the $5-HT_1$ receptor were found to be inactive in the raphe-lesioned circling model (Blackburn et al., 1984b). Thus the identity of the receptor responsible for the circling is in doubt.

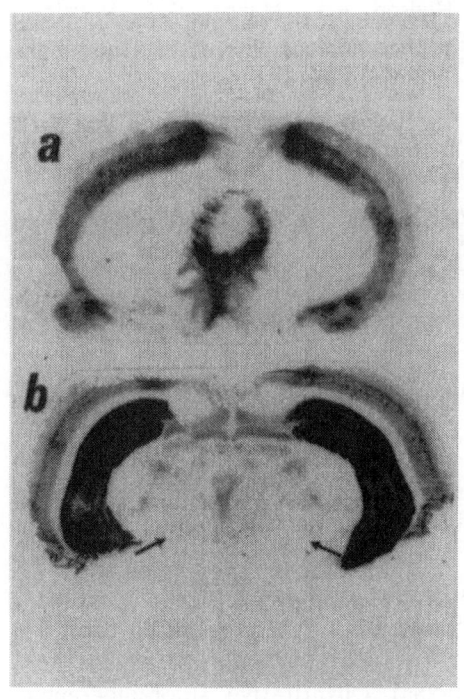

Fig. 2.1
Autoradiographs showing [^{3}H]8-OH-DPAT binding to coronal sections of brain from rats with a DRN lesion. Note the absence of binding in caudate-putamen (a) and substantia nigra (arrowed in b).

This doubt is increased by the present autoradiographic study which demonstrated a complete absence of high affinity [^{3}H]8-OH-DPAT binding in the basal ganglia and substantia nigra on either side of the brain in raphe lesioned rats.

It was proposed that 5-HT receptors in the rat basal ganglia and substantia nigra are of the 5-HT$_{1B}$ subtype (Pazos & Palacios, 1985). The question therefore arises as to why 8-OH-DPAT causes circling behaviour. One explanation could be that the 5,7-DHT lesion may have caused the appearance of 5-HT$_{1A}$ receptors in the substantia nigra or elsewhere in the basal ganglia which may have such low density of low affinity that they escape detection on [^{3}H]8-OH-DPAT autoradiographs.

References

Arnt, J., Hyttel, J. & Larsen, J.J. (1984). The citalopram/5-HTP-induced head shake syndrome is correlated to 5-HT$_2$ receptor affinity and also influenced by other transmitters. Acta Pharmacol. Toxicol., **55**, 363-372.

Blackburn, T.P., Foster, G.A., Heapy, C.G. & Kemp, J.D. (1980). Unilateral 5,7-dihydroxytryptamine lesions of the dorsal raphe nucleus (DRN) and rat rotational behaviour. Eur. J. Pharmacol., **67**, 427-438.

Blackburn, T.P., Bowery, N.G., Cox, B., Hudson, A.L., Martin, D.A. & Price, G.W. (1984a). Lesions of the dorsal raphe nucleus increases the nigral concentration of 5-HT receptors. Br. J. Pharmacol., **82**, 203P.

Blackburn, T.P., Kemp, J.D., Martin, D.A. & Cox, B. (1984b). Evidence that 5-HT agonist induced rotational behaviour in the rat is mediated via 5-HT$_1$ receptors. Psychopharmacology, **83**, 163-165.

Deshmukh, P.P., Yamamura, H. Woods, L. & Nelson, L.D. (1983). Computer-assisted autoradiographic localization of subtypes of serotonin receptors in rat brain. Brain Res **288**, 338-343.

Konig, J.F.R. & Klippel, R.A. (1963). The Rat Brain. The Williams & Wilkins Co., New York.

Leysen, J.E., Awouters, F., Kennis, L., Laduron, P.M., Vandenburg, J. & Janssen, P.A.J. (1981). Receptor binding profile of R41,468. A novel antagonist at 5-HT$_2$ receptors. Life Sci., **28**, 1015-1022.

Marcinkiewicz, M., Verge, D., Gozlan, H., Pichat, L. & Hamon, M. (1984). Autoradiographic evidence for the heterogeneity of 5-HT$_1$ sites in the rat brain. Brain Res., **29**, 159-163.

Pazos, A. & Palacios, J.M. (1985). Quantitative autoradiographic mapping of serotonin receptors in the rat brain. I. Serotonin-1 receptors. Brain Res., **346**, 205-230.

Pazos, A., Cortes, R. & Palacios, J.M. (1985a). Quantitative autoradiographic mapping of serotonin receptors in the rat brain. II. Serotonin-2 receptors. Brain Res., **346**, 231-249.

Pazos, A., Hoyer, D. & Palacios, J.M. (1985b). The binding of serotonergic ligands to the porcine choroid plexus: characterisation of a new type of serotonin recognition site. Eur. J. Pharmacol., **106**, 539-546.

Pedigo, N.W., Yamamura, H.I. & Nelson, D.L. (1981). Discrimination of multiple [^{3}H]5-hydroxytryptamine binding sites by the neuroleptic spiperone in the rat brain. J. Neurochem., **36**, 220-226.

Peroutka, S.J. & Snyder, S.H. (1979). Multiple serotonin receptors: differential binding of [^{3}H]-5-hydroxytryptamine, [^{3}H]-lysergic acid diethylamide and [^{3}H]-spiroperidol. Mol. Pharmacol., **16**, 687-699.

Slater, P. (1980). Circling produced by serotonin and dopamine agonists in raphe lesioned rats: a serotonin model. Pharmacol. Biochem. Behav., **13**, 817-821.

Schnellmann, R.G., Waters, S.J. & Nelson, D.L. (1984). [^{3}H]-5-Hydroxytryptamine binding sites: species and tissue variation. J. Neurochem., **42**, 65-70.

Weissmann, D., Mach, E., Oberlander, C., Demassey, Y. & Pujol, J-F. (1986).

Evidence for hyperdensity of 5-HT$_{1B}$ binding sites in the substantia nigra of the rat after 5,7-dihydroxytryptamine intraventricular injection. Neurochem. Int., **9**, 191-200.

SEROTONIN–MEDIATED BEHAVIOUR FOLLOWING CENTRAL OR PERIPHERAL ADMINISTRATION OF THE SELECTIVE NK-3 TACHYKININ AGONIST SENKTIDE

A.J.Stoessl, C.T.Dourish, S.D.Iversen
Merck Sharp & Dohme Research Laboratories, Neuroscience Research Centre, Terlings Park, Harlow, Essex, U.K. CM20 2QR

Although an anatomical relationship between substance P (SP) and serotonin (5-HT) has been recognized for almost a decade (Chan-Palay et al., 1978; Hokfelt et al., 1978; Neekers et al., 1979; Johansson et al., 1981; Magoul et al., 1986; Pretel & Ruda, 1986), its functional significance is less clearly defined. SP is known to release 5-HT both in vivo and in vitro (Reubi et al., 1978; Forchetti et al., 1982; Reisine et al., 1982) and it has been suggested that this is mediated via an inhibitory action on 5-HT terminal autoreceptors (Mitchell & Fleetwood-Walker, 1981). Previous studies examining the interactions of these two transmitters on unconditioned behaviour have pointed to a modulatory effect of each on behaviours induced by the other (Hylden & Wilcox, 1983; Ogren et al., 1985; Fasmer et al., 1987).

We have recently had the opportunity to study the behavioural effects of senktide, a synthetic peptide which is a highly selective and potent agonist at the NK-3 tachykinin receptor site (Wormser et al., 1986). Senktide, when administered intracisternally (ics) to the mouse (Stoessl et al., 1987) or rat, was found to produce a 5-HT behavioural syndrome that was attenuated by specific 5-HT antagonists. More recently, we have observed the same behavioural responses following subcutaneous (sc) administration in both species.

Intracisternal injections in the mouse were performed under light ether anaesthesia. A cannula was introduced to a depth of 3 mm and peptides were administered in a volume of 5 ξl from a hand-held 10 ξl Hamilton syringe. Rats were anaesthetized with N_2O (70%) and halothane (4%) and placed in a stereotaxic frame (Kopf instruments). The head was hyperflexed and a cannula was introduced into the cisterna magna to a depth of 5 mm. Senktide or its vehicle (1% dimethylsulfoxide in 0.9% saline, v/v) was administered in a volume of 50 µl from a hand-held 50 µl Hamilton syringe. Subcutaneous injections were performed in a volume of 1 mg/kg (rats) or 10 ml/kg (mice) in the back of the neck.

Head twitches were counted for a ten minute period immediately following ics senktide, and for 30 min following sc administration in mice. In rats, wet dog shakes (WDS) were counted for 60 min following senktide administration via either route. Forepaw treading (FPT) was scored on a 0-4 scale following ics senktide in the mouse. In the other studies, the duration (s) of FPT was recorded using a keyboard linked to a microcomputer. Pressing the appropriate keys allowed the automated recording of latency, frequency and duration of up to eight categories of behaviour.

In mice, ics senktide elicited head twitches. The maximal response rate was approximately 50/10 min, at doses between 01. - 0.6 nmol. At higher doses, the head twitch rate declined as other behaviours emerged (see below).

Head twitches following ics senktide were significantly (p<0.05 or better) attenuated by the following antagonists given i.p. 30 min prior to senktide: ketanserin (1.25 mg/kg), ritanserin (1.0 mg/kg), methysergide (10 mg/kg) and prazosin 1.0 mg/kg) (Stoessl et al., 1987).

Forepaw treading following ics senktide in the mouse was attenuated by methysergide and (–)–pindolol (4.0 mg/kg), but unaffected by (+)–pindolol (4.0 mg/kg) or ritanserin. Other behaviours seen following ics senktide in the mouse included hindlimb splaying and tail rattling. These were not noticeably affected by any of the antagonists tested (Stoessl et al., 1987).

Subcutaneous injection of senktide resulted in an almost identical profile of behaviour in mice. The maximal response rated for both head twitching and forepaw treading were seen at 1.2 µmol/kg, which was the highest dose tested.

In rats, senktide elicited WDS with a bell-shaped dose response curve. The doses eliciting the maximal response rat were 0.1 nmol ics (108 WDS/h) and 1.2 µmol/kg sc (50 WDS/h). The duration of FPT continued to increase with higher doses of senktide, and was more pronounced following ics administration (700 s/h at 1.2 nmol) than sc injection (140 s/h at 2.4 µmol/kg). Other behaviours observed following either route of administration included flat body posture, hindlimb splaying, and in some animals, Straub tail and head weaving.

These observations suggest that activation of NK-3 tachykinin receptors results in a pharmacologically significant degree of serotonergic stimulation. The interaction appears to be selective for NK-3 receptors since ics administration of SP and a variety of other tachykinins to the mouse did not result in these behaviours, although they were also elicited by L-363,851, another synthetic tachykinin agonist with selectivity for the NK-3 site (Stoessl et al., 1987).

The mechanism by which this interaction might take place is as yet undetermined. Two major possibilities include (i) an interaction between NK-3 tachykinin and 5-HT receptors or (ii) tachykinin mediated release of 5-HT. Agnati and colleagues (1983) have demonstrated increased binding of [^{3}H] 5-HT to brain homogenates in the presence of SP, but this mechanism does not appear to explain the behavioural effects of senktide, since the latter does not displace the binding of either [^{3}H] 8–OH–DPAT or [^{3}H] ketanserin to rat frontal cortex, not does it shift the equilibrium binding curves for these agents at concentrations up to 100 nM (Stoessl, unpublished observations).

Tachykinin-induced 5-HT release thus appears likelier, and recent work by Solti and Bartfai (1987) has shown that neurokinin B (the endogenous mammalian agonist at NK-3 sites) increases basal and K^+-evoked release of [^{3}H] 5-HT from rat frontal cortical slices in vitro.

Recently, we examined the effects of senktide on brain 5-HT turnover ex vivo. Mice received senktide (0.6 µmol/kg sc) or its vehicle (0.5% DMSO) and were killed by stunning and decapitation 1, 5, 15 or 30 min later.

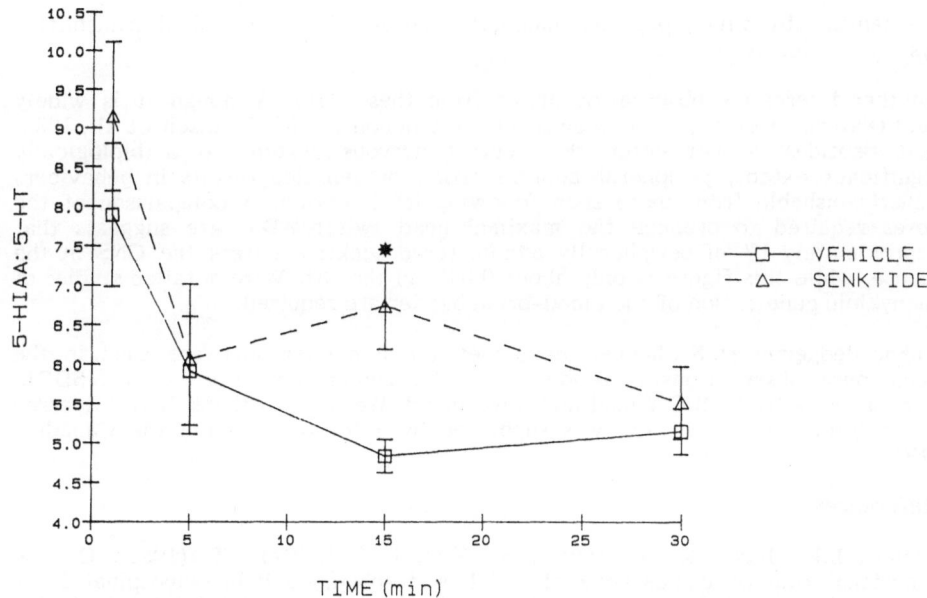

*Fig. 3.1 The ratio of 5-HIAA to 5-HT (both measured in pmol/mg wet tissue weight) following senktide (0.6 µmol/kg sc) or its vehicle (0.5% DMSO in saline, v/v). Each point is the mean (±SEM) of 6 mice. *p<0.03.*

The brains were rapidly removed and dissected on ice, and the brainstem stored at −70°C pending biochemical analysis. 5-HT and 5-HIAA were measured by high performance liquid chromatography with electrochemical detection. Analysis revealed an increase in the ratio of 5-HIAA to 5-HT 15 min after senktide compared to the vehicle treated animals, but not at other time points (Fig. 3.1) ($p<0.03$, 2-way ANOVA). There was no significant difference in absolute tissue levels of 5-HT of 5-HIAA between the two groups. These observations support increased turnover of 5-HT in mouse brainstem following senktide, with a time course that approximates that of the behavioural effects.

Presynaptic mediation of the effects of senktide on 5-HT behaviour is also suggested by our recent finding that senktide-induced wet dog shakes and forepaw treading in the rat were markedly attenuated by a course of p-chlorophenylalanine which resulted in 90% depletion of whole brain 5-HT (Stoessl et al., unpublished).

At present, the localization and functional significance of this tachykinin: 5-HT interaction within the nervous system are unknown. Further work examining the mechanisms by which peptides modulate the activity of classical transmitter systems is clearly needed.

Another interesting observation arises from these data. Although it is widely held (Pardridge et al., 1981; Meisenberg & Simmons, 1983; Ermisch et al., 1985) that peptides do not enter the central nervous system to a biologically significant extent, peripheral administration of senktide results in behaviours indistinguishable from those seen following ics injection. A comparison of the doses required to produce the maximal head twitch/WDS rate suggests that approximately 1% of peripherally administered senktide enters the CNS in the mouse, while this figure is only about 0.04% in the rat. More detailed studies of tachykinin penetration of the blood-brain barrier are required.

Acknowledgements: S.Channell developed the computer software used in the behavioural observations. Senktide was synthesized and characterized at MSDRL, Terlings Park by Dr.S.C.Yound and S.Raymond. We wish to thank Dr.L.L.Iversen for helpful comments. AJS was supported by a fellowship from the Canadian MRC.

References

Agnati, L.F., Fuxe, K., Benfenati, F., Zini, I. & Hokfelt, T. (1983). On the functional role of coexistence of 5-HT and substance P in bulbospinal 5-HT neurons. Substance P reduces affinity and increases density of 3H-5-HT binding sites. Acta Physiol. Scand., 117, 299-301.

Chan-Palay, V., Jonsson, G. & Palay, S.L. (1978). Serotonin and substance P coexist in neurons of the rat's central nervous system. Proc. Natl. Acad. Sci. USA, 75, 1582-1586.

Ermisch, A., Ruhle, H-J., Landgraf, R. & Hess, J. (1985). Blood-brain barrier and peptides. J. Cerebral Blood Flow Metab., 5, 350-357.

Fasmer, O.B., Post, C. & Hole, K. (1987). Increased sensitivity to intrathecal substance P following chronic administration of zimelidine. Neurosci. Lett., 74, 81-84.

Forchetti, C.M., Marco, E.J. & Meek, J.L. (1982). Serotonin and gamma-aminobutyric acid turnover after injection into the median raphe of substance P and D-Ala-Met-Enkephalin amide. J. Neurochem., 38, 1336-1341.

Hokfelt, T., Ljungdahl, A., Steinbusch, H., Verhofstad, A., Nilsson, G., Brodin, E., Pernow, B. & Goldstein, M. (1978). Immunohistochemical evidence of substance P-like immunoreactivity in some 5-hydroxytryptamine-containing neurons in the rat central nervous system. Neurosci., 3, 517-538.

Hylden, J.L.K. & Wilcox, G.L. (1983). Intrathecal serotonin in mice: analgesia and inhibition of a spinal action of substance P. Life Sci., 33, 789-795.

Johansson, O., Hokfelt, T., Pernow, B., Jeffcoate, S.L., White, N., Steinbusch, H.W.M., Verhofstad, A.A.J., Emson, P.C. & Spindel, E. (1981).

Immunohistochemical support for three putative transmitters in one neuron: coexistence of 5-hydroxytryptamine, substance P- and thyrotropin releasing hormone-like immunoreactivity in medullary neurons projecting to the spinal cord. Neuroscience, 6, 1857-1881.

Magoul, R., Onteneinte, B., Oblin, A. & Calas, A. (1986). Inter- and intracellular relationship of substance P-containing neurons with serotonin and GABA in the dorsal raphe nucleus: combination of autoradiographic and immunocytochemical techniques. J. Histochem. Cytochem., 34, 735-742.

Meisenberg, G., & Simmons, W.H. (1983). Peptides and the blood-brain barrier. Life Sci., 32, 2611-2623.

Mitchell, R. & Fleetwood-Walker, S. (1981). Substance P., but not TRH, modulates the 5-HT autoreceptor in ventral lumbar spinal cord. Eur. J. Pharmacol. 76, 119-120.

Neekers, L.M., Schwartz, J.P., Wyatt, R.J. & Speciale, S.G. (1979). Substance P afferents from the habenula innervate the dorsal raphe nucleus. Exp. Brain Res., 37, 619-623.

Ogren, S-O, Fuxe, K. & Agnati, L.F. (1985). Evidence for a modulatory influence of centrally administered substance P on a 5-hydroxytryptamine mediated behavioural response in the male rat. Acta Physiol. Scand., 125, 743-745.

Pardridge, W.M., Frank, H.J.L., Cornford, E.M., Braun, L.D., Crane, P.D. & Oldendorf, W.H. (1981). Neuropeptides and the blood-brain barrier. In: Neurosecretion and brain peptides. (Eds. Martin, J.B., Reichlin, S. & Bick, K.L.) pp 321-328. New York, Raven Press.

Pretel, S., Ruda, M.A. (1986). A comparison of noradrenalin, enkephalin and substance P input to the caudal raphe nuclei of the cat. Anat. Rec., 214 104 A.

Reisine, T., Soubrie, P., Artaud, F. & Glowinski, J. (1982). Application of L-glutamic acid and substance P to the substantia nigra modulates in vivo [^{3}H]serotonin release in the basal ganglia of the cat. Brain Res. 236: 317-317.

Reubi, J.C., Emson, P.C., Jessell, T.M. & Iversen, L.L. (1978). Effects of GABA, dopamine, and substance P on the release of newly synthesized ^{3}H-5-hydroxytryptamine from rat substantia nigra in vitro. Naunyn-Schmiedeberg's Arch. Pharmacl. 304, 271-275.

Solti, M., Bartfai, T., (1987). Tachykinin regulation of serotonin release: enhancement of [^{3}H]serotonin release from rat cerebral cortex by neuromedin K and substance P acting at distinct receptor sites. Brain Res. 401, 377-380.

Stoessl, A.J., Dourish, C.T., Young, S.C., Williams, B.J., Iversen, S.D. & Iversen, L.L. (1987). Senktide, a selective neurokinin B-like agonist, elicits serotonin-mediated behaviour following intracisternal administration in the mouse. Neurosci. Letters, 80, 321-326.

Wormser, U., Laufer, R., Hart, Y., Chorev, M., Gilon, C. & Selinger, Z. (1986). Highly selective agonists for substance P receptor subtypes. The EMBO Journal, 5, 2805–2808.

5-HT AND SEXUAL BEHAVIOUR.

P. Bevan, Department of Pharmacology, Duphar B.V., P.O.Box 2, 1380 AA Weesp, The Netherlands

In this section, we will examine the role of 5-HT neurotransmission in male and female sexual behaviour in rats. A pre-requisite of such studies is an easily observable and, preferably, quantifiable behaviour. In females, this presents no real problem, as the display of lordosis is generally regarded as an indicator of sexual receptivity in the female rat. This might be regarded as a measure of "motivation" in the broad sense, although it is always induced by the administration of hormone, for the convenience of the experimenter.

Measurement of male sexual behaviour is inevitably rated according to performance, although motivational aspects are undoubtedly important. Measurement of performance is convenient because both facilitation and inhibition can be easily quantified and thus the effect of drugs observed. Measurement of motivational aspects is more elusive, but the possibility of using castrated males will be discussed.

5-HT has traditionally been regarded as a purely inhibitory transmitter, a dogma which has been repeatedly challenged in the literature by electrophysiologists for more than 20 years. Similarly, 5-HT had been ascribed only an inhibitory role in the modulation of male and female sexual behaviour. The evidence for this was based on observations that manipulations which would lead to increased 5-HT levels would decrease sexual behaviour whereas 5-HT depletion would increase sexual activity. Behavioural pharmacologists now have some of the tools available to examine the role of 5-HT in sexual behaviour more thoroughly. However, the tools presently available are not optimal, neither in the behavioural nor in the molecular pharmacological sense, and this will be touched on in the chapters which follow.

Despite the limitations of the pharmacological tools available, there can be no doubt that 5-HT neurotransmission can exert a far more subtle influence on sexual behaviour than previously thought. In female rats, then, all the evidence points to a role of $5-HT_1$, $5-HT_2$ and even $5-HT_3$ receptors in the modulation of oestrogen-induced lordosis. Drugs acting at $5-HT_{1A}$ receptors inhibit lordosis whereas $5-HT_2$ receptor agonists facilitate this behaviour. Drugs acting at $5-HT_{1B}$ receptors appear to facilitate sexual behaviour in female rats, although inhibition has been reported to result from the administration of higher doses. $5-HT_3$ antagonists facilitate lordosis suggesting that $5-HT_3$ agonists would inhibit this behaviour.

Male rats, to my surprise, react differently to serotonergic drugs than do females. $5-HT_2$ receptors seem to play no part in modulating the sexual activity of the male rat. At least, both d-LSD and DOM are inactive. Moreover, 8-OH-DPAT facilitates male sexual performance and $5-HT_{1B}$ agonists generally decrease sexual activity in males. No $5-HT_3$ drugs have been tested on male sexual activity to my knowledge. Nevertheless, it would appear that serotonergic drugs have opposite effects on the sexual behaviour of male and female rats.

The effects of serotonergic drugs on male sexual behaviour illustrate some of the difficulties in interpreting effects of drugs on complex behaviours. For example, 8-OH-DOAT facilitates performance in male rats, and the interval to the first ejaculation is reduced. However, the latency to the first intromission is increased. This suggests possible effects at the sensory-motor level and motivational level respectively, and these issues will receive attention.

There is little doubt, then, that serotonergic drugs can both increase and decrease sexual behaviour in rats, both male and female. Some progress has also been made in the pharmacological characterisation of the receptors involved. The localisation of the receptors has also been investigated. Whereas their morphological localisation i.e. presynaptic or postsynaptic, remains speculative in my view, the anatomical distribution seems fairly clear, at least in the case of those receptor sub-types mediating female sexual behaviour.

Inhibitory effects would appear to be mediated via receptors located in the ventromedial nucleus of the hypothalamus whereas facilitatory effects are mediated by $5-HT_2$ receptors in the median eminence. A role for 5-HT receptors in the hypothalamus is also implied in male sexual behaviour. Naturally, the importance of 5-HT receptors located in other brain regions, particularly fore-brain regions, must be considered too.

Almost all behaviours are complex events involving multi-pathway multi-transmitter systems, and this point is addressed by our contributors here. It is possible therefore that some of the effects described subsequently in this section may be due to the modulation by 5-HT pathways of other neurotransmitter events resulting in, for example, functional antagonism. Indeed, some serotonergic drugs may be acting on non-serotonergic nerve terminals to modify the release of transmitters other than 5-HT.

The discussions which follow do much to enhance our understanding of the physiology and pharmacology of sexual behaviour in the laboratory rat. An extension of this knowledge to cover other species is eagerly awaited. This is of particular importance, not least because of the potential clinical application of many of the drugs being discussed.

NEW ASPECTS ON THE SEROTONERGIC MODULATION OF MALE RAT SEXUAL BEHAVIOUR

Sven Ahlenius[1,2], Knut Larsson[1]
1)Department of Neuropharmacology,
Astra Research Centre, S–151 85 Södertälje, and 2)Department of Psychology University of Göteborg, P.O.Box 14158, S–400 20 Göteborg, Sweden

Introduction
It is generally assumed that the neurotransmitter 5-hydroxytryptamine (5-HT, serotonin) has an inhibitory role in the mediation of male rat sexual behaviour. This assumption is based on the observation that an increase in brain 5-HT, as produced by inhibitors of monoamine oxidase or the 5-HT precursor, 5-hydroxytryptophan (5-HTP), is accompanied by an inhibition of the behaviour. Conversely, a decrease in brain 5-HT, as produced by the inhibition of tryptophan hydroxylase by p-chloro-phenylalanine (PCPA), chemical or mechanical lesions of 5-HT neurons, facilitates male rat sexual behaviour. The experimental evidence has been summarized in a number of recent reviews (e.g. Meyerson and Malmnäs, 1978; Crowley and Zemlan, 1981; Larsson and Ahlenius, 1985; Bitran and Hull, 1987). In this presentation we will focus on results obtained by means of the new serotonergic agent 8-OH-2-(di-n-propylamino) tetralin (8-OH-DPAT) (Arvidsson et al., 1981).

This compound, 8-OH-DPAT, pharmacologically characterized as a centrally active 5-HT agonist (Hjorth et al., 1982), produced totally unexpected effects on the male rat sexual behaviour (Ahlenius et al., 1981). In fact, the firmly established picture on the role of 5-HT in male rat sexual behaviour, as outlined above, made us postulate unique properties of this compound on brain serotonergic neurotransmission. Additional biochemical experiments have shown that the compound is an excellent serotonergic receptor ligand in vivo and in vitro (e.g. Hamon et al., 1987; Palacios et al., 1987), and results from receptor binding experiments have suggested that it has high affinity for the 5-HT receptor subtype 5-HT$_{1A}$ (Deshmukh et al., 1983; Middlemiss and Fozard, 1983).

It should also be noted, however, that interpretation of the behavioural physiology of this proposed receptor is limited by the lack of selective antagonists. Certain ß-blocking agents like (-)pindolol and (-)alprenolol do have some specificity in this regard, but so far it has not been possible to separate ß-blocking from serotonergic blocking properties. It is interesting to note that clinically, a new group of anxiolytic compounds represented by buspirone and gepirone, has tentatively been identified as 5-HT$_{1A}$ agonists (e.g. McMillan et al., 1987), and these compounds share the effects of 8-OH-DPAT on male rat sexual behaviour (Glaser et al., 1987). If both these effects indeed are due to 5-HT$_{1A}$ receptor stimulation, this indicates that 8-OH-DPAT may have potential as an anxiolytic compound.

The male rat sexual behaviour
Since the male rat sexual behaviour is observed under laboratory conditions, a few words on equipment and the behavioural items observed, may be appropriate.

The animals used in this laboratory are adult male rats of the Wistar strain (Möllegaard Breeding laboratories, Vejle, Denmark). Light-dark cycle is artificially maintained (dark 11.00 - 23.00 h), and observations of the animals are performed between 13.00 - 17.00 h, in a semicircular arena with a perspex front. Before used experimentally, the animals have gained sexual experience in 2-4 pretests. Stimulus females of the same strain are brought into estrous by sequential estrogen-progesterone treatment.

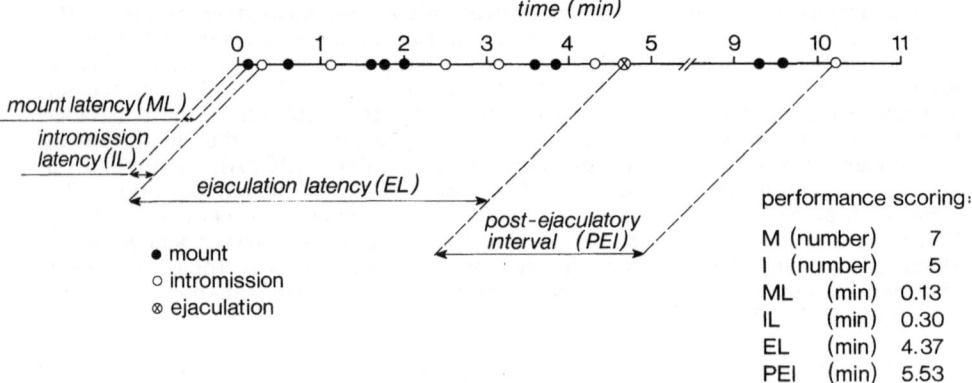

Fig. 4.1 Schematic presentation of the normal male rat sexual behaviour

The behavioural items observed are schematically presented in Fig. 4.1. Thus, the behaviour is recorded until the first ejaculation, and terminated when the male resumes copulation by an intromission. For further details on apparatus and behavioural observations see Larsson (1979).

The sexual behaviour is usually considered to have motivational, as separated from performance, components (Beach, 1956; Sachs, 1978). When discussing effects on performance of the male rat sexual behaviour we have used the terms "facilitation" and "inhibition" as explained in Fig. 4.2 (cf. Fig. 4.1).

By facilitation we mean that ejaculation is achieved after fewer mounts and intromissions, and within a shorter time than normally. Inhibitory effects are noted when the time to ejaculation is prolonged, usually accompanied by an increase in number of intromissions and mounts.

Effects of 8-OH-DPAT on male rat sexual behaviour

The administration of 8-OH-DPAT to male rats produces a marked decrease in the number of intromissions preceding ejaculation, and in the ejaculation latency, as well as a significant decrease in the post-ejaculatory interval (Fig. 4.3).

"Facilitation" by pCPA

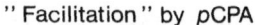

"Inhibition" by 5-HTP

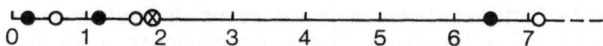

"Marked facilitation" by 8-OH-DPAT

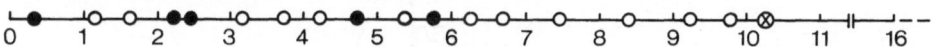

● mount
○ intromission
⊗ ejaculation

time (min)

Fig. 4.2 *Schematic presentation of drug-induced changes in the male rat sexual behaviour*

It should be noted that spontaneous ejaculations are not triggered by 8-OH-DPAT administration (Renyi, 1985). In addition, when given to castrated, sexually inactive, male rats there is a resumption of sexual activity, including ejaculatory behaviour (Ahlenius et al., 1981). Thus, there appears to be effects on motivational aspects, as well as on the performance of the sexual behaviour. Similar effects were found by use of the 8-OMe analog of 8-OH-DPAT. This compound, 8-OMe-DPAT, was also available in its (+) and (-) enantiomers.

However, we could not find evidence for stereoselectivity of the compound by use of the resolved material. In this first series of experiments we also compared the effects of 8-OH-DPAT with effects produced by the hallucinogenic serotonergic compounds d-LSD and 2,5-dimethoxy-4-methylamphetamine (DOM). These latter compounds, however, were devoid of effects on the male rat sexual behaviour. It is also interesting to note that drug discrimination studies indicate that d-LSD and 8-OH-DPAT give rise to different internal cues as experienced by the animals (Glennon, 1986; Cunningham et al., 1987). In fact, up to this point we had not observed effects in this laboratory, nor were there reports in the literature of effects, comparable to those obtained by 8-OH- or 8-OMe-DPAT. The effect is, of course, highly abnormal and any comparison with human behaviour would be highly speculative. If we owned King Solomon's ring, we would surely hear complaints about ejaculatio praecox!

In an effort to put the 8-OH-DPAT-induced effects in context, and to investigate its mechanism of action we found that structurally related ergot drugs like lisuride and quinpirole facilitated the male rat sexual behaviour in a manner similar to 8-OH-DPAT. Results from these experiments will be described next.

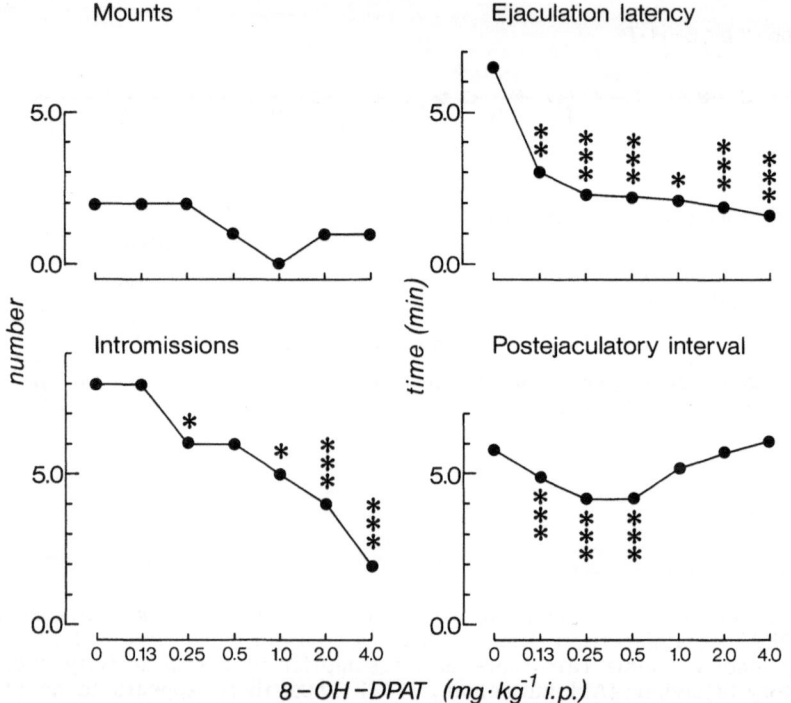

Fig. 4.3 *Effect of 8-OH-DPAT on male rat sexual behaviour. 8-OH-DPAT was administered in doses as indicated in the figure, 15 min before behavioural observations. Repeated measurements were made on the same animals (n=17) using a change-over design. The results are presented as medians in the figure. Statistical comparisons with the saline control session was made by means of the Wilcoxon matched-pairs signed-ranks test (from Ahlenius et al., 1981).*
* $p<0.05$ **<0.02 ***$p<0.01$

Effects of lisuride, quinpirole and some other ergot drugs on male rat sexual behaviour
Our initial experiments indicated that lisuride and pergolide, but not bromocriptine, produce a similar decrease in the number of intromissions, and in time to ejaculation, as seen after 8-OH-DPAT administration. In the case of lisuride, we also found that male rats, made inactive by castration, initiated sexual behaviour, including the ejaculatory pattern (Ahlenius et al., 1980; 1982).

Somewhat suprising, continued experiments revealed that quinpirole (or its racemic form LY 141865), like 8-OH-DPAT, also facilitated male rat sexual behaviour. Quinpirole has been characterized as a selective DA D_2 receptor agonist (Tsuruta et al., 1981), and thus our results suggest a possible dopaminergic involvement in the effects produced by 8-OH-DPAT. We have recently observed that the increase in striatal DA synthesis produced by reserpine can be antagonized by treatment with 8-OH-DPAT, indicating agonist actions at central DA receptors (Ahlenius et al., 1988), and, on closer examination, the initial pharmacological characterization of 8-OH-DPAT indeed indicated results in the same direction (Hjorth et al. 1982). It should be noted that lisuride, which is similar to 8-OH-DPAT, as regards effects on male rat sexual behaviour, also is an agonist at brain 5-HT as well DA receptors (Kehr, 1977). Thus, in our continued experiments, we first turned to the problem of separating serotonergic from dopaminergic mechanisms in the effects produced by 8-OH-DPAT.

Effects of serotonergic and catecholaminergic antagonists on the effects produced by 8-OH-DPAT, lisuride and quinpirole on male rat sexual behaviour
As described above, we find a facilitation of the male rat sexual behaviour by the administration of 8-OH-DPAT, lisuride, pergolide or quinpirole (or its racemic form LY 141865).

In order to investigate a possible role of DA we pretreated the animals with the DA receptor blocking agent haloperidol (Andén et al., 1970). However, we were unable to antagonise the effects produced by 8-OH-DPAT, lisuride or quinpirole by this pretreatment (Ahlenius and Larsson 1984a). A further observation is worth mentioning. The DA agonists apomorphine (Andén et al., 1967), or 5-OH-DPAT (Arvidsson et al., 1981) also facilitate the male rat sexual behaviour, although much less dramatically than 8-OH-DPAT (Butcher et al., 1969; Tagliamonte et al., 1974; Paglietti et al., 1978; Ahlenius et al., 1981). This effect, however, in contrast to the effect produced by 8-OH-DPAT, is readily antagonized by haloperidol (0.16 mg.kg^{-1} i.p.) (Ahlenius and Larsson, 1987). The comparison between effects produced by 5-OH- and 8-OH-DPAT is particularly illuminating. The actions on the sexual behaviour of these two compounds, which are characterized as preferentially active as central DA and 5-HT agonists, respectively (Arvidsson et al., 1981), were thus separated by the use of haloperidol. Taken together, results from these experiments indicate the involvement of 5-HT in the effects produced not only by 8-OH-DPAT and lisuride, but also possitly in the effects produced by pergolide and quinpirole.

In a following series of experiments we investigated the effects of the 5-HT antagonists metergoline, methiothepin, and pirenperone on the 8-OH-DPAT-induced facilitation of male rat sexual behaviour. None of these agents, however, with demonstrated effects on 5-HT$_1$ as well as 5-HT$_2$ receptors could block the 8-OH-DPAT induced effects. The same drugs, in the same doses, completely blocked 5-HTP-induced effects on male rat sexual behaviour (Ahlenius and Larsson, 1984b).

It has been reported that certain β-blocking agents, including (-)pindolol and (-)alprenolol, also block 5-HT$_1$ receptors, and the subgroup designated 5-HT$_{1A}$, in particular (Green and Grahame-Smith, 1976; Middlemiss et al.,

1977; Tricklebank et al., 1984). Pretreatment with either of these two compounds, produced at least a partial antagonism of the 8-OH-DPAT-induced facilitation of male rat sexual behaviour (see Ahlenius and Larsson, 1987).

Thus, summarizing the above evidence there is probably no involvement of central DA receptors in the effect produced by 8-OH-DPAT, at least not of DA receptors sensitive to haloperidol. Of the centrally active 5-HT antagonists, (-)pindolol and (-)alprenolol were the only compounds with some effects. A word of caution is in place, however, since (-)pindolol as well as (-)alprenolol produced a dose-dependent inhibition of the male rat sexual behaviour by themselves.
The negative or ambiguous results with the antagonists as described above made us look into the possibility that 8-OH-DPAT could have 5-HT antagonistic properties.

8-OH-DPAT as a 5-HT antagonist
Biochemical and behavioural observations indicate that lisuride is a centrally active 5-HT agonist (Kehr, 1977; Rosenfeld and Makman, 1981; Rogawski and Aghajanian, 1979; Silbergeld and Hruska, 1979). It is also well known, however, that lisuride, like d-LSD and a number of other ergot compounds, possess 5-HT receptor antagonistic properties as observed in peripheral models of serotonergic activity (Podvalova and Dlabac, 1972; see Müller-Schweinitzer and Weidmann, 1978). In view of the great similarity between lisuride and 8-OH-DPAT, as regards effects on male rat sexual behaviour, we considered the possibility that 8-OH-DPAT also could have 5-HT antagonistic actions, and that these actions could be responsible for the behavioural effects described here. Our negative results with the different catecholamine antagonists also greatly influenced the planning of experiments in this direction.

In order to investigate a possible 5-HT antagonistic action of 8-OH-DPAT or lisuride, we pretreated animals with 5-HTP, which results in a characteristic inhibition of the male rat sexual behaviour (cf. Fig. 4.2). The 5-HTP-induced effects were dose-dependently antagonised by 8-OH-DPAT or lisuride treatment, and these experiments gave us the most clear indication of a mechanism of action of these compounds (Ahlenius and Larsson, 1985). Thus, it appears that both lisuride and 8-OH-DPAT can act as 5-HT antagonists at central sites. That the effects by 8-OH-DPAT, on male rat sexual behaviour, indeed are centrally mediated has been demonstrated by Svensson and Hansen (1984), who showed that the effects were elicited by intracerebroventricular or intrathecal application of 8-OH-DPAT in µg quantities.

Together with other results (e.g. Hjorth et al., 1982), this demonstrates mixed agonist/antagonist actions of 8-OH-DPAT and lisuride and suggests the possibility that these drugs are partial 5-HT agonists. It is interesting to note that recently Gerber et al. (1985) provided other evidence for central serotonergic antagonistic actions of lisuride and 8-OH-DPAT, although comparatively high doses were needed of the latter compound. There is further evidence that 8-OH-DPAT is a partial agonist. Thus, PCPA pretreatment enhances the behavioural response to 8-OH-DPAT (Engel et al., 1984), and the increase in male rats of plasma prolactin produced by the 5-HT agonist 5-MeODMT is significantly attenuated by concomitant 8-OH-DPAT treatment (Carlsson and Eriksson, 1986). In fact, partial agonist properties may explain the particular pharmacological profile of these compounds. A case in point is the characteristic pharmacological profile exhibited by some recent dopaminergic partial agonist like (-)3-PPP and trans-dihydrolisuride (see Clark et al., 1985a). On the assumption that different receptor populations display different sensitivity, as determined e.g. by normal agonist receptor occupancy, a partial agonist may show up as an agonist or antagonist depending on the preparation used (see Carlsson, 1983).

Effects of bromo-lisuride, alone and in combination with 8-OH-DPAT, on male rat sexual behaviour

A few facts regarding structure-activity relationships, like the important differences between 8-OH- and 5-OH-DPAT as well as the similarities in the effects of the stereoisomers of 8-OMe-DPAT, have been mentioned in passing above. In this section we will discuss some further observations that may be of relevance in this connection.

On the face it, the results from the experiment, shown in Fig. 4.4, were, by and large, negative; i.e. bromo-lisuride had few effects of its own on the male rat sexual behaviour, and there was no antagonism of the effects produced by 8-OH-DPAT. Another ergot, bromocriptine, also does not affect the male rat sexual behaviour (Ahlenius et al., 1982). Both compounds, in contrast to lisuride, pergolide and quinpirole, are substituted in the 2-position of the ergoline ring skeleton by a bromide. If we assume that the facilitation of the male rat sexual behaviour, brought about by the administration of lisuride, pergolide or quinpirole, somehow is linked to the serotonergic system, it is possible that brom in the 2-position eliminates the affinity to serotonergic receptors. At least there is no effect on the male rat sexual behaviour, comparable to 8-OH-DPAT, nor was the 8-OH-DPAT-induced effect antagonized by bromo-lisuride pretreatment. The affinity to the dopamine receptor, however, is not affected. Bromo-lisuride has been shown to possess DA receptor blocking properties (Wachtel et al., 1983), and bromocriptine, like the other ergots tested here, is a DA receptor agonist (Pieri et al., 1975). The DA receptor blocking properties by bromo-lisuride may explain the slight reduction in number of intromissions (statistically significant) and a tendency to a decrease in time to ejaculation. These latter effects are comparable to the effects obtained by the use of haloperidol (Ahlenius and Larsson, 1984c).

Comparison of pharmacological profile of 8-OH-DPAT and RDS-127: Relevance for mechanism of action

2-N,N-di-n-Propylamino-4,7-dimethoxyindane (RSD-127) has been shown to be a centrally active DA agonist with some selectivity for dopaminergic autoreceptors (Bhatnagar et al., 1982; Arneric et al., 1983; Pinnock, 1984). In a series of experiments, this compound has been reported to produce effects on male rat sexual behaviour similar to those observed after 8-OH-DPAT administration, including ejaculation on the first intromission and induction of copulatory behaviour in castrated animals (Clark et al., 1982; 1983). Furthermore, these effects can be obtained by intracerebroventricular administration of the drug (Clark et al., 1985b), demonstrating that the effect is centrally elicited. Interestingly, the effects of RDS-127 on male rat sexual behaviour were not antagonized by pretreatment with pimozide or metergoline

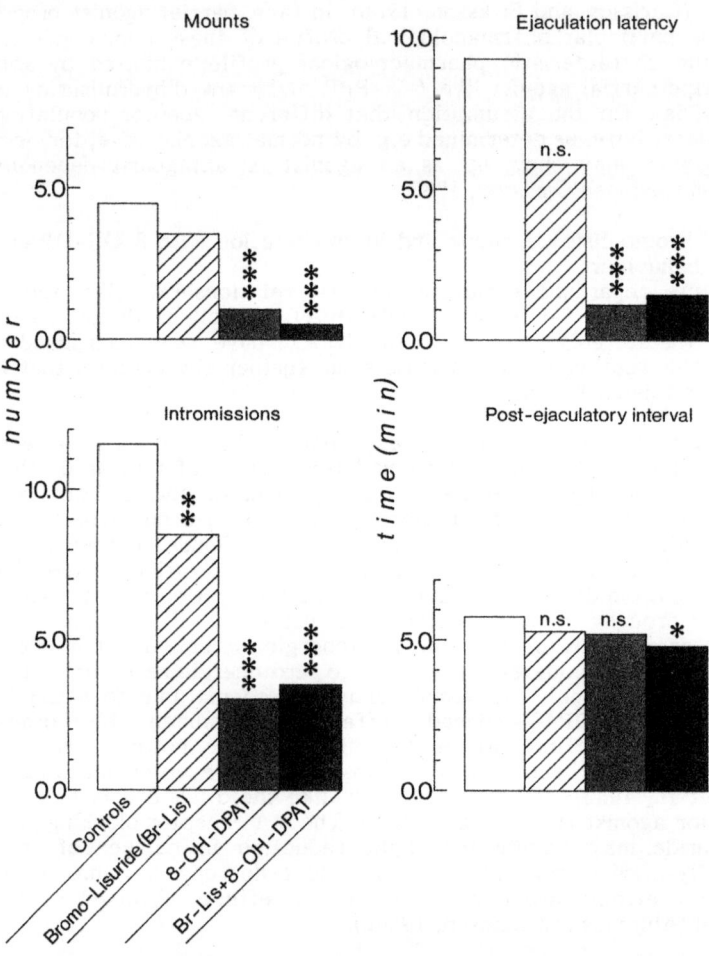

Fig. 4.4 Effects of bromo-lisuride, alone and in combination with 8-OH-DPAT, on the male rat sexual behaviour.
Bromo-lisuride, 0.2 mg.kg^{-1} i.p., was administered 30 min, and 8-OH-DPAT, 0.25 mg.kg^{-1} i.p., 15 min before behavioural observations started. Repeated measurements were made on the same animals (n=18) using a change-over design. Thus, the animals served as their own controls, and statistical comparisons with the control condition were made by means of the Wilcoxon matched-pairs signed-rank test. There were at least 2 days between successive drug treatments. The results are presented as medians in the figure.
n.s. p>0.05 *p<0.05 **p<0.02 ***p<0.01

(Clark and Smith, 1986), blocking DA and 5-HT receptors, respectively (Ferrini and Glässer, 1965; Andén et al., 1970; Fuxe et al., 1975). As described above, we failed to antagonize the 8-OH-DPAT-induced effects by use of metergoline or haloperidol, the latter compound being similar to pimozide as a DA receptor blocking agent.

Furthermore, it was demonstrated that RSD-127 potently displaces [^{3}H]8-OH-DPAT binding in rat brain cortex, indicating a common site of action, possibly the 5-HT$_{1A}$ receptor (Clark et al., 1985b). These results then, together with results from this laboratory, demonstrate the possibility of separating serotonergic from dopaminergic mechanisms in the effects of 8-OH-DPAT and RSD-127 by means of observations of male rat sexual behaviour.

Quinpirole, as well as the more potent and selective later development LY 163502 (Bymaster et al., 1986), both facilitate male rat sexual behaviour in a manner similar to 8-OH-DPAT and RSD-127 (Ahlenius and Larsson, 1984a; Foreman and Hall, 1987). However, affinity to central 5-HT receptors remains to be demonstrated for these agents, characterized as highly selective for the D2 subtype of brain DA receptors (Bymaster et al., 1986; Tsuruta et al., 1981).

Experiments aimed at elucidating the nature of the effects of 8-OH-DPAT on the male rat sexual behaviour
The compound RSD-127, discussed above, facilitates ejaculation in male rats, not only in copula but also ex copula, in intact as well as spinal animals (Stefanick et al., 1982). This latter effect, not observed with 8-OH-DPAT (Renyi, 1985), could possibly be attributed to DA agonist properties of this compound. The most prominent effect of 8-OH-DPAT is a marked reduction of the number of intromissions preceding ejaculation, accompanied by a decrease in time to ejaculation, sometimes to the point that the animals ejaculate on the first intromission. We thus speculated that the effect could be some sensory-motor "short-cut" at the spinal or possibly supraspinal level. If so, some simple measurements of sensory-motor reactivity, like the startle response, would be decisive. Indeed, 8-OH-DPAT as well as lisuride, enhanced the startle amplitude (but not the startle latency) (Svensson and Ahlenius, 1983; Svensson, 1985). However, there are several indications that this effect is not directly linked to the effects of 8-OH-DPAT on the male rat sexual behaviour; (1) 5-HTP and 8-OH-DPAT produced qualitatively the same effect; (2) The effects of 8-OH-DPAT and lisuride are readily blocked by metitepine; and (3) quinpirole or pergolide lack effect on the startle amplitude (but do affect startle latency) (Svensson, 1985).

These observations fit the original characterization of 8-OH-DPAT as a 5-HT receptor agonist, producing the effects seen e.g. by administration of 5-HTP (Hjorth et al., 1982).Finally, it should be noted that the doses of 8-OH-DPAT required to affect the acoustic startle are at least 10 times the doses affecting the male rat sexual behaviour. We also tested the effects of 8-OH-DPAT on sensory-motor reactivity in another situation, the respiratory response to acoustic stimulation (Criborn 1969). PCPA produces a decrease in this response, an effect completely antagonized by 5-HTP treatment (in combination with benserazide). In this behavioural situation 8-OH-DPAT, like PCPA, produced a decrease in the respiratory response (Ahlenius et al., 1985). Finally, we have investigated the importance of sensory feed-back from the penis via the pudendal nerve for the 8-OH-DPAT-induced facilitation of male rat sexual behaviour. Since the number of intromissions preceding ejaculation are reduced by 8-OH-DPAT treatment, it was speculated that sensory feed-back from the penis was essential for the effects observed. Pudendectomy produces a characteristic increase in the number of mounts and prolongation in time to ejaculation (see Dahlöf and Larsson, 1976). This effect by pudendectomy was evident also in the present experiments. There were no signs, however, of antagonism by pudendectomy of 8-OH-DPAT-induced effects on the male rat sexual behaviour (Fig. 4.5).

In fact, there is a dramatic normalization of the behaviour in lesioned animals, and at least the sensory afferent supply via the pendundal nerve is apparently not necessary for the effects produced by 8-OH-DPAT (Dahlöf et al., 1988). It is interesting to compare these results with the effects obtained by the administration of PCPA to pudendectomized rats. Also in this case there was a normalization of the sexual behaviour by the drug treatment (Dahlöf, 1980).

Taken together, there is some evidence for increased reactivity to sensory stimulation by 8-OH-DPAT. On the other hand, the respiratory response to acoustic stimulation was decreased, and it is not clear how these observations are related to the effects of 8-OH-DPAT on the male rat sexual behaviour.

The sensory feed-back via the pudendal nerve is not critically involved in these effects by 8-OH-DPAT, and it remains to be shown to what extent other forms of sensory input is influenced by 8-OH-DPAT treatment. PCPA also restores the suppression of sexual behaviour in anosmic male rats (Dahlöf, unpublished observations). As mentioned above, we have previously noted several instances of similarities in effects produced by PCPA and 8-OH-DPAT, suggesting a decrease in central 5-HT neurotransmission as a common denominator in their mechanisms af action. If this applies also here, we have evidence from different modalities, suggesting that sensory input does not greatly influence the effects produced by 8-OH-DPAT on male rat sexual behaviour. The characteristic motor pattern of the pelvis accompaning mounts and intromission, has been shown not to be affected by 8-OH-DPAT treatment (Morali and Larsson, 1984). Thus, the effects of 8-OH-DPAT on male rat sexual behaviour appears to be not only central of origin, but to a large extent uninfluenced by the sensory input, and the compound does not greatly influence motor aspects influence motor aspects of the behaviour.

Selective actions of 8-OH-DPAT, and related compounds, in limbic brain areas
8-OH-DPAT clearly produces effects due to stimulation of postsynaptic 5-HT receptors, like hindlimb extension and forepaw treading (Hjorth et al., 1982; Smith and Peroutka, 1986). There is also evidence for stimulation of presynaptic 5-HT receptors on serotonergic cell bodies in brain stem raphe nuclei (Dourish et al., 1986). This latter finding indicates that the unexpected effects of 8-OH-DPAT on body temperature (Ahlenius et al., 1985; Hjorth, 1985; Goodwin et al., 1985; cf. Jacob and Girault, 1979) feeding (Dourish et al., 1985) and sexual behaviour (Ahlenius et al., 1981) could be due to 5-HT autoreceptor activation. The evidence in favour of 8-OH-DPAT as a partial 5-HT agonist, explaining its effects in terms of regional differences in receptor sensitivity and regulation, is

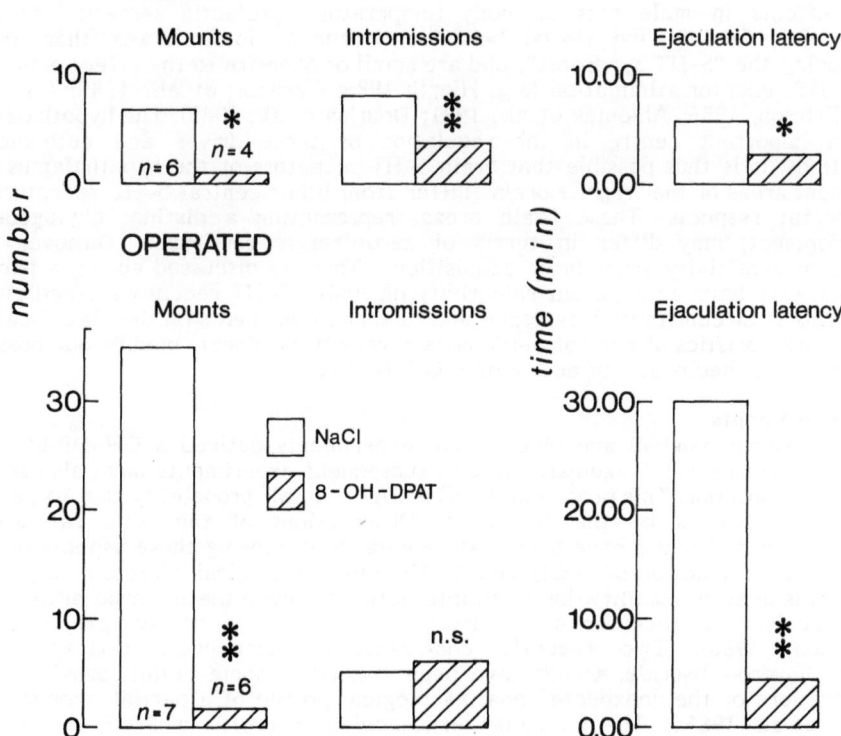

Fig. 4.5 Effects of 8-OH-DPAT on the sexual behaviour in pudendectomized male rats.
Three weeks following sensory deafferentation by means of cutting the dorsal penile nerve, or sham operation, the animals were given physiological saline or 8-OH-DPAT as indicated in the figure. 8-OH-DPAT was administered s.c. in a dose of 0.25 mg.kg^{-1}, 20 min before behavioural observations started. The results are presented as medians in the figure. Statistical comparisons between drug treated and control animals were made by means of the Mann-Whitney U-test (from Dahlöf et al., 1988).
n.s. p>0.05 *p<0.05 **p<0.01

in general agreement with such a conclusion.

The effects in male rats on body temperature, prolactin secretion and, in particular, feeding and sexual behaviour occur at lower doses than those producing the "5-HT syndrome", and are small or opposite to the effect expected by 5-HT receptor stimulation (e.g. Hjorth 1985; Carlsson et al., 1985; Carlsson and Erikson, 1986; Ahlenius et al., 1981; Dourish et al., 1985). The hypothalamus is an important centre in the regulation of these drives and autonomous functions. It is thus possible that brain 5-HT receptors of the hypothalamus and adjacent areas of the "reptile brain" differ from other central 5-HT receptors in important respects. These brain areas, representing a distinct phylogenetic development, may differ in terms of serotonergic feed-back, turnover and receptor sensitivity from later acquisitions. Thus, as discussed above, a partial agonist may have an apparent selectivity on limbic 5-HT receptors. Needless to say, this is circumstantial evidence and further experiments on the classification and characteristics of central 5-HT receptors will, no doubt, modify our present views on the mechanism of action of 8-OH-DPAT.

Final comments

Initial pharmacological and biochemical experiments defined 8-OH-DPAT as a centrally acting 5-HT agonist, whereas subsequent experiments have also shown antagonist actions. This profile of 8-OH-DPAT is in all probability due to partial agonist properties of the compound. Observations of the male rat sexual behaviour in particular have been instrumental in disclosing these aspects of the mechanisms of action of 8-OH-DPAT. The pharmacological effects of a partial agonist is in all probability due to an interaction between the intrinsic activity of the agonist and receptor sensitivity, pre- as well as postsynaptically (see Carlsson, 1983). Two recently characterized compounds, (-)3-PPP and trans-dihydro- lisuride, which have been studied in some detail, provide good illustrations of the unexpected pharmacological profile of a partial agonist (see Clark et al., 1985a). The particular pharmacological profile of 8-OH-DPAT, and some related compounds, as described above, may thus be explained by an apparent selectivity for certain populations of 5-HT receptors.

There are regional differences in brain 5-HT turnover (e.g. Fuenmayor and Garcia, 1984), and this may very well be associated with differences in neuronal feed-back and receptor regulation. On the basis of the functions particularly sensitive to 8-OH-DPAT treatment, there appears to be a preference for limbic 5-HT receptors.

Thus, we certainly have a new type of 5-HT receptor agonist in 8-OH-DPAT, as initially proposed from experiments on male rat sexual behaviour, but not necessarily a new type of 5-HT receptor. The effects obtained by 8-OH-DPAT in our experiments on male rat sexual behaviour could very well be due to a blockade of postsynaptic receptors, although stimulation of presynaptic receptors contribute to the effect obtained. This reasoning is based on the assumption of a high degree of competition at the particular postsynaptic sites of interest. Admittedly, this is at present an assumption for which direct experimental support is lacking.

As has been discussed in detail above, the male rat sexual behaviour is a function particularly sensitive to 8-OH-DPAT treatment. The effect can probably not be reduced to a simple sensory-motor facilitation, and may rather have a motivational component, affecting the initiation as well as performance of the behaviour. It is interesting to note that another conspicuous effect by 8-OH-DPAT is overeating in satiated rats (Dourish et al., 1985). Depicting the effect of 8-OH-DPAT in terms of motivation has the advantage of not focussing on the particular behavioural item, but it should be noted that 8-OH-DPAT and lisuride have a "normal" 5-HT agonist profile on female rat sexual behaviour, and the lordosis response (Ahlenius et al., 1986; Fernandez-Guasti et al., 1987; Hlinak, 1987). Furthermore, there is no effect, or inhibition, of the male mouse sexual behaviour (Svensson et al., 1987). Needless to say, effects of 8-OH-DPAT on male rat sexual behaviour are interesting in their own right, and may further our understanding of the physiology of sexual behaviour.

Acknowledgements: We are indebted to Ms Madelene Kröning for her skills in preparing the figures, and to Ms Ann Fritz for patience with the manuscript. The studies summarized in this review were made possible by grants from the Swedish Humanitites and Social Sciences Research Council, The Bank of Sweden Tercentenary Foundation, and Astra Alab AB.

References

Ahlenius, S., Criborn, C.-O. & Henriksson, Ch. (1985). Central 5-HT and the respiratory response to acoustic stimulation in awake rats: effects of PCPA, 5-HTP and 8-OH-DPAT. J. Neural Transm., 63, 285-295.

Ahlenius, S., Engel, J., Larsson, K. & Svensson, L. (1982). Effects of pergolide and bromocriptine on male rat sexual behaviour. J. Neural Transm., 54, 165-170.

Ahlenius, S., Fernandez-Guasti, A., Hjorth, S. & Larsson, K. (1986). Suppression of the lordosis behaviour by the putative 5-HT receptor agonist 8-OH-DPAT in the rat. Eur. J. Pharmacol., 124, 361-363.

Ahlenius, S., Hillegaart, V. & Wijkström, A. (1988). Region-selective effects by 8-OH-DPAT on rat brain dopamine synthesis in vivo. Neurochem. Int., 13, (S1), 80.

Ahlenius, S. & Larsson, K. (1984a). Lisuride, LY-141865 and 8-OH-DPAT facilitate male rat sexual behaviour via a non-dopaminergic mechanism. Psychopharmacol., 83, 330-334.

Ahlenius, S. & Larsson, K. (1984b). Failure to antagonize the 8-hydroxy-2-(di-n-propylamino)- tetralin-induced facilitation of male rat sexual behaviour by the administration of 5-HT receptor antagonists. Eur. J. Pharmacol., 99, 279-286.

Ahlenius, S. & Larsson, K. (1984c). Apomorphine and haloperidol-induced effects on male rat sexual behaviour: no evidence for actions due to stimulation of central dopamine autoreceptors. Pharmacol. Biochem. Behav., 21, 463-466.

Ahlenius, S. & Larsson, K. (1985). Antagonism by lisiride and 8-OH-DPAT of 5-HTP-induced prolongation of the performance of male rat sexual behavior. Eur. J. Pharmacol. 110, 379-381.

Ahlenius, S. & Larsson, K. (1987). Evidence for a unique pharmacological profile of 8-OH-DPAT by evaluation of its effects on male rat sexual behaviour. In: Pharmacology of Central 5-HT$_{1A}$ Receptors (Eds. Dourish C.T., Ahlenius S., Hutson P.), pp. 185-198. Ellis Horwood, Chichester.

Ahlenius, S., Larsson, K. & Svensson, L. (1980). Stimulating effects of lisuride on masculine sexual behaviour of rats. Eur. J. Pharmacol., 64, 47-51.

Ahlenius, S., Larsson, K., Svensson, L., Hjorth, S., Carlsson, A., Lindberg, P., Wikström, H., Sanchez, D., Arvidsson, L-E., Hacksell, U. & Nilsson, J.L.G. (1981). Effects of a new type of 5-HT receptor agonist on male rat sexual behaviour. Pharmacol. Biochem. Behav., 15, 785-792.

Anden, N.-E., Butcher, S.G., Corrodi, H., Fuxe, K. & Ungerstedt, U. (1970). Receptor activity and turnover of dopamine and noradrenaline after neuroleptics. Eur. J. Pharmacol., 11, 303-314.

Anden, N.-E., Rubenson, A., Fuxe, K. & Hökfelt, T. (1967). Evidence for dopamine receptor stimulation by apomorphine. J. Pharm. Pharmacol., 19, 627.

Arneric, S.P., Long, J.P., Williams, M., Goodale, D.B., Mott, J., Lakosi, J.M. & Gebhart, G.L. (1983). RSD-127 (2-di-n-propylamino-4,7-dimethoxyindane): central effects of a new dopamine receptor agonist. J. Pharmacol. Exp. Ther., 224, 161-170.

Arvidsson, L.-E., Hacksell, U., Nilsson, J.L.G., Hjorth, S., Carlsson, A., Lindberg, P., Sanchez, D. & Wikström, J. (1981). 8-Hydroxy-2(di-n-propylamino) tetralin, a new centrally acting 5-hydroxytryptamine receptor agonist. J. Med. Chem., 24, 921-923.

Beach, F.A. (1956). Characteristics of masculine "sex drive". In: The Nebraska Symposium on Motivation (Ed. Jones, M.R.), pp. 1-32. University of Nebraska Press, Lincoln.

Bitran, D. & Hull, E.M. (1987). Pharmacological analysis of male rat sexual behaviour. Neurosci. Biobehav. Rev., 11, 365-389.

Bhatnagar, R.K., Arneric, S.P., Cannon, J.G., Flynn, J. & Long, J.P. (1982). Structure activity relationships of presynaptic dopamine receptor agonists. Pharmacol. Biochem. Behav., 17 (Suppl. 1), 11-19.

Butcher, L.L., Butcher, S.G. & Larsson, K. (1969). Effects of apomorphine, (+)-amphetamine, and nialamide on tetrabenazine-induced suppression of sexual behaviour in the male rat. Eur. J. Pharmacol., 7, 283-288.

Bymaster, F.P., Reid, L.R., Nichols, C.L., Kornfeld, E.C. & Wong, D.T. (1986). Evaluation of acetylcholine levels in striatum of rat brain by LY163502, trans-(-)-5,5a,6,7,8,9a,10-octahydro-6-propylpyrimido <4,5-g>quinolin-2-amine dihydrochloride, a potent and stereospecific dopamine (D_2) agonist. Life Sci., 38, 317-322.

Carlsson, A. (1983). Intrinsic activity vs. state of the receptor. J. Neural Transm., 57, 309-315.

Carlsson, M. & Eriksson, E. (1986). A central serotonin receptor agonist, 8-hydroxy-2-(di-n-propylamino) tetralin, has different effects on prolactin secretion in male and female rats. Pharmacol. Toxicol., 58, 297-302.

Carlsson, M., Svensson, K., Eriksson, E. & Carlsson, A. (1985). Rat brain serotonin: biochemical and functional evidence for a sex difference. J. Neural Transm., 62, 297-313.

Clark, D., Hjorth, S. & Carlsson, A. (1985a). Dopamine-receptor agonists: mechanisms underlying autoreceptor selectivity. I. Review of the evidence. J. Neural Transm., 62, 1-52.

Clark, J.T., Peroutka, S.J., Ciarenello, R.D. & Smith, E.R. (1985b). Central effects of RSD-127: sexual behaviour after intracerebroventricular administration and in vitro binding studies. Behav. Brain Res., 18, 251-260.

Clark, J.T. & Smith, E.R. (1986). Failure of pimozide and metergoline to antagonize the RSD-127-induced facilitation of ejaculatory behavior. Physiol. Behav., 37, 47-52.

Clark, J.T., Smith, E.R., Stefanick, M.L., Arneric, S.P., Long, J.P. & Davidson, J.M. (1982). Effects of a novel dopamine-receptor agonist RSD-127 (2-N,N-di-n-propylamino-4,7-dimethoxyindane), on hormone levels and sexual behaviour in the male rat. Physiol. Behav., 29, 1-6.

Clark, J.T., Stefanick, M.L., Smith, E.R. & Davidson, J.M. (1983). Further studies on alterations in male rat copulatory behaviour induced by the dopamine-receptor agonist RSD-127. Pharmacol. Biochem. Behav., 19, 781-786.

Criborn, C-O. (1969). Recordings of the respiratory process in mice without strain on the respiratory organs. Life Sci., 8 (I), 1351-1358.

Crowley, W.R. & Zemlan, F.P. (1981). The neurochemical control of mating behaviour. In: Neuroendocrinology of Reproduction (Ed. Adler, N.T.), pp. 451-484. Plenum Publishing Corp., New York.

Cunningham, K.A., Callahan, P.M. & Appel, J.B. (1987). Discriminative stimulus properties of 8-hydroxy-2-(di-n-propylamino) tetralin (8-OH-DPAT): implications for understanding the actions of novel anxiolytics. Eur. J. Pharmacol., 138, 29-36.

Dahlöf, L.-G. (1980). PCPA and the sexual behaviour of the pudendectomized male rat. Biol. Behav., 5, 211-218.

Dahlöf, L.-G., Ahlenius, S. & Larsson, K. (1988). Copulatory performance of penile desensitized male rats following the administration of 8-OH-DPAT. Physiol. Behav. 43: (In Press.)

Dahlöf, L.-G. & Larsson, K. (1976). Interactional effects of pudendal nerve section and social restriction on male rat sexual behaviour. Physiol. Behav., 16, 757-762.

Deshmukh, P.P., Yamamura, H.I., Woods, L. & Nelson, D.L. (1983). Computer-assisted autoradiographic localization of subtypes of serotonin-1 receptors in rat brain. Brain Res., 288, 338-343.

Dourish, C.T., Hutson, P.H. & Curzon, G. (1985). Low doses of the putative serotonin agonist 8-hydroxy-2-(di-n-propylamino) tetralin (8-OH-DPAT) elicit feeding in the rat. Psychopharmnacol., 86, 197-204.

Dourish, C.T., Hutson, P.H., Kennett, G.A. & Curzon, G. (1986). 8-OH-DPAT-induced hyperphagia: its neural basis and possible therapeutic relevance. Appetite, 7 (Suppl), 127-140.

Engel, J.A., Hjorth, S., Svensson, K., Carlsson, A. & Liljequist, S. (1984). Anticonflict effect of the putative serotonin receptor agonist 8-hydroxy-2-(di-n-propylamino) tetralin (8-OH-DPAT). Eur. J. Pharmacol., 105, 365-368.

Fernandez-Guasti, A., Ahlenius, S., Hjorth S. & Larsson, K. (1987). Separation of dopaminergic and serotonergic inhibitory mechanisms in the mediation of estrogen-induced lordosis behaviour in the rat. Pharmacol. Biochem. Behav., 27, 93-98.

Ferrini, R. & Glässer, A. (1965). Antagonism of central effects of tryptamine and 5-hydroxytryptophan by 1,6-dimethyl-8 beta-carbobenzyloxyaminomethyl -10-α-ergoline). Psychopharmacol., 8, 271-276.

Foreman, M.M., Hall, J.L. (1987). Effects of D_2 dopaminergic receptor stimulation on male rat sexual behaviour. J. Neural Transm., 68, 153-170.

Fuenmayor, L.D. & Garcia, S. (1984). The effect of fasting on 5-hydroxytryptamine metabolism in brain regions of the albino rat. Br. J. Pharmacol., 83, 357-362.

Fuxe, K., Agnati, L. & Everitt, B. (1975). Effects of methergoline on central monoamine neurons. Evidence for a selective blockade of central 5-HT receptors. Neurosci. Lett., 1, 283-290.

Gerber, R., Barbaz, B.J., Martin, L.L., Neale, R., Williams, M. & Liebman, J.M. (1985). Antagonism of L-5-hydroxytryptamine-induced head twitching in rats by lisuride: a mixed 5-hydroxytryptamine agonist-antagonist? Neurosci. Lett., 60, 207-213.

Glaser, T., Dompert, W.U., Schuurman, T., Spencer, D.G. & Traber, J. (1987).

Differential pharmacology of the novel 5-HT$_{1A}$ receptor ligands 8-OH-DPAT, BAT R 1531 and ipsapirone. In: Pharmacology of Central 5-HT$_{1A}$ Receptors (Eds. Dourish, C.T., Ahlenius, S., Hutson, P.), pp. 106-119. Chichester: Ellis Horwood,

Glennon, R.A. (1986). Discriminative stimulus properties of the 5-HT$_{1A}$ agonist 8-hydroxy-2-(di-n-propylamino) tetralin (8-OH-DPAT). Pharmacol. Biochem. Behav., 25, 135-139.

Goodwin, G.M., De Souza, R.J., Green, A.R. (1985). The pharmacology of the hypothermic response in mice to 8-hydroxy-2-(di-n-propylamino) tetralin (8-OH-DPAT). Neuropharmacol., 24, 1187-1194.

Green, A.R. & Grahame-Smith, D.G. (1976) (-)Propranolol inhibits the behavioural responses of rats to increased 5-hydroxytryptamine in the central nervous system. Nature, 262, 594-596.

Hamon, M., Emerit, M.B., El Mestikawy, S., Verge, D., Daval, G., Marquet, A. & Gozlan, H. (1987). Pharmacological, biochemical and functional properties of 5-HT$_{1A}$ receptor binding sites labelled by [^{3}H]8-hydroxy-2-(di-n-propylamino) tetralin in the rat brain. In: Pharmacology of Central 5-HT$_{1A}$ Receptors (Eds. Dourish, C.T., Ahlenius, S., Hutson, P.), pp. 34-51. Chichester: Ellis Horwood.

Hjorth, S. (1985). Hypothermia in the rat induced by the potent serotonergic agent 8-OH-DPAT. J. Neural. Transm., 61, 131-135.

Hjorth, S., Carlsson, A., Lindberg, P., Sanchez, D., Wikström, H., Arvidsson, L-E, Hacksell, U. & Nilsson, J.L.G. (1982). 8-hydroxy-2-(di-n-propylamino) tetralin (8-OH-DPAT), a potent and selective simplified ergot congener with central 5-HT receptor stimulating activity. J. Neural. Transm., 55, 163-188.

Hlinak, Z. (1987). Lisuride inhibits temporarily sexual behaviour in female rats. Pharmacol. Biochem. Behav., 27, 211-215.

Jacob, J.J. & Girault, M-M.T. (1979). 5-Hydroxytryptamine. In: Body Temperature - Regulation. Drug Effects and Therapeutic Implications. (Eds. Lomax, P., Schoenbaum, E.), pp. 183-230. New York: Marcel Dekker.

Kehr, W. (1977). Effect of lisuride and other ergot derivatives on monoaminergic mechanisms in rat brain. Eur. J. Pharmacol., 41, 261-273.

Larsson, K. (1979). Features of the neuroendocrine regulation of masculine sexual behaviour. In: Endocrine Control of Sexual Behavior. (Ed. Beyer, C.), pp. 77-163. New York: Raven Press.

Larsson, K. & Ahlenius, S. (1985). Masculine sexual behaviour and brain monoamines. In: Psychopharmacology of Sexual Disorders. (Ed. Segal, M.), pp. 15-32. London and Paris: John Libbey.

McMillan, B.A., Scott, S.M., Williams, H.L. & Sanghera, M.K. (1987). Effects of gepirone, an aryl-piperazine anxiolytic drug, on aggressive behaviour and brain monoaminergic neurotransmission. Naunyn Schmiedeberg's Arch Pharmacol., 335, 454-464.

Meyerson, B.J. & Malmnäs, C-O. (1978). Brain monoamines and sexual behaviour. In: Biological Determinates of Sexual Behavior. (Ed. Hutchinson, J.B.), pp. 521-554. London: Wiley.

Middlemiss, D.N., Blakeborough, L. & Leather, S.R. (1977). Direct evidence for an interaction of β-adrenergic blockers with the 5-HT receptor. Nature (Lond.), 267, 289-290.

Middlemiss, D.N. & Fozard, J.R. (1983). 8-hydroxy-2-(di-n-propylamino) tetralin discriminates between subtypes of the $5-HT_1$ recognition site. Eur. J. Pharmacol., 90, 151-153.

Morali, G. & Larsson, K. (1984). Differential effects of a new serotoninomimetic drug, 8-OH-DPAT, on copulatory behaviour and pelvic thrusting pattern in the male rat. Pharmacol. Biochem. Behav., 20, 185-187.

Müller-Schweinitzer, E. & Weidman, H. (1978). Basic pharmacological properties. In: Ergot Alkaloids and Related Compounds, (Eds. Berde, B., Schild, H.O.), Handbook Exp. Pharmacol., 49, 87-232. Berlin: Springer.

Paglietti, E., Pellegrini-Quarantotti, B., Mereu, G., Gessa, G.L. (1978). Apomorphine and L-Dopa lower ejaculation threshold in the male rat. Physiol. Behav., 20, 559-562.

Palacios, J.M., Pazos, A. & Hoyer, D. (1987). Characterization and mapping of $5-HT_{1A}$ sites in the brain of animals and man. In: Pharmacology of Central $5-HT_{1A}$ Receptors, pp. 67-81. Chichester: Ellis Horwood.

Pieri, M., Pieri, L., Sander, A., da Prada, M. & Haefely, W. (1975). A comparison of drug-induced rotation in rats lesioned in the medial forebrain bundle with 5,6-dihydroxytryptamine or 6-hydroxydopamine. Arch. Int. Pharmacodyn., 217, 118-130.

Pinnock, R.D. (1984). Action of putative dopamine receptor agonists, TL-99,3-PPP and RSD-127 on substantia nigra neurons. Brain Res., 292, 190-193.

Podvalova, I. & Dlabac, A. (1972). Lysenyl, a new antiserotonin agent. Res. Clin. Stud. Headache, 3, 325-334.

Renyi, L. (1985). Ejaculations induced by p-chloroamphetamine in the rat. Neuropharmacol., 24, 697-704.

Rogawski, M.A., Aghajanian, G.K. (1979). Response of central monoaminergic neurons to lisuride: Comparison with LSD. Life Sci., 24, 1289-1298.

Rosenfeld, M. & Makman, M.H. (1981). The interaction of lisuride, an ergot derivative, with serotonergic and dopaminergic receptors in rabbit brain. J. Pharmacol. Exp. Ther., 216, 526–531.

Sachs, B.D. (1978). Conceptual and neural mechanisms of masculine copulatory behaviour. In: Sex and Behavior: Status and Prospectus. (Eds. McGill, T.E., Dewsbury, D.A., Sachs, B.D.), pp. 267–295. New York: Plenum Press.

Silbergeld, E.K. & Hruska, R.E. (1979). Lisuride and LSD: Dopaminergic and serotonergic interactions in the "seronergic syndrome". Psychopharmacol, 65, 233–237.

Smith, L.M., Peroutka, S.J. (1986). Differential effects of 5-hydroxytryptamine $_{1A}$ selective drugs on the 5-HT behavioural syndrome. Pharmacol. Biochem. Behav., 24, 1513–1519.

Stefanick, M.L., Smith, E.R., Clark, J.T. & Davidson,. J.M. (1982). Effects of a potent dopamine receptor agonist, RSD-127, on penile reflexes and seminal emission in intact and spinally transected rats. Physiol. Behav., 29, 973–978.

Svensson, K., Ahlenius, S., Larsson, K., Arvidsson, L.-E. & Carlsson A. (1978). Evidence for a facilitatory role of central 5-HT in male mouse sexual behaviour. In: Pharmacology of central 5-HT$_{1A}$ receptors. (Eds. Dourish, C.T., Ahlenius, S., Hutson, P.), pp. 199–210. Chichester: Ellis Horwood.

Svensson, L. (1985). Effects of 8-OH-DPAT, lisuride and some ergot-related compounds on the acoustic startle response in the rat. Psychopharmacol., 85, 469–475.

Svensson. L. & Ahlenius, S. (1983). Enhancement by the putative 5-HT receptor agonist 8-OH-2-(di-n-propylamino)tetralin of the acoustic startle response in the rat. Psychopharmacol, 79, 104–107.

Svensson, L. & Hansen, S. (1984). Spinal monoaminergic modulation of masculine copulatory behaviour in the rat. Brain Res., 302, 315–321.

Tagliamonte, A., Fratta, W., del Fiacco, M. & Gessa, G.L. (1974). Possible stimulatory role of brain dopamine in the copulatory behaviour of male rats. Pharmacol. Biochem. Behav., 2, 257–260.

Tricklebank, M.D., Forler, C. & Fozard, J.R. (1984). The involvement of subtypes of the 5-HT$_{1A}$ receptor and catecholaminergic systems in the behavioural response to 8-hydroxy-2-(di-n-propylamino) tetralin in the rat. Eur. J. Pharmacol., 106, 271–282.

Tsuruta, K., Frey, E.A., Grewe, C.W., Cote, T.E.D., Eskay, R.L. & Kebabian, J.W. (1981). Evidence that LY 141865 specifically stimulates the D-2 dopamine receptor. Nature, 292, 463–465.

Wachtel, H., Kehr, W. & Sauer, G. (1983). Central antidopaminergic properties of 2-bromolisuride, an analogue of the ergot dopamine agonist lisuride. Life Sci., 33, 2583–2597.

DIFFERENTIAL ROLES OF 5-HT RECEPTOR SUBTYPES IN FEMALE SEXUAL BEHAVIOUR

Scott D.Mendelson, Boris B.Gorzalka
University of British Columbia, Vancouver, British Columbia V6T 1W5, Canada

Introduction
The evaluation of the effects of serotonergic drugs on the sexual behaviour of the female rat began with the work of Meyerson, who in the early 1960's set forth the hypothesis that serotonergic activity inhibits the expression of lordosis behaviour (Meyerson, 1964a,b,c). Lordosis is the downward flexion of the back and lifting of the rump and tail that may be displayed by a female rat in response to the mounting and pelvic thrusting of a male. The display of lordosis is generally regarded as an indicator of sexual receptivity in the female rat.

Much of the data gathered by Meyerson and others appeared consistent with serotonergic inhibition of lordosis (Meyerson, Malmnas & Everitt, 1985). However, many inconsistencies have been found in the literature (see Mendelson & Gorzalka, 1985). These inconsistencies may partially reflect differential roles of subtypes of 5-HT receptors in the control of lordosis. Over recent years there has been an accumulation of evidence for the existence of subtypes of central 5-HT receptors (e.g., Roberts & Straughn, 1967; Haigler & Aghajanian, 1974a; Von Hungen, Roberts & Hill, 1975). In 1979, by analysing the binding characteristics of serotonin, LSD and spiperone, and comparing the manner in which serotonin and spiperone displaced [^{3}H]LSD from serotonergic binding sites in rat cortical tissue, Peroutka and Snyder were able to identify two pharmacologically distinct populations of serotonin binding sites. One population of binding sites displayed high affinity binding of [^{3}H]serotonin and was designated as the 5-HT$_1$ subtype, whereas the second population displayed high affinity binding of [^{3}H]spiperone and was designated as the 5-HT$_2$ subtype. The 5-HT$_1$ class of binding sites itself has since been divided into subtypes designated as 5-HT$_{1A}$, 5-HT$_{1B}$ and 5-HT$_{1C}$ (Pedigo, Yamamura & Nelson, 1981; Pazos, Hoyer & Palacios, 1984).

Although still controversial, there is growing evidence that the subtypes of central serotonergic binding sites as characterized by in vitro binding analyses represent functional 5-HT receptors. There is evidence that 5-HT$_{1A}$ receptors act as somato-dendritic autoreceptors on serotonergic cell bodies (Sprouse & Aghajanian, 1986) and as the receptors that mediate the postsynaptic stimulation of adenylate cyclase (Markstein, Hoyer & Engel, 1986). 5-HT$_{1B}$ receptors appear to act as prejunctional autoreceptors at serotonergic cell terminals (Engel, Gothert, Hoyer, Schlicker & Hillenbrand, 1986). 5-HT$_{1C}$ receptors have been found to mediate serotonergic stimulation of phosphoinositide hydrolysis in the choroid plexus (Sanders-Bush & Conn, 1986). Evidence indicates that 5-HT$_2$ receptors mediate the neural excitatory effects of serotonin in brain tissue (Peroutka & Snyder, 1979).

The discovery of subtypes of central 5-HT receptors complements the earlier characterization of the D and M subtypes of peripheral 5-HT receptors (Gaddum & Picarelli, 1957). The D receptor is now thought to be similar, if not identical, to the $5\text{-}HT_2$ receptor (Bradley et al., 1986). The M receptor appears distinct from the $5\text{-}HT_1$ and $5\text{-}HT_2$ subtypes, and it has been suggested that it be designated as the $5\text{-}HT_3$ receptor (Bradley et al., 1986).

The following is a review and re-evaluation of the effects of 5-HT antagonists and agonists on lordosis behaviour. Although none of the drugs discussed bind exclusively to 5-HT receptors, their effects appear to be mediated primarily by serotonergic systems. Therefore, we have considered the effects of these drugs only in terms of probable action at specific subtypes of 5-HT receptors. In this review it becomes apparent that serotonin can produce either inhibitory or facilitatory effects on lordosis behaviour, as has been suggested in the recent dual role hypotheses (Mendelson & Gorzalka 1985; Wilson & Hunter, 1985). It is concluded that inhibitory effects of serotonin are mediated primarily by post-synaptic $5\text{-}HT_{1A}$ receptors, whereas facilitatory effects are mediated by $5\text{-}HT_2$ receptors. It is also concluded that activity at somato-dendritic $5\text{-}HT_{1A}$ autoreceptors may facilitate lordosis.

Finally, on the basis of preliminary evidence, it is suggested that $5\text{-}HT_3$ receptors may indicate inhibitory, and prejunctional $5\text{-}HT_{1B}$ autoreceptors facilitatory effects of serotonin on lordosis behaviour.

5-HT receptor antagonists

The classical 5-HT antagonists bind to both $5\text{-}HT_1$ and $5\text{-}HT_2$ receptors. However, these drugs tend to show varying degrees of preference for $5\text{-}HT_2$ receptors (Peroutka, Lebovitz & Snyder, 1981). Recent work indicates that differences may also exist in the degree to which a 5-HT antagonist binds to the various subtypes of the $5\text{-}HT_1$ receptor. Even drugs considered to be quite selective $5\text{-}HT_2$ antagonists tend to bind with high affinity to $5\text{-}HT_{1C}$ receptors. Moreover, it appears that most 5-HT antagonists bind with a markedly higher affinity to $5\text{-}HT_{1A}$ than to $5\text{-}HT_{1B}$ sites.

The effectiveness of the classical 5-HT antagonists at $5\text{-}HT_2$ receptors appears well established. However, while it is clear that these drugs bind to the various subtypes of the $5\text{-}HT_1$ receptor, there remains some doubt as to their effectiveness as antagonists at these sites (Haigler & Aghajanian, 1974a). There is in fact evidence that some classical 5-HT antagonists may act as weak partial agonists at $5\text{-}HT_1$ receptors (Haigler & Aghajanian, 1974a; Peroutka et al., 1981). However, the newer, highly selective $5\text{-}HT_2$ antagonists possess little, if any, agonist activity (Janssen, 1985).

In studies investigating the role of serotonin in lordosis, methysergide has been the most commonly employed receptor antagonist. Methysergide is one of the less selective antagonists, binding with high affinity to $5\text{-}HT_2$, $5\text{-}HT_{1A}$ and $5\text{-}HT_{1C}$ receptors, and with somewhat lower affinity to $5\text{-}HT_{1B}$ sites (Dr.S.J.Peroutka, personal communication).

The administration of methysergide directly into the hypothalamus, hippocampus, or amygdala facilitates lordosis in estrogen-primed females (Zemlan et al., 1973; Ward et al., 1975; Foreman and Moss, 1978; Franck and Ward, 1981). In one instance, inhibition was observed following injection of methysergide into the preoptic area (Clemens, 1978).

Peripherally administered methysergide also facilitates lordosis in estrogen-primed, ovariectomized females (Ward et al., 1975; Hendrik and Gerall, 1976; Davis and Kohl, 1978; Foreman and Moss, 1978; Rodriguez-Sierra and Davis, 1979; Franck and Ward, 1981; Hunter, Hole and Wilson, 1985; Mendelson & Gorzalka, 1986a; Ulibarri & Yahr, 1987). However, the maximal facilitatory effects of peripherally administered methysergide have been reported to occur 2 to 6 h after treatment (Zemlan et al., 1973; Davis and Kohl, 1978).

Pharmacokinetic data indicates that the maximal antiserotonergic effects of intraperitoneally administered methysergide can occur in 30 min and may decline within 1 hr (Sofia & Vassar, 1975). When evaluated 30 min to 1 hr after its peripheral administration, methysergide has been found to be ineffective in females primed with a low dose of estrogen (Mendelson & Gorzalka, 1985), and to inhibit lordosis in females primed with a high dose of estrogen, or with estrogen and progesterone (Meyerson & Eliasson, 1977; Sietnieks, 1985; Mendelson & Gorzalka, 1986a). Together, these data suggest that at the times when it is most effective as a 5-HT antagonist, peripherally administered methysergide inhibits lordosis behaviour. It is conceivable that the lordosis-facilitating effect of methysergide after 2 hr is due to the action of a metabolite.

Cinanserin is one of the classical antagonists more selective for $5-HT_2$ receptors (Leysen & Tollenaere, 1982). When administered into the medial preoptic or posterior areas of the hypothalamus, cinanserin facilitated lordosis in estrogen-primed females (Zemlan et al., 1973; Ward et al., 1975). However, when administered peripherally to estrogen-primed females, 25 mg/kg cinanserin did not facilitate lordosis (Everitt et al., 1975). In females primed with estrogen and progesterone, peripheral administration of 5 mg/kg cinanserin appeared ineffective (Sietnieks, 1985), whereas 10 mg/kg cinanserin substantially inhibited lordosis (Hunter et al., 1985). Hunter et al. (1985) also reported that 10 mg/kg cinanserin produced a slight increase in lordosis behaviour in estrogen-primed females with very low baseline levels of receptivity.

However, because of the arbitrary placement of animals into "receptive" and "non-receptive" groups prior to statistical analysis, we suspect that this apparent facilitation is merely an artifact of the experimental design, that is, a regression toward the mean (see Raible & Gorzalka, 1986). Nonetheless, these data do indicate that while cinanserin may inhibit lordosis, it does not eliminate this behaviour.

Metergoline, like methysergide, is somewhat non-selective in its binding to the various 5-HT receptor subtypes (Hoyer, Engel & Kalkman, 1985). In females treated chronically with estrogen, 0.05 mg/kg metergoline was found to facilitate and doses over 0.5 mg/kg to inhibit lordosis (Fuxe et al., 1976). In females treated either with estrogen (Hunter et al., 1985), or with estrogen and progesterone (Hunter et al., 1985; Sietnieks, 1985), 5 mg/kg metergoline inhibited lordosis.

Other classic antagonists that have been evaluated for their effects upon lordosis behaviour are methiothepin, mianserin, cyproheptadine, and pizotefin. Methiothepin is somewhat nonselective in its serotonergic binding (Hoyer et al., 1985), whereas mianserin, cyproheptadine and pizotefin bind preferentially to

5-HT$_2$ receptors (Lysen & Tollenaere, 1982). When administered systematically to females primed with estrogen and progesterone, methiothepin (Mendelson and Gorzalka, 1986b), and cyproheptadine (Mendelson & Gorzalka, 1986b; Sietnieks, 1985) inhibited lordosis. At a dose lower than our effective one (Mendelson and Gorzalka, 1986b), methiothepin did not inhibit lordosis (Fernandez-Guasti, Ahlenius, Hjorth & Larsson, 1987). The peripheral administration of mianserin inhibited lordosis in females treated either with estrogen, or with estrogen and progesterone (Hunter et al., 1985; Sietnieks, 1985). Pizotefin also inhibited lordosis in females primed with estrogen and progesterone (Mendelson and Gorzalka, 1986b).

Recently, highly selective 5-HT$_2$ antagonists have become available for evaluation. In two studies, peripheral administration of the 5-HT$_2$-selective antagonist pirenperone inhibited lordosis in steroid-primed females (Mendelson and Gorzalka, 1985; Sietnieks, 1985). Intraventricular administration of pirenperione also inhibited lordosis (Mendelson and Gorzalka, unpublished data). In a recent paper, a single dose of pirenperone (0.25 mg/kg) was reported to be ineffective in females primed with estrogen and progesterone (Fernandez-Guasti et al., 1987). However, while this dose of pirenperone was higher than those found effective in our own study (Mendelson and Gorzalka, 1985), it was lower than that found minimally effective by Sietnieks (1985). In at least three studies, ketanserin (1 - 10 mg/kg), a 5-HT$_2$ antagonist related in structure to pirenperone, has been found to inhibit lordosis behaviour (Mendelson & Gorzalka, 1985, 1986b; Hunter et al., 1985). However, the 5-HT$_2$ selective antagonist altanserin did not inhibit lordosis in steroid-primed rats at doses up to 0.2 mg/kg (Sietnieks, 1985).

The ergoline derivative LY53857 has recently been reported to be a potent, and highly selective 5-HT$_2$ antagonist. Unlike most 5-HT$_2$ antagonists, LY53857 is relatively inactive at α_1 adrenergic and dopaminergic receptors (Cohen, Fuller & Kurz, 1983). We have found that LY53857 inhibits lordosis (Mendelson & Gorzalka, unpublished). Ritanserin, another 5-HT$_2$ antagonist with relatively little α-adrenergic activity (Janssen, 1985), also inhibits lordosis (Mendelson & Gorzalka, unpublished data). These data suggest that whether or not 5-HT$_2$ antagonists act at α_1 adrenergic or dopaminergic receptors, the blockade of activity at 5-HT$_2$ receptors is sufficient to inhibit lordosis behaviour.

Although the 5-HT$_3$ receptor has been characterized as a peripheral receptor, recent evidence indicates the existence of 5-HT$_3$ binding sites in brain tissue (Kilpatrick, Jones & Tyers, 1987). The possibility that these binding sites represent functional 5-HT$_3$ receptors is suggested by the recent report that intrahypothalamic administration of the selective 5-HT$_3$ antagonist ICS 205-930 facilitates gastric emptying in the guinea-pig (Costall et al., 1986). We have recently observed that the peripheral administration of 5 mg/kg ICS 205-930 facilitates lordosis in estrogen-primed females (Mendelson & Gorzalka, unpublished data).

Studies employing central administration of ICS 205-930 are in progress in our laboratory. Interestingly, we have as yet observed no indication of facilitatory effects of the 5-HT$_3$ antagonist MDL 72222 (Fozard, 1984). This could be a reflection of the differential effectiveness of these drugs on the subtypes of

5-HT$_3$ receptors (Richardson et al., 1985). However, pending the evaluation of other 5-HT$_3$ selective drugs, this suggestion must be regarded as highly speculative.

Serotonin receptor agonists
Before discussing the effects of drugs that mimic 5-HT, we must note that the administration of 5-HT itself into pre-optic and hypothalamic areas inhibited lordosis in steroid-primed females (Foreman & Moss, 1978; Clement, 1978). However, the injection of 10 µg 5-HT into the third ventricle produced no effect, while injection of 100 µg 5-HT into the lateral ventricle significantly facilitated lordosis (Wilson & Hunter, 1985). These findings suggest that 5-HT can either inhibit or facilitate lordosis, depending on which areas of the brain receive treatment.

LSD was the first 5-HT agonist to be evaluated for effects on lordosis. In vitro binding data indicate that the binding of LSD to the various subtypes of 5-HT receptors is relatively nonselective. It binds with roughly equal high affinities to 5-HT$_2$, 5-HT$_{1A}$ and 5-HT$_{1C}$ receptors, and with slightly lower affinity to 5-HT$_{1B}$ sites (Engel et al., 1986).

Peripheral administration of LSD had inhibited lordosis in females treated with estrogen and progesterone (Eliasson, Mechanek & Meyerson, 1972; Meyerson, Carrer & Eliasson, 1974; Eliasson & Meyerson, 1976, 1977; Sietnieks & Meyerson, 1980). Although one laboratory has reported inhibitory effects of LSD in females treated with estrogen alone (Everitt et al., 1975), another failed to confirm this even with relatively high doses of the drug (Sietnieks & Meyerson, 1980).

The inhibitory effects of LSD have been taken as evidence of serotonergic inhibition of lordosis. However, this interpretation ignores the fact that LSD may act as either an agonist or an antagonist, depending, perhaps, on the subtype of 5-HT receptor. The ability of LSD to reduce the rate of firing of neurons in the dorsal raphe (Haigler & Aghajanian, 1974b), and to inhibit the release of serotonin from neuron terminals (Middlemiss, 1982) suggests that LSD may act as an agonist at 5-HT$_{1A}$ and 5-HT$_{1B}$ receptors, respectively.

That the discrimination of LSD from saline can be blocked by selective 5-HT$_2$ antagonists (Janssen, 1983) suggest that LSD may act as at least a partial agonist at 5-HT$_2$ receptors. However, LSD has been found to block the excitatory, and most likely, 5-HT$_2$ mediated (Peroutka et al., 1981) effects of serotonin in the cortex (Roberts & Straughn, 1967) and reticular formation (Boakes et al., 1970). We suspect that the inhibitory effects of LSD are mediated primarily by postsynaptic 5-HT$_1$ receptors. However, in view of the effects of the classic and selective 5-HT$_2$ antagonists, it is tempting to suggest that the lordosis-inhibiting effect of LSD may be partially due to blockade of activity at specific populations of central 5-HT$_2$ receptors.

It has been suggested that the inhibition of lordosis by LSD is due to an increase in activity at 5-HT$_2$ receptors (Sietnieks, 1985). This conclusion was reached following the finding that the inhibitory effects of LSD were reduced by cinanserin, cyproheptadine, and pirenperone (Peroutka et al., 1981; Leysen & Tollenaere, 1982; Hoyer et al., 1985). Finally, like LSD, cyproheptadine, pirenperone, methysergide, metergoline, and mianserin inhibited lordosis in Sietnieks' (1985) study. Thus, it seems unlikely that Sietnieks' data provide evidence that LSD inhibits lordosis by increasing activity at 5-HT$_2$ receptors.

LSD in very low doses (5-20 µg/kg) appears to facilitate lordosis in estrogen-treated females (Everitt et al., 1975; Sietnieks & Meyerson, 1983). These effects of LSD have been attributed to inhibition of serotonergic activity through action upon autoreceptors in the dorsal raphe. However, whereas a reduction of neuronal activity in the dorsal raphe might contribute to the lordosis-facilitating effect of LSD, the facilitation of lordosis by LSD cannot be due entirely to this mechanism. The reduction of neuronal activity in the raphe following LSD treatment is a relatively short-lived phenomenon. Indeed raphe neurons may begin to recover their normal patterns of firing within 5 minutes after the intravenous administration of LSD (Aghajanian, Foote & Sheard, 1968). In contract, the facilitation of lordosis induced by LSD may persist undiminished for as long as 3 hr after treatment (Sietnieks & Meyerson, 1983).

Recently, it has been found that very low doses of LSD (5-10 µg/kg) enhance the ability of serotonin to facilitate the glutamate-induced excitation of motor neurons in the rat facial nucleus (McCall & Aghajanian, 1980). This effect of serotonin appears to be mediated by 5-HT$_2$ receptors (Penington & Reiffenstein, 1986b). The enhancement of serotonergic activity in the facial nucleus by LSD was found to persist for over 4 hr, a time course similar to that observed in the facilitation of lordosis by LSD (Sietnieks & Meyerson, 1983). These data suggest that postsynaptic enhancement, rather than presynaptic inhibition of serotonergic activity may be responsible for the prolonged lordosis-facilitating effect of low doses of LSD.

The hallucinogenic phenylalkylamines 2,5-dimethoxy- 4-methylamphetamine (DOM), 2,4,5-trimethoxyamphetamine (TMA), 2,5-dimethoxy-4-methyl-phenyl-ethylamine (DOMPE), and mescaline have also been found to facilitate lordosis at low doses, and to inhibit lordosis at higher doses (Everitt & Fuxe, 1977). As with LSD, the facilitatory effects of the phenylalkylamines have been attributed to reductions in serotonergic activity (Everitt & Fuxe, 1977). However, neither mescaline (Haigler & Aghajanian, 1973) nor DOM (Penington and Reiffeinstein, 1986a) has significant effects on serotonergic autoreceptors in the dorsal raphe. Interestingly, like LSD, mescaline (McCall & Aghajanian, 1980) has been found to enhance, and DOM (Penington & Reiffenstein, 1986b) to mimic the 5-HT$_2$ receptor-mediated excitatory effect of serotonin on neurons of the facial motor nucleus.

Of further interest, the affinities of DOM and TMA for 5-HT$_2$ sites have been found to be 30-fold higher (Shannon et al., 1984), and the affinity of mescaline 12-fold higher (Leysen & Tollenaere, 1982) than their affinities for 5-HT$_1$ sites. The range of doses in which these drugs produced facilitation and inhibition of lordosis (Everitt & Fuxe, 1977) seems to reflect the drugs' relative affinities for 5-HT$_2$ and 5-HT$_1$ sites.

Low doses of the hallucinogenic tryptamine derivatives N,N,dimethyltryptamine (DMT), 5-methoxy-dimethyl- tryptamine (5-MeODMT), and psilocybin also facilitate lordosis in estrogen-primed females. High doses of these drugs are inhibitory in females primed with estrogen or with estrogen and progesterone (Everitt & Fuxe, 1977). Tryptamine derivatives have been thought to bind with highest affinity to 5-HT_1 receptors (Peroutka et al., 1981). Within this class of receptors, 5-MeODMT and DMT show a marked selectivity for 5-HT_{1A} sites (Peroutka, 1986). Thus the facilitatory and inhibitory effects of the N-methylated tryptamines could be at least partially due to activity at somato-dendritic autoreceptors and postsynaptic 5-HT_{1A} receptors, respectively. However, at doses comparable to those that facilitate lordosis, psilocin (the active metabolite of psilocybin) enhances, and DMT mimics the neural excitatory effect of serotonin in the facial motor nucleus (McCall & Aghajanian, 1980). Thus it appears that the lordosis-facilitating effects of hallucinogenic tryptamines could be at least partially due to activity at 5-HT_2 receptors.

The piperazine derivate quipazine binds with moderately high affinity to 5-HT_{1C}, 5-HT_{1B}, 5-HT_2 receptors, and with slightly lower affinity to 5-HT_{1A} receptors (Hoyer et al., 1985). Quipazine was first reported to inhibit lordosis in females primed with estrogen and progesterone (Rodriguez-Sierra & Davis, 1979). However, in later studies comparable doses of quipazine were found to be ineffective in females treated with estrogen and progesterone (Arendash & Gorski, 1982; Mendelson & Gorzalka, 1985), and to facilitate lordosis in females treated with estrogen alone (Hunter et al., 1985). Interestingly, quipazine has also been found to facilitate lordosis, to a limited degree, in spinal rats (Kow, Zemlan & Pfass, 1979). We have subsequently observed that low doses of quipazine facilitate, whereas doses over 9 mg/kg may inhibit lordosis in estrogen-primed females (Mendelson & Gorzalka, unpublished data).

Quipazine is active at somato-dendritic 5-HT_{1A} autoreceptors (Blier & de Montigny, 1983), thus it may facilitate lordosis by reducing the activity of lordosis-inhibiting serotonergic pathways. At higher doses, quipazine may inhibit lordosis by activating postsynaptic 5-HT_{1A} receptors, or by enhancing the release of 5-HT in certain areas through its action as a weak antagonist at prejunctional 5-HT_{1B} autoreceptors (Martin & Sanders-Bush, 1982). However, quipazine appears to act primarily as a 5-HT_2 agonist. Quipazine elevates serum corticosterone levels, and this is reversed by treatment with the selective 5-HT_2 antagonist LY535857 (Cohen et al., 1983). In stimulus generalization studies, animals trained to respond to the 5-HT_2 agonist DOM will also respond to quipazine (Glennon, Young & Rosencrans, 1983). Perhaps most importantly, quipazine attenuates the lordosis-inhibiting effects of the 5-HT_2 antagonists pirenpirone, ketanserin, methysergide, cyproheptadine, and pizotefin (Mendelson & Gorzalka, 1985, 1986b). These findings suggest that quipazine facilitates lordosis by stimulating 5-HT_2 receptors, and attenuates the effects of 5-HT_2 antagonists by restoring activity to these receptors. In view of the much higher affinity of antagonists for 5-HT_2 receptors, the suggestion that quipazine could complete effectively with these drugs for 5-HT_2 binding sites is surprising. Interestingly, recent data indicate that a subpopulation of 5-HT_2 binding sites possesses a conformational state with high affinity for 5-HT_2 agonists (Lyon, Davis & Titeler, 1987). Quipazine has been found to bind with quite high affinity to the agonist binding state of the 5-HT_2 receptor (Lyon et al., 1987). At moderately high doses, quipazine might be expected to displace 5-HT_2 antagonists from these sites. It is worth noting that DMT, 5-MeODMT, and phenylalkylamines similar to DOM bind with very high affinity to the agonist binding state of the 5-HT$_2$ receptor (Lyon et al., 1987).

The piperazine MK 212 facilitates lordosis, and this effect has been attributed to stimulation of 5-HT$_2$ receptors (Wilson & Hunter, 1985). MK 212 produced the head-twitch response (Clineschmidt, McGuffin & Pflueger, 1977) in a manner typical of 5-HT$_2$ agonists and it substitutes for quipazine in drug discrimination studies (Lucot, 1984). However, binding studies show a very low affinity of MK 212 for 5-HT$_2$ and other 5-HT receptors (Engel et al., 1986). It would be of interest to determine the affinity of MK 212 for the agonist binding state of the 5-HT$_2$ receptor.

Recently, drugs with very high selectivity for 5-HT$_{1A}$ receptors have become available. The highly selective 5-HT$_{1A}$ agonist 8-OH-DPAT, and the slightly less selective partial agonists buspirone, ipsapirone, and gepirone inhibit lordosis in females primed either with estrogen, or with estrogen and progesterone (Ahlenius et al., 1986; Mendelson & Gorzalka, 1986c,d). Extremely small doses of the somewhat selective 5-HT$_{1A}$ agonist lisuride (Peroutka, 1986) also inhibit lordosis in females primed with estrogen and progesterone (Sietnieks, 1985). At lower doses, ipsapirone and gepirone facilitate lordosis in estrogen-primed females (Mendelson & Gorzalka, 1986d). Because both drugs reduce the activity of serotonergic neurons in the dorsal raphe (Dourish, Hutson & Curzon, 1986; Eison et al., 1986) the lordosis-facilitating effects of these drugs may be due to activity at somatodendritic 5-HT$_{1A}$ autoreceptors.

It is of interest to note that doses of buspirone, ipsapirone, and gepirone that either facilitate lordosis or are ineffective in females primed with estrogen alone inhibit lordosis in females primed with both estrogen and progesterone (Mendelson & Gorzalka, 1986d). The lordosis-inhibiting effects of the selective 5-HT$_{1A}$ agonists 8-OH-DPAT (Mendelson & Gorzalka, unpublished data) and lisuride (Hlinak, 1987; Sietnieks, 1985), the 5-HT$_{1A}$ active agonist LSD (Sietnieks & Meyerson, 1980) and the uncharacterized 5-HT agonist α-methyltryptamine (Espino, Sano & Wade, 1975) also appear to be either enhanced by or dependent upon treatment with progesterone. These data support our recent suggestion that progesterone enhances the effects of activity at 5-HT$_{1A}$ receptors (Mendelson & Gorzalka, 1986d).

The 5-HT agonist 1-(3-trifluoromethylphenyl)piperazine (TFMPP) has a serotonergic binding profile similar to that of quipazine. However, TFMPP appears to act primarily as an agonist at 5-HT$_{1B}$ receptors. Unlike quipazine, TFMPP inhibits the K$^+$-induced release of 5-HT from hypothalamic synaptosomes (Martin & Sanders-Bush, 1982). In stimulus generalization studies, animals trained to respond to TFMPP will respond to the 5-HT$_{1B}$ agonists m-chlorophenylpiperazine (MCPP) and RU24969, but not to quipazine, the 5-HT$_2$ agonist DOM, or the selective 5-HT$_{1A}$ agonist 8-OH-DPAT (Cunningham & Appel, 1986; Glennon, McKenny & Young, 1984). Low, but not high, doses of peripherally administered TFMPP or MCPP facilitate lordosis in estrogen-primed females (Mendelson & Gorzalka, unpublished data). We have very recently observed that 0.03 mg/kg of the putative 5-HT$_{1B}$ agonist CGS 12066B (Neale et al., 1987) facilitates lordosis in estrogen-primed females (Mendelson & Gorzalka, unpublished data). Together, these data suggest that prejunctional $5-HT_{1B}$ autoreceptors mediate lordosis-facilitating effects of 5-HT. In apparent contradiction to this possibility, RU 24696 has been reported to inhibit lordosis (Hunter & Wilson, 1985). However, whereas RU 24696 has often been characterized as a selective $5-HT_{1B}$ agonist, the drug binds with nearly equal high affinity to $5-HT_{1A}$ receptors (Tricklebank, Middlemiss and Neill, 1986). Indeed, at doses that inhibit lordosis (Wilson & Hunter, 1985), RU 24969 mimics the hypothermic effect of the $5-HT_{1A}$ selective agonist 8-OH-DPAT (Tricklebank et al., 1986). Nevertheless, there is recent evidence suggesting that $5-HT_{1B}$ receptors may exist post- as well as pre-synaptically (Kennett, Dourish & Curzon, 1987). Thus we recognize the possibility that higher doses of $5-HT_{1B}$ agonists could produce lordosis-inhibiting effects through action at certain postsynaptic $5-HT_{1B}$ sites.

Conclusions

In reviewing the effects of 5-HT antagonists and agonists it becomes apparent that serotonin can either inhibit or facilitate the expression of lordosis behaviour in the female rat. It appears that the lordosis-inhibiting effects of serotonin are mediated primarily by post-synaptic $5-HT_{1A}$ receptors.

Moreover, we suggest that the $5-HT_{1A}$ receptors that mediate these effects exist primarily in the forebrain. This conclusion is consistent with the variety of reports (cited in Mendelson & Gorzalka, 1985) indicating that simple depletion of forebrain serotonin levels, by either chemical or surgical means, facilitates lordosis. On the basis of preliminary evidence, we further suggest that forebrain $5-HT_3$ receptors may mediate some of the lordosis-inhibiting effects of serotonin.

The lordosis-facilitating effects of serotonin appear to be mediated primarily by $5-HT_2$ receptors. Moreover, it is tempting to suggest that the subpopulation of $5-HT_2$ receptors possessing a high affinity agonist binding state may play a particularly important role in mediating these effects. Of course, we might again note that drugs selective for $5-HT_2$ receptors tend also to bind with high affinity to $5-HT_{1C}$ receptors. We therefore cannot eliminate the possibility that activity at certain populations of $5-HT_{1C}$ receptors enhances lordosis.

There is evidence that stimulation of the medullary reticular formation facilitates lordosis (Cohen, Schwartz-Giblin & Pfaff, 1987). Moreover, its appears that $5-HT_2$ receptors mediate neural excitatory effects of serotonin in this area (Haigler & Aghajanian, 1974a; Peroutka et al., 1981). Very recently we have found that the administration of small doses of the $5-HT_2$ agonist LY 53857 directly into the medullary reticular formation inhibits lordosis, whereas the administration of quipazine into this area is facilitatory (Mendleson & Gorzalka, unpublished). In view of these results, and of the many reports of facilitation of lordosis following depletion of forebrain serotonin, we hypothesize that the lordosis-facilitating effects of serotonin are mediated primarily by neural excitatory activity at $5-HT_2$ receptors in the brainstem.

It is tempting to offer an additional mechanism by which activity at $5-HT_2$ receptors might facilitate lordosis. It has been suggested that in areas of the brain where $5-HT_1$ and $5-HT_2$ receptors co-exist, $5-HT_2$ receptors may serve to modulate the activity of $5-HT_1$ receptors. In some areas, activity at

5-HT$_2$ appears to diminish the effects of activity at 5-HT1 receptors (Aghajanian, Sprouse & Rasmussen, 1987). It may be that selective stimulation of 5-HT$_2$ receptors can facilitate lordosis indirectly by attenuating the lordosis-inhibiting effects of activity at adjacent 5-HT$_{1A}$ receptors in the forebrain. Conversely, 5-HT$_2$ antagonists would inhibit lordosis by freeing forebrain 5-HT$_{1A}$ receptors from the modulating effects of activity at 5-HT$_2$ receptors. It is interesting to consider that by this mechanism, the apparent effectiveness of 5-HT$_2$ agonists and antagonists in affecting lordosis might vary as a function of baseline levels of serotonergic activity.

Evidence suggests that activity at somato-dendritic 5-HT$_{1A}$ autoreceptors may also facilitate lordosis behaviour. Ostensibly, this would be due to reductions in the activity of lordosis-inhibiting serotonergic pathways ascending to the forebrain. We further suggest that activity at prejunctional autoreceptors of the 5-HT$_{1B}$ type facilitates lordosis activity. As with the activation of autoreceptors of the 5-HT$_{1A}$ type, the activation of prejunctional 5-HT$_{1B}$ receptors would result in the reduction of activity in certain lordosis-inhibiting serotonergic pathways.

In summary, we hypothesize that serotonin can either inhibit or facilitate lordosis behaviour. We suggest that the inhibitory effects of serotonin are mediated primarily by post-synaptic 5-HT$_{1A}$ receptors in the forebrain, whereas the facilitatory effects are mediated by 5-HT$_2$ receptors in the brain stem. We further suggest that activity at somato-dendritic 5-HT$_{1A}$ autoreceptors in the raphe nuclei may facilitate lordosis. Finally, on the basis of preliminary evidence, we suggest that 5-HT$_3$ receptors may mediate inhibitory, and prejunctional 5-HT$_{1B}$ autoreceptors facilitatory effects of serotonin on lordosis behaviour.

Acknowledgement: This research was supported by a Natural Sciences and Engineering Council of Canada grant to B.B. Gorzalka.

References

Aghajanian, G.K., Foote, W.E. & Sheard, M.H. (1968). Lysergic acid diethylamide: sensitive neuronal units in the midbrain raphe. Science, **161**, 706-708.

Aghajanian, G.K., Sprouse, J.S. & Rasmussen, K. (1987). Physiology of the midbrain serotonin system. (Ed. Meltzer, H.Y.). Psychopharmacology, the third generation of progress. New York: Raven Press. (In press.)

Ahlenius, S., Fernandez-Guasti, A., Hjorth, S. & Larsson, K. (1986). Suppression of lordosis behaviour by the putative 5-HT receptor agonist 8-OH-DPAT in the rat. Eur. J. Pharmacol. **124**, 361-363.

Blier, P. & de Montigny C. (1983). Effects of quipazine on pre- and postsynaptic serotonin receptors: single cell studies in the rat CNS. Neuropharmacol., **22**, 495-499.

Boakes, R.J., Bradley P.B., Briggs, I. & Dray, A. (1970). Antagonism of

5-hydroxytryptamine by LSD 25 in the central nervous system: a possible neuronal basis for the actions of LSD 25. Br. J. Pharmacol., **40**, 202-218.

Bradley, P.B. & Briggs, I. (1974). Further studies on the mode of action of psychotomimetic drugs: antagonism of the excitatory actions of 5-hydroxytryptamine by methylated derivatives of tryptamine. Br. J. Pharmac., **50**, 345-354.

Bradley, P.B., Engel G., Fenuik, W., Fozard J.R., Humphrey P.P.A., Middlemiss, D.N. Mylecharane, E.J. Richardson, B.P. & Saxena, P.R. (1986). Proposals for the classification and nomenclature of functional receptors for 5-hydroxytryptamine. Neuropharmacol., **25**, 563-576.

Cerrito, F. & Raiteri M. (1979). Serotonin release is modulated by presynaptic autoreceptors. Eur. J. Pharmacol., **57**, 427-430.

Clemens L.G. (1978). Neural plasticity and feminine sexual behaviour in the rat. In: Sex and Behaviour (Eds. McGill, T.E., Dewsbury, D.A. & Sachs, B.D.), pp. 243-266. New York: Plenum Press.

Cohen, M.L., Fuller, R.W. & Kurz, K.D. (1983). LY53857, a selective and potent serotonergic (5-HT$_2$) receptor antagonist, does not lower blood pressure in the spontaneously hypertensive rat. J. Pharmacol. Exp. Ther., **227**, 327-332.

Cohen, M.S., Schwartz-Giblin, S. & Pfaff, D.W. (1987). Brainstem reticular stimulation facilitates back muscle motoneuronal responses to pudendal nerve input. Brain Res., **405**, 155-158.

Costall, B., Kelly M.E., Naylor, R.J., Tan, C.C.W. & Tattersall, F.D. (1986). 5-hydroxytryptamine M-receptor antagonism in the hypothalamus facilitates gastric emptying in the guinea-pig. Neuropharmacol., **25**, 1293-1296.

Cunningham, K.A. & Appel, J.B. (1986). Possible 5-hydroxytryptamine$_1$ (5-HT$_1$) receptor involvement in the stimulus properties of 1-(m-trifluoromethylphenyl) piperazine (TFMPP) J. Pharmacol. Exp. Ther., **237**, 369-377.

Davis, G.A. & Kohl, R.L. (1978). Biphasic effects of the antiserotonergic methysergide on lordosis in rats. Pharmacol. Biochem. Behav., **9**, 487-491.

Dourish, C.T., Hutson P.H. & Curzon G. (1986). Putative anxiolytics 8-OH-DPAT, buspirone and TVX Q7821 are agonists at 5-HT$_{1A}$ autoreceptors in the raphe nuclei. Trends Pharmacol. Sci., **7**, 212-214.

Eison, A.S., Eison M.S., Stanley, M. & Riblet, L.A. (1986). Serotonergic mechanisms in the behavioural effects of buspirone and gepirone. Pharmacol. Biochem. Behav., **24**; 701-708.

Eliasson, M., Michanek, A. & Meyerson B.J. (1972). A differential inhibitory action of LSD and amphetamine on copulatory behaviour in the female rat. Acta Pharmacol. Toxicol., **31** (suppl. 1), 22.

Engel, G., Gothert, M., Hoyer, D., Schlicker, E. & Hillenbrand, K. (1986). Identity of inhibitory presynaptic 5-hydroxytryptamine (5-HT) autoreceptors in the rat brain cortex with 5-HT$_{1B}$ binding sites. Naunyn-Schmiedeberg's Arch. Pharmacol., **332**, 1-7.

Espino, C., Sano, M. & Wade G.N. (1975). Alpha-methyltryptamine blocks facilitation of lordosis by progresterone in spayed, estrogen-primed rats. Pharmacol. Biochem. Behav., **3**, 557-559.

Everitt, B.J. & Fuxe K. (1977). Serotonin and sexual behaviour in female rats. Effects of hallucinogenic indolealkylamines and phenylethylamines. Neurosci. Lett., **4**, 215-220.

Everitt, B.J., Fuxe K., Hokfelt, T. & Jonsson G. (1975). Role of monoamines in the control by hormones of sexual receptivity in the female rat. J. Comp. Physiol. Psychol., **89**, 556-572.

Fernandez-Guasti, A., Ahlenius, S., Hjorth, S. & Larsson, K. (1987). Separation of dopaminergic and serotonergic inhibitory mechanisms in the mediation of estrogen-induced lordosis behaviour in the rat. Pharmacol. Biochem. Behav., **27**, 93-98.

Foreman, M.M. & Moss, R.L. (1978). Role of hypothalamic serotonergic receptors in the control of lordosis behaviour in the female rat. Horm. Behav., **10**, 97-106.

Fozard, J.R. (1984). MDL 72222, a potent and highly selective antagonist at neuronal 5-hydroxytryptamine receptors. Naunyn-Schmiedeberg's Arch. Pharmac., **326**, 36-44.

Franck, J.E. & Ward I.L. (1981). Intralimbic progesterone and methysergide facilitate lordotic behaviour in estrogen-primed female rats. Neuroendocrinology, **32**, 50-56.

Fuxe, K., Everitt, B.J., Agnati, L., Fredholm, B. & Jonsson, G. (1976). On the biochemistry and pharmacology of hallucinogens. In: Schizophrenia Today (Eds. Kemali, D., Bartholini, G. & Richter, D.), pp. 135-157. Oxford: Plenum Press.

Gaddum, J.H. & Picarelli, Z.P. (1957). Two kinds of tryptamine receptor. Br. J. Pharmac. Chemother., **12**, 323-328.

Glennon, R.A., McKenny J.D., Young, R. (1984). Discriminative stimulus properties of the serotonin agonist 1-(3-trifluoromethyphenyl)piperazine (TFMPP). Life Sci., **35**, 1475-1480.

Glennon, R.A., Young, R. & Rosencrans, J.A. (1983). Antagonism of the effects of the hallucinogen DOM and the purported 5-HT agonist quipazine by 5-HT$_2$ antagonists. Eur. J. Pharmacol., **91**, 189-196.

Haigler, H.J. & Aghajanian, G.K. (1973). Mescaline and LSD: direct and indirect effects on serotonin-containing neurons in brain. Eur. J. Pharmacol., **21**, 53-60.

Haigler, H.J. & Aghajanian, G.K. (1974a). Peripheral 5-HT antagonists: Failure to antagonize serotonin in brain areas receiving a prominent serotonergic input. J. Neural. Trans. 35, 257-273.

Haigler, H.J. & Aghajanian, G.K. (1974b). Lysergic acid diethylamine and serotonin: A comparison of effects on serotonergic neurons and neurons receiving a serotonergic input. J. Pharmacol. Exp. Ther., 188, 688-699.

Hendrik, E. & Gerall, A.A. (1976). Facilitation of receptivity in estrogen-primed rats during successive mating tests with progestins and methysergide, J. Comp. Physiol. Psych., 90, 590-600.

Hlinak, Z. (1987). Lisuride inhibits temporarily sexual behaviour in female rats. Pharmacol. Biochem. Behav., 27, 211-215.

Höyer, D., Engel, G. & Kalkman H.O. (1985). Molecular pharmacology of $5-HT_1$ and $5-HT_2$ binding sities in rat and pig brain membranes: Radioligand binding studies with [^{3}H]5-HT, [^{3}H]8-OH-DPAT, (-) [^{125}I]iodocyano-pindolol, [^{3}H]mesulergine and [^{3}H]ketanserin. Eur. J. Pharmacol., 118, 13-23.

Hunter, A.J., Hole, D.R. & Wilson, C.A. (1985). Studies into the dual effects of serotonergic pharmacological agents on female sexual behaviour in the rat: preliminary evidence that endogenous 5-HT is stimulatory. Pharmacol. Biochem. Behav., 22, 5-13.

Janssen, P.A.J. (1983). $5-HT_2$ receptor blockade to study serotonin-induced pathology. Trends Pharmacol. Sci., 4, 198-206.

Janssen, P.A.J. (1985). Pharmacology of potential and selective S_2-serotonergic antagonists. J. Cardiovasc. Pharmacol., 7, S2-S11.

Kennett, G.A., Dourish, C.T. & Curzon, G. (1987). $5-HT_{1B}$ agonists induce anorexia at a postsynaptic site. Eur. J. Pharmacol., 141, 429-435.

Kilpatrick, G.J., Jones, B.J. & Tyers, M.B. (1987). Identification and distribution of $5-HT_3$ receptors in rat brain using radioligand binding. Nature, 330 (6150), 746-748.

Kow, L-M, Zemlan, F.P. & Pfaff, D.W. (1979). Attempts to reinstate lordosis Reflex in estrogen-primed spinal female rats with monoamine agonists. Horm. Behav., 13, 232-240.

Lakoski, J.M. & Aghajanian, G.K. (1985). Effects of ketanserin on neuronal responses to serotonin in the prefrontal cortex, lateral geniculate, and dorsal raphe nucleus. Neuropharmacol., 24, 265-273.

Leysen, J.E. & Tollenaere, J.P. (1982). Biochemical models for serotonin receptors. Ann. Rep. Med. Chem., 17, 1-10.

Lyon, R.A., Davis, K.H. & Titeler, M. (1987) ^{3}H-DOB

(4-bromo-2,5-dimethoxyphenylisopropylamine) labels a guanyl nucleotide sensitive state of cortical 5-HT$_2$ receptors. Mol. Pharmacol., 31, 194-199.

Markstein, R., Hoyer, D. & Engel G. (1986). 5-HT$_{1A}$-receptors mediate stimulation of adenylate cyclase in rat hippocampus. Naunyn-Schmiedeberg's Arch. Pharmacol., 333, 335-341.

Martin, L.L. & Sanders-Bush, E. (1982). Comparison of the pharmacological characteristics of 5-HT$_1$ and 5-HT$_2$ binding sites with those of serotonin autoreceptors which modulate serotonin release. Naunyn Schmiederberg's Arch. Pharmacol., 321, 165-170.

McCall, R.B. & Aghajanian, G.K. (1980). Hallucinogens potentiate responses to serotonin and norepinephrine in the facial motor nucleus. Life Sci., 26, 1149-1156.

Mendelson, S.D. & Gorzalka, B.B. (1985). A facilitatory role for serotonin in the sexual behaviour of the female rat. Pharmacol. Biochem. Behav., 22, 1025-1033.

Mendelson, S.D. & Gorzalka, B.B. (1986a). Methysergide inhibits and facilitates lordosis behaviour in a time-dependent manner. Neuropharmacol., 25, 749-755.

Mendelson, S.D. & Gorzalka, B.B. (1986b). Serotonin type 2 antagonists inhibit lordosis behaviour in the female rat: Reversal with quipazine. Life Sci., 38, 33-39.

Mendelson, S.D. & Gorzalka, B.B. (1986c). 5-HT$_{1A}$ receptors: Differential involvement in female and male sexual behaviour in the rat. Physiol. Behav., 37, 345-351.

Mendelson, S.D. & Gorzalka, B.B. (1986d). Effects of 5-HT$_{1A}$ selective anxiolytics on lordosis behaviour: Interactions with progesterone. Eur. J. Pharmacol., 132, 323-326.

Meyerson, B.J. (1964a). The effect of neuropharmacological agents on hormone-activated estrous behaviour in ovariectomized rats. Arch. Int. Pharmacodyn., 150, 4-33.

Meyerson, B.J. (1964b). Estrous behaviour in spayed rats after estrogen or progesterone treatment in combination with reserpine or tetrabenazine. Psychopharmacol., 6, 210-218.

Meyerson, B.J. (1964c). Central nervous monoamines and hormone induced estrus behaviour in the spayed rat. Acta Physiol. Scand., 63, Supp. 241, 3-32.

Meyerson, B.J., Carrer H. & Eliasson, M. (1974). 5-hydroxytryptamine and sexual behaviour in the female rat. In: Advances in Biochem. Psychopharmacol. (Eds. Costa, E., Gessa, G.L. & Sandler, M.), Vol. 11, pp. 229-242. New York: Raven Press.

Meyerson, B.J. & Eliasson, M. (1977). Pharmacological and hormonal control of reproductive behaviour. In: Handbook of Psychopharmacol (Eds. Iversen L.L., Iversen S.D. & Snyder S.H.), Vol. 8, pp. 159-232. New York: Plenum Press.

Meyerson, B.J. , Malmnas, C.O. & Everitt, B.J. (1985). Neuropharmacology, neurotransmitters and sexual behaviour in mammals. In: Handbook of Behav. Neurobiol.(Eds. Adler, N., Pfaff, D. & Goy, R.W.), Vol. 7, pp. 495-536. New York: Plenum Press.

Middlemiss, D.N. (1982). Multiple 5-hydroxytryptamine receptors in the central nervous system of the rat. In: Presynaptic receptors: mechanism and function (Ed. De Belleroche J.), pp. 46-74. Chichester: Ellis Horwood.

Neale, R.F., Fallon, S.L., Boyar, W.C., Wasley, J.W.F., Martin, L.L., Stone, G.A., Glaeser, B.S., Sinton C.M. & Williams, M. (1987). Biochemical and pharmacological characterization of CGS 12066B, a selective 5-HT_{1B} agonist. Eur. J. Pharmacol., 136, 1-9.

Pazos, A., Hoyer, D., Palacios, J.M. (1984). The binding of serotonergic ligands to the porcine choroid plexus: Characterization of a new type of serotonin recognition site. Eur. J. Pharmacol., 106, 539-546.

Pedigo, N.W., Yamamura, H.I. & Nelson D.L. (1981). Discrimination of multiple [^{3}H]5-hydroxy-tryptamine binding sites by the neuroleptic spiperone in rat brain. J. Neurochem., 36: 220-226.

Penington, N.J. & Reiffenstein R.J. (1986a). Direct comparison of hallucinogenic phenethylamines and d-amphetamine on dorsal raphe neurons. Eur. J. Pharmacol., 122, 373-377.

Penington, N.J. & Reiffenstein R.J. (1986b). Possible involvement of serotonin receptors in the facilitatory effect of a hallunicogenic phenethylamine on single facial motoneurons. Can. J. Physiol. Pharmacol., 64, 1302-1309.

Peroutka, S.I., Lebovitz, R.M. & Snyder S.H. (1981). Two distinct central serotonin receptors with different physiological functions. Science, 212, 827-829.

Peroutka, S.I. & Snyder S.H. (1979). Multiple serotonin receptors: Differential binding of (^{3}H) 5-hydroxytryptamine, (^{3}H) lysergic acid diethylamide and (^{3}H) spiroperidol. Mol. Pharmacol., 16, 687-699.

Peroutka, S.I. (1986). Selective labeling of 5-HT_{1A} and 5-HT_{1B} binding sites in bovine brain. Brain Res., 344, 167-171.

Raible, L.H. & Gorzalka, B.B. (1986). Short and long term inhibitory actions of alpha-melanocyte stimulating hormone on lordosis in rats. Peptides, 7, 581-586.

Richardson, B.P., Engel, G. Donatsch, P. & Stadler, P.A. (1985). Identification of serotonin M-receptor subtypes and their specific blockade by a new class of drugs. Nature, 316, 126-131.

Roberts, M.H.T. & Straughan, D.W. (1967). Excitation and depression of cortical neurons by 5-hydroxytryptamine. J. Physiol., 193, 269-294.

Rodriguez-Sierra, J.F. & Davis, G.A. (1979). Tolerance to the lordosis-facilitating effects of progesterone or methysergide. Neuropharmacol., 18, 335-339.

Rodriguez-Sierra, J.F., Naggar, A.N. & Komisaruk, B.R. (1979). Monoaminergic mediation of masculine and feminine copulatory behaviour in female rats. Pharmacol. Biochem. Behav., 5, 457-463.

Sanders-Bush, E. & Conn, P.J. (1986). Effector systems coupled to serotonin receptors in brain: Serotonin stimulated phosphoinositide hydrolysis. Psychopharmacol. Bull., 22, 829-836.

Shannon, M., Battaglia, G., Glennon, R.A. & Titeler, M.(1984). $5-HT_1$ and $5-HT_2$ binding properties of the hallucinogen 1-(2,5-dimethoxyphenyl) -2-aminopropane (2,5-DMA). Eur. J. Pharmacol., 102, 23-29.

Sietnieks, A. (1985). Involvement of $5-HT_2$ receptors in the LSD- and 5-HTP induced suppression of lordotic behaviour in the female rat. J. Neural. Transmission, 61, 81-94.

Sietnieks, A. & Meyerson B.J. (1980). Enhancement by progesterone of lysergic acid diethylamide inhibition of the copulatory response in the female rat. Eur. J. Pharmacol., 63, 57-64.

Sietnieks, A. & Meyerson B.J. (1983). Progesterone enhancement of lysergic acid diethylamide and levo-5-hydroxytryptophan Stimulation of the copulatory response in female rat. Neuroendocrinol., 36, 462-467.

Sills, M.A., Wolfe, B.B. & Frazer, A. (1986). Determination of selective and nonselective compounds for the $5-HT_{1A}$ and $5-HT_{1B}$ receptor subtypes in rat frontal cortex. J. Pharmacol. Exper. Ther., 231, 480-487.

Sofia, R.D. & Vassar H.B. (1975). The effect of ergotamine and methysergide on serotonin metabolism in the rat brain. Arch. Int. Pharmacodyn. Ther., 216, 40-50.

Sprouse, J.S. & Aghajanian, G.K. (1986). (-) Propranolol blocks the inhibition of serotonergic dorsal raphe cell firing by $5-HT_{1A}$ selective agonists. Eur. J. Pharmacol., 128, 295-298.

Tricklebank, M.D., Middlemiss, D.N. & Neill, J. (1986). Pharmacological analysis of the behavioural and thermoregulatory effects of the putative $5-HT_1$ receptor agonist, RU 24969, in the rat. Neuropharmacol., 25, 877-886.

Tyers, M.B. (1988). The anxiolytic activities of $5-HT_3$ antagonists in laboratory animals. In: Behavioural Pharmacology of 5-HT (Eds. Bevan, P.,

Archer, T., & Cools, L.). New York: Lawrence Erlbaum.

Ulibarri, C. & Yahr, P. (1987). Poly-A+ mRNA and defeminization of sexual behaviour and gonadotropin secretion in rats. Physiol. Behav., 39, 767-774.

Von Hungen, K., Roberts, S. & Hill, D.F. (1975). Serotonin-sensitive adenylate cyclase activity in immature rat brain. Brain Res., 8, 257-267.

Ward, I.L., Crowley, W.R., Zemlan, F.P. & Margulus, D.L. (1975). Monoaminergic mediation of female sexual behaviour. J. Comp. Physiol. Psych., 88, 53-61.

Wilson, C.A. & Hunter A.J. (1985). Progesterone stimulates sexual behaviour in female rats by increasing 5-HT activity on 5-HT$_2$ receptors. Brain Res., 333, 223-229.

Zemlan, F.P., Ward, I.L., Crowley W.R. & Margules, D.L. (1973). Activation of lordotic responses in female rats by suppression of serotonergic activity. Science, 179, 1010-1011.

HYPOTHALAMIC SITES OF ACTION OF THE DUAL EFFECT OF 5-HT ON FEMALE SEXUAL BEHAVIOUR IN THE RAT

M.D.James, S.M.Lane, D.R.Hole, C.A.Wilson
Department Obstetrics & Gynaecology, St. George's Hospital Medical School, London SW17 ORE. United Kingdom

The presence of gonadal steroids is essential for the occurrence of female sexual receptivity and the cyclical release of the gonadotrophin surge in the rat. In ovariectomised animals, receptivity can be induced 48-56 hours after either a single injection of a high dose of oestradiol benzoate (e.g. 50 µg OB) or a low dose (e.g. 2 µg OB) followed 2 days later by progesterone (P). The latter steroid treatment also induces the release of a gonadotrophin surge.

In the presence of steroids, 5-HT can modulate female sexual behaviour and gonadotrophin release. In both cases 5-HT exerts a dual effect (Wilson et al., 1985; Walker 1983) and recent findings indicate that the inhibitory effect on sexual behaviour occurs via $5-HT_1$ receptors and the stimulatory action via $5-HT_2$ receptors. This was concluded from the facts that selective $5-HT_1$ agonists e.g. 8-OH-DPAT and RU 24969, reduce female sexual activity, while putative $5-HT_2$ agonists (MK 212 and quipazine) and antagonists (ketanserin and pirenperone) were stimulatory and inhibitory, respectively (Wilson & Hunter, 1985; Mendelson & Gorzalka, 1985; 1986a; 1986b, Ahlenius et al., 1986).

We have further investigated the receptor sub-types involved, using some of the new pharmacological agents now available.

There are many reports in the literature showing that both oestrogen (O) and P can alter 5-HT activity within the central nervous system (CNS) and we have suggested that the stimulatory effect of P on both sexual behaviour and LH release in oestrogen-primed animals is exerted via increasing hypothalamic 5-HT activity (Walker & Wilson, 1983; Wilson & Hunter, 1985). We have carried out experiments to confirm these findings and elucidate the specific site(s) where P might be exerting its effect.

Methods

Experiment I
Wistar rats (bred at St.George's Hospital Medical School) were ovariectomised and three weeks later, groups of rats were primed with various steroid regimes that would affect sexual receptivity and plasma LH concentration. These regimes included: –
1) 5µg OB followed 48 hours later by 0.5 mg P
2) 5 µg OB or
3) 50 µg OB alone
4) 2 µg OB plus 0.1 mg P at 48h and
5) a control group receiving 0.1 ml corn oil. The steroids (Sigma & Co. Ltd., Poole, Dorset) were given in 0.1 ml corn oil subcutaneously (s.c.).

Tests for receptivity were carried out 54-56 hours after the OB treatment

(Hunter et al., 1985) and then the rats were autopsied by decapitation. The brains were removed, frozen immediately and stored at –70°C. Blood was collected from the cervical cut, spun at 400g for 15 minutes at 4°C and the plasma stored at –20°C until assayed for LH by radioimmunoassay. Within one month of autopsy the following hypothalamic areas were dissected by the micropunch technique: preoptic area (POA), suprachiasmatic nucleus (SCN), arcuate nucleus (ARC) median eminence (ME), ventromedial nucleus (VMN). All the samples were assayed for 5–HT and 5–hydroxyindole acetic acid (5–HIAA) after high–performance liquid chromatography by electrochemical detection. All the results are shown as means ± standard errors.

Experiment II
The design of this experiment has been reported in detail (Hunter et al., 1985). Four groups of 20 ovariectomised rats were primed with 5µg OB s.c., which in these animals induced receptivity in a half to one third of the rats. 48 hours later half of each group received one of the serotonergic agents (see results) as a solution or suspension intraperitoneally (i.p.) in 1 ml/kg saline. The other half of the group received saline alone. Lordotic activity (LQ) was tested one hour later.

Significances were assessed by the Student 't' test or the Wilcoxon matched–pair test.

Results

Experiment I: The effect of steoids on sexual behaviour, plasma LH and hypothalamic 5–HT activity
Administration of oil or 5 µg OB had no effect on sexual receptivity in ovariectomised animals (LQ = 0% in both cases): 5 µg OB reduced plasma LH levels (oil group: LH = 23.0 ± 4.0 ng/ml, 5 µg OB group: LH = 15.2 ± 2.0 ng/ml; $p<0.05$). When 5 µg OB was followed by 0.5 mg P, all the animals were highly receptive and all showed a significant increase in plasma LH. The ratio of 5–HIAA: 5–HT (taken as a measure of 5–HT turnover) was significantly increased in the ME and significantly reduced in the VMN in the O plus P treated animals, compared to the oil treated controls (see Table 6.1). No other area measured showed any significant differences.

A submaximal steroid treatment of 2 µg OB plus 0.1 mg P induced receptivity in 11/20 rats and when the plasma LH levels and ratio of 5–HIAA: 5–HT were compared in the non–receptive (NR) and receptive (R) animals, there was no significant difference in their LH levels but the 5HIAA: 5HT ratio was significantly higher in the ME of the R rats compared to the NR group. There was a tendency for a lower ratio in the VMN of the R rats. Other areas showed no significant differences. The steroid regime used in this experiment induced raised LH plasma concentrations in 14/20 rats. When the group was sub–divided according to this criterion (< or > 20 ng/ml) their sexual activity was similar in both groups and so was the 5–HIAA: 5–HT ratio in all the hypothalamic areas.

Treatment with 50 µg OB also induced sexual receptivity in approximately half the animals (9/17), but there was no difference in the indole ratio in any hypothalamic area in the NR and R groups (see Table 6.1).

Table 6.1 Effect of steroid treatment on sexual activity, plasma LH and 5-HIAA: 5-HT ratio in hypothalamic areas

Treatment			LQ% ± SEM	Plasma LH ng/ml ± SEM	5-HIAA: 5-HT ± SEM in the	
					ME	VMN
Oil	NR[a]	(12)	0%	23.0 ± 4.0	0.5 ± 0.18	0.81 ± 0.16
	R	(0)	-	-	-	-
5µg OB+0.5mg P	NR	(0)	-	-	-	-
	R	(7)	100%	91.5 ± 20.0	1.8 ± 0.44**	0.39 ± 0.07*
2µg OB+0.1mg P	NR	(9)	2.2 ± 2.2%	64.3 ± 12.4	0.44 ± 0.04	0.87 ± 0.06
	R	(11)	87.5 ± 5.5%	48.8 ± 8.3	0.67 ± 0.07+	0.63 ± 0.08
2µg OB+0.1mg P	NS[b]	(6)	62.8 ± 15.2%	16.9 ± 3.9	0.48 ± 0.4	0.75 ± 0.2
	S	(14)	64.3 ± 11.1%	72.3 ± 6.0	0.56 ± 0.05	0.70 ± 0.4
50µg OB	NR	(8)	6.9 ± 3.6%	23.2 ± 6.4	0.56 ± 0.24	0.48 ± 0.03
	R	(9)	65.0 ± 5.9%	25.1 ± 4.5	0.40 ± 0.09	0.55 ± 0.04

a NR - non-receptive; R - receptive
b NS - no LH surge; S - LH surge present
*p<0.05; **p<0.01 compared to oil controls
+p<0.05 compared to NR rats on same treatment

Experiment II: The effect of 5-HT agonists in Sexual Behaviour
8-OH-DPAT (a selective 5-HT$_{1A}$ agonist) at 0.2 and 0.5 mg/kg reduced the LQ to 30% of the saline control responses (p<0.001) in R rats. 0.1 mg/kg was ineffective. 2 mg/kg MCPP (a selective 5-HT$_{1B}$ agonist) reduced the LQ to 52% (p<0.01) of controls in R rats.

1 mg/kg was ineffective. Neither compound affected NR rats. The selective 5-HT$_2$ agonist 1-(2.5-dimethoxyphenyl-4-iodo)-2-aminopropane (DOI) stimulated behaviour in NR rats increasing lordosis by 49% and 52% at 0.1 and 0.5 mg/kg. It did not effect R animals.

b)The effect of 5-HT antagonists on Sexual Behaviour.
Previously we have shown that a selective 5-HT$_2$ antagonist (ketanserin) has a soley inhibitory effect on receptivity (Hunter et al., 1985). A variety of 5-HT antagonists known to effect 5-HT$_1$ receptors were now tested and found to reduce the LQ in R rats e.g. 0.5 and 0.1 mg/kg mesulergine (5-HT$_{1C}$; LQ reduced to 50 and 60%), 1 mg/kg pindolol (5-HT$_1$; LQ reduced to 51%), 0.5 mg/kg methiopine (5-HT$_1$ and $_2$; LQ reduced to 54%). There was a tendency for an increase in lordosis in NR rats after pindolol and mesulergine and the partial agonist isaperone at 2 mg/kg. Three 5-HT$_3$ antagonists were found to significantly increase sexual behaviour in NR rats and had no effect in R rats. At 0.2 and 0.5 mg/kg GR 38032F, MDL 72222 and BRL 43694 all increased LQ by 50% over the saline response.

Discussion

These results show that in OB primed animals, P can increase sexual activity, LH release and a rise in 5-HT activity in the ME. By employing a submaximal treatment of O and P, we have shown that there is no correlation between sexual activity and LH release confirming the findings of Södersten & Eneroth (1987) and that the increase in 5-HT activity in the ME noted previously by others (Vitale et al., 1984) is not correlated with LH release, but is correlated with sexual activity. When sexual activity was induced by OB alone, there was no difference in 5-HT activity in the R and NR rats in any hypothalamic area. This confirms our previous finding that it is likely that P exerts its stimulatory effect via a 5-HT system while O does not (Wilson & Hunter, 1985).

Luine et al. (1983) have shown that 5-HT has an inhibitory effect on sexual behaviour in the VMN. After P treatment we found a decrease in 5-HT activity in the VMN of the R rats. Johnson & Crowley (1986) also showed a reduction in this area after OB plus P treatment. This indicates that in order to stimulate sexual behaviour P reduces the activity of an inhibitory system in the VMN concomitantly enhancing a stimulatory system in the ME.

Although further experiments are required, the data using selective agonists and antagonists indicate that 5-HT inhibits behaviour via $5-HT_1$ receptors and stimulates via $5-HT_2$ receptors as has been suggested before (Wilson & Hunter, 1985; Mendelson & Gorzalka, 1985). We have also shown that $5-HT_3$ antagonists have a significant stimulatory effect on behaviour indicating that they can act at the level of the CNS and that $5-HT_3$ receptors may have an inhibitory effect on sexual behaviour.

Based on the results of this report one can hypothesise that 5-HT exerts its inhibitory effect via $5-HT_1$ or $5-HT_3$ receptors in the VMN and its stimulatory efect via $5-HT_2$ receptors in the ME.

References

Ahlenius, S., Fernandez-Guasti, A., Hjorth, S. & Larsson, K. (1986). Suppression of lordosis behaviour by the putative 5-HT receptor agonist 8-OH-DPAT in the rat. Eur. J. Pharmac., **124**, 361-363.

Hunter, A.J., Hole, D.R. & Wilson, C.A. (1985). Studies into the dual effects of serotonergic pharmacological agents on female sexual behaviour in the rat: preliminary evidence that endogenous 5-HT is stimulatory. Pharmac. Biochem. Behav., **22**, 5-13.

Johnson, M.D. & Crowley, W.R. (1986). Role of central serotonin systems in the stimulatory effects of ovarian hormones and naloxone on luteinizing hormone release in female rats. J. Endocrinol., **118**, 1180-1186.

Luine, V.N., Frankfurt, M., Rainbow, T.C., Biegon, A. & Azmitia, E. (1983). Intrahypothalamic 5,7, dihydroxytryptamine facilitates feminine sexual behaviour and decreases [^{3}H] imipramine binding and 5-HT uptake. Brain Res., **264**, 344-348.

Mendelson, D.S. & Gorzalka, B.B. (1985). A facilitatory role for serotonin in the sexual behaviour of the female rat. Pharmac. Biochem. Behav., **22**, 1025-1033.

Mendelson, D.S. & Gorzalka, B.B. (1986a). 5-HT$_{1A}$ receptors: Differential involvement in female and male sexual behaviour in the rat. Physiol. Behav. **37**, 345-351.

Mendelson, D.S. & Gorzalka, B.B. (1986b). Serotonin type 2 antagonists inhibit lordosis behaviour in the female rat: Reversal with quipazine. Life Sci., **38**, 33-39.

Sodersten, P. & Eneroth, P. (1987). Dissociation between the ovarian factors controlling sexual receptivity and preovulatory secretion of LH in cyclic female rats. J. Endocr. **112**, 133-138.

Vitale, M.L., de las Nieves Parisi, M., Chiocchio, S.R. & Tramezzani, J.H. (1984). Median eminence serotonin involved in the proestrus gonadotrophin release. Neuroendocrinol., **39**, 136-141.

Walker, R.F. (1983). Quantitative and temporal aspects of serotonin's facilitatory action on phasic secretion of luteinizing hormone in female rats. Neuroendocrinol., **36**, 468-474.

Walker, R.F. & Wilson, C.A. (1983). Changes in hypothalamic serotonin associated with amplification of LH surges by progesterone in rats. Neuroendocrinol., **37**, 200-205.

Wilson, C.A. & Hunter, A.J. (1985). Progesterone stimulates sexual behaviour in female rats by increasing 5-HT activity on 5-HT$_2$ receptors. Brain Res. **333**, 223-229.

5-HT$_{1A}$ AND 5-HT$_{1B}$ AGONISTS PRODUCE OPPOSITE EFFECTS ON THE EJACULATORY RESPONSE INDUCED BY AMPHETAMINE IN RATS

Lucy Rényi & Tommy Lewander
Biochemical Neuropharmacology, Astra Alab AB, Research & Development Laboratories, S–151 85 Södertälje, Sweden

d-Amphetamine, when given in large dosages, releases not only the catecholamines, noradrenaline (NA), and dopamine (DA) but also 5-hydroxytryptamine (5-HT) (Sloviter et al., 1978). Consequently, it produces a behavioural syndrome which includes e.g. hindlimb abduction, forepaw treading, the wet dog shake response (WDSR) (Bedard & Pycock, 1983), and the ejaculatory response (ER). The ER induced by the nonselective endogenous neurotransmitter, 5-HT, and by the nonselective postsynaptic 5-HT agonist, 5-methoxy-N,N-dimethyltryptamine (5-MeODMT) was recently introduced as a model for detecting changes in functions mediated by central 5-HT receptors (Rényi, 1985). The ER like the WDSR is a multitransmitter response. Only a nonselective agonism of the 5-HT receptors results in the ER and WDSR since they require both 5-HT$_1$ and 5-HT$_2$ receptors (Green et al., 1981; Yap & Taylor, 1983; Rényi, 1987). Indeed, none of these behaviours could be induced by the 5-HT$_{1A}$ agonist, 8-hydroxy-2-(di-n-propylamino)-tetralin (8-OH-DPAT) (Middlemiss & Fozard, 1983), or by the 5-HT$_{1B}$ agonist, 5-methoxy-3(1,2,3,6-tetrahydro- pyridin-4-yl)1H indole (RU 24969) (Hunt & Oberlander, 1983), even at large doses (Rényi, 1985; Tricklebank et al., 1985, 1986). Furthermore, both the ER and the WDSR is facilitated by NA and inhibited by DA (Bedard & Pycock, 1977; Handley & Brown, 1982; Dickinson & Curzon, 1983; Rényi, 1985). Thus, these behaviours can be influenced in many ways. The aim of the present study was to separate the 5-HT$_1$ receptors functionally and to investigate some connections between the 5-HT and the catecholamine systems.

Method
Sprague–Dawley rats, about 4 months old, were housed in groups of 5 in plastic cages, 55x35x19 cm, with sawdust flooring. The tests were performed between 8 and 12 a.m. in a soft-lighted quiet room. The 5-HT$_1$ agonists were injected s.c. 5 min, raclopride i.p. 30 min before d-amphetamine. The seminal material, which was wholly or partially hidden within the sheath, was collected 10–25 min thereafter by drawing back the foreskin, exposing the penis and removing the compact seminal material. The plug was then placed on filter paper, dried on each side, and weighed (Rényi, 1985).

Results
Three 5-HT$_{1A}$ agonists, 8-OH-DPAT, 1-(m-trifluormethyl-phenyl) -4-(p-amino-phenylethyl) piperazine (PAPP, LY 165163), and 3-dipropylamino-5-hydroxychroman (NDO-008) (Thorberg et al., 1987) decreased dose-dependently the ER induced by d-amphetamine (10 mg.kg i.p.) in that order of potency (Table 7.1). Significant effects were obtained at 0.03–0.3 mg/kg dosages. The compounds did not produce the ER by themselves even at high doses (8.0, 6.0, 6.0 mg/kg s.c., respectively; data not shown).

The 5-HT$_{1B}$ agonist, RU 24969, and the D$_2$ antagonist, raclopride, increased the ER induced by d-amphetamine. The lowest significant doses were at 0.3-0.5 mg/kg (Table 7.1). The ER was also enhanced by 3.0 mg/kg of RU 24969 (462%, n=10, ***p<0.001; Mann-Whitney U-test) when it was induced by a high dose of 5-MeODMT (8.0 mg/kg i.p.), whereas RU 24969 did not affect the ER when it was induced by a lower dose of 5-MeODMT (3.0 mg/kg i.p.).

Table 7.1 Effects of various 5-HT$_1$ agonists and raclopride on the ejaculatory response induced by d-amphetamine (10 mg/kg i.p.)

Treatment	Dose mg/kg	n	Number of rats with ejaculation	Weight of the seminal material mg	(min-max)	Ejaculatory response bases on seminal weight (% of saline)
Saline		30	23	118	(2-278)	
8-OHDPAT	0.01	10	6	130	(4-194)	110
	0.025	10	6	48	(2-164)**	41
	0.1	10	2	31	(6-57)***	26
	0.6	10	5	16	(14-20)***	14
PAPP	0.025	10	8	97	(3-170)	82
	0.1	10	5	24	(2-80)***	20
	0.5	10	1	6	(-)***	5
NDO-008	0.1	10	7	101	(6-220)	86
	0.3	10	2	23	(20-27)***	19
	1.5	10	5	3	(2-8)***	3
RU 24969	0.1	10	10	131	(12-290)	111
	0.25	10	10	273	(167-448)**	231
	3.0	10	10	416	(272-690)***	352
Raclopride	0.25	10	7	122	(52-234)	103
	0.5	10	9	212	(73-353)**	180
	1.0	10	10	267	(52-381)**	226

The 5-HT$_1$ agonists were given s.c. 5 min, raclopride i.p. 30 min before d-amphetamine. Differences from saline **p<0.01, ***p<0.001 (Mann-Whitney U-test).

Discussion

As shown previously, the 5-HT$_{1A}$ agonist 8-OH-DPAT surprisingly produced a dose-dependent decrease in the ER and WDSR induced by the nonselective 5-HT agonists 5-MeODMT and quipazine, respectively (Rényi, 1987). This finding was confirmed in the present study, since three 5-HT$_{1A}$ agonists also inhibited the ER when it was induced by the nonselective 5-HT agonist, 5-HT itself, or by d-amphetamine. The mechanism by which the 5-HT$_{1A}$ agonists antagonize another postsynaptic 5-HT behaviour is not quite clear.

This is the case even at high doses during which at least two $5-HT_{1A}$ mediated behaviours, flat body posture and forepaw treading, are distinctly visible. One explanation may be that these drugs "occupy" the essential postsynaptic $5-HT_{1A}$ receptors in such a way that these receptors no longer are available for the nonselective 5-HT agonists with affinity for both $5-HT_{1A}$ and $5-HT_2$ receptors.

Sprouse & Aghajanian (1986) also reported somehow different effects of $5-HT_{1A}$ agonists and nonselective 5-HT agonists on dorsal raphe neurons. Another possibility is that central catecholamine systems may be involved in functional responses to $5-HT_{1A}$ agonists (Hjorth, 1985; Tricklebank et al., 1985).

Likewise, for the present we can only speculate about the mechanisms by which the $5-HT_{1B}$ agonist, RU 24969, enhances the d-amphetamine induced ER. Hyperlocomotion in the rat, induced by RU 24969, is a proposed $5-HT_{1B}$ behaviour. This effect clearly depends on intact catecholamine systems, however, the drug has no significant affinity for either dopamine D_2 sites or α_1-adrenoceptors (Tricklebank et al., 1986). The similar effect of RU 24969 and the D_2 antagonist, raclopride, on the ER indicates that the $5-HT_{1B}$ receptors may be involved in behaviour modulated by DA (Oberlander et al., 1987). Since DA has an inhibitory effect on the ER (Rényi, 1985), RU 24969 may have a direct or indirect inhibitory effect on some DA activities. The finding that RU 24969 also potentiated the ER when it was induced by a high dose of 5-MeODMT supports this hypothesis. At this dose, as with d-amphetamine, 5-MeODMT shows some DA-ergic activity. Still another explanation may be that RU 24969 releases NA (Cousty et al., 1986) a neurotransmitter, which in contrast to DA, facilitates the ER (Rényi, 1985).

In conclusion, the results obtained corroborate previous findings, namely that the ER induced by nonselective 5-HT-agonists requires a balance on the one hand between different subtypes of 5-HT receptors and on the other hand between the 5-HT and catecholamine systems.

References

Bedard, P., Pycock, C.J. (1977). "Wet-dog" shake behaviour in the rat: a possible quantitative model of central 5-hydroxytryptamine activity. Neuropharmacol., 16, 663-670.

Cousty, D., Hunt, P. & Pujol, J.F. (1986). Effet d'un agoniste serotonergique, le RU 24969, sur la liberation de noradrenaline in vitro. J. Pharmac. Paris. (In press.)

Dickinson, S.L. & Curzon, G. (1983). Roles of dopamine and 5-hydroxytryptamine in stereotyped and nonstereotyped behaviour. Neuropharmacol., 22, 805-812.

Green, A.R., Hall., J.E., & Rees, A.R. (1981). A behavioural and biochemical study in rats of 5-hydroxytryptamine agonists and antagonists with observations on structure-activity requirements for agonists. Br. J. Pharmacol., 73, 703-719.

Handley, S.L. & Brown, J. (1982). Effects of the 5-hydroxytryptamine-induced headtwitch of drugs with selective actions on α_1 and α_2 adrenoceptors. Neuropharmacol., 21, 507-510.

Hjorth, S. (1985). Hypothermia in the rat induced by the potent serotonergic agent 8-OHDPAT. J.Neural Trans., 61, 131-135.

Hunt, P.J. & Oberlander, C. (1981). The interaction of indole derivatives with the serotonin receptor and non-dopaminergic circling behaviour. Serotonin-current aspects of neurochemistry (Ed: Haber, B.), pp. 547-562. New York: Plenum Press.

Middlemiss, D.N. & Fozard, J.R. (1983). 8-Hydroxy-2-(di-n-propylamino)-tetralin discriminates between subtypes of the 5-HT$_1$ recognition site. Eur. J. Pharmacol., 90, 151-153.

Oberlander, C., Demassey, Y., Verdu, A., Van de Velde, D. & Bardelay, C. (1987). Tolerance to the serotonin 5-HT$_1$ agonist RU 24969 and effects on dopaminergic behaviour. Eur. J. Pharmacol., 139, 205-214.

Rényi, L. (1985). Ejaculations induced by p-chloroamphetamine in the rat. Neuropharmacol., 24, 697-704.

Rényi, L. (1987). The involvement of 5-HT$_1$ and 5-HT$_2$ receptors in different components of the 5-HT syndrome in the rat. Neurosci. (Suppl.), 22, 225.

Sloviter, R.S., Dust, E.G. & Connor, J.D. (1978). Evidence that serotonin mediates some behavioural effects of amphetamine. J. Pharmacol. Exp. Ther., 206, 248-352.

Sprouse, J.S. & Aghajanian, G.K. (1986). (-)-Propranolol blocks the inhibition of serotonergic dorsal raphe cell firing by 5-HT$_{1A}$ selective agonists. Eur. J. Pharmac., 128, 295-298.

Thorberg, S,-O., Hall, H., Akesson, CH., Svensson, K. & Nilsson, J.L.G. (1987). Aminochromans: potent agonists as central dopamine and serotonin receptors. Acta Pharmac. Suec., 24, 169-182.

Tricklebank, M.D., Forler, C. & Fozard, J.R. (1985). The involvement of subtypes of the 5-HT$_1$ receptor and of catecholamonergic systems in the behavioural response to 8-hydroxy-2-(di-n-propylamino)tetralin in the rat. Eur. J. Pharmacol., 106, 271-282.

Tricklebank, M.D., Middlemiss, D.N. & Neill, J. (1986). Pharmacological anaylsis of the behavioural and thermoregulatory effects of the putative 5-HT$_1$ receptor agonist, RU 24969, in the rat. Neuropharmacol., 25, 877-886.

Yap, C.Y. & Taylor, D.A. (1983). Involvement of 5-HT$_2$ receptors in the wet-dog shake behaviour induced by 5-hydroxy-tryptophan in the rat. Neuropharmacol., 22, 801-804.

FUNCTIONAL INTERPLAY OF SEROTONIN (5-HT)-RECEPTOR SUBTYPES

H.H.G.Berendsen, R.J.M. Smets, C.L.E.Broekkamp
Dept. CNS Pharmacology, Organon International B.V., P.O.Box 20, 5340 BH Oss, The Netherlands

Introduction

The synthesis of serotonin (5-HT) receptor subtype selective agonists and/or antagonists has given the behavioural pharmacologist the opportunity to discover the functions for these receptors. A great variety of changes in rodent behaviour by drugs has been related to serotonin receptor activation. At present there is sufficient information to suggest that drug-induced penile erection is due to activation of the $5-HT_{1B}$ receptor (Berendsen and Broekkamp 1987) and that drug induced head shakes are mediated by the $5-HT_2$ receptor (Yap and Taylor, 1983). Goodwin et al. (1987) suggested that activation of the presynaptic $5-HT_{1A}$ receptor mediates hypothermia in mice and rats whereas Tricklebank et al. (1985) found that activation of the postsynaptic $5-HT_{1A}$ receptor mediates forepaw treading. The $5-HT_{1A}$ agonist 8-OH-N,N-dipropyl-2 aminotetralin (8-OH-DPAT) also induces hyperactivity in rats (Dourish et al., 1985), hypoactivity in mice and a recently discovered symptom which we described as lower lip retraction (Berendsen et al., 1988). Compounds with activity on more than one subtype of 5-HT receptor do not always induce the above mentioned behaviour. For example the compound RU 24969, (5 methoxy-3(1,2,3,6-tetra-hydropiridin-4-yl) 1H indole) which has both $5-HT_{1B}$ and $5-HT_{1A}$ activity in vitro (Doods et al., 1985), does not induce the $5-HT_{1B}$ related penile erections (Berendsen and Broekkamp, 1987) or the $5-HT_{1A}$ related forepaw treading, but we do observe $5-HT_{1A}$ related lower lip retraction. On the other hand 5-methoxy-N,N-dimethyltryptamine (5-MeODMT), a compound with $5-HT_{1A}$ activity and some $5-HT_2$ and $5-HT_{1B}$ activity (Engel et al., 1986), does not induce lower lip retraction but does induce forepaw treading (Berendsen et al., unpublished).

Therefore it might be possible that activation of a particular receptor subtype prevents or masks the effects mediated by other 5-HT receptor subtypes. The location of the receptor pre- or post synaptically seems to be important in this respect. Here we report experiments which actually demonstrate that activation of one particular 5-HT receptor subtype influences the expression of the behaviour related to activation of another 5-HT receptor subtype.

Methods

Naive adult Wistar rats (HSD/cpb: WU, Harlan Spraque Dawley, Zeist, The Netherlands) weighing 200-320 grams and naive male mice (Crl: CD-1 (ICR) BR, from Charles River, Germany) weighing 24-26 grams were used. All animals were housed under a 12 hour light-dark cycle (lights on 06.00 hrs) and were allowed free access to standard food pellets and tap water. In the experiments in which rats were used the animals were placed in individual observation cages immediately after treatment. Injections were made subcutaneously in a dose volume of 5 ml/kg body weight in rats and of 10 ml/kg body weight in mice. Control animals received an equivalent volume of vehicle. Experiments were performed between 09.00 and 13.00 hrs.

Lower lip retraction in rats is defined as a complete retraction of the lower lip so that the teeth of the lower jaw are completely visible. Lower lip retraction was induced by 8-OH-DPAT (0.46 mg/kg). A dose of the 5-HT$_{1B}$-receptor agonist 1-3'chlorophenyl-4-iodophenyl)-2-amino propane (DOI) (Shannon et al., 1984) was injected at the same time (N=10 per group). After 15, 30 and 45 min the retraction of the lower lip was rated as follows: 0=absent; 0.5=present but not complete; 1=complete retraction.

Hypothermia in mice was induced by 8-OH-DPAT 0.25 mg/kg. Ten minutes after pretreatment with a dose of mCPP or DOI rectal temperature of the animals was measured (9 mice per group). Immediately thereafter 8-OH-DPAT was injected and 10, 20, 30 and 40 minutes later the temperature of the mice was measured again.

Locomotor activity in mice was measured in small cages (11x11x16 cm) equipped with photoelectric cells. Immediately after treatment with a dose of mCPP or DOI plus 8-OH-DPAT (0.5 mg/kg) the mice were placed in these cages and their activity was measured for 30 minutes (N=12 mice per group).

In rats forepaw treading was measured using a time sampling method. Rats were injected with 8-OH-DPAT 0.22 mg/kg together with a dose of mCPP or DOI (8 rats per group). The rats were scored for presence or absence of forepaw treading every 30 seconds during 15 minutes starting 15 minutes after injection of the compounds.

In rats penile erections were measured as described before (Berendsen and Broekkamp, 1987). After injection of mCPP (0.46 mg/kg) plus a dose of 8-OH-DPAT or DOI the number of penile erections were counted during a 30 minute observation period (N=10 rats per group).

In rats head shakes were measured by counting the total number of head shakes that occurred during 30 minutes after injection of DOI (0.22 mg) plus a dose of mCPP or 8-OH-DPAT (N=8 rats per group).

Results
The results are summarized in table 8.1.

8-OH-DPAT (0.46 mg/kg) induced complete lower lip retraction. This was dose dependently antagonised by mCPP and DOI with ED$_{50}$'s of 0.2 mg/kg and 0.05 mg/kg respectively.

8-OH-DPAT (0.25 mg/kg) lowered the body temperature in mice by approximately 2°C 20 minutes after treatment. Both mCPP and DOI antagonised this hypothermic effect of 8-OH-DPAT dose dependently. The ED$_{50}$'s were 0.2 and 0.03 mg/kg respectively. 0.46 mg/kg of mCPP or 0.1 mg/kg of DOI completely prevented the hypothermic effect.

8-OH-DPAT induced a dose dependent hypoactivity in mice. 0.5 mg/kg of 8-OH-DPAT caused an 80% reduction in activity if compared to placebo during the first 10 minutes after treatment. mCPP and DOI antagonise this hypoactivity with estimated ED$_{50}$'s of 0.5 and 0.1 mg/kg respectively.

8-OH-DPAT (0.22 mg/kg) induced forepaw treading in rats was antagonised by mCPP (ED_{50} = 0.9 mg/kg) but was potentiated by DOI (ED_{150}=0.08 mg/kg).

Penile erections in rats induced by mCPP (0.46 mg/kg) were inhibited by 8-OH-DPAT (ED_{50}=0.03 mg/kg). DOI did not affect the mCPP induced penile erections up to 1 mg/kg. DOI by itself induced ejaculations but no penile erections.

Head shakes induced by DOI (0.22 mg/kg) were antagonised by mCPP only in relatively high doses (ED_{50}=0.9 mg/kg). 8-OH-DPAT was more potent with an ED_{50} of 0.03 mg/kg.

Table 8.1 Influence of 5-HT-subtype specific agonists on each others effect. Given are the ED_{50} ± S.E. in mg/kg

		effect of 5-HT_{1A} agonist (8-OH-DPAT)	effect of 5-HT_{1B} agonist (mCPP)	effect of 5-HT_2 agonist (DOI)
8-OH-DPAT (5-HT_{1A}) induced	lower lip retraction 0.46 mg/kg in rats		antagonism ED_{50}=0.2±0.04	antagonism ED_{50}=0.05±0.01
	hypothermia 0.25 mg/kg in mice		antagonism ED_{50}=0.2±0.04	antagonism ED_{50}=0.03±0.01
	hypoactivity 0.5 mg/kg in mice		antagonism ED_{50}=0.5[1]	antagonism ED_{50}=0.1[1]
	forepaw treading 0.22 mg/kg in rats		antagonism ED_{50}=0.9±0.6	antagonism ED_{150}=0.08±0.02
mCPP (5-HT_{1B}) induced	penile erections 0.46 mg/kg in rats	antagonism ED_{50}=0.03±0.007		no effect up to 1.0 mg/kg
DOI (5-HT_2) induced	head shakes 0.22 mg/kg in rats	antagonism ED_{50}=0.03±0.01	antagonism ED_{50}=0.9±0.7	

1) These are estimated ED_{50}'s

Discussion

The hypothermia, hypoactivity and lower lip retraction induced by the 5-HT_{1A}-receptor agonist 8-OH-DPAT are antagonised by the 5-HT_{1B}-receptor agonist mCPP and by the 5-HT_2-receptor agonist DOI. The antagonistic effect of DOI in this respect is stronger than that of mCPP (Table 8.1). However, forepaw treading which is also induced by 8-OH-DPAT is influenced by mCPP and DOI in a different manner. mCPP antagonizes the forepaw treading whereas DOI potentiates this behaviour. Since 8-OH-DPAT induced hypothermia and lower lip retraction are supposed to be presynaptically mediated (Goodwin et al., 1985; Berendsen et al., 1988) whereas forepaw treading induced by this compound is post-synaptically mediated (Tricklebank et al. 1985) the data suggest that pre-synaptically induced 5-HT_{1A} effects are antagonised by both 5-HT_{1A} agonists and that post-synaptically induced

$5-HT_{1A}$ effects are antagonised by $5-HT_{1B}$ agonists and potentiated by $5-HT_2$ agonists. This hypothesis can be used to account for a number of anomalous findings. For example, the failure of 5-MeODMT to induce lower lip retraction may be due to its agonist effects on $5-HT_2$ and $5-HT_{1B}$ sites. Similarly, the introduction of fore-paw treading may be due to the synergistic effects at $5-HT_{1A}$ and $5-HT_2$ sites. Furthermore it does not induce head shakes because of its agonistic effects on $5-HT_{1A}$ and $5-HT_{1B}$ sites.

Our data also show that penile erections induced by the $5-HT_{1B}$ agonist mCPP are potently antagonised by 8-OH-DPAT indicating that $5-HT_{1A}$ activity antagonises $5-HT_{1B}$ induced behaviour. This may explain why a compound with a mixed $5-HT_{1B}$ and $5-HT_{1A}$ agonist profile such as RU 24969 (Doods et al., 1985) fails to induce penile erections (Berendsen and Broekkamp, 1987). Indeed, as RU 24969 induces lower lip retraction, $5-HT_{1A}$ receptor activation may well dominate the pharmacological profile of this compound.

These experiments show that behaviour induced by a 5-HT-subtype specific agonist can sometimes be overruled or masked by an effect on another 5-HT receptor subtype. Care is therefore needed when attempting to draw conclusions from functional measurements when the compound to be investigated is not a very selective one.

References

Berendsen, H.G. & Broekkamp C.L.E. (1987). Drug induced penile erections in rats: indications of serotonin$_{1B}$ receptor mediation. Eur. J. Pharmacol., 135, 279-287.

Doods, H.N., Kalkman, H.O., De Jonge, A., Thoolen, M.J.M.C., Wilffert B., Timmermans, P.B.M.W.M. & Van Zwieten, P.A. (1985). Differential selectivities of RU 24969 and 8-OH-DPAT for the purported $5-HT_{1A}$ and $5-HT_{1B}$ binding sites. Correlation between $5-HT_{1A}$ affinity and hypotensive activity. Eur. J. Pharmacol., 112, 363-370.

Dourish, C.T., Hutson, P.H. & Curzon, G., (1985). Low doses of the putative serotonin agonist 8-hydroxy-2-(di-n-propylamino)tetralin (8-OH-DPAT) elicit feeding in the rat. Psychopharmacol. 86, 197-204.

Engel, G., Göthert, M., Hoyer, D., Schlicher E., & Hillebrand, K. (1986). Identity of inhibitory presynaptic 5-hydroxytryptamine (5-HT) autoreceptors in the rat brain cortex with $5-HT_{1B}$ binding sites. Naunyn-Schmiedeberg's Arch. Pharmacol., 332, 1-7.

Goodwin, G.M., De Souza, R.J. & Heal, D.J. (1987). The pharmacology of the behavioural and hypothermic responses of rats to 8-hydroxy-2-(di-n-propylamino) tetralin (8-OH-DPAT). Psychopharmacol., 91, 506-511.

Middlemiss, D.N. & Fozard, J.R. (1983). 8-hydroxy-2-(di-n-propylamino) tetralin discriminates between subtypes of the $5-HT_1$ recognition site. Eur. J. Pharmacol., 90, 151-153.

5-HT & AGGRESSION

P.Bevan. Department of Pharmacology, Duphar B.V., P.O.Box 2, 1380 AA Weesp, Holland.

Aggression is a behaviour displayed by almost all animal species and at first sight would appear to be a relatively simple phenomenon. In fact, aggression is a complex set of behaviours displayed in different conflict situations including attack, or offense, defense and flight. Moreover, in carnivores a phenomenon is identified which could loosely be described as predatory aggression. Competition among contemporaries may also involve aggressive displays. Aggression may be considered a response to the environment. Environment plays an important role in initiating aggression and one wonders, then, to what extent the artificial laboratory environment may influence the phenomena being examined. Ethologists, of course, try to recreate the natural situation as far as is practicable, even to the extent of using rats trapped in the wild for their studies.

To the naive observer, different expressions of aggression can clearly be differentiated, for example attack and defense. Within a given species, however, both forms may use the same elements of the animals' behavioural repertoire: Attack and defense differ primarily in the frequency and temporal sequence of performing the elements. The term "agonistic behaviour" is thus often used to cover all events associated with aggressive behaviours and is the term favoured by ethologists.

As one reviews the literature on aggression, it quickly becomes apparent that aggression always seems to be reduced by the drug under study, and only rarely increases. Certainly, this is true for drugs influencing the 5-HT system. This suggests that the paradigms selected are more suitable for detecting decreases in aggressive behaviour. Indeed, in most models used, for example isolated induced aggression in mice, aggression is first "kindled" and then subjected to psychopharmacological manipulation. There is also a question of specificity in that general stimulation or sedation can render proper determination of aggression levels impossible. There remains the possibility that many of the drugs tested, certainly those found to be inactive in reducing aggression, may in fact stimulate aggressive behaviours in appropriate circumstances. This needs to be examined before a definitive statement on the role of 5-HT in aggressive behaviours can be made.

From a pharmacological point of view, many of the studies reported in the literature leave a lot to be desired. Scant attention has been paid to the doses used or to the confounding effects that other properties of the drug can have on the behaviour being examined. For example, it is well known that 5-HT-related drugs can influence appetite, vigilance as well as nociception. Furthermore, some of the effects reported are difficult to reconcile suggesting some degree of artefact, perhaps due to one or more of the factors just mentioned. For instance, MAO inhibitors and 5-HT uptake inhibitors all decrease aggression in the isolation-induced aggression model in mice. However, 5-HT depleters or synthesis inhibitors do the same. Whilst there may be an explanation for this contradiction, it is not readily apparent. Of course, this might say more about the model than about the experiments and it stresses the need to take a closer look at the behavioural specificity of the drug effects. Considering the range of drugs, serotonergic and non-serotonergic, which influences isolation-induced aggression, it may be that this is a behaviour very sensitive to psychoactive drugs rather than an indicator of specific anti-aggressive properties. This may be true of other commonly used models too.

5-HT seems to be attributed with attenuating aggression, a feature consistent irrespective of the species under consideration or the type of aggression studied. This is an example of remarkable evolutionary resiliance. It is also a piece of pharmacological good fortune, for the conclusion seems to hold despite the fact that the pharmacological tools available in the past have lacked specificity. Moreover, most of the behavioural observations have, by modern standards, been quite crude.

The reader could be forgiven a certain scepticism of the specificity of some of the drug-induced behavioural effects reported. Sleeping rats rarely display aggression. Such scepticism is ill-founded, however, as sophisticated ethological observations of the behavioural effects of the newer, more selective agents merely confirm the old data. Indeed, the latest data show that 5-HT agonists can attenuate agonistic behaviour with remarkable specificity, leaving other behavioural elements of the animal's repertoire intact. Such studies have provided strong evidence of the involvement of $5-HT_{1B}$ receptors in the mediation of aggressive behaviours. Selective effects on aggression are only recorded for drugs which act as agonists at this receptor sub-type. This conclusion is clearly provisional though, and much work still needs to be done. None of the drugs available for testing are particularly pure $5-HT_{1B}$ agonists, all having affinity for other 5-HT receptor sub-types. Furthermore, the hypothesis can only be confirmed once antagonism studies have been performed. Once again we have a dilemma in that there are currently no good specific antagonists available.

Aggression seems to be unique amongst the behaviours reported in this volume in that there appears to be no functional antagonism between the different 5-HT receptor sub-types. For example, whereas $5-HT_{1B}$ agonists reduce aggression, aggression is not induced or increased by $5-HT_{1A}$ agonists. As mentioned earlier, this may be more a feature of the behavioural paradigms used than a reflection of the physiological substrate.

The reviews which follow serve two purposes. First, the uninitiated will enjoy an introduction to aggression research, or the study of agonistic behaviour, which considers both evolutionary and ethological issues. Secondly, the role of 5-HT and 5-HT systems are considered and form a neat pharmacological overlay of the behavioural background.

MODULATORY ACTION OF SEROTONIN IN AGGRESSIVE BEHAVIOUR

Berend Olivier, Jan Mos, Martin Tulp, Jacques Schipper and Paul Bevan.
Department of Pharmacology, Duphar B.V., P.O.Box 2, 1380 AA Weesp, The Netherlands.

Introduction

Serotonin (5-HT) has been suggested to play an important role in the control of aggressive behaviour in animals (Valzelli, 1981) and man (Mühlbauer, 1985). This does not imply, however, that serotonin is the most important neurotransmitter involved in aggression, because a number of other neurotransmitters have also been suggested to modulate aggressive behaviour, e.g. adrenaline and noradrenaline (Bell and Hepper, 1987; Eichelman 1987), acetylcholine (Bell et al., 1985), and GABA (Eichelman, 1987). Although multiple-transmitter modulation is more realistic for complex behaviours like aggression, serotonin may still play a key role although many aspects of its precise role still remain to be elucidated (Miczek, 1987; Olivier et al., 1987; Valzelli, 1984).

Support for a simple relationship between neural 5-HT activity and aggression in animals is not unequivocal. This is largely due to the lack of direct correlations between the spontaneous neurophysiological activity of subsets of 5-HT neurones in the CNS and measures of ongoing aggressive behaviour. Part of the available evidence is the correlation between levels of 5-HT and 5-hydroxyindoleacetic acid (5-HIAA) (i.e. 5-HT-turnover) in certain specific brain areas with some forms of aggressive behaviour (c.f. Daruna, 1978). More evidence stems from different pharmacological manipulations of the 5-HT system and the effects on some parameters of aggression (for a review, see Miczek, 1987). Thus it was observed that depleting 5-HT in the brain facilitated or elicited various kinds of aggressive behaviour in the rat, e.g. predation (Applegate, 1980; Valzelli et al., 1981; Vergnes and Kempf, 1981, 1982) and shock-induced fighting (defence) (Kantak et al., 1981; Knutson et al., 1979; Sheard and Davis, 1976).

In a resident-intruder paradigm in rats, Vergnes et al. (1986) showed that depletion of 5-HT induced by parachlorophenylalanine (pCPA), enhanced offence when the resident was treated, but did not affect defensive behaviour of the intruder. This suggests that 5-HT plays a role in modulating offence, rather than defence.

Early work on 5-HT and aggression suggested that 5-HT activation decreased aggression whereas an overall inactivation of 5-HT (achieved by various manipulations) enhanced aggression (e.g. Valzelli, 1981). However, the increased knowledge about the neuroanatomy, neurochemistry and neuropharmacology of the serotonergic systems in the CNS on the one hand, and the different types of aggressive behaviours on the other hand, hint at a much more complicated pattern of the involvement of 5-HT in agonistic behaviour. The neuroanatomical distribution of 5-HT receptors and localization of cell groups of 5-HT in the CNS and their differential projections (c.f. Pazos and Palacios, 1985; Pazos et al., 1985; Steinbusch, 1981) strongly indicate that it is no longer justified to speak about an unidimensional role of 5-HT in any aggression paradigm.

Moreover, the recent differentiations in 5-HT receptor subtypes or binding sites (Arvidsson et al., 1986; Hoyer et al., 1985; Peroutka, 1986; Richardson and Engel, 1986) and their distinct functional roles in other behavioural mechanisms (e.g. feeding, fear, depression) point to a further differentiation in the 5-HT systems in the CNS with regard to different kinds of agonistic behaviour.

Aggressive behaviour, as observed in a variety of aggression models, consists of an extensive number of separate components with complex functions and organized in typical patterns and sequences resulting in species-specific behaviour. Depending on the animal model used (i.e. the environmental situation), each model delivers its own set of species-specific behaviours. The distinction between the different types of aggression occurring in different models is mainly based on the frequency distribution and sequence of the possible behavioural acts (repertoire) of the species involved.

A defending and offending rat make use of the same set of behaviour elements, but because of a different frequential and sequential distribution of the elements, the behaviour can be classified into offensive or defensive. It is current terminology to refer to aggressive interactions between animals by the term "agonistic behaviour", which refers to all elements of behaviour present in situations of conflict, including offence, defence and flight.

The present contribution tries to tackle this problem by using several serotonergic compounds with different mechanism of action and testing these drugs in several animal models of aggression. Because drugs can inhibit aggression in diverse ways (e.g. sedation, motor disturbances, psycho-stimulation) it is important to describe drug effects carefully, preferably by using ethological set-ups, in which not only aggression is measured but simultaneously all other behaviours are recorded in detail too.

Although studying the behavioural effects of serotonergic compounds in different aggression paradigms may lead to clues about the involvement of the various putative 5-HT receptors in agonistic behaviour, there are several shortcomings at this moment to optimally gain from this approach. First, the availability of potent and specific pharmacological tools is problematic. Although we seem to have reasonably specific agonists for $5-HT_{1A}$, $5-HT_{1C}$ and $5-HT_3$ receptors, those for $5-HT_2$ and $5-HT_{1B}$ are lacking. For 5-HT antagonists a more or less reverse picture holds; reasonably specific antagonists for $5-HT_2$, $5-HT_3$ and $5-HT_{1C}$ receptors but not for $5-HT_{1A}$ and $5-HT_{1B}$ receptors. Therefore most hypotheses on functional involvements of 5-HT have to be formulated carefully, and should be retested upon emergence of new and specific tools. Secondly, most 5-HT drugs are not sufficiently specific, not only for the different 5-HT receptors, but neither for other receptors. Thirdly, psychoactive drugs also affect peripherally located receptors which may lead to factors interfering with the proper performance of behaviour. Furthermore, pharmacokinetic and metabolic factors may complicate interpretations. Another pitfall is that the different 5-HT receptor subtypes do not seem to be similar across species, although this data is largely based on receptor binding and not on functional studies. This implies however that functional coupling of aggression to a certain receptor subtype should preferably be based on more than one species and, not just rodents only. Notwithstanding all these problems and pitfalls, at this moment applying drugs to animals and observing the resulting behavioural changes it seems the most direct way to study the involvement of certain receptor subtypes in agonistic behaviour.

Neurochemical profile of 5-HT drugs used
The following section firstly provides a short description of the neurochemical profiles of the prototypic serotonergic drugs. Apart from data on receptor affinities, some functional indicators are given, namely 5-HT agonistic activity in vitro on the presynaptic autoreceptor, modulation of 5-HT turnover in vivo and inositolphosphate production in the choroid plexus of the pig, as a $5-HT_{1C}$ model.

Receptor binding
Of the drugs tested in behavioural studies of aggression several compounds have a high affinity for $5-HT_{1A}$ binding sites relative to $5-HT_{1B}$, $5-HT_{1C}$ or $5-HT_2$ (5-MeODMT, 8-OH-DPAT, buspirone, ipsapirone and flesinoxan). Others have about equal affinity for $5-HT_{1A}$, $5-HT_{1B}$ and $5-HT_{1C}$ sites (RU 24969, eltoprazine), whereas TFMPP and quipazine (Hoyer and Neyt, 1987; Kilpatrick et al., 1987) exert some preference for $5-HT_{1C}$ and $5-HT_3$ sites respectively. In general, affinity of these compounds for $5-HT_2$ receptors is relatively low. Ritanserine exerts the highest preference for the $5-HT_{1C}$ and $5-HT_2$ receptor and is an antagonist. Surprisingly, DOI, besides its relatively strong affinity for $5-HT_2$ receptors (agonist), is also a strong agonist on the $5-HT_{1C}$ receptor (Schipper and Tulp, unpublished). Of course, several compounds also have affinity for other receptors and in some cases (e.g. buspirone–D_2) it is likely that such affinities cause contribute to the biological activity.

Autoreceptor modulation of 5-HT release
5-HT release from nerve terminals is subjected to a negative feedback via presynaptic 5-HT receptors.

Various authors (Engel et al., 1986; Middlemiss, 1984) proposed that these autoreceptors are of the $5-HT_{1B}$ type. Several 5-HT agonists inhibit the 5-HT release in vitro as measured by the K^+- or electrically-stimulated release of pre-stored (^3H) 5-HT from brain slices (Göthert, 1980). As indicated by the pD_2 values in table 9.2, 5-HT, RU 24969, 5-MeODMT, TFMPP and eltoprazine act as agonists at these presynaptic autoreceptors. Compounds with a high affinity for the $5-HT_{1A}$ binding site, such as 8-OH-DPAT, buspirone and ipsapirone, do not affect 5-HT release, indicating a lack of agonistic activity on the 5-HT autoreceptor. The phenylpiperazines TFMPP and eltoprazine show activity comparable to 5-HT, but their intrinsic activities classify them as partial agonists.

5-HT turnover in vivo
After inhibition of the aromatic-L-amino-acid decarboxylase by NSD 1015, 5-hydroxytryptophan (5-HTP) accumulates in brain regions containing 5-HT terminals. The rate of accumulation of 5-HTP has been used as an index of 5-HT neuronal activity (Carlsson et al., 1972). 5-HT agonists inhibit 5-HTP accumulation and it has been suggested that this is due to a negative feedback mechanism (Arvidsson et al., 1986). The 5-HT agonists that inhibit 5-HT release

Table 9.1 Affinity of serotonin and several serotonergic compounds for 20 receptor (sub)types expressed as pK_i's ($-\log K_i$). All data are based on at least three measurements. For experimental details and literature on methods used see Olivier et al. (1987). '-' denotes: $pK_i < 5.0$ and 'n.t.' denotes: not tested.

Receptor (sub)type	5-HT$_{1A}$ agonists				mixed 5-HT$_1$ agonists					5-HT$_{1C}$ agonist	5-HT$_2$/5-HT$_{1C}$ antagonists		5-HT$_3$ antagonists			mixed 5-HT antagonists		miscellaneous			
	8-OH-DPAT	Flesinoxan	Buspirone	Ipsapirone	Serotonin	5-Me-O-DMT	RU 24969	TFMPP	Eltoprazine	DOI	Mianserine	Ritanserine	Quipazine	MDL 72222	GR 38032F	Methysergide	dl-propranolol	Fluprazine	Berperide	Fluvoxamine	Fenfluramine
5-HT$_{1A}$	8.6	8.8	7.8	8.3	8.4	8.2	8.1	6.7	7.4	–	6.1	6.1	5.6	–	–	7.7	6.7	6.4	7.7	–	5.7
5-HT$_{1B}$	5.8	6.3	5.5	5.5	8.6	7.1	8.2	7.3	7.3	5.7	5.9	5.8	6.2	–	5.4	7.0	6.3	5.4	6.3	–	5.1
5-HT$_{1C}$	5.2	–	6.1	–	8.4	7.6	7.3	7.9	7.1	8.2	8.5	9.3	7.1	–	5.2	8.3	6.0	5.6	6.7	–	6.1
5-HT$_2$	–	5.4	6.0	5.6	5.9	5.6	5.8	6.1	5.8	6.7	8.0	8.5	6.0	–	–	7.8	n.t.	5.8	7.4	5.9	–
alpha$_1$	5.6	6.4	6.2	6.6	–	5.5	5.9	5.9	6.1	5.1	7.1	7.1	5.4	–	5.4	–	n.t.	5.8	8.7	5.4	6.3
alpha$_2$	6.5	–	–	5.6	–	5.5	5.6	5.8	5.5	5.6	6.8	6.5	5.9	–	–	5.5	5.2	–	–	–	–
beta$_{1/2}$	–	–	–	–	–	5.3	6.4	6.3	6.7	6.5	5.0	–	5.9	–	–	5.2	8.9	–	–	–	–
DA$_1$	–	–	–	5.2	5.4	5.6	–	–	–	–	5.9	6.2	–	–	–	n.t.	–	–	6.3	–	–
DA$_2$	5.7	6.9	7.4	6.4	–	–	5.9	6.1	6.0	–	6.0	7.9	–	–	–	6.8	–	5.6	6.7	–	–
muscarinic	5.7	–	–	–	–	n.t.	–	5.6	–	–	6.3	5.9	–	5.8	–	–	–	5.6	5.6	–	–
his-H$_1$	–	6.1	6.0	5.9	n.t.	6.1	–	5.6	5.4	6.1	8.7	7.9	5.9	6.0	–	n.t.	n.t.	5.6	6.0	–	–
mu-opiate	–	5.0	5.2	–	–	–	5.2	–	–	5.3	–	5.6	–	–	5.6	–	–	5.1	5.3	–	–
kappa-opiate	–	5.3	5.6	5.1	–	–	–	–	–	–	–	–	–	–	–	–	–	–	–	–	–
delta-opiate	–	–	–	–	–	–	–	–	–	–	–	–	–	–	–	–	–	–	–	–	–
BDZ	–	–	–	–	–	–	–	–	–	–	–	–	5.3	–	–	n.t.	n.t.	–	–	–	–
glycine	–	–	–	–	–	–	–	–	–	–	–	–	5.6	–	–	n.t.	n.t.	–	–	–	–
GABA	–	–	–	–	–	–	–	–	–	–	–	–	5.1	5.5	–	n.t.	–	–	–	–	–
TRH	–	–	–	–	–	–	–	–	–	–	–	–	–	–	–	n.t.	n.t.	–	–	–	–
CCK	–	–	–	–	–	–	–	–	–	–	–	–	–	–	–	n.t.	n.t.	–	–	–	n.t.

in vitro also reduce 5-HTP accumulation in vivo. Interestingly, 8-OH-DPAT and flesinoxan also reduce 5-HTP accumulation. The in vitro data suggest that it is unlikely that the 5-HT$_{1A}$ compounds reduce 5-HT neuronal activity via presynaptic receptors mediating 5-HT release.

It is possible that another presynaptic feedback mechanism exists via 5-HT$_{1A}$ receptors which modulates synthesis rather than the release mechanism. On the other hand, one cannot exclude the possibility that a postsynaptic 5-HT$_{1A}$ receptor is involved, resulting in a "long-loop" feedback mechanism as known for e.g. dopaminergic neurones. Further research will be necessary to unravel the mechanisms involved in the modulation of 5-HT neuronal activity via 5-HT receptor subtypes in vivo.

5-HT$_{1C}$-phosphoinositol production in the choroid plexus of the pig
Phosphatidyl inositol turnover in pig choroid plexus was measured according to Conn et al. (1986) with some slight modifications. 5-HT elicited a 10-fold increase in [^{3}H]-IP formation. The maximal response occurred at 10^{-5}M and the pD$_2$ was 6.8. Several drugs exert weak 5-HT$_{1C}$-agonistic partial activity, like RU 24969, 5-MeODMT, TFMPP, quipazine and DOI, whereas mianserine, ritanserine and eltoprazine have 5-HT$_{1C}$ antagonistic activity.

SEROTONERGIC DRUGS AND OFFENSIVE AGGRESSION

Isolation induced aggression in mice
The most frequently used manipulation to induce aggression of animals is isolation for some time. Male laboratory mice show attack behaviour after several weeks of isolation (Valzelli, 1969). This isolation-induced aggression model is one of the most frequently used aggression models in behavioural pharmacology. Because isolated male mice show a full repertoire of agonistic behaviours against an intruder in their home cage (Krsiak, 1974; Miczek and Krsiak, 1979), this isolation-induced aggression model can also be used in an ethological way, giving the possibility to detect very specific drug effects (c.f. Olivier and Van Dalen, 1982). The model mainly represents offensive aspects of agonistic behaviour, but in some animals defensive aspects prevail, so that this behavioural model can be elegantly used to differentiate between several drugs that influence agonistic behaviour (Janssen et al., 1960; Krsiak, 1975, 1979). Although this paradigm has been suggested to model a psychopathology (Valzelli, 1973), more recent studies, assessing ethological, pharmacological and endocrinological features, strongly suggest that such isolated males are highly similar to adult male mice in nature defending their territory (Blanchard et al., 1979; Brain, 1975; Miczek and O'Donnell, 1978).

A comparable model is the intermale aggression model in rats and mice. In this model, a relatively short isolation period is used to increase the probability of aggression in a short test against a male conspecific in a neutral test arena (Olivier, 1981; Olivier and Van Dalen, 1982). In this test, a very diverse pattern of activities is present because such a situation delivers a mixture of offensive-defensive behaviour.

This model can differentiate more specifically between different putative anti-aggressive drugs and is more interesting in this respect because it shows properties of compounds which are revealed only partially, or not at all by common pharmacological test models e.g isolation-induced aggression (c.f. Miczek and Krsiak, 1979; Olivier, 1981).

Table 9.2 Effects of serotonergic compounds on in vitro 5-HT release (5-HT_{1B} model), in vivo 5-HTP accumulation and in vitro PI hydrolysis (5-HT_{1C} model)

	5-HT release in cortex[1]			5-HTP formation[2]	PI hydrolysis[3]		
	pD_2	α	pA_2	ED_{50} mg/kg	pD_2	α	pIC_{50}
5-HT	7.7	1.0			6.8	1.0	
RU24969	8.0	0.8		1.0	6.1	0.3	
5-MeODMT	5.5	0.9		<10	5.4	0.3	
TFMPP	7.7	0.7		1.9	6.4	0.6	
Eltoprazine	7.6	0.4		0.3		0	5.0
Fluprazine	<5.5		<5.5		<6.0		<5.5
8-OH-DPAT		0	6.2	0.2	<5.0		<5.5
Quipazine		0	5.5	>10	5.7	0.8	
Buspirone		0	6.0	5.4	<6.0		<6.0
Ipsapirone	<6.0		<6.0	1.8	<5.5		<5.5
Flesinoxan				0.5	<6.0		<6.0
DOI					7.0	0.7	
Methysergide	<6.0			>10			
Mianserine		0	6.0	>30		0	7.6
MDL 72222	<5.5		<5.5		<5.5		<5.5
GR 38032F	<5.5		<5.5		<5.5		<5.5
Ritanserine	<5.5		<5.5			0	7.7

1 Effects on K^+ (20 mM)-evoked ^3H-5-HT release from rat cortex slices. Neuronal 5-HT uptake was blocked during superfusion by fluvoxamine (10 μmol/l). Compounds were added to the superfusion medium 15 min before stimulation. pD_2 values, intrinsic activity (α) and pA_2 values are calculated from the concentration-response curves. For details see Engel et al. (1986).
2 Effects on 5-HTP formation in striatum of rats. Compounds were given 60 min and NSD 1015 (100 mg/kg IP) 30 min before decapitation. 5-HTP levels were determined by use of HPLC with electrochemical detection. Doses necessary to induce half-maximal inhibition of 5-HTP levels were determined from the dose-response curves. All compounds were given orally except 5-Me-O-DMT and 8-OH-DPAT which were given intraperitoneally.
3 Effects on inositol phosphate accumulation in slices of pig choroid plexus in the presence of 5 mM Li^+. For details see Conn et al. (1986). Agonist activity was determined by pD_2 values and intrinsic activity (α). Antagonist activity was determined by the neg. logarithm of concentration needed to inhibit 50% of the effect of 1 μM 5-HT (pIC_{50}).

The effects of serotonergic drugs were studied in two set-ups. First, in a relatively simple screening test, the ED_{50} for inhibition of isolation-induced aggression was determined using prototypic serotonergic drugs. Although the results given in table 9.3 show that a substantial number of drugs inhibit aggression it is unclear from such values which behavioural inhibitory mechanisms are involved. Therefore, more detailed ethological studies were performed, which, by taking into account a broader repertoire of behaviours, may indicate more specifically how aggression was reduced.

The methodology used has been described in detail in Olivier and Van Dalen (1982) and Olivier et al. (1986b). Briefly, the behaviour of the isolated animal was recorded in a neutral cage using 16 categories to describe the ongoing behaviour (during 5 min confrontations between a 4-8 week isolated male mouse and a group-housed male conspecific). This approach permits a direct, simultaneous measurement of drug effects on aggressive and non-aggressive behaviours, revealing the specificity of drug action (c.f. Miczek, 1987; Miczek and Krsiak, 1979). Figure 9.1 summarizes how eleven serotonergic drugs affect such behaviour.

Table 9.3 ED_{50}-values of serotonergic drugs in isolation-induced aggression in male mice. Drugs were given orally 60 min before testing suspended in 1% tragacanth, except for 8-OH-DPAT and ritanserine.

Drug	ED_{50} (mg/kg PO)
RU24969	0.7
5-Me-O-DMT	4.2
TFMPP	0.2
Eltoprazine (DU 28853)	0.4
Fluprazine	1.2
8-OH-DPAT	0.3 (IP)
Quipazine	>38
Buspirone	>20
Ipsapirone	>20
Fluvoxamine	70
Fenfluramine	10
Methysergide	>10
Ritanserine	>10 (IP)
MDL72222	>10
GR 38032F	>10
DOI	>10
Flesinoxan	1.1
Befiperide	1.4
Mianserine	6.8
dl-Propranolol	>20

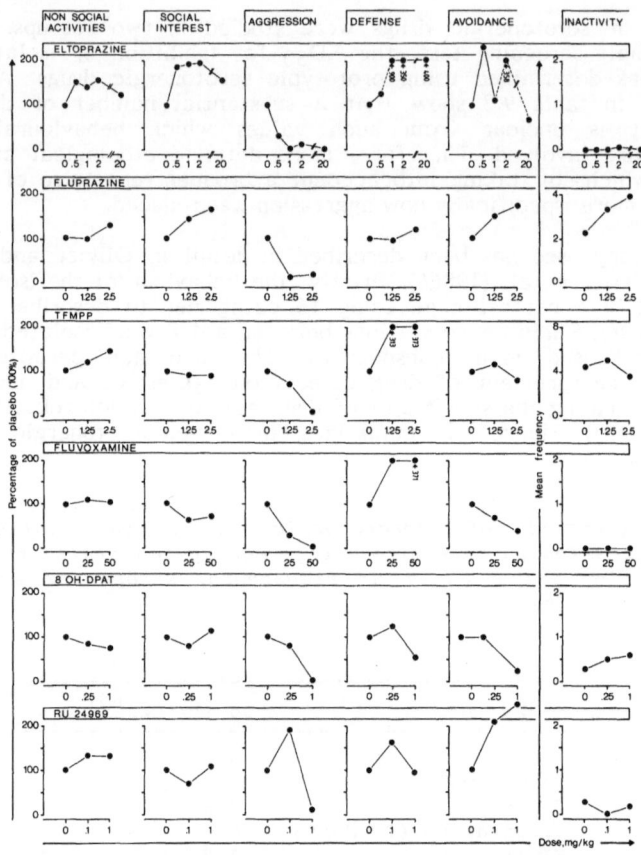

Legends chapter 9

Fig. 9.1 Effects of eltoprazine (DU 28853), fluprazine, TFMPP, fluvoxamine, 8-OH-DPAT, RU 24969 (left page), befiperide, flesinoxan, buspirone, ipsapirone and GR 38032F (right page) on six behavioural categories in intermale aggression in mice.

Eltoprazine dose-dependently decreases aggression with a concomitant increase in social interest, whereas non social activities (mainly exploration) were also somewhat enhanced. It is especially noteworthy that no sedation was evident even at the dose of 20 mg/kg, which is 40 times the ED_{50} of isolation-induced aggression.

Although defence seems to be dramatically enhanced, it has to be remembered that under control conditions, defence hardly occurs, whereas at doses where aggression is no longer present (1 mg/kg and higher) animals defend (including avoidance) themselves against obtrusive intruders.

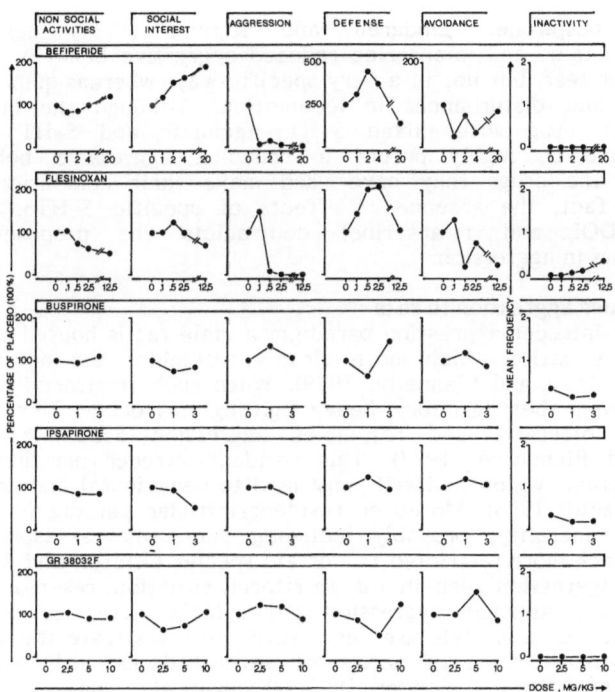

A more detailed pattern for eltoprazine's behavioural effects is given in Olivier et al. (1987). A rather similar pattern emerges for fluprazine and TFMPP (for full descriptions see Olivier et al., 1986d). Fluvoxamine, a specific 5-HT reuptake blocker (Claassen, et al., 1979) also decreases aggression, but in a less specific way because it is accompanied by decreases in social interest and avoidance. 8-OH-DPAT, a specific 5-HT_{1A}-agonist (Middlemiss and Fozard, 1983) reduces aggression but also decreases non-social activities somewhat, and reduces defence (at 1 mg/kg) and avoidance.

RU 24969, a mixed 5-HT_1-agonist (Tricklebank, 1985) reduces aggression at the highest dose tested, concomitantly increasing avoidance.

Befiperide, a mixed 5-HT-agonist (Van der Heyden, this volume), had a similar pattern to eltoprazine. Flesinoxan, a specific 5-HT_{1A}-agonist, reduces aggression potently, but this is associated with decreases in non-social activities and avoidance. This pattern strongly resembled that after 8-OH-DPAT. At the highest dose used (12.5 mg/kg) some slight sedation occurred. In contrast, buspirone and ipsapirone, also 5-HT_{1A}-agonists, had no effects on aggression at the doses used, which are certainly in the psychoactive range. GR 38032F, a 5-HT_3 antagonist, had no clearcut effects up to 10 mg/kg.

McMillan et al. (1987) confirmed the antiaggressive activities of 8-OH-DPAT and TFMPP, but in contrast to what we found, they also reported anti-aggressive effects for buspirone. Lindgren and Kantak (1987) described that 5-methoxytryptamine and mianserine reduced aggressive behaviour in a mouse social behaviour test, but not in a very specific way, whereas quipazine reduced aggression without disturbances in locomotion. Although the latter authors predict that a drug with mixed $5-HT_1$-agonistic and $5-HT_2$-antagonistic properties would be highly potent in reducing aggressive behaviour, the specificity of the drugs they have used make their conclusions somewhat premature. In fact, the absence of effects of specific $5-HT_2$ agonists and antagonists (DOI and ritanserine) contradicts the proposed role for $5-HT_2$-receptors in aggression.

Resident-intruder aggression in rats

In the resident-intruder aggression paradigm a male rat is housed with a female conspecific, a situation which more closely resembles the natural situation (Barnett, 1975; Lore and Flannelly, 1979). When such territorial males meet a strange intruder in their territory, heavy fighting may occur. This aggression is considered as offensive and resembles aggression occurring in the wild (Blanchard and Blanchard, 1977). This resident-intruder paradigm needs no prolonged isolation, which in itself may lead to behavioural and neurochemical anomalies (Valzelli, 1973). Moreover resident-intruder paradigms have a very wide species generality, probably including man, whereas isolation-induced aggression is much more restricted to certain species (Miczek and Krsiak, 1979). The offensive aggression seen in the territorial situation, resembles to a great extent that seen in intermale aggression (although the latter has more defensive components), and both models are very usefull to investigate the specificity of and the behavioural mode of action of several drugs with both anti- and pro-aggressive qualities (c.f. Olivier, 1981; Olivier et al., 1984).

Details of the methods are provided in Olivier (1981) and Olivier et al. (1986a). Briefly, male rats (residents) living together with a female in a large territory cage, were confronted during 10 min with a strange male intruder. Behaviour was recorded using ethograms as described before (Olivier, 1981). Drugs were given 30 min (i.p.) or 60 min (p.o.) before testing. Animals were tested once a week and treatments (vehicle and 3 doses of a drug) were randomized according to a Latin square design. The results for eltoprazine, TFMPP, 5-MeODMT, fluvoxamine, buspirone and befiperide are shown in fig. 9.2. Behavioural effects are shown on four main behavioural categories: aggression, social interest, exploration and inactivity.

Each category comprises several behavioural elements (c.f. Olivier et al., 1984, 1986d) and adequately reflects the general effects of drugs on social behaviour. Eltoprazine dose-dependently reduces aggression without a concomitant decrease in social interest. Exploration is not changed or even somewhat enhanced. Although inactivity is not significantly enhanced, it tends to increase at the highest dose, presumably replacing the time normally spent on aggression. A similar pattern occurs after TFMPP and fluprazine (not shown). 5-MeODMT reduces aggression in a nonspecific way, indicated by a simultaneous decrease in social interest and an increase in inactivity.

Fluvoxamine has no dramatic effects at the doses used but the general pattern indicates a non-specific inhibitory effect on aggression. Buspirone nonspecifically reduces aggression as the compound is quite sedative at doses suppressing attacks (c.f. Olivier et al., 1984) in such tests. Befiperide reduced aggression without a concomitant decrease in social interest. However, some sedation is indicated by a decrease in exploration and an increase in inactivity at the highest dose.

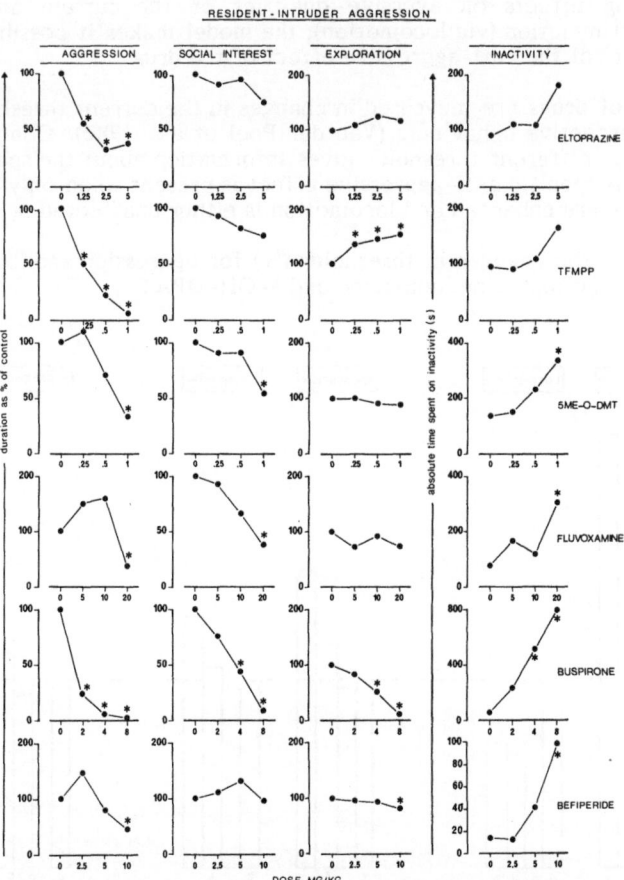

Fig. 9.2 The effects of eltoprazine, TFMPP, 5-MeODMT, fluvoxamine, buspirone and befiperide on four behavioural categories in resident-intruder aggression in rats.

Hypothalamically-induced behaviour in rats

Aggression similar to that of offensive territorial males can be induced by electrical stimulation in the hypothalamus of male and female rats (Kruk et al., 1979; Mos et al., 1987). Overt attack behaviour is used as the behavioural characteristic; the current threshold is used to judge the anti-aggressive potency of the compound. Since stimulation of the same electrode can be used to measure drug effects on aversive qualities of the current and to measure sedation or stimulation (via locomotion), the model makes it possible to estimate the specificity of the anti-aggressive effect of the drug.

The effects of drugs are measured in changes in the current thresholds needed to evoke the respective behaviours (Van der Poel et al., 1982). Comparison of the effects on the different thresholds gives information about the specificity of the drug effect. A specific anti-aggressive effect is present when only the thresholds for aggression are enhanced and locomotion is either unaffected or decreased.

Fig. 9.3 shows the changes in thresholds (%) for aggression and locomotion after treatment with eltoprazine, quipazine and 8-OH-DPAT.

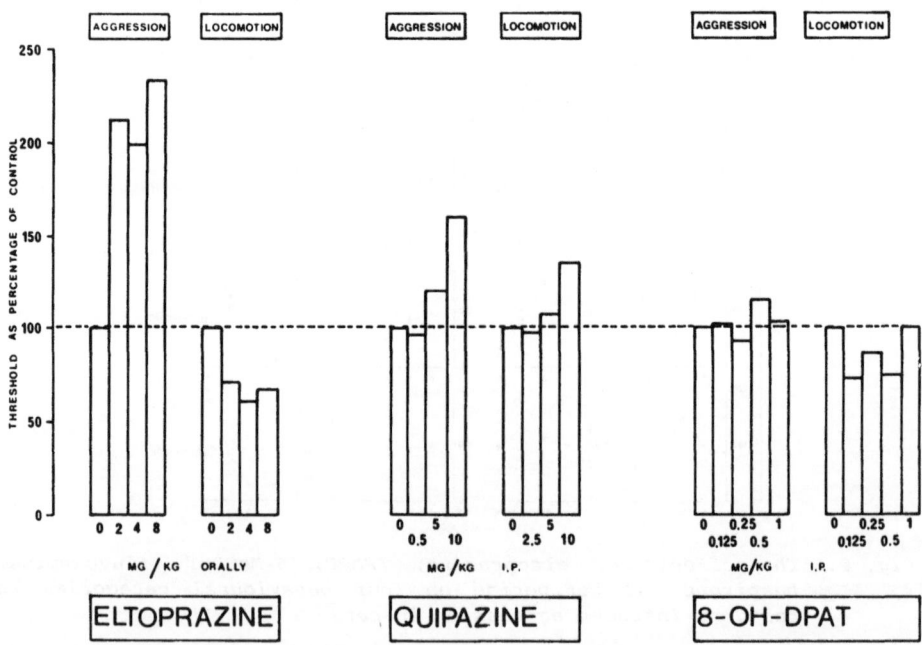

Fig. 9.3 *The effects of eltoprazine, fluprazine, 8-OH-DPAT and fluvoxamine on the threshold currents (in % of vehicle) to induce aggression or locomotion in the hypothalamus of male rats.*

Eltoprazine markedly enhanced the threshold currents for aggression whereas the threshold for locomotion was even decreased, indicating the specificity of action on aggression. Fluprazine (not shown) had a comparable effect on aggression but hardly influenced locomotion (see Van der Poel et al., 1982). 8-OH-DPAT has no effect on aggression but reduces locomotion thresholds whereas quipazine decreases aggression but this parallels an increase in locomotion.

Recent evidence (Kruk et al., 1987) showed that this EBS-induced behaviour paradigm in rats shows a quite specific profile for serenics; enhancement of thresholds for aggression and teeth-chattering, no effect or even a decrease on locomotion thresholds and no effect on switch-off behaviour, a measure for the interference of a drug with the aversive qualities also resulting from electrical brain stimulation. Kruk et al. (1987) also found that quipazine, a nonspecific 5-HT-agonist and $5-HT_3$ antagonist, had a quite nonspecific action in EBS-behaviours: increases in all thresholds, indicating its behavioural nonspecificity. These authors also found that 8-OH-DPAT, a specific $5-HT_{1A}$-agonist, had, at doses between 0.05 and 0.2 mg/kg i.p., no effects, whereas TFMPP, a more specific $5-HT_{1B}$-agonist, and a putative metabolite of fluprazine, exerted a specific effect at 0.5 - 2 mg/kg (i.p.) on aggression without interference with locomotion. Interestingly, TFMPP enhanced switch-off thresholds indicating at least that the anti-aggressive action is not caused by fear-induction. dl-Propranolol, a β-adrenergic blocker (at 5-20 mg/kg i.p.) and a $5-HT_1$ antagonist also had a specific anti-aggressive profile: it inhibited aggression but had no influence on locomotion.

Maternal aggression in rats

Maternal aggression is a female aggression paradigm, in which a lactating animal with young behaves very aggressively towards a large variety of intruders. In rats, maternal aggression is most pronounced during the first part of the lactating period (Erskine et al., 1978a,b; Olivier and Mos, 1986a). The aggression performed by a lactating female can be considered as offensive (Van der Poel et al., 1984).

The effects a number of serotonergic compounds were studied in a maternal aggression paradigm, specifically developed in our laboratory to test psychoactive drugs (Olivier and Mos, 1986a,b; Olivier et al., 1985, 1986a,b). The lactation period (3-12 days after birth) appears a relatively stable time to perform aggression tests using the females as their own controls (Olivier et al., 1985, 1986a; Olivier and Mos, 1986a; Mos and Olivier, 1986).

Several drugs were tested using this paradigm, during test periods of 5 minutes. Each female was repeatedly tested on alternate days between post-partum days 3 to 12. Testing occurred every other day with vehicle or drug (several doses) alternated with "wash-out" days. For the sake of comparison the mean number of bite attacks/minute is presented, a measure which is corrected for the latency to the first attack (Fig. 9.4).

This representation of the data shows that most serotonin-modulating drugs tested thusfar inhibit aggression, although there are vast differences between drugs. 8-OH-DPAT has a somewhat peculiar profile in that aggression suddenly decreases without a clear dose-response distribution. Behaviourally this

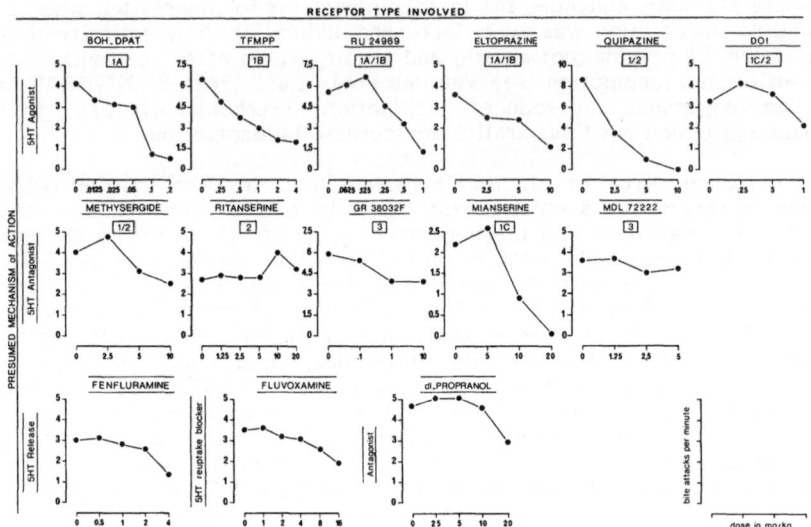

Fig. 9.4 The mean number of bite attacks/minute (± SEM) is shown for fourteen serotonergic drugs in the maternal aggression test in rats.

inhibition is also manifest in a sudden drop in interest for the intruder and a high increase in inactivity (see Fig. 9.5).

Flesinoxan potently reduced aggression without interference with social interest, exploration and pup care, but strongly reduced body care and enhanced inactivity, indicating that 5-HT_{1A} agonists differ in their mode of anti-aggressive action. Other 5-HT agonists all have regular dose-dependent, inhibitory effects on aggression (Fig. 9.4), except DOI.

However, more refined behavioural analyses reveal substantial differences between the drugs involved (Fig. 9.5). Quipazine, a nonspecific 5-HT-agonist, clearly has a very nonspecific behavioural profile: decreases in aggression and social interest and a strong increase in inactivity (c.f. Olivier and Mos, 1986a). The 5-HT_{1B}-agonist, TFMPP and both mixed 5-$HT_{1A/B}$ agonists RU 24969 and eltoprazine (and also the weak 5-HT agonist fluprazine (Bradford et al., 1984)), have a similar profile in this female paradigm namely reducing aggression whilst increasing social interest, exploration, pup care and inactivity.

The 5-HT antagonists methysergide, ritanserine, GR 38032F, MDL 72222 and dl-propranolol did not affect aggression, but mianserine, a 5-HT_{1C} antagonist, had a potent anti-aggressive effect, presumably caused by interactions of this drug on other receptor types (see Table 9.1). Fenfluramine and fluvoxamine both reduced aggression but not in a very specific way (Fig. 9.5).

SEROTONERGIC DRUGS AND DEFENSIVE AGGRESSION

Defensive aggression models

In the foregoing models the offensive aspects of aggression were emphasized. In contrast, defensive aggression lacks active approach, and no wounds (or incidental wounds on the snout) are made on the attacker. A much used aggression-model in pharmacology is foot-shock or pain-induced aggression, which is nowadays conceived as a defensive aggression model (Blanchard et al., 1977a). Electrical shock to the hindpaws of a pair of rats or mice can evoke defensive behaviour (Ulrich and Azrin, 1962), consisting of mutual defensive upright postures and squealing. These reactions are well integrated, but no complete sequences of fighting and no signs of offensive (threat) displays are present at least in rats. An interfering factor in such a model is that the behaviour-releasing stimulus (pain) can be masked by analgesic properties of psychoactive drugs. The latter fact and the very restricted behavioural repertoire in this paradigm, limits the use of this defensive model to test anti-aggressive properties of drugs, although it has been (Valzelli, 1978) used extensively to assess such qualities and still is used.

A more natural defensive model is the resident-intruder paradigm, in which the intruder is attacked by a resident male or lactating female and has to defend itself (Miczek and Krsiak, 1981; Olivier and Mos, 1986b; Rodgers, 1981). In this situation a defending rat displays all behavioural elements occurring in natural situations as defensive upright postures, freeze-crouch postures, full submissive postures, fleeing and vocalisations (sonic and ultrasonic). This model gives the opportunity to record effects of psychotropic agents on the complete defensive behavioural repertoire.

Foot-shock-induced defence (SID) in mice and rats

Several studies indicate that 5-HT agonists may reduce foot-shock induced defence (SID). Olivier et al. (1987) showed that fluprazine, but not eltoprazine, inhibited SID in mice. Quipazine has frequently been reported to inhibit SID (Sheard, 1981; Rolinski and Herbut, 1981). Also decreases in SID have been reported after treatments with 5-HT-antagonists cyproheptadine, metergoline and methysergide (Rolinski and Herbut, 1981). As potential analgesic effects may heavily interfere with the evoking of this kind of behaviour, and serotonin is also widely implicated in pain perception (see Berge et al., this volume), it is difficult to delineate from these kind of data an exact role of 5-HT in defensive behaviour.

Defensive behaviour of animals when attacked

Those forms of agonistic behaviour in which elements of initiative and approach prevail belong to offensive aggression. This offence contrasts with defence, in which fighting is merely a response to being attacked, without initiative and essentially "reactive". Flight and submission is behaviour aimed at escaping or preventing further agonistic interactions (c.f. Dixon and Kaesermann, 1987).
Some of the drugs known to suppress aggression effectively have highly undesirable effects: e.g., neuroleptics decrease activity, including social interest, whereas low doses of benzodiazepines may even increase aggression (Olivier et al., 1985).

104 Olivier et al.

However, aggression is not always detrimental. Ideally, drugs should inhibit aggression but leave animals competent to deal with situations that require initiative and adequate defence and flight in response to threat and danger.

To test the effects of drugs on this aspect of defensive/flight behaviour, we tested drug-treated male intruders in an aggression paradigm, where they will be heavily attacked by lactating females and are strongly dependent on their own defensive capabilities (Olivier and Mos, 1986; Olivier et al., 1987). Fluprazine and eltoprazine had no dramatic effects on the defensive and flight capabilities of drug-treated intruders (results not shown; see Olivier et al., 1987). This suggest that, at least $5-HT_{1A/B}$ agonists, do not interfere with aspects of behaviour of animals that are of utmost importance for species survival. Unfortunately, this paradigm has not been used for several other serotonergic compounds, but data on e.g. d-amphetamine and haloperidol (Olivier and Mos, 1986b) indicate that these compounds do interfere with this defence/flight repertoire.

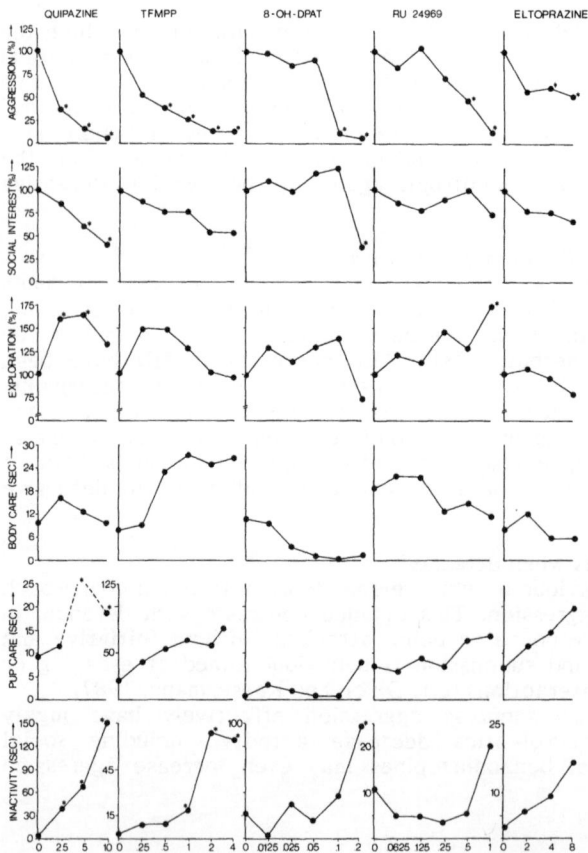

Fig. 9.5 Effects of nine serotonergic drugs on the mean duration of six behavioural categories in maternal aggression in rats. Data are expressed either as % of vehicle (0 mg/kg) or as mean duration (sec).

Table 9.4 Effects of serotonergic drugs on muricide in rats. Data are expressed as Lowest Effective Dose (LED) required to inhibit mouse killing.

Drug	LED (mg/kg IP) in males	LED (mg/kg IP) in females	Specificity of inhibiting effect *
Fluprazine	8	40	+
TFMPP	2	0.5	+
8-OH-DPAT	>5	>5	-
RU24969	1	2	+/-
Eltoprazine	5	20	+
Quipazine	4	4	-
Fenfluramine	2	4	-
Fluvoxamine	20	10	+/-
Methysergide	>20	>30	0
Ritanserine	>20	>20	0
5-Me-O-DMT	1	5	+
Buspirone	10	15	-
Ipsapirone	>10	>10	0
GR 38032F	>10	>10	0
Mianserine	20	nt	-
dl-Propranolol	10	nt	+
Befiperide	10	nt	+
DOI	nt	nt	nt

* + means behaviourally specific effects, - means nonspecific inhibition by e.g. sedation or motoric disturbances; 0 no inhibition. nt = not tested

Another interesting paradigm to study defence/flight behaviour is brain stimulation induced escape behaviour (Graeff et al., 1986; Schütz et al., 1985; see also this volume). These authors showed that microinjection of serotonin or 5-MeODMT into the dorsal central grey of rats enhanced the thresholds for inducing escape behaviour. Zimeldine, a 5-HT uptake blocker, was able to potentiate the effect of 5-HT and enhanced the threshold also when given alone, thus indicating that during brain-stimulation endogenous 5-HT is released. Graeff et al. (1986) found evidence that primarily $5-HT_2$ receptors in the central grey were involved in the anti-escape activity of 5-HT. Although more work is clearly needed to further verify the hypothesis that $5-HT_2$ receptors are involved in the modulation of defence/flight, it may be quite attractive to speculate upon a more refined hypothesis in which $5-HT_{1B}$ receptors modulate offence and $5-HT_2$-receptors defence/flight.

SEROTONERGIC DRUGS AND PREDATORY BEHAVIOUR

Predatory aggression
A much used aggression model is mouse-killing by rats (muricide). This predatory or interspecific aggression model can be perceived as predatory attack on a prey.

The muricide model has been used extensively to assess anti-aggressive qualities of drugs, but is of limited value because there is a lot of dispute whether its main motivational background stems from aggressive or feeding behaviour (Baenninger, 1978), or from both.

In this study rats of the TMD-S3 strain were used, a strain which has a high spontaneous mouse killing frequency (Walsh, 1982). Experienced rats kill a mouse immediately upon confrontation; all killing latencies are reliably smaller than 1 min. By measuring during a 30 min test, the killing latency after giving the drug 30 min. prior to the test (i.p.), one can estimate the lowest effective dose (LED) which significantly inhibits muricide.

Table 9.4 shows the LED's of several serotonergic compounds. Serenics (fluprazine, eltoprazine), TFMPP, 5-MeODMT and dl-propranolol inhibit muricidal behaviour in a specific way. 8-OH-DPAT does not inhibit muricide. RU 24969 effectively inhibits muricide, although some stimulatory action of the compound may interfere with the killing behaviour. Quipazine, fenfluramine, mianserine and fluvoxamine inhibit muricide but not in a behaviourally specific way. Methysergide, ritanserine and GR 38032F (5-HT antagonists) had no inhibiting effects on mouse-killing. Buspirone nonspecifically inhibited mouse-killing only at higher doses, whereas ipsapirone, another 5-HT$_{1A}$ agonist had no inhibiting effects.

DISCUSSION

The present contribution outlines the behavioural effects of several serotonergic drugs with different mechanisms of action on diverse aggression paradigms using mice and rats. This approach has several pitfalls, e.g. it is assumed that the effects of drugs can be solely explained by their actions on receptors in the brain, although 5-HT receptors are also located abundantly in the periphery (see Richardson and Engel, 1986). Moreover, after peripheral administration of drugs pharmacokinetic and metabolic processes may lead to unexpected pharmacodynamic results. It is therefore quite conceivable that data obtained with in vitro receptor binding do not directly predict actual in vivo (behavioural) effects. This should be kept in mind when interpreting behavioural effects although as long as clearly-distinct behavioural effects are noted, one is apt to neglect the above-mentioned considerations.

The data obtained in this paper are related to the hypothesis that the behavioural changes induced by the drugs are in some way related to their in vitro pharmacological profile (see Tables 9.1 and 9.2 for a short summary). Table 9.5 summarizes, in a very schematic manner, how the different 5-HT drugs tested exert their behavioural effects in several aggression paradigms in mice and rats.

The last column indicates the most likely serotonergic mechanism of action based on our own research and on the available literature on binding studies (e.g. Arvidsson et al., 1986) and several behavioural studies including drug discrimination (e.g. Arvidsson et al., 1986; Glennon, 1986; Glennon et al., 1982, 1984).

Although we still lack specific drugs with certain 5-HT properties, e.g. 5-HT$_2$ and 5-HT$_3$-agonists and 5-HT$_{1A/B}$ antagonists, the available data

Table 9.5 Summary of the effects of serotonergic drugs on several aggression paradigms in mice (m) and rats (r).

Drug	isolation induced aggression (m)	intermale aggression (m)	footshock induced defence (m)	resident-intruder aggression (r)	maternal aggression (r)	EBS (r)	muricide (r)	Putative 5-HT mechanism of action. The sequence indicates the decreasing potency in the different mechanisms.
8-OH-DPAT	↓	↓	-	-	↓	o	o	1A-agonist
Buspirone	o	o	-	↓	↓	-	↓	1A-agonist
Ipsapirone	o	o	-	-	↓	-	o	1A-agonist
Flesinoxan	↓	↓	-	-	↓	-	↓	1A-agonist
Befiperide	↓	↓	-	↓	-	↓	↓	1A,2-agonist
RU24969	↓	Ⓞ	-	↓	↓	-	↓	1A,1B-agonist, weak 1C-agonist
Eltoprazine	↓	Ⓞ	o	Ⓞ	Ⓞ	Ⓞ	Ⓞ	1A,1B-agonist, weak 1C-antagonist
5-Me-O-DMT	↓	-	-	↓	-	-	↓	1A,1C,1B-agonist
TFMPP	↓	Ⓞ	-	Ⓞ	Ⓞ	Ⓞ	Ⓞ	1C,1B-agonist, weak 1A-agonist
DOI	o	-	-	-	o	-	-	1C-agonist, weak 2-agonist
Fluprazine	↓	Ⓞ	Ⓞ	Ⓞ	Ⓞ	Ⓞ	Ⓞ	weak 1A,2,1C,1B agonist
Mianserine	↓	-	-	-	↓	-	-	1C,2-antagonist
dl-Propranolol	↓	-	-	↓	o	↓	↓	weak 1-antagonist
Methysergide	o	-	-	-	o	o	o	1,2-antagonist
Ritanserine	o	-	-	-	o	-	o	1C-antagonist, weak 2-agonist
MDL 72222	o	-	-	-	o	-	-	3-antagonist
GR 38032F	o	o	-	↓	-	-	o	3-antagonist
Quipazine	o	-	-	↓	↓	↓	o	3-antagonist, weak 1C,2-agonist
Fluvoxamine	↓	Ⓞ	-	↓	↓	↓	↓	reuptake blocker
Fenfluramine	↓	-	-	-	↓	-	↓	release

Ⓞ :specific behavioural decrease; ↓:nonspecific behavioural decrease; - o effect; -: not tested. EBS=Electrical brain stimulation-induced aggression.

summarized in table 9.5 suggests that the 5-HT$_{1B}$ binding site plays a specific modulatory role in aggressive and more precisely offensive behaviour.

The 5-HT$_{1A}$ site does not seem to play a specific role because 8-OH-DPAT, 5-MeODMT, buspirone, ipsapirone and flesinoxan have either no anti-aggressive activity or display a nonspecific behavioural effect. Moreover, the 5-HT$_2$-site also seems not to be directly involved in the modulation of aggressive behaviour, although the lack of specific 5-HT$_2$-agonists (besides DOI) means that this statement cannot be verified. A similar argument may hold for the 5-HT$_{1C}$ and 5-HT$_3$-sites, although it is still a matter of discussion whether there is a central location where 5-HT$_3$ ligands may bind (see Tyers, this volume).

Of course, almost every compound will eventually cause an inhibition of aggressive behaviour, but it is assumed that such an effect must be specific on aggression and should not involve effects on other behavioural systems, like exploration, motor activity, defensive capabilities and social interest. The use of ethological methodology shows that it is possible to simultaneously measure these several aspects of agonistic behaviour occurring between competing conspecifics, whether they are mice or rats (see Olivier et al., 1984, 1986b,c). On the basis of this simultaneous measurement of all ongoing behaviour, a new class of psychoactive drugs termed "serenics" has been developed (Olivier et al.,

1986c; Bradford et al., 1984). Serenics, represented here by fluprazine and eltoprazine, appear to have their main effects on 5-HT-neuronal systems, especially via the $5-HT_{1B}$ binding sites, although a contribution from the $5-HT_{1A}$ site cannot be completely ruled out. Therefore, the general hypothesis that 5-HT activity is inversely correlated with aggression, seems no longer tenable. A more refined hypothesis should specify a role for $5-HT_{1B}$ receptors. Before this can be done, a considerable amount of work has to be performed and new pharmacological tools (especially specific $5-HT_{1A}$ and $5-HT_{1B}$ antagonists and $5-HT_2$, $5-HT_3$ agonists) have to be available.

The presence of $5-HT_{1B}$ receptors in several areas known to be involved in the modulation of agonistic behaviour (Albert et al., 1982) lends support to a direct regulation of such behaviour by a $5-HT_{1B}$-related mechanism. The different 5-HT receptor types have been shown to be differentially localized in the CNS using quantitative autoradiography (Pazos and Palacios, 1985; Pazos et al., 1985). $5-HT_{1B}$ sites were present in high density in the globus pallidus, dorsal subiculum, substantia nigra and the olivary pretectal nucleus, whereas the highest density of $5-HT_{1A}$ sites was found in the dentate gyrus of the hippocampus and the lateroseptal nucleus. The neocortex and the hypothalamus also showed high concentration of both types of receptors. In contrast, the density of $5-HT_1$ receptors in the brainstem and spinal cord was low.

Finally, it should not be forgotten that 5-HT is very widely involved in many other kinds of behavioural and motoric systems (see reviews by Green, 1985; Osborne, 1982; Soubrie, 1986). Recent evidence shows that 8-OH-DPAT enhances food intake (Dourish et al., 1985) and sexual behaviour (Ahlenius et al., 1981). In contrast, more specific $5-HT_{1B}$ agonists like TFMPP decrease food intake and sexual behaviour (Olivier and Mos, 1987). This also holds for the mixed $5-HT_{1A/B}$ agonists eltoprazine and RU 24969 (Olivier and Mos, unpublished).

It is clear that further studies await the development of more specific, 5-HT-pharmacological tools and localized brain studies which will facilitate the unravelling of the complex modulation by 5-HT receptor subtypes of agonistic and other behaviours.

Acknowledgement: We thank Marijke Mulder for typing the manuscript, Ruud van Oorschot for technical support and Dr.J.A.M. van der Heyden for supplying some of the data used.

References

Ahlenius, S., Larsson, K., Svensson, L., Hjorth, A., Carlsson, A., Lindberg, P., Wikstrom, H., Sandchen, D., Arvidsson, L.E., Hacksell, U. & Nilsson, J.L.G. (1981). Effects of a new type of 5-HT receptor agonist on male rat sexual behaviour. Pharmacol. Biochem. Behav., 15, 785-792.

Albert, D.J. & Walsh, M.L. (1982). The inhibitory modulation of agonistic behaviour in the rat brain: a review. Neurosci. Biobehav. Rev., 6, 125-143.

Applegate, C.D. (1980). 5.7-Dihydroxytryptamine-induced mouse killing and behavioural reversal with ventricular administration of serotonin in rats. Behav. Neural. Biol., 30, 178-190.

Arvidsson, L.E., Hacksell, U. & Glennon, R.A. (1986). Recent advances in central 5-hydroxytryptamine receptor agonists and antagonists. Progress in Drug Research., Vol 30, pp.365-471. Basel, Boston, Stuttgart: Birkhaüser Verlag.

Baenninger, R. (1978). Some aspects of predatory behaviour. Aggr. Behav., 4, 287-311.

Barnett, S.A. (1975). The Rat. A Study in Behavior. Chicago and London: The University of Chicago Press.

Bell, R. & Hepper, P.G. (1987). Catecholamines and aggression in animals. Behav. Brain Res., 23, 1-21.

Bell, R., Warburton, D.M. & Brown, K. (1985). Drugs as research tool in psychology: cholinergic drugs and aggression. Neuropsychobiol., 14, 181-192.

Blanchard, R.J. & Blanchard, D.C. (1977). Aggressive behaviour in the rat. Behav. Biol., 21, 197-224.

Blanchard, R.J., Blanchard, D.C. & Takahashi, L.K. (1977). Reflexive fighting in the albino rat: aggressive or defensive behaviour? Aggr. Behav., 3, 145-155.

Blanchard, R.J., O'Donnell, V. & Blanchard. D.C. (1979). Attack and defensive behaviours in the albino mouse (Mus musculus). Aggr. Behav., 5, 341-352.

Bradford, L.D., Olivier, B., Van Dalen, D. & Schipper, J. (1984). Serenics: the pharmacology of fluprazine and DU 28412. In: Ethopharmacological Aggression Research. (Eds. Miczek, K.A., Kruk, M.R., Olivier, B.), pp. 191-207. New York: Alan R. Liss.

Brain, P.F. (1975). What does individual housing mean to a mouse? Life Sci., 16, 187-200.

Carlsson, A., Davis, J.N., Kehr, W., Lindquist, M. & Atack, C.V. (1972). Simultaneous measurement of tyrosine and tryptophan hydroxylase activities in brain in vivo using an inhibitor of the aromatic amino acid decarboxylase. Naunyn-Schmiedeberg's Arch. Pharmacol., 275, 153-168.

Claassen, V.C., Davies, J.E., Hertting, G. & Placheta, P. (1979). Fluvoxamine, a specific 5-hydroxytryptamine uptake inhibitor. Br. J. Pharmacol., 60, 505-516.

Conn, P.J,. Sanders-Bush, E., Hoffman, B.J. & Hartig. P.R. (1986). A unique serotonin receptor in the choroid plexus is linked to phosphatidyl inositol turnover. Proc. Natl. Acad., USA, 83, 4086-4088.

Daruna, J.H. (1978). Patterns of brain monoamine activity and aggressive behaviour. Neurosci. Biobehav. Rev., 2, 101-113.

Dixon, A.K. & Kaesermann, H.P. (1987). Ethopharmacology of flight behaviour. In: Ethopharmacology of agonistic behaviour in humans and animals. (Eds. Olivier, B., Mos, J. & Brain, P.F.), pp. 46-79. Dordrecht: Martinus Nijhoff.

Dourish, C.T., Hutson, P.H. & Curzon, G. (1985). Characteristics of feeding induced by the serotonin agonist 8-hydroxy-2-(Di-n-propylamino)-tetralin (8-OH-DPAT). Brain Res. Bull., **15**, 377-384.

Eichelman, B. (1987). Neurochemical and psychopharmacologic aspects of aggressive behaviour. In: Psychopharmacology: the third generation of progress. (Ed. Meltzer, H.Y.), pp. 697-704. New York: Raven Press.

Engel, G., Göthert, M., Hoyer, D., Schlicker, E. & Hillenbrand, K. (1986). Identity of inhibitory presynaptic 5-hydroxytryptamine. (5-HT) autoreceptors in the rat brain cortex with 5-HT_{1B} binding sites. Naunyn-Schmiedeberg's Arch. Pharmacol., **332**, 1-7.

Erskine, M.S., Barfield, R.J. & Goldman, B.S. (1978a). Intraspecific fighting during late pregnancy and lactation in rats and effects of litter removal. Behav. Biol., **23**, 206-218.

Erskine, M.S., Denenberg, V.H. & Goldman, B.D. (1978b). Aggression in the lactating rat: effects of intruder age and test arena. Behav. Biol., **23**, 52-66.

Glennon, R.A. (1986). Site selective serotonin agonists as discriminative stimuli. Psychopharmacol., **89** (S1): 135.

Glennon, R.A., McKenney, J.D.& Young, R. (1984). Discrimination stimulus properties of the serotonin agonist 1-(3-trifluoromethylphenyl) piperazine. (TFMPP) Life Sci., **35**, 1475-1480.

Glennon, R.A., Rosecrans, J.A. & Young, R. (1982). The use of the drug discrimination paradigm for studying hallucinogenic agents. In: Drug discrimination: Applications in CNS pharmacology. (Eds. Colpaert, F.C. & Slangen, J.L.), pp. 69-98. Amsterdam: Elsevier Biomedical Press.

Göthert, M. (1980). Serotonin receptor mediated modulation of Ca^{2+} dependent 5-hydroxytryptamine release from neurones of the rat brain cortex. Naunyn-Schmiedeberg's Arch. Pharmacol., **314**, 223-230.

Graeff, F.G., Brandao, M.L., Audi, E.A., Schütz, M.T.B. (1986). Modulation of the brain aversive system by GABAergic and serotonergic mechanisms. Behav. Brain Res., **22**, 173-180.

Green, A.R. (1985). Neuropharmacology of Serotonin, pp. 1-436. Oxford: Oxford University Press.

Hoyer, D., Engel, G. & Kalkman, H.O. (1985). Molecular pharmacology of 5-HT_1 and 5-HT_2 recognition sites in rat and pig brain membranes. Radioligand binding studies with [^{3}H] 5-HT, [^{3}H] 8-OH-DPAT, (-) [^{125}I] iodocyanopindolol, [^{3}H] mesulergine, and [^{3}H] ketanserin. Eur. J. Pharmacol., **118**, 13-23.

Hoyer, D. & Neyt, H.C. (1987). Identifcation of serotonin 5-HT_3 recognition sites by radioligand binding in NG108-15 neuroblastoma-glioma cells. Eur. J.

Pharmacol., 143, 291-292.

Janssen, P.A.J., Jagenau, A.H.M. & Schellekens, K.H.J. (1960). Chemistry and pharmacology of compounds related to 4-(4-hydroxy-4-phenyl-piperidino) - butyrophenone. Part IV. Influence of haloperidol (R 1625) and of chlorpromazine on the behaviour of rats in an unfamiliair "open field" situation. Psychopharmacologia, 1, 389-392.

Kantak, K.M., Hegstrand, L.R. & Eichelman, B. (1981). Facilitation of shock-induced fighting following intraventricular 5,7-dihydroxytryptamine and 6-hydroxy dopa. Psychopharmacol., 74, 157-160.

Kilpatrick, G.J., Jones, B.J. & Tyers, M.B. (1987). Identification of 5-HT$_3$ receptors in rat brain using radioligand binding. Nature, 330, 746-748.

Knutson, J.F., Kane, N.L., Schlosberg, A.J., Fordyce, D.J.& Simansky, K.J. (1979). Influence of PCPA, shock level, and home cage conditions on shock-induced aggression. Physiol. Behav., 23, 897-907.

Krsiak, M. (1974). Behavioural changes and aggressivity evoked by drugs in mice. Res. Comm. Chem. Pathol. Pharmacol., 7, 253-257.

Krsiak, M. (1975). Timid singly-housed mice: their value in prediction of psychotropic activity of drugs. Br. J. Pharmacol., 55, 141-150.

Krsiak, M. (1979). Effects of drugs on behaviour of aggressive mice. Br. J. Pharmacol., 65, 525-533.

Kruk, M.R., Van der Poel, A.M. & De Vos-Frerichs, T.P. (1979). The induction of aggressive behaviour by electrical stimulation in the hypothalamus of male rats. Behav., 70, 292-322.

Kruk, M.R., Van der Poel, A.M., Lammers, J.H.C.M., Hagg, T., De Hey, A.M.D.M. & Oostvegel, S. (1987). Ethopharmacology of hypothalamic aggression in the rat. In: Ethopharmacology of agonistic behaviour in animals and humans. (Eds. Olivier, B., Mos, J. & Brain, P.F.), pp. 33-45. Dordrecht: Martinus Nijhoff.

Lindgren, T. & Kantak, K.M. (1987). Effects of serotonin receptor agonists and antagonists on offensive aggression in mice. Aggr. Behav., 13, 87-96.

Lore, R. & Flannelly, K. (1979). Rat societies. Sci. Am., 236, 106-116.

McMillan, B.A., Scott, M.A., Williams, H.L. & Sanghera, M.K. (1987). Effects of gepirone, an aryl-piperazine anxiolytic drug, on aggressive behaviour and brain monoaminergic neurotransmission. NS Arch. Pharmacol., 335, 454-464.

Miczek, K.A. (1987). The psychopharmacology of aggression. In: Handbook of Psychopharmacology. (Eds. Iversen, L.L., Iversen, S.D. & Snyder, S.H.), Behavioural Pharmacology, Vol. 19, pp. 183-327. New York: Plenum Press.

Miczek, K.A. & Krsiak, M. (1979). Drug effects on agonistic behaviour. In:

Advances in Behavioural Pharmacology. (Eds. Thompson, T. & Dews, P.B.), Vol. 2, pp. 87–162. New York: Academic Press.

Miczek, K.A. & Krsiak, M. (1981). Pharmacological analysis of attack and flight. In: Multidisciplinary approaches to aggression research. (Eds. Brain, P.F. & Benton, D.), pp. 341–354. Amsterdam: Elsevier/North–Holland Biomedical Press.

Middlemiss, D.N. (1984). 8–hydroxy–2–(di–n–propylamino)–tetralin is devoid of activity at the 5–hydroxytryptamine autoreceptor in rat brain. NS Arch. Pharmacol., 327, 18–22.

Middlemiss, D.N. & Fozard, J.R. (1983). 8–hydroxy–2–(di–n–propylamino)–tetralin discriminates between subtypes of the $5-HT_1$ recognition sites. Eur. J. Pharmacol., 90, 151–153.

Mos, J. & Olivier, B. (1986). RO 15–1788 does not influence post–partum aggression in lactating female rats. Psychopharmacol., 90, 278–280.

Mos, J., Olivier, B., Lammers, J.H.C.M., Van der Poel, A.M., Kruk, M.R., Zethof, T. (1987). Pregnancy and lactation do not interact with current thresholds for brain stumulation induced aggression in female rats. Brain Res., 404, 263–266.

Mühlbauer, (1985). Human aggression and the role of central serotonin. Pharmacopsychiat., 18, 218–221.

Olivier, B. (1981). Selective anti–aggressive properties of DU 27725: ethological analysis of intermale and territorial aggression in the male rat. Pharmacol. Biochem. Behav., 14, (S1), 61–77.

Olivier, B. & Mos, J. (1986a). A female aggression paradigm for use in psychopharmacology: maternal agonistic behaviour in rats. In: Cross–Disciplinary studies on Aggression. (Eds. Brain, P.F. & Martin Ramirez, J.M.), pp. 3–111. University of Seville Press.

Olivier, B., Mos, J. & Schipper, J. (1986b). Serotonin and Aggressive behaviour in the rat. Psychopharmacol., 89, 26.

Olivier, B. & Mos, J. (1986b). Serenics and aggression. Stress Med., 2, 197–209.

Olivier, B. & Mos, J. (1988). Serotonergic drugs and male sexual behaviour. In: Sonesta symposium proceedings 1988. (In press.)

Olivier, B., Mos, J., Schipper, J., Tulp, M.T.M., Van der Heyden, J.A.M., Berkelmans, B. & Bevan, P. (1987). Serotonergic modulation of agonistic behaviour. In: Ethopharmacological analysis of agonistic behaviour in animals and man. (Eds. Olivier, B., Mos, J. & Brain, P.F.), pp. 162–186. Dordrecht: Martinus Nijhoff.

Olivier, B., Mos, J. & Van Oorschot, R. (1985). Maternal aggression in rats: effects of chlordiazepoxide and fluprazine. Psychopharmacol., 86, 68–76.

Olivier, B., Mos, J. & Van Oorschot, R. (1986a). Maternal aggression in rats: lack of interaction between chlordiazepoxide and fluprazine. Psychopharmacol., **88**, 40-43.

Olivier, B., Van Aken, H., Jaarsma, I., Van Oorschot, R., Zethof, T. & Bradford, L.D. (1984). Behavioural effects of psychoactive drugs on agonistic behaviour of male territorial rats. (resident-intruder paradigm) In: Ethopharmacological Aggression Research. (Eds. Miczek, K.A., Kruk, M.R. & Olivier, B.), pp. 137-156. New York: Alan R. Liss Inc.

Olivier, B. & Van Dalen, D. (1982). Social behaviour in rats and mice: an ethologically based model for differentiating psychoactive drugs. Aggr. Behav., **8**, 163-168.

Olivier, B., Van Dalen, D. & Hartog, J. (1986c). A new class of psychoactive drugs: Serenics. Drugs Future, **11**, 473-499.

Osborne, N.N. (1982). Biology of Serotonergic Transmission. Chichester: John Wiley and Sons.

Pazos, A. & Palacios, J.M. (1985). Quantitative autoradiographic mapping of serotonin receptors in rat brain. I. Serotonin-1 receptors. Brain Res., **346**, 205-230.

Pazos, A., Cortes, R. & Palacios, J.M. (1985). Quantitative autoradiographic mapping of serotonin receptors in the rat brain. II. Serotonin-2 receptors. Brain Res., **346**, 231-249.

Peroutka, S.J. (1986). Pharmacological differentiation and characterization of $5-HT_{1A}$, $5-HT_{1B}$, and $5-HT_{1C}$ binding sites in rat frontal cortex. J. Neurochem., **47**, 529-540.

Richardson, B.P. & Engel, G. (1986). The pharmacology and function of $5-HT_3$ receptors. TINS, 424-428.

Rodgers, R.J. (1981). Drugs, aggression and behavioural methods. In: Multidisciplinary approaches to aggression research. (Eds. Brain, P.F. & Benton, D.), pp. 325-340. Amsterdam, New York, Oxford: Elsevier/North-Holland Biomedical Press.

Rolinski, Z. & Herbut, M. (1981). The role for the serotonergic system in foot-shock induced behaviour in mice. Psychopharmacol., **73**, 246-251.

Schütz, M.T.B, De Aguiar, J.C., Graeff, F.G. (1985). Antiaversive role of serotonin in the dorsal periaqueductal grey matter. Psychopharmacol., **85**, 340-345.

Sheard, M.H. (1981). Shock-induced fighting. (SIF): psychopharmacological studies. Aggr. Behav., **7**, 41-49.

Sheard, M.H. & Davis, M. (1976). Shock-elicited fighting in rats: importance of intershock interval upon the effect of p-chlorophenylalanine. (PCPA) Brain Res., 111, 433-437.

Soubrie, P. (1986). Reconciling the role of central serotonin neurons in human and animal behaviour. Behav. Brain Sci., 9, 319-364.

Steinbusch, H.W.M. (1981). Distribution of serotonin-immunoreactivity in the central nervous system of the rat - cell bodies and terminals. Neurosci., 6, 557-618.

Tricklebank, M.D. (1985). The behavioural response to 5-HT receptor agonists and subtypes of the central 5-HT receptor. Trends Pharmacol. Sci., 403-407.

Ulrich, R.E. & Azrin, N.H. (1962). Reflexive fighting in response to aversive stimulation. J. Exp. Anal. Behav., 5, 511-520.

Valzelli, L. (1969). The "isolation syndrome" in mice. Psychopharmacologia, 31, 305-320.

Valzelli, L. (1973). Aggressive behaviour induced by isolation. In: Aggressive Behaviour. (Eds. Garattini, S. & Sigg, E.B.), pp. 70-76. New York: John Wiley and Sons.

Valzelli, L. (1978). Psychopharmacology of aggression. Basel: Karger.

Valzelli, L. (1981). Psychopharmacology of aggression: an overview. Int. Pharmacopsychiat., 16, 39-48.

Valzelli, L. (1984). Reflections on experimental and human pathology of aggression. Progr. Neuropsychopharmacol. Biol. Psychiat., 8, 311-325.

Valzelli, L., Garattini, S., Bernasconi, S. & Sala, A. (1981). Neurochemical correlates of muricidal behaviour in rats. Neuropsychobiol., 7, 172-178.

Van der Poel, A.M., Olivier, B., Mos, J., Kruk, M.R., Meelis, W. & Van Aken, J.H.M. (1982). Anti-aggressive effect of a new phenylpiperazine compound (DU 27716) on hypothalamically induced behavioural activities. Pharmacol. Biochem. Behav., 17, 147-153.

Van der Poel, A.M., Mos, J., Kruk, M.R. & Olivier, B. (1984). A motivational analysis of ambivalent actions in the agonistic behaviour of rats in tests used to study effects of drugs on aggression. In: Ethopharmacological Aggression Research. (Eds. Miczek, K.A., Kruk, M.R. & Olivier, B.), pp. 115-135. New York: Alan R. Liss Inc.

Vergnes, M., DePaulis, A. & Boehrer, A. (1986). Parachlorophenylalanine-induced serotonin depletion increases offensive but not defensive aggression in male rats. Physiol. Behav., 36, 653-658.

Vergnes, M. & Kempf, E. (1981). Tryptophan deprivation: Effects on mouse-killing and reactivity in the rat. Pharmacol. Biochem. Behav., 14, 19-23.

Vergnes, M. & Kempf, E. (1982). Effect of hypothalamic injections of 5,7-dihydroxy-tryptamine on elicitation of mouse-killing in rats. Behav. Brain Res., 5, 387-397.

Walsh, L.L. (1982). Strain and sex differences in mouse killing by rats. J. Comp. Physiol. Psychol., 96, 278-283.

BRAIN 5-HT SYSTEM AND INHIBITION OF AGGRESSIVE BEHAVIOUR

K.A.Miczek, P. Donat
Department of Psychology, Tufts University, Medford, Massachusetts 02155, U.S.A.

Introduction
Are aggressive behaviours inhibited similar to the way tryptophan hydroxylase is inhibited by para-chlorophenylalanine (PCPA) or raphe neurons are inhibited by microiontophoresed lysergic acid diethylamide (LSD) (e.g., Koe and Weissman 1966, Aghajanian et al., 1968)? How useful is the extrapolation of an inhibitory process from the molecular and cellular level to the behavioural level? When Brodie and Shore (1957) suggested an inhibitory function for brain 5-HT on behaviour and autonomic nervous system activity, they were guided by the concept of W.R.Hess (1954) who proposed a trophotrophic subcortical system guiding recuperative, sedative behaviours that are associated with increased parasympathetic output. 5-HT was postulated to serve as the critical neurotransmitter in this system. An opposing ergotrophic system was proposed to subserve behavioural arousal accompanied by sympathetic activation. The concept of serotonin as a trophotrophic neurotransmitter has been applied to the neural control of aggressive behaviour up to the present (e.g., Valzelli, 1982; Eichelman, 1979; Pucilowski and Kostowski, 1983).

How do the initial observations with the currently developed ligands for 5-HT receptor subtypes fit with the classic concept of serotonin's inhibitory control of aggressive behaviour? In order to appreciate the significance and predictive value of the results obtained with these new pharmacological manipulations, it is important to examine briefly past efforts with the most often used tests of animal aggression, and subsequently consider evolutionary and ethological issues of 5-HT and aggressive behaviour. It will become quickly apparent that the distal and proximal antecedents, the execution of complex and intricate patterns of aggressive, defensive, submissive and predatory behaviour as well as the various functions that these behaviours serve, cannot be readily linked to any specific biochemical aspect of the anatomically separate pools of 5-HT containing neurons or their receptor subtypes.

5-HT and animal aggression as assessed by traditional laboratory tests
During the 1960's and 1970's the merging tools for the study of 5-HT and aggression included procedures that impaired 5-HT neurotransmission such as synthesis inhibition, electrolytic or neurotoxic lesions of the raphe nuclei, maintenance on tryptophan-free diet, depletion of vesicular 5-HT stores or, alternatively, procedures that enhanced 5-HT neurotransmission such as administration of 5-HT precursors, inhibition of the enzymatic degradation of 5-HT by MAO inhibitors or by re-uptake blockade, promotion of 5-HT release. Current interest in 5-HT derives primarily from the biochemical identification of 5-HT receptor subtypes and the increased selectivity and specificity of receptor agonists and antagonists. Olivier et al. (1988) are the first to examine several of these newer agents for their effects on animal aggression.

The earlier neurochemical and neuropharmacological studies have relied mostly on traditional tests of animal aggression such as (1) pain- or shock-induced aggression, mostly in rats, (2) isolation-induced aggression, mostly in mice and (3) mouse-killing (muricide) by rats.

The evidence on the effects of neuropharmacological manipulations of 5-HT on aggressive behaviour by <u>isolated</u> mice is summarized in Fig. 10.1. Nearly every pharmacological manipulation of 5-HT neurotransmission results in a more or less specific decrease of isolation-induced aggression. Whether or not mice are maintained on tryptophan-free diets, administered with the 5-HT synthesis inhibitor para-chlorophenylalanine, with storage depleting agents such as reserpine, with non-selective receptor antagonists, or, alternatively, with the precursor 5-HTP, the releasing agent fenfluramine, MAO inhibitors or reuptake blockers - all of these treatment decreased the proportion of isolated mice displaying aggressive behaviour or reduced the frequency of aggressive behaviour. The only exceptions are increases in aggressive behaviour of isolated mice after administration with a very low 5-HTP or pargyline dose.

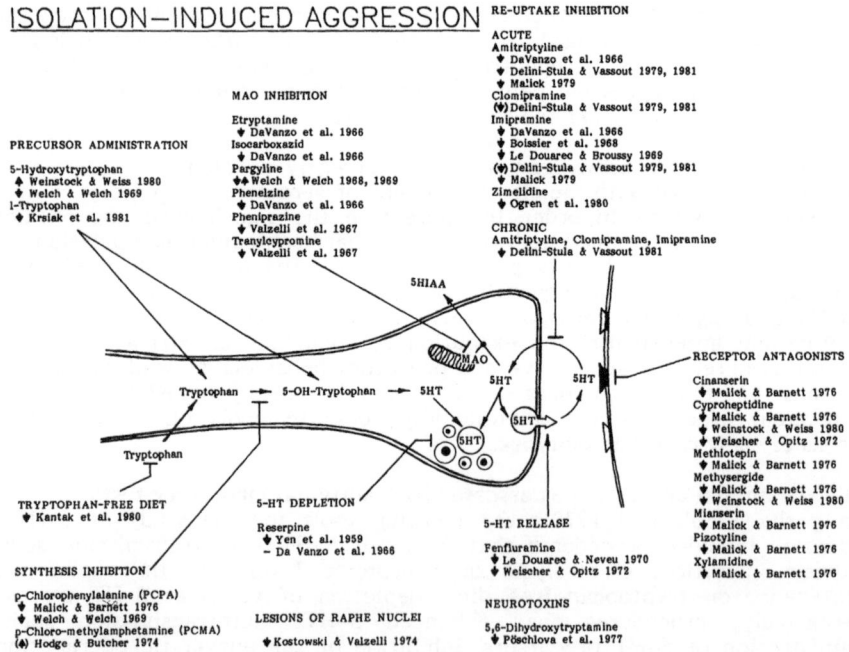

Fig. 10.1 Schematic representation of a brain 5-HT presynaptic terminal and postsynaptic receptors with potential sites for drugs that enhance (top) or impair (bottom) serotonergic neurotransmission. Beneath each drug name, several key references are listed that report either an increase (upward arrow), decrease (downward arrow) in aggressive behaviour by isolated mice or no reliable change in this behaviour (dash).

Shock- or pain-induced aggression is an experimental preparation that entails presentation of electric shock pulses to pairs of rats; as a result of this stimulation the rats exhibit mainly defensive reactions, namely upright postures, audible vocalizations, front paw movements and occasionally bites at the snout and front paws (e.g., Blanchard et al., 1978). As summarized in Figure 10.2, experimental manipulations that impair 5-HT such as inhibition of tryptophan hydroxylase, electrolytic or neurotoxic lesions of serotonin-containing neurons or receptor blockade with certain antagonists may increase defensive responses in reaction to shock pulses. Yet, this increase in defensive behaviour only occurs with a subset of electric shock parameters and with specific pharmacological agents.

More consistently, heightened defense is seen often after chronic administration of monoamine reuptake blockers and MAO inhibitors such as imipramine, iprindol, mianserine, or maprotiline (Allikmets and Lapin, 1967; Eichelman and Barchas, 1975; Eichelman, 1979; Mogilnicka and Przewlocka, 1981; Mogilnicka et al., 1983; Prasad and Sheard, 1983a,b). By contrast, acute administration of these and similarly acting drugs decreases pain- or shock-induced defensive reactions (see review by Miczek, 1987). However, acute and chronic treatments with antidepressants do not always produce opposite effects on aggressive responses.
For example, chronic blockade of 5-HT reuptake with clomipramine in isolated mice may amplify the acute anti-aggressive effects (Delini-Stula and Vassout, 1981). It remains to be determined whether the intensified aggressive of defensive responses in chronically treated animals are based on changes in noradrenergic or serotonergic receptor sensitivity or uptake processes.

In summary, any interference with normal 5-HT functions may be detrimental to attack and threat behaviour of isolated mice. Manipulations that impair 5-HT or chronic antidepressant treatment may facilitate defensive responses.
Isolated, grouped or crowded housing in mice may lead to substantial changes in whole brain 5-HT content as well as in regional 5-HT turnover (e.g., Welch and Welch, 1969; Modigh, 1973; Raad and Deisz, 1975; Valzelli and Bernasconi, 1979). Since isolated housing in mice on its own may result in aggressive or timid or social interactions with another mouse (e.g. Miczek and Krsiak, 1979), it is difficult to predict in isolated individuals from a selected index of brain 5-HT activity to a certain level of aggressive behaviour or vice versa.

5-HT and killing
A mysterious fascination appears to surround the killing of an animal by another animal, usually of a different species. Whether or not killing by an omnivorous species such as rats or by carnivorous animals such as ferrets or cats should be considered aggressive as distinct from predatory behaviour, has been a matter of considerable debate (e.g., Huntingford, 1976; Rossi, 1975). A pragmatic alternative, but conceptually unsatisfactory, is the use of the term "predatory aggression". The induction of killing responses in animals that have not previously shown this behaviour has been examined with various pharmacological, electrolytic or neurotoxic insults of brain, 5-HT mainly in laboratory rats confronting a mouse. As summarized in Figure 10.2, neurotoxic destruction of 5-HT neurons with 5,6- or 5,7-dihydroxytryptamine (5,6-DHT or 5,7-DHT), large electrolytic lesions of raphe nuclei, omission of l-tryptophan from the diet, or

inhibiton of tryptophan hydroxylase with para-chlorophenylalanine induce mouse-killing behaviour in a larger proportion of rats than is seen in control animals. Contrary to this pattern of effects is the absence of any killing-inducing effects by 5-HT receptor antagonists.

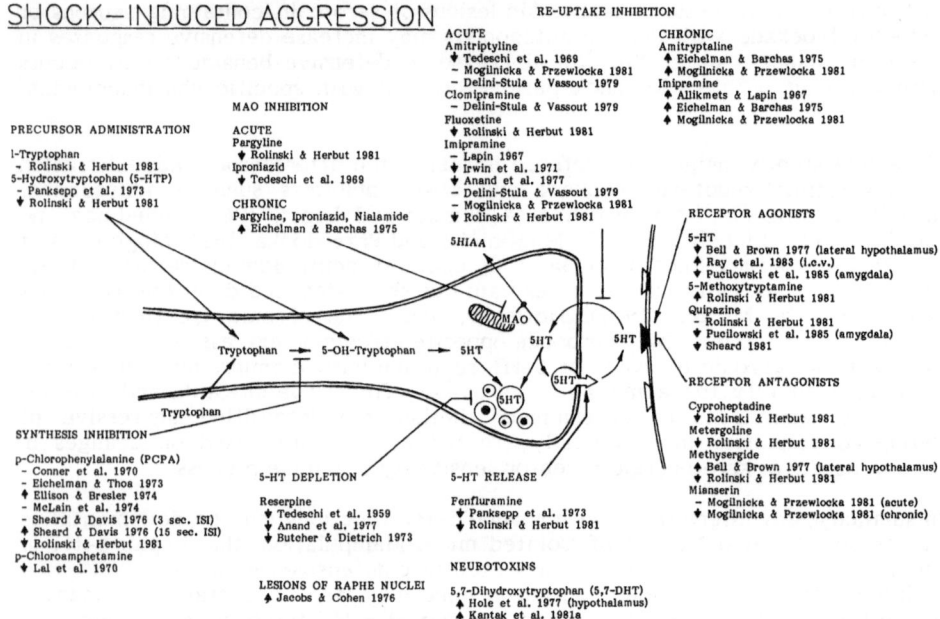

Fig. 10.2 *Schematic representation of a brain 5-HT presynaptic terminal and postsynaptic receptors with potential sites for drugs that enhance (top) or impair (bottom) serotonergic neurotransmission. Beneath each drug name, several key references are listed that report either an increase (upward arrow), decrease (downward arrow) in pairs of mice or rats exposed to electric shock pulses or no reliable change in this behaviour (dash).*

While the induction of mouse-killing is a relatively consistent observation after insults to 5-HT systems, it is, however, not an all-or-none phenomenon. Many severely 5-HT depleted rats fail to become so-called "killer" rats, and others again kill without showing any discernable abnormality in 5-HT (e.g., Miczek et al., 1975; Salama and Goldberg, 1973).

It has not been possible to identify any critical threshold value of 5-HT depletion that is necessary or sufficient to induce mouse-killing behaviour in so-called non-killer rats, or alternatively, to establish a systematic dose-effect relationship between the degree of 5-HT depletion and the probability of killing behaviour. It is very difficult to induce mouse-killing behaviour pharmacologically in rats that have been exposed to the potential prey before the pharmacological treatment (e.g., Marks et al., 1977; Vergnes et al., 1977; Vergnes and Kempf, 1981). Moreover, once induced, mouse-killing persists in the absence of any detectable change in 5-HT levels, synthesis or metabolism (e.g., Vergnes and Kempf, 1981). The suppressive influence of habituation on the one hand, and the reinforcing effect of the killing act on the other may involve mechanisms that override the impact of detrimental manipulations of the 5-HT systems.

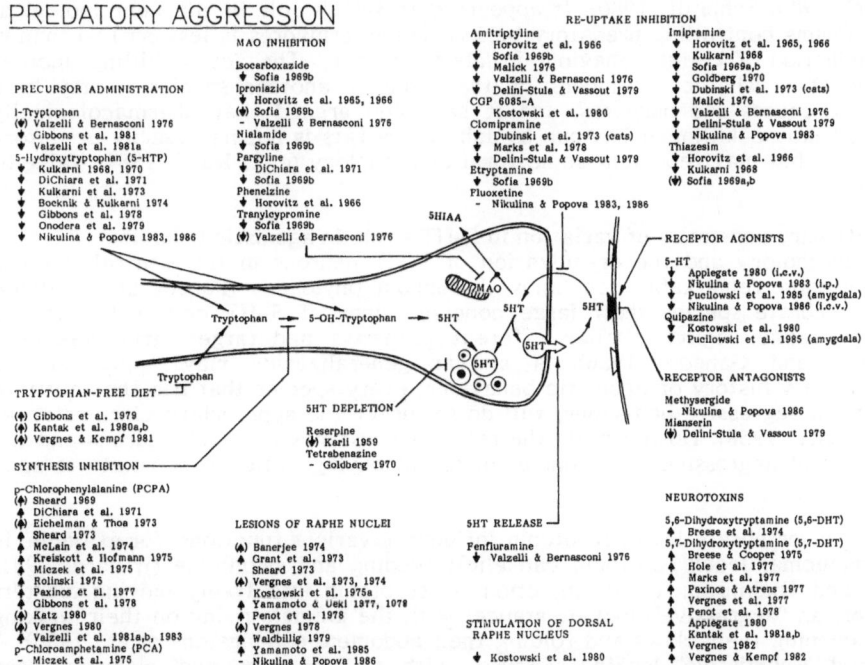

Fig. 10.3 Schematic representation of a brain 5-HT presynaptic terminal and postsynaptic receptors with potential sites for drugs that enhance (top) or impair (bottom) serotonergic neurotransmission. Beneath each drug name, several key references are listed that report either an increase (upward arrow), decrease (downward arrow) in predatory aggression, most often mouse-killing behaviour by laboratory rats, or no reliable change in this behaviour (dash).

A particularly troubling issue concerns the contrast between carnivorous species and the omnivorous rat with regard to the proposed inhibitory control of 5-HT over killing behaviour.

When ferrets, grasshopper mice or cats are subjected to treatments that impair 5-HT, no facilitation of killing behaviour is seen. For example, McCarty et al. (1976) found decreased predatory behaviour in grasshoper mice (Onychomys torridus) toward crickets after 5-HT-depleting treatment with PCPA (50 mg/kg/day for 5 days). A frequently used screen for drugs with antidepressant properties in industrial laboratories involves the blockade of mouse-killing behaviour (e.g. Howard and Pollard, 1983). However, antidepressants that block re-uptake of 5-HT and that inhibit mouse-killing in rats, fail to block predatory behaviour by such carnivores as ferrets or cats (Leaf et al., 1978; Schmidt and Meierl, 1980; Schmidt, 1980). It appears that 5-HT's role in the neurobiological mechanisms controlling predatory behaviour by carnivores is less critical than in the induction of killing behaviour in laboratory rats. The act of killing another individual — whether members of the same or another species — may here multiple causes and purposes, and it has been argued that pharmacologically induced mouse-killing behaviour by laboratory rats is not necessarily predatory in nature, but may represent some form of pathology (e.g., Karli, 1981; Valzelli, 1985).

Evolutionary constancy of variation in 5-HT's role in agonistic behaviour
The morphology and the organization of 5-HT neurons in the central nervous system show remarkable constancy throughout phylogeny (Parent et al., 1984). All vertebrate species show large concentrations of 5-HT cell bodies in the midline raphe region, similar efferent pathways and target structures (e.g. Azmitia and Gannon, 1986). A similar generalization may apply to the evolutionary history of agonistic behaviour; every species that has the ability to engage in aggressive behaviour, will do so, under the appropriate circumstances (e.g. Scott, 1958). However, do the relative constancy of 5-HT neurons and the presence of aggressive behaviour at different phylogenetic levels relate to each other?

At the invertebrate level, serotonin influences various functions; for example, in the medicinal leech serotonin can elicit feeding and swimming (Willard, 1981; Lent and Dickinson, 1984). Injection of serotonin into freely moving lobsters triggers an "aggressive"-looking stance, with the animals rising on their walking legs, opening their claws and folding their abdomens (Livingstone et al., 1980). A lowered "submissive"-looking posture, with walking legs and claws pointed forward, is evoked by octopamine injections, the invertebrate counterpart to norepinephrine. Kravitz and coworkers (1981, 1986) have identified two sites of action for serotonin in the lobster, a priming action directly at the peripheral neuromuscular junctions for flexor and extensor muscle pairs, and a more integrative action in the central nervous system, i.e. the central nerve cord. At this latter site, serotonin is released from nerve endings in neuropil regions and may interact with command circuitry which in turn leads to the "read out of a motor programme causing flexion" (Kravitz, 1986). The flexed posture, normally shown by an aggressive animal, is generated by a pattern of excitatory and inhibitory motoneurons. It is then quite likely that serotonin-containing cells in the thoracic and abdominal regions of the central nerve cord of the lobster are important in generating an entire portion of the behavioural repertoire such as the "aggressive"-looking posture.

In ants (Formica rufa), increased concentrations of 5-HT are seen after performance of heightened aggressive behaviour toward each other or toward a beetle (Kostowski et al., 1975b). Administration of 5-HTP (0.3 µg/mg) increased the percentage of ants fighting among each other, although the number of ants attacking a beetle declined (Kostowski and Tarchalska, 1972). These observations in ants, like those in lobsters, suggest a facilitative or enabling role of serotonin in the initiation and/or performance of certain aggressive acts and postures – a function in complete opposition to the concept of 5-HT as trophotrophic neurotransmitter.

In the South American weakly electric fish (Gymnotidea), Bullock (1969) described a type of electric organ discharge – named "chirp" – that is associated with aggressive behaviour, and these short chirps may serve as aggressive signals that are instrumental in establishing dominance. Maler and Ellis (1987) found that intraventricular injections of 0.1 µg norepinephrine or i.m. desmethylimipramine increased the frequency of aggressive chirping, whether occurring spontaneously or in response to an experimentally stimulated electric organ discharge. By contrast, intraventricular injection of 1 µg 5-HT, i.m. fluoxetine, or methiothepin led to a prolonged decrease in "aggressive" chirping while still preserving the jamming avoidance response. These results, although limited to a small range of pharmacological and behavioural conditions, suggest that in this species brain 5-HT is critical for the suppression of aggressive signals. It will be interesting to learn about the dynamics of 5-HT synthesis, release and metabolism as well as the regulation of the various 5-HT receptors in the moment-to-moment control of aggressive behaviour as well as in the long-term disposition to engage in aggressive behaviour in this species.

In hamsters, Payne and coworkers (1981, 1984, 1985) found indications of increased 5-HT "turnover" in the hypothalami of isolated and group-housed animals that were assayed immediately after a 10-min fighting episode. While isolated housing increases aggressive behaviour in hamsters, it failed to result in any detectable changes in mesencephalic or hypothalamic levels of 5-HT of 5-HIAA nor in alterations in MAO activity after pharmacological blockade.

As in mice (see Fig. 10.1), pargyline decreased fighting behaviour of isolated hamsters. Neither the isolated resident hamster's attack and threat behaviour nor the group-housed intruder's flight reactions led to differential changes in 5-HT of 5-HIAA. It was evident that 5-HT or 5-HIAA are not related in any direct way to a certian level of aggressive or flight behaviour in hamsters, and Payne and associates suggested that the alterations in this amine and its metabolite after fighting may represent part of a more pervasive stress response.

Offensive and defensive behaviour of tree-shrews (Tupaia belangeri) is vigorous and intense. Single unit activity of dorsal raphe neurons during offensive behaviour and defensive reactions, transmitted via telemetry from tungsten microelectrodes, shows dramatic changes in confrontations between tree-shrews. Walletschek and Raad (1982) recorded 3- to 4-fold increases in firing rate of neurons in the dorsal raphe nucleus when the combatant engaged in immobile defensive postures (Fig. 10.4).

In one of the first behavioural single unit recording studies, Adams et al. (1968) has recorded from neurons in the central grey area dorsal to the raphe that selectively increased their discharge rate during defensive reactions of a cat that was attacked by an opponent. Whether or not these central grey neurons contained 5-HT or localized 5-HT receptors is uncertain. The marked increase in activity in selective neurons in dorsal raphe n. and in the central grey region suggests a critical role of these midline structures in the initiation and execution of defensive reactions, at least in cats and tree-shrews.

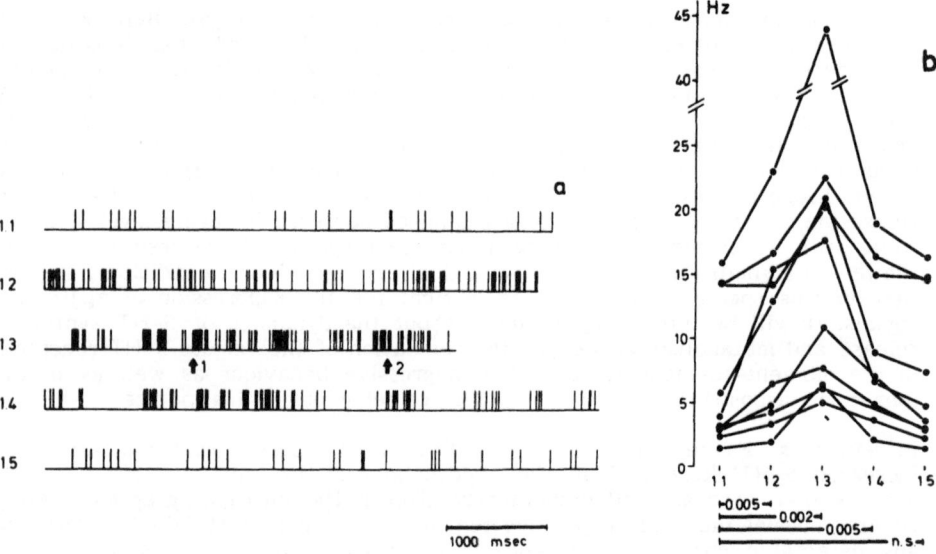

Fig. 10.4 Typical neuronal activity of dorsal raphe neurons the interaction of tree-shrew with the experimenter. 1.1. tree shrew rests in the sleeping box; 1.2 experimenter enters the chamber and approaches the sleeping box; 1.3 experimenter opens the sleeping box and inserts his hand (1), removes it (2) and closes the box; 1.4 experimenter leaves the chamber; 1.5 experimenter has left for at least two minutes. b. Firing rates of the dorsal raphe neurons during the interactions of the tree-shrew with the experimenter. The numbers 1.1; 1.2; 1.3; 1.4; 1.5 indicate the same behavioural items as in a. Firing rates for each behavioural item during 20 interaction sequences are presented. The numbers in the bars between the items indicate the p-values. Deviation from resting (item 1.1) was evaluated statistically (From Walletschek and Raab, Physiology and Behavior, 1982).

The role of brain 5-HT in aggressive behaviour among non-human primates has been studied in squirrel monkeys (Saimiri sciureus), talapoin monkeys (Miopithecus talapoin), rhesus macaques (Macaca mulatta), stumptail macaques (Macaca arctoides), and vervet monkeys (Cercopithecus aethiops). In addition to pharmacological manipulations, correlative biochemical measurements were obtained that reveal a complex pattern of results. Raleigh, McGuire and associates (1981, 1983, 1984) reported consistent high levels of whole blood 5-HT in high ranking vervet monkeys. Only when the dominant vervet monkey actively occupies the dominant position within the group, the level of whole blood 5-HT is about one third higher than that measured in subdominant group members. When the dominant animal is removed from the group, the whole blood 5-HT level decreases to that characteristic of subdominant animals (Raleigh et al., 1984). The relevance of whole blood 5-HT measurements for the various 5-HT cell bodies, pathways and receptors in brain needs to be established.

In several primate species, concentrations of 5-HIAA in cerebrospinal fluid (CSF) have been measured in individuals that display different amounts of aggressive behaviour. Overall, the correlations between social status - as defined by success in dyadic agonistic interactions - and 5-HIAA in CSF have been disappointing. Although some preliminary data have been presented suggesting a significantly higher concentration of 5-HIAA in CSF of dominant vervet monkeys (Raleigh et al., 1983), similar studies in squirrel monkeys (Green et al. personal communication), talapoin monkeys (Yodyingyad et al., 1985) and rhesus macaques (Kraemer, 1985) detected no reliable correlation between CSF 5-HIAA and aggressive behaviour toward group members. For example, during the phase of group formation, talapoin monkeys that emerged as dominant tended to decrease CSF 5-HIAA and those that became subordinate showed an increase in this parameter. However, the day-to-day variation in number of attacks or in number of threats did not correlate with CSF 5-HIAA (Fig. 10.5). Subordinate males featured high concentrations of CSF 5-HIAA on days with high, moderate or no aggressive behaviour (Yodyingyuad et al., 1985). When alpha and beta members of pairs of squirrel monkeys were initially examined for monoamine metabolites in CSF, subordinate males showed higher levels of MHPG, but not 5-HIAA or HVA. In a second phase the alpha and beta male pairs were housed with females. Elevations in CSF MHPG, but not in 5-HIAA of HVA characterized the pairs with the highest level of aggressive interactions (Table 10.1; Green, Coe, Faull, Levine and Barchas, personal communication). In a methodologically very demanding preparation in pairs of laboratory rats, Sahakian et al. (1986) predicted from elevated CSF tryptophan concentrations, but not from 5-HT turnover estimates, the individual which exhibited most neck and body bites towards his opponent.

The absence of any reliable correlation between the number of aggressive responses in selected primate species and CSF metabolites should not be altogether surprising, since it is unclear how the various anatomically distinctive 5-HT neuronal pools contribute to CSF metabolite concentration and how the initiation, execution and termination of aggressive interactions relates to CSF outflow of 5-HIAA. Of course, this problem applies also to the highly publicized clinical data that attempt to correlate low values of 5-HIAA in human CSF with indices of a behavioural history of violent behaviour, impulsive reactions or suicide attempts (e.g. Asberg et al., 1976, 1987, Brown et al., 1979, 1982,

Linnoila et al., 1983).

At present, it is not possible to develop a cogent evolutionary argument for a direct relationship between any specific marker of 5-HT activity and a specific type of aggressive or defensive response that is consistent across several species, including primates. As intriguing as for example the data from lobsters and three-shrews may appear, they involve aggressive versus defensive responses, and indices of 5-HT activity that change in opposite direction.

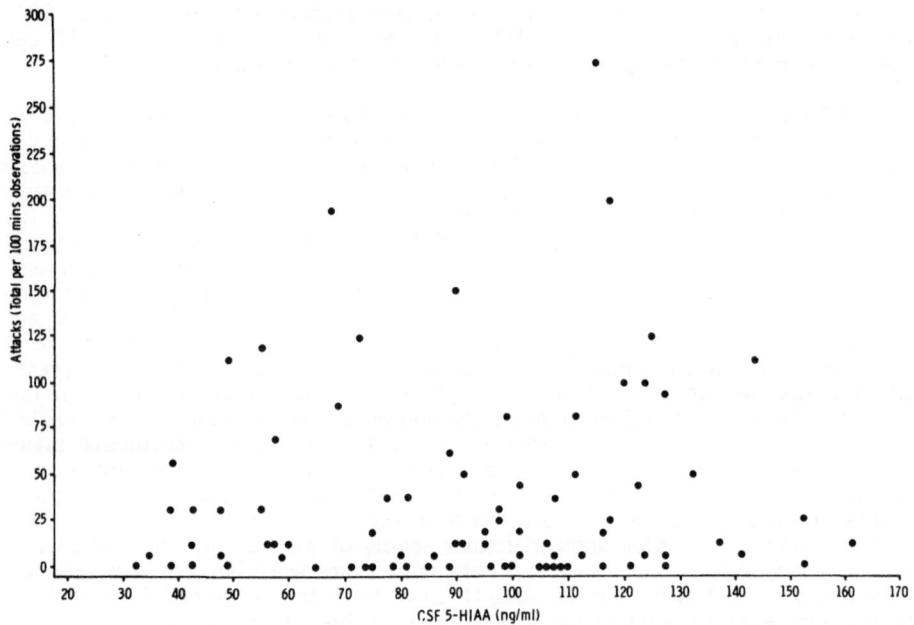

Fig. 10.5 Daily variation in frequency of attacks fails to correlate with the same day levels of 5-HIAA in the cerebrospinal fluid of grouped male talapoin monkeys (From Yodyingyuad et al., Neuroscience, 1985).

Ethopharmacological study of aggression
Particularly in the past decade, an ethological approach has begun to influence what kind of aggressive behaviour is studied and in which way the behavioural measurements are acquired and analyzed, that is suitable for neuropharmacological investigations under controlled laboratory conditions (Miczek, Kruk and Olivier, 1984). An initial contribution of this approach led to a shift away from the isolated or electrically shocked animal and away from the single behavioural score or measure. Experimental preparations were developed

for several animal species, including laboratory rodents, that focus on adaptive forms of aggression such as seen in confrontations between a resident and an intruder, the aggressive and submissive interactions that characterize the establishment and maintenance of social groups or the aggressive behaviour by a lactating female.

As illustrated in Figure 10.6, videoanalysis of aggressive behaviour, aided by microprocessors, enables encoding of latency, frequency and duration of each salient act, posture, display, gesture and movement that is characteristic of aggressive, defensive, submissive and flight behaviour. These detailed analyses show that agonistic behaviour is composed of many behavioural elements, that each movement and posture as well as their temporal and sequential patterns are typical of a given species, that this behaviour is interactive involving at least two individuals, and that it has an episodic pattern with periods of intense fighting alternating with periods of relative quiescence.

Table 10.1 CSF monoamine metabolite concentrations in group-housed squirrel monkeys.

Group condition	Social status	Cisternal CSF concentrations (nmol/l)		
		MHPG	HVA	5-HIAA
Agonistic (n=3)	α	104 ± 9**	322 ± 196	168 ± 101
	β	162 ± 7**	218 ± 96	93 ± 29
Stable (n=4)	α	114 ± 16	515 ± 382	182 ± 125
	β	99 ± 9	221 ± 139	94 ± 55

Values are means ± s.e.; **$p<0.01$. From Green, Coe, Faull, Levine and Barchas (1986).

An initial example of ethopharmacological research serves to demonstrate a pervasive decrease in aggressive behaviour, but not defensive and escape reactions, after 5-HT receptor blockade (Miczek, 1981). In resident mice confronting an intruder in their home cage, methysergide decreased dose-dependently all salient elements of aggressive behaviour such as attack bites, sideways threats, pursuits and tail rattles; concurrently, non-aggressive activities such as walking, rearing and self-directed grooming were also decreased by methysergide in parallel dose-effect curves. By contrast, when the intruder mouse was administered with methysergide, its defensive upright postures and escape reactions remained unaltered by this treatment and audible vocalizations actually tended to increase with higher doses of methysergide. Propranolol, a beta-receptor blocker, selectively decreased attack and threat components of the resident mouse' behaviour without significantly altering other aspects in the behavioural repertoire. Synthesis inhibition by PCPA can, however, increase a resident rat's aggressive behaviour toward an intruder significantly (Sheard, 1973a). This set of observations illustrates again the

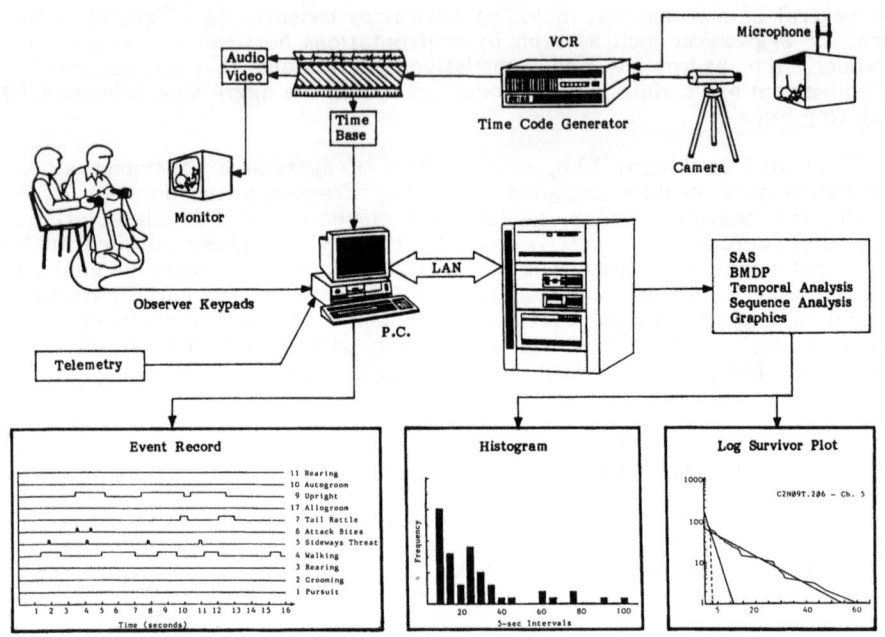

Fig. 10.6 Schematic representation of the system for detailed recording and analysis of behavioural events by observation, as used in ethological studies of aggressive behaviour.

important differential significance of pre- versus post-synaptic events in 5-HT neurotransmission for aggressive behaviour. In female mice, however, PCPA and methysergide exert no differential effects on aggressive behaviour toward a male opponents near the nest site early during the lactation period (Mann and Svare, unpublished, in Svare and Mann, 1983, Fig. 10.7).

Mianserin, methiotepin, methysergide, single or daily administration of PCPA significantly reduces the proportion of lactating mice that exhibit aggressive behaviour and decreases the frequency of attacks (Mann and Svare, unpublished; Ieni and Thurmond, 1985). Gonadal steroids also reduce fighting behaviour of lactating mice as well as reduce 5-HT activity. However, similar to male aggression, both enhancement and impairment of 5-HT activity reduce maternal aggression. These observations question whether 5-HT exerts a critical role in the mechanisms mediating aggressive behaviour and maternal care responses of females. However, it appears possible to achieve a higher degree of behavioural specificity in decreasing intense maternal aggressive behaviour with more selective 5-HT agonists (e.g. Olivier et al., 1988).

The proposed disinhibiting effects of manipulations that impair 5-HT systems may be examined in animals that are habituated to a provocative stimulus. When exposed to an intruder animal, a resident mouse will initially attack, threaten and pursue the intruder with high frequency and intensity, and with repeated confrontations in short intervals, the resident's aggressive behaviour will decline exponentially (Winslow and Miczek, 1984). This habituation process affords the opportunity to study alterations in aggressive behavoir that range from high to low levels. While amphetamine and apomorphine substantially increase the low rates of aggressive behaviour that are characteristic for the late phase of the habituation process, no such increased or "disinhibited" levels of aggressive behaviour are seen after treatment with methysergide or lisuride (Fig. 10.8; Winslow and Miczek, 1983).

Similarly, ethological studies on primate behaviour detected either no changes in social interactions or in attacks and threats after 5-HT synthesis inhibition by daily PCPA administration nor changes that paralleled those of 5-HT agonists (Redmond et al., 1971; Maas et al., 1973; Raleigh et al., 1980). Moreover, the modest decrease in aggressive behaviour after tryptophan in non-human and human primates is frequently difficult to dissociate from the lethargic and sedative effects of this treatment (e.g., Chamove, 1983; Morand et al., 1983;

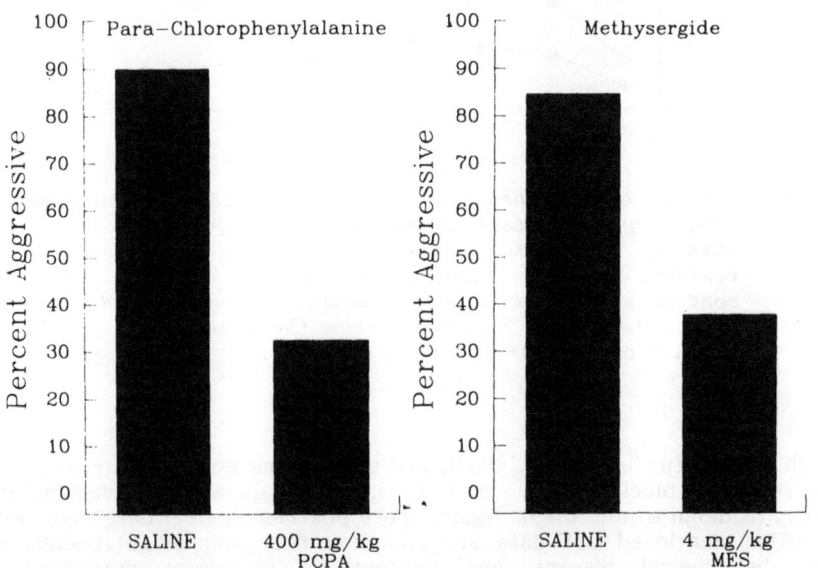

Effects of pharmacological alteration
of serotonin function on postpartum aggression

Fig. 10.7 *Effects of PCPS or methysergide on the percentage of lactating female mice exhibiting aggressive behaviour toward a stimulus male (From Svare, personal communication).*

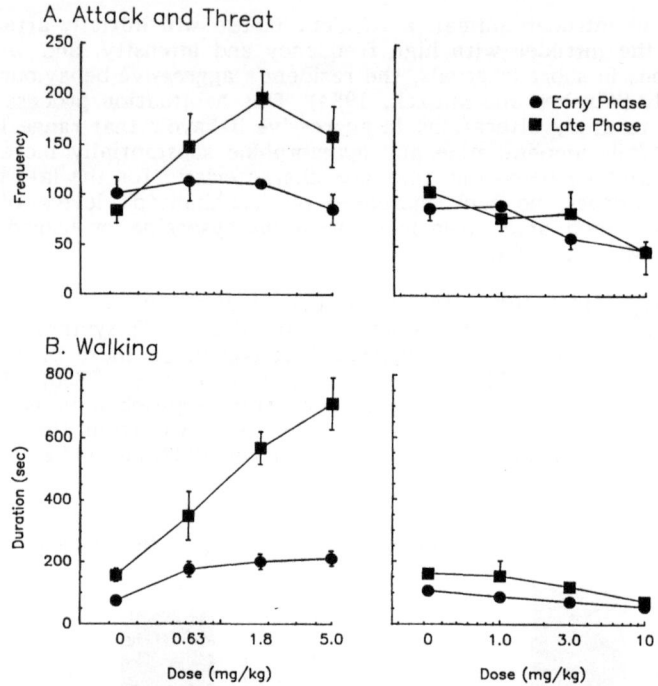

Fig. 10.8 Effects of d-amphetamine and methysergide on the cumulative frequency of attack bites and sideways threats (top) and walking duration (bottom) during the initial resident-intruder confrontations and during the later confrontations of a 10 consecutive 5-min trial sequence, each trial being separated from the next by a 5 min interval (Data from Winslow and Miczek, 1983).

Raleigh, 1987; Yuwiler et al., 1981), and under some conditions tryptophan and 5-HT reuptake blockers may actually increase social intiatives in vervet monkeys (Raleigh et al., 1985). Again, these pharmacological data together with the earlier mentioned CSF data are complex; they emphasize species, social status, behavioural history, and context as important determinants of behavioural drug effects, and render any simplistic generalizations about the relationship between 5-HT and primate aggressive behaviour as premature.

Conclusions
5-HT systems are a significant part of the neurobiological mechanisms mediating different aggressive, defensive and flight behaviours. Serotonergic mechanisms appear important in the induction (e.g., mouse killing), execution (e.g., hypothalamic attack behaviour), and consequences of different kinds of

aggressive behaviour. Manipulations that enhance 5-HT activity consistently decrease aggressive and defensive reactions. Recently developed 5-HT agonists with higher degree of selectivity for receptor subtypes promise to act more specifically as antiaggressive drugs. However, whether or not the antiaggressive effects of these newer substances are due to their 5-HT receptor agonist properties has to await confirmation with appropriate antagonists. Unfortunately, the behavioural specificity of these manipulations to aggressive, defensive or predatory behaviour often remains unclear. So far, biochemical studies of 5-HT, 5-HIAA or metabolic enzymes in blood, CSF, whole brain or discrete brain structures have been performed in individuals under conditions that may have resulted in aggressive behaviour (i.e. isolated housing) or in individuals that had shown the aggressive behaviour some time ago (e.g., mouse killing in rats, aggressive displays in monkeys, or violence in humans). It remains unclear how the dynamic changes in 5-HT synthesis, release, or turnover in discrete anatomical systems are related to the dynamics of aggressive interactions on a moment-to-moment basis.

So-called "disinhibitory" effects by manipulations that impair 5-HT activity are limited to certain defensive reactions in selected species, and to the mouse-killing response of laboratory rats. In spite of the impressive evolutionary constance in morphology of 5-HT systems in brain, no parallel constancy in function can be established, at least with regard to aggressive behaviour. Early research attempted to identify 5-HT as a pervasive "trophotrophic" or behaviourally inhibitory neurotransmitter. Current efforts are directed toward determining the role of 5-HT receptor subtypes in the neurobiological mechanisms for the multiple distal and proximal causes for aggressive behaviour, their varied purpose, including dispersive and cohesive functions, and the indicate behavioural patterns that characterize conflict situations.

Acknowledgements: Preparation of this chapter was supported by U.S.P.H.S. research grants AA05122 and DA02632. The excellent technical assistance of J.T.Sopko is gratefully acknowledged. We are grateful to Dr.J.F.DeBold for his helpful comments.

References

Adams, D.B. (1968). Cells related to fighting behaviour recorded from midbrain central gray neuropil of cat. Sci., 159, 894-896.

Aghajanian, G.K., Foote, W.E. & Sheard, M.H. (1968). Lysergic acid diethylamide: sensitive neuronal units in the midbrain raphe. Sci., 161, 706-708.

Allikmets, L.K., Lapin, I.P. (1967). Influence of lesions of the amygdaloid complex on behaviour and on effects of antidepressants in rats. Int. J. Neuropharmacol., 6, 99-108.

Anand, M., Gupta, G.P. & Bhargava, K.P. (1977). Modification of electroshock fighting by drugs known to interact with dopaminergic and noradrenergic neurons in normal and brain lesioned rats. J. Pharm. Pharmacol., 29, 437-439.

Applegate, C.D. (1980). 5,7-Dihydroxytryptamine-induced mouse killing and behavioural reversal with ventricular administration of serotonin in rats. Behav. Neural Biol., 30, 178-190.

Asberg, M., Schalling, D., Traskman-Bendz, L. & Wagner, A. (1987). Psychobiology of suicide, impulsivity, and related phenomena. In: Psychopharmacology: the Third Generation of Progress. (Ed. Meltzer H.Y.), pp. 655-668. New York: Plenum Press.

Asberg, M., Thoren, P. & Traskman, L. (1976). "Serotonin depression" - a biochemical subgroup within the effective disorders? Sci., 191, 478-480.

Azmitia, E.C. & Gannon, P.J. (1986). The primate serotonergic system: a review of human and animal studies and a report on Macace fascicularis. In: Advance in Neurology. Myoclonus. (Ed. Fahn, S.), pp. 407-468. New York: Raven Press.

Banerjee, U. (1974). Modification of the isolation-induced abnormal behaviour in male Wistar rats by destructive manipulation of the central monoaminergic systems. Behav. Biol., 11, 573-579.

Bell, R. & Brown, K. (1977). Lateral hypothalamus and shock-induced aggression: serotonergic inhibition. IRCS Medical Science; Nervous System: Pharmacology. Physiol. Psychol. Psychiat., 5, 221.

Blanchard, R.J., Blanchard, D.C. & Takahashi, L.K. (1978). Pain and aggression in the rat. Behav. Biol., 23, 291-305.

Bocknik, S.E. & Kulkani, A.S. (1974). Effect of a decarboxylase inhibitor (RO 4-4602) on 5-HTP induced muricide blockade in rats. Neuropharmacol., 13, 279-281.

Boissier, J.R., Grasset, G. & Simon, P. (1968). Effect of some psychotropic drugs on mice from a spontaneously aggressive strain. J. Pharm. Pharmacol., 20, 972-973.

Breese, G.R. & Cooper, B.R. (1975). Behavioural and biochemical interactions of 5,7-dihydroxytryptamine with various drugs when administered intracisternally to adult and developing rats. Brain Res., 98, 517-527.

Breese, G.R., Cooper, B.R., Grant, L.D. & Smith, R.D. (1974). Biochemical and behavioural alterations following 5,6-dihydroxytryptamine administration into brain. Neuropharmacol., 13, 177-187.

Brodie, B.B. & Shore, P.A. (1957). A concept for a role of serotonin and norepinephrine as chemical mediators in the brain. Ann. NY Acad. Sci., 66, 631-642.

Brown, G.L., Ebert, M.H., Goyer, P.F., Jimerson, D.C., Klein, W.J., Bunney, W.E. & Goodwin, F.K. (1982). Aggression, suicide, and serotonin-relationships to CSF amine metabolites. Am. J. Psychiat., 139, 741-746.

Brown, G.L., Goodwin, F.K., Ballenger, J.C. Goyer, P.F. & Major, L.F. (1979).

Aggression in humans correlates with cerebrospinal fluid amine metabolites. Psychiat. Res., 1, 131-139.

Bullock, T.H. (1969). Species differences in effect of electroreceptor input on electric organ pacemakers and other aspects of behaviour in electric fish. Brain, Behav. and Evol., 2, 85-5118.

Butcher, L.L. & Dietrich, A.P. (1973). Effects on shock-elicited aggression in mice of preferentially protecting brain monoamines against the depleting action of reserpine. Arch. Pharmacol., 277, 61-70.

Chamove, A.S. (1983). Dietary effects of rhesus social behaviour: Altered amino acid diets. Dev. Psychobiol., 16, 505-509.

Conner, R.L., Stolk, J.M., Barchas, J.D., Dement, W.C. & Levine, S. (1970). The effect of parachlorophenylalanine (PCPA) on shock-induced fighting behaviour in rats. Physiol. Behav., 5, 1221-1224.

DeVanzo, J.P., Daugherty, M., Ruckart, R. & Kang, L. (1966). Pharmacological and biochemical studies in isolation-induced fighting mice. Psychopharmacologia, 9, 210-219.

Delini-Stula, A. & Vassout, A. (1979). Differential effects of psychoactive drugs on aggressive responses in mice and rats. In: Psychopharmacology of Aggression. (Ed. Sandler, M.), pp. 41-60. New York: Raven Press.

Delini-Stula, A. & Vassout, A. (1981). The effects of antidepressants on aggressiveness induced by social deprivation in mice. Pharmacol. Biochem. Behav., 14 (S1), 33-41.

Di Chiara, G., Camba, R. & Spano, P.F. (1971). Evidence for inhibition by brain serotonin of mouse killing behaviour in rats. Nature, 233, 272-273.

Dubinsky, B., Karpowicz, J.K. & Goldberg, M.E.. (1973). Effects of tricyclic antidepressants on attack elicited by hypothalamic stimulation: relation to brain biogenic amines. J. Pharmacol. Exper. Ther., 187, 550-557.

Eichelman, B. (1979). Role of biogenic amines in aggressive behaviour. In: Psychopharmacology of Aggression. (Ed. Sandler, M.), pp. 61-93. New York: Raven Press.

Eichelman, B. & Barchas, J. (1975). Facilitated shock-induced aggression following anti-depressive medication in the rat. Pharmacol. Biochem. Behav., 3, 601-604.

Eichelman, BS. Jr. & Thoa, N.B. (1973). The aggressive monoamines. Biol. Psychiat., 6, 143-164.

Ellison, G.D. & Bresler, D.E. (1974). Tests of emotional behaviour in rats following depletion of norepinephrine, of serotonin, or of both. Psychopharmacologia, 34, 275-288.

Gibbons, J.L., Barr, G.A. Bridger, W.H. & Leibowitz, S.F. (1978). Effects of parachlorophenylalanine and 5-hydroxytryptophan on mouse killing behaviour in killer rats. Pharmacol. Biochem. Behav., **9**, 91-98.

Gibbons, J.L., Barr, G.A. Bridger, W.H. & Leibowitz, S.F. (1981). L-Tryptophan's effects on mouse killing, feeding, drinking, locomotion, and brain serotonin. Pharmacol. Biochem. Behav., **15**, 201-206.

Gibbons, J.L., Barr, G.A. Bridger, W.H. & Leibowitz, S.F. (1979). Manipulations of dietary tryptophan: Effects on mouse killing and brain serotonin in the rat. Brain Res., **169**, 139-153.

Goldberg, M.E. (1970). Pharmacologic activity of a new class of agents which selectively inhibit aggressive behaviour in rats. Arch. int. de Pharmacodynamie et de Therapie, **186**, 287-297.

Grant, L.D., Coscina, D.V., Grossman, S.P. & Freedman, D.X. (1973). Muricide after serotonin depleting lesions of midbrain raphe nuclei. Pharmacol. Biochem. Behav., **1**, 77-80.

Hess, W.R. (1954). Das Zwischenhirn. Basel: Benno Schwabe & Co.

Hodge, G.K. & Butcher, L.L. (1974). 5-Hydroxytryptamine correlates of isolation-induced aggression in mice. Eur. J. Pharmacol., **28**, 326-337.

Hole, K., Johnson, G.E. & Berge, O.G. (1977). 5,7-Dihydroxytryptamine lesions of the ascending 5-hydroxytryptamine pathways: habituation, motor activity and agonistic behaviour. Pharmacol. Biochem. Behav., **7**, 205-210.

Horovitz, Z.P., Piala, J.J., High, J.P., Burke, J.C. & Leaf, R.C. (1966). Effects of drugs on the mouse-killing (muricide) test and its relationship to amygdaloid function. Int. J. Neuropharmacol., **5**, 405-411.

Horovitz, Z.P., Ragozzino, P.W. & Leaf, R.C. (1965). Selective block of rat mouse-killing by antidepressants. Life Sci., **4**, 1909-1912.

Howard, J.L. & Pollard, G.T. (1983). Are primate models of neuropsychiatric disorders useful to the pharmaceutical industry? In: Ethopharmacology: Primate Models of Neuropsychiatric Disorders. (Ed. Miczek, K.A.), pp. 307-312. New York: Alan R. Liss Inc.

Huntingford, F.A. (1976). The relationship between inter- and intra-specific aggression. Animal Behav., **24**, 485-497.

Ieni, J.R. & Thurmond, J.B. (1985). Maternal aggression in mice: effects of treatments with PCPA, 5-HTP and 5-HT receptor antagonists. Eur. J. Pharmacol., **111**, 211-220.

Irwin, S., Kinohi, R., Van Sloten, M. & Workman, M.P. (1971). Drug effects on distress-evoked behaviour in mice: methodology and drug class comparisons. Psychopharmacologia, **20**, 172-185.

Jacobs, B.L. & Cohen, A. (1976). Differential behavioural effects of lesions of the median or dorsal raphe nuclei in rats: open field and pain-elicited aggression. J. Comp. Physiol. Psychol., 90, 102-108.

Kantak, K.M. (1981). Facilitation of shock-induced fighting following intraventricular 5,7-dihydroxy-tryptamine and 6-hydroxy-dopa. Psychopharmacol., 74, 157-160.

Kantak, K.M., Hegstrand, L.R. & Eichelman, B. (1980a). Dietary tryptophan modulation and aggressive behaviour in mice. Pharmacol. Biochem. Behav., 12, 675-679.

Kantak, K.M., Hegstrand, L.R. & Eichelman, B. (1981a). Dietary tryptophan reversal of septal lesion and 5,7-DHT lesion elicited shock-induced fighting. Pharmacol. Biochem. Behav., 15, 343-350.

Kantak, K.M., Hegstrand, L.R. & Eichelman, B. (1981b). Facilitation of shock-induced fighting following intraventricular 5,7-dihydroxy-tryptamine and 6-hydoxy-dopa. Psychopharmacol., 74, 157-160.

Kantak, K.M., Hegstrand, L.R., Whitman, J. & Eichelman, B. (1980b). Effects of dietary supplements and a tryptophan-free diet on aggressive behaviour in rats. Pharmacol. Biochem. Behav., 12, 173-179.

Karli, P. (1959). Action de substances dites "tranquillisantes" sur l'aggressivite interspecifique Rat-Souris. Comptes Rendus de Societe de Biologie, 153, 467-469.

Karli, P. (1981). Conceptual and methodological problems associated with the study of brain mechanisms underlying aggressive behaviour. In: The Biology of Aggression. (Eds. Brain, P.F. & Benton, D.), pp. 322-361. Rockville, MD: Sijthoff and Noordhoff.

Katz, R.J. (1980). Role of serotonergic mechanisms in animal models of predation. Progress Neuro-Psychopharmacol., 4, 219-231.

Koe, B.K., Weissman, A. (1966). p-Chlorophenylalanine: a specific depletor of brain serotonin. J. Pharmacol. Exper.Ther., 154, 499-516.

Kostowski, W., Czlonkowski, A., Markowska, L., Markiewicz, L. (1975a). Intraspecific aggressiveness after lesions of midbrain raphe nuclei in rats. Pharmacol., 13, 81-85.

Kostowski, W., Pucilowski, O. & Plaznik, A. (1980). Effect of stimulation of brain serotonergic system on mouse-killing behaviour in rats. Physiol. Behav., 25, 161-165.

Kostowski, W., & Tarchalska, B. (1972). The effects of some drugs affecting brain 5-HT on the aggressive behaviour and spontaneous electrical activity of the central nervous system of the ant, Formica rufa. Brain Res., 38, 143-149.

Kostowski, W., Tarchalska-Krynska, B., Markowska, L. (1975b). Aggressive behaviour and brain serotonin and catecholamines in ants (Formica rufa). Pharmacol. Biochem. Behav., 3, 717-719.

Kostowski, W. & Valzelli, L. (1974). Biochemical and behavioural effects of lesions of raphe nuclei in aggressive mice. Pharmacol. Biochem. Behav., 2, 277-280.

Kraemer, G.E. (1985). The primate social environment, brain neurochemical changes and psychopathology. Trends Neurosci., 8, 339-340.

Kravitz, E.A. (1986). Serotonin, octopamine, and proctolin: Two amines and a peptide, and aspects of lobster behaviour. In: Fast and Slow Chemical Signalling in the Nervous System. (Eds. Iversen, L.L. & Goodman, E.), pp. 244-259. Oxford Science Publication.

Kravitz, E.A., Glusman, S., Livingstone, M.S. & Harris-Warrick, R.M. (1981). Serotonin and octopamine in the lobster nervous system: mechanism of action at neuromuscular junctions and preliminary behavioural studies. In: Serotonin Neurotransmission and Behavior. (Ed. Gelperin, A.), pp. 189-210. Cambridge, Mass.: MIT Press.

Kreiskott, H. & Hofmann, H.P. (1975). Stimulation of a specific drive (predatory behaviour) by p-chlorophenylalanine (pCPA) in the rat. Pharmakopsychiat. - Neuro Psychopharmakol., 8, 136-140.

Krsiak, M., Sulcova, A., Tomasikova, Z., Dlohozkova, N., Kosar, E., Masek, K. (1981). Drug effects on attack, defense and escape in mice. Pharmacol. Biochem. Behav., 14 (S1), 47-52.

Kulkarni, A.S. (1968). Muricidal block produced by 5-hydroxytryptophan and various drugs. Life Sci., 7, 125-128.

Kulkarni, A.S. (1970). Decarboxylase inhibitor on 5-HTP induced blockade of mouse killing. Pharmacologist (Abstr.), 12, 207.

Kulkarni, A.S., Rahwan, R.G. & Bocknik, S.E. (1973). Muricidal block induced by 5-hydroxytryptophan in the rat. Arch. int. de Pharmacodynamie et de Therapie, 201, 308-313.

Lal, H., DeFeo, J.J. & Thut, P. (1970). Prevention of pain-induced aggression by parachloroamphetamine. Biol. Psychiat., 2, 205-206.

Lapin, I.P. (1967). Simple pharmacological procedures to differentiate antidepressants and cholinolytics in mice and rats. Psychopharmacologia, 11, 79-87.

Le Douarec, J.C. &, Broussy, L. (1969). Dissociation of the aggressive behaviour in mice produced by certain drugs. In: Aggressive Behavior. Eds. Garattini, S. & Sigg, E.B.), pp. 281-295. Amsterdam: Excerpta Medica Foundation.

Le Douarec, J.C. & Nuveu, C. (1970). Pharmacology and biochemistry of fenfluramine. In: Amphetamine and related compounds. (Eds. Costa, E. & Garattini, S.), pp. 75-105. New York: Raven Press.

Leaf, R.C., Wnek, D.J. & Lamon, S. (1978). Despite various drugs, cats continue to kill mice. Pharmacol. Biochem. Behav., 9, 445-452.

Lent, C.M. & Dickinson, M.H. (1984) Serotonin integrates the feeding behaviour of the medicinal leech. J. Comp. Physiol. A, 154, 457-471.

Linnoila, M., Virkkunen, M., Scheinin, M., Nuutila, A., Rimon, R. & Goodwin, F.K. (1983). Low cerebrospinal fluid 5-hydroxyindoleacetic acid concentration differentiates impulsive from nonimpulsive violent behaviour. Life Sci., 33, 2609-2614.

Livingstone, M.S., Harris-Warrick, R.M. & Kravitz, E.A. (1980). Serotonin and octopamine produce opposite postures in lobsters. Sci., 208, 76-79.

Maas, J.W., Redmond, D.E. Jr., Gauen, R. (1973). Effects of serotonin depletion on behaviour in monkeys. In: Serotonin and Behavior. (Eds. Barchas, J. & Usdin, E.), pp. 351-356. New York: Academic Press.

Maler, L., Ellis, W.G. (1987). Inter-male aggressive signals in weakly electric fish are modulated by monoamines. Behav. Brain Res., 25, 75-81.

Malick, J.B. (1976). Pharmacological antagonism of mouse-killing behaviour in the olfactory bulb lesion-induced killer rat. Aggr. Behav., 2, 123-130.

Malick, J.B. (1979). The pharmacology of isolation-induced aggressive behaviour in mice. In: Current Developments in Psychopharmacology. (Eds. Essman, W.B. & Valzelli, L.), pp. 1-27. New York: SP Medical and Scientific Books.

Malick, J.B. & Barnett, A. (1976). The role of serotonergic pathways in isolation-induced aggression in mice. Pharmacol. Biochem. Behav., 5, 55-61.

Marks, P.C., O'Brien, M. & Paxinos, G. (1977). 5,7-DHT-induced muricide: inhibition as a result of preoperative exposure of rats to mice. Brain Res., 135, 383-388.

Marks, P.C., O'Brien, M. & Paxinos, G. (1978). Chlorimipramine inhibition of muricide: the role of the ascending 5-HT projection. Brain Res., 149, 270-273.

McCarty, R.C., Whitesides, G.H. & Tomosky, T.K. (1976). Effects of para-chlorophenylalanine on the predatory behaviour of Onychomys tossidus. Pharmacol. Biochem. Behav., 4, 217-220.

McLain, W.C., Cole, B.T., Schrieber, R. & Powell, D.A. (1974). Central catechol- and indolamine systems and aggression. Pharmacol. Biochem. Behav., 2, 123-126.

Miczek, K.A. (1981). Differential antagonism of d-amphetamine effects on motor activity and agonistic behaviour in mice. Neurosci. Abstr., 7, 343.

Miczek, K.A. (1987). The psychopharmacology of aggression. In: Handbook of Psychopharmacology. (Eds. Iversen, L.L., Iversen, S.D. & Snyder, S.H.), Vol. 19: New Directions in Behavioural Pharmacology, pp. 183-328. New York: Plenum Press.

Miczek, K.A., Altman, J.L. Appel, J.B. & Boggan, W.O. (1975). Para-chlorophenylalanine, serotonin and killing behaviour. Pharmacol. Biochem. Behav., 3, 355-361.

Miczek, K.A. & Krsiak, M. (1979). Drug effects on agonistic behaviour. In: Advances in Behavioral Pharmacology. (Eds. Thompson, T. & Dews, P.B.), pp. 87-162. New York: Academic Press Inc.

Miczek, K.A., Kruk, M.R. & Olivier, B. (1984). Ethopharmacological Aggression Research. New York: Alan R. Liss Inc.

Modigh, K. (1973). Effects of isolation and fighting in mice on the rate of synthesis of noradrenaline, dopamine and 5-hydroxytryptamine in the brain. Psychopharmacologia, 33, 1-17.

Mogilnicka, E., Boissard, C.G., Waldmeier, P.C. & Delini-Stula, A. (1983). The effects of single and repeated doses of maprotiline, oxaprotiline and its enantiomers on foot-shock induced fighting in rats. Pharmacol. Biochem. Behav., 19, 719-723.

Mogilnicka, E., & Przewlocka, B. (1981). Facilitated shock-induced aggression after chronic treatment with antidepressant drugs in the rat. Pharmacol. Biochem. Behav., 14, 129-132.

Morand, C., Young, S.N. & Ervin, F.R. (1983). Clinical response of aggressive schizophrenics to oral tryptophan. Biol. Psychiat., 18, 575-578.

Nikulina, E.M. & Popova, N.K. (1983). Ob uchastii serotonina v proiavlenii chiccnnicheskoi agressii u myshei (On serotonin participation in mice predatory aggression). Zhurnal vysshei nervnoi deiatelnosti imeni IP Pavlova, 33, 737-742.

Nikulina, E.M. & Popova, N.K. (1986). Serotonin's influence on predatory behaviour of highly aggressive CBA and weakly aggressive DD strains of mice. Aggr. Behav., 12, 277-283.

Ogren, S.O., Holm, A.C., Renyi, A.L. & Ross, S.B. (1980). Anti-aggressive effect of zimelidine in isolated mice. Acta pharmacologia et toxicologia, 47, 71-74.

Onodera, K., Kisara, K. & Ogura, Y. (1979). Effect of 5-hydroxytryptophan on muricide response induced by thiamine deficiency. Arch. int. de Pharmacodynamie et de Therapie, 240, 220-227.

Panksepp, J., Zolovick, A.J., Jalowiec, J.E., Stern, W.C. & Morgane, P.J. (1973). Fenfluramine: effects on aggression. Biol. Psychiat., 6, 181-186.

Parent, A., Poitras, D. & Dube, L. (1984). Comparative anatomy of central monoaminergic systems. In: Handbook of Chemical Neuroanatomy. (Eds. Bjorklund, A. & Hokfelt, T.), Vol. 2: Classical Transmitters in the CNS, Part I, pp. 409-439. Elsevier Science Publishers B.V.

Paxinos, G. & Atrens, D.M. (1977). 5,7-Dihydroxytryptamine lesions: effects on body weight, irritability, and muricide. Aggr. Behav., 3, 107-118.

Paxinos, G., Burt, J., Atrens, D.M. & Jackson, D.M. (1977). 5-hydroxytryptamine depletion with parachlorophenylalanine: effects on eating, drinking, irrability, muricide, and copulation. Pharmacol. Biochem. Behav., 6, 439-447.

Payne, A.P., Andrews, M.J. & Wilson, C.A. (1984). Housing, fighting and biogenic amines in the midbrain and hypothalamus of the golden hamster. In: Ethopharmacological aggression research. (Eds. Miczek, K.A., Kruk, M. & Olivier, B.), pp. 227-247. New York: Alan R. Liss Inc.

Payne, A.P., Andrews, M.J. & Wilson, C.A. (1985). The effects of isolation, grouping and aggressive interactions on indole- and catecholamine levels and apparent turnover in the hypothalamus and midbrain of the male golden hamster. Physiol. Behav., 34, 911-916.

Payne, A.P. & Wilson, C.A. (1981). Effects of housing conditions and aggressiveness on neural biogenic amine metabolism in male golden hamsters. In: Biology of aggression. (Eds. Brain, P.F. & Benton, D.), pp. 147-154. Alphen a/d Rijn: Sijthoff and Noordhoff.

Penot, C., Vergnes, M., Mack, G. & Kempf, E. (1978). Comportement d'aggression interspecifique et reactivite chez le chat: etude comparative des effets de lesions electrolytiques du raphe et d'injections intraventriculaires de 5,7 DHT. Biol. Behav., 3, 71-85.

Poschlova, N., Masek, K. & Krsiak, K. (1977). Amphetamine-like effects of 5,6-dihydroxytryptamine on social behaviour in the mouse. Neuropharmacol., 16, 317-321.

Prasad, V. & Sheard, M.H. (1983a). Synergistic effect of propranolol and quipazine on desimipramine enhanced shock-elicited fighting in rats. Pharmacol. Biochem. Behav., 19, 419-421.

Prasad, V. & Sheard, M.H. (1983b). Time course of chronic desipramine on shock-elicited fighting in rats. Agressol., 24, 15-17.

Pucilowski, O. & Kostowski, W. (1983). Aggressive behaviour and the central serotonergic systems. Behav. Brain Res., 9, 33-48.

Pucilowski, O., Plaznik, A. & Kostowski, W. (1985). Aggressive behaviour inhibition by serotonin and quipazine injected into the amygdala in the rat. Behav. Neural Biol., 43, 58-68.

Raad, A. & Deisz, R. (1975). Male and female mice living in differently-sized groups. II. Serotonin metabolism in discrete brain areas, open-field activity and corticoid release. J. Comp. Physiol., **99**, 165-175.

Raleigh, M.J. (1987). Differential behavioural effects of tryptophan and 5-hydroxy-tryptophan in vervet monkeys: influence of catecholaminergic systems. Psychopharmacol., **93**, 44-50.

Raleigh, M.J., Brammer, G.L. & McGuire, M.T. (1983). Male dominance, serotonergic systems, and the behavioural and physiological effects of drugs in vervet monkeys (Cercopithecus aethiops sabaeus). In: Ethopharmacology: primate models of neuropsychiatric disorders. (Ed. Miczek, K.A.), pp. 185-198. New York: Alan R. Liss Inc.

Raleigh, M.J., Brammer, G.L., McGuire, M.T. & Yuwiler, A. (1985). Dominant social status facilitates the behavioural effects of serotonergic agonists. Brain Res., **348**, 274-282.

Raleigh, M.J., Brammer, G.L., Yuwiler, A., Flannery, J.W., McGuire, M.T. & Geller, E. (1980). Serotonergic influences on the social behaviour of vervet monkeys (Cercopithecus aethiops sabaeus). Exp. Neurol., **68**, 322-334.

Raleigh, M.J., McGuire, M.T., Brammer, G.L. & Yuwiler, A. (1984). Social and environmental influences on blood serotonin concentrations in monkeys. Arch. Gen. Psychiat. **41**, 405-410.

Raleigh, M.J., Yuwiler, A., Brammer, G.L., McGuire, M.T. & Geller, E. & Flannery, J.W. (1981). Peripheral correlates of serotonergically-influenced behaviours in vervet monkeys (Cercopithecus aethiops sabaeus). Psychopharmacol., **72**, 241-246.

Ray, A., Sharma, K.K., Alkondon, M., Sen, P. (1983). Possible interrelationship between the biogenic amines involved in the modulation of footshock aggression in rats. Arch. int. de Pharmacodynamie et de Therapie, **265**, 36-41.

Redmond, D.E., Maas, J.W., Kling, A., Graham, C.W. & Dekirmenjian, H. (1971). Social behaviour of monkeys selectively depleted of monoamines. Sci., **174**, 428-431.

Rolinski, Z. (1975). Interspecies aggressiveness of rats towards mice after the application of p-chlorophenylalanine. Polish J. Pharmacol. Pharm., **27**, 223-229.

Rolinski, Z. & Herbut, M. (1981). The role of the serotonergic system in foot shock-induced behaviour in mice. Psychopharmacol., **73**, 246-251.

Rossi, A.C. (1975). The "mouse-killing" rat: Ethological discussion on an experimental model of aggression. Pharmacol. Res. Comm., **7**, 199-216.

Sahakian, B.J., Sarna, G.S., Kantamaneni, D.B., Jackson, A., Hutson, P.H., Curzon, G. (1986). CSF tryptophan and transmitter amine turnover may predict social behaviour in the normal rats. Brain Res., **399**, 162-166.

Salama, A.I., Goldberg, M.E. (1973). Temporary increase in forebrain norepinephrine turnover in mouse-killing rats. Eur. J. Pharmacol., 21, 372-374.

Schmidt, W.J. (1980). Unlike rats, ferrets do kill under antidepressants. Naturwissenschaften, 67, 262-263.

Schmidt, W.J. & Meierl, G. (1980). Antidepressants and the control of predatory behaviour. Physiol. Behav., 25, 17-19.

Scott, J.P. (1958). Aggression. Chicago: The University of Chicago Press.

Sheard, M.H. (1969). The effect of p-chlorophenylalanine on behaviour in rats: relation to brain serotonin and 5-hydroxyindoleacetic acid. Brain Res., 15, 524-528.

Sheard, M.H. (1973a). Aggressive behaviour: modification by amphetamine, p-chlorophenylalanine and lithium in rats. Agressol., 14, 323-326.

Sheard, M.H. (1973b). Brain serotonin depletion by p-chlorophenylalanine or lesions of raphe neurons in rats. Physiol. Behav., 10, 809-811.

Sheard, M.H. (1981). Shock-induced fighting (SIF): Psychopharmacological studies. Aggr. Behav., 7, 41-49.

Sheard, M.H. & Davis, M. (1976). Shock elicited fighting in rats: importance of intershock interval upon the effect of p-chlorophenylalanine (PCPA). Brain Res., 111, 433-437.

Sofia, R.D. (1969a). Effects of centrally active drugs on four models of experimentally-induced aggression in rodents. Life Sci., 8, 705-716.

Sofia, R.D. (1969a). Structural relationship and potency of agents which selectively block mouse killing (muricide) behaviour in rats. Life Sci., 8, 1201-1210.

Svare, B.B. & Mann, M.A. (1983). Hormonal influences on maternal aggression. In: Hormones and Aggressive Behavior. (Ed. Svare, B.B.), pp. 91-104. New York: Plenum Press.

Tedeschi, D.H., Fowler, P.J. Miller, E.B. & Macko, E. (1969). Pharmacological analysis of footshock-induced fighting behaviour. In: Aggressive Behaviour. (Eds. Garattini, S. & Sigg, E.B.), pp. 245-252. Amsterdam: Excerpta Medica Foundation.

Tedeschi, R.E., Tedeschi, D.H., Mucha, A., Cook, L., Mattis, P.A. & Fellows, E.J. (1959). Effects of various centrally acting drugs on fighting behaviour of mice. J. Pharmacol. Exper. Ther., 125, 28-34.

Valzelli, L. (1967). Drugs and aggressiveness. Adv. Pharmacol., 5, 79-108.

Valzelli, L. (1982). Serotonergic inhibitory control of experimental aggression.

Pharmacol. Res. Comm., 14, 1-13.

Valzelli, L. (1985). Animal models of behavioural pathology and violent aggression. Methods/Findings Exper. Clin. Pharmacol., 7, 189-193.

Valzelli, L. & Bernasconi, S. (1976). Psychoactive drug effect on behavioural changes induced by prolonged socia-environmental deprivation in rats. Psychol. Med., 6, 271-276.

Valzelli, L. & Bernasconi, S. (1979). Aggressiveness by isolation and brain serotonin turnover changes in differtent strains of mice. Neuropsychobiol., 5, 129-135.

Valzelli, L., Bernasconi, S. & Dalessandro, M. (1981a). Effect of tryptophan administration on spontaneous and P-CPA-induced muricidal aggression in laboratory rats. Pharmacol. Res. Comm., 13, 891-897.

Valzelli, L., Bernasconi, S. & Dalessandro, M. (1983). Time-course of p-cpa-induced depletion of brain serotonin and muricidal aggression in the rat. Pharmacol. Res. Comm., 15, 387-395.

Valzelli, L., Bernasconi, S. & Garattini, S. (1981b). p-chlorophenyl-alanine-induced muricidal aggression in male and female laboratory rats. Neuropsychobiol., 7, 315-320.

Valzelli, L., Giacalone, E. & Garattini, S. (1967). Pharmacological control of aggressive behaviour in mice. Eur. J. Pharmacol., 2, 144-146.

Vergnes, M. (1978). Interspecific aggression and reactivity in rats: effects of selective raphe lesions and additional olfactory bulb ablation. Aggr. Behav., 4, 207-218.

Vergnes, M. (1980). Induction du comportement d'aggression rat-souris par la p-chlorophenylalanine: role de l'amygdale. Physiol. Behav., 25, 353-356.

Vergnes, M. (1982). Serotonergic inhibition of mouse-killing behaviour in the rat: localization of the brain structures involved. Aggr. Behav., 8, 208-211.

Vergnes, M. & Kempf, E. (1981). Tryptophan deprivation: effects on mouse-killing and reactivity in the rat. Pharmacol. Biochem. Behav., 14 (S1),: 19-23.

Vergnes, M. & Kempf, E. (1982). Effect of hypothalamic injection of 5,7-dihydroxytryptamine on elicitation of mouse-killing in rats. Behav. Brain Res., 5, 387-397.

Vergnes, M., Mack, G. & Kempf, E. (1983). Lesions du raphe et reaction d'agression interspecifique rat souris effect comportementaux et biochimiques. Brain Res., 57, 67-74.

Vergnes, M., Mack, G. & Kempf, E. (1974). Controle inhibiteur du comportement d'aggression interspecifique du rat: systeme serotoninergique du raphe et afferences olfactives. Brain Res., 70, 481-491.

Vergnes, M., Penot, C., Kempf, E. &Mack, G. (1977). Lesion selectives des neurones serotonergiques du raphe par 5,7-dihydroxytryptamine: effets sur le comportement d'aggression interspecifique du rat. Brain Res., 133, 167-171.

Waldbillig, R.J. (1979). The role of the dorsal and median raphe in the inhibition of muricide. Brain Res., 160, 341-346.

Walletschek, H. & Raad, A. (1982). Spontaneous activity of dorsal raphe neurons during defensive and offensive encounters in the tree-shrew. Physiol. Behav., 28, 697-705.

Weinstock, M., Weiss, C. (1980). Antagonism by propranolol of isolation-induced aggression in mice: correlation with 5-hydroxytryptamine receptor blockade. Neuropharmacol., 19, 653-656.

Weischer, M-L. & Opitz, K. (1972). Einfluss von Fenfluramin, Chlorphentermin und verwandten Verbindungen auf das Verhalten von aggressiven Mausen. Arch. int. de Pharmacodynamie et de Therapie, 195, 252-259.

Welch, B.L. & Welch, A.S. (1968). Rapid modification of isolation-induced aggression behaviour and elevation of brain catecholamines and serotonin by the quick-acting monoamine-oxidase inhibitor pargyline. Comm. Behav. Biol., 1, 347-351.

Welch, B.L. & Welch, A.S. (1969). Aggression and the biogenic amino neurohumors. In: Aggressive Behavior. (Eds. Garattini, S. & Sigg, E.B.), pp. 188-202. Amsterdam: Excerpta Medica Foundation.

Willard, A.L. (1981). Effects of serotonin on the generation of the motor program for swimming by the medicinal leech. J. Neurosci., 1, 936-944.

Winslow, J.T. & Miczek, K.A. (1983). Habituation of aggression in mice: pharmacological evidence of catecholaminergic and serotonergic mediation. Psychopharmacol., 81, 286-291.

Winslow, J.T. & Miczek, K.A. (1984). Habituation of aggressive behaviour in mice: a parametric study. Aggr. Behav., 10, 103-113.

Yamamoto, T., Araki, H., Abe, Y. & Ueki, S. (1985). Effects of chronic LiCi and RbCl on muricide induced by midbrain raphe lesions in rats. Pharmacol. Biochem. Behav., 22, 559-563.

Yamamoto, T. & Ueki, S. (1977). Characteristics in aggressive behaviour induced by midbrain raphe lesions in rats. Physiol. Behav., 19, 105-110.

Yamamoto, T. & Ueki, S. (1978). Effects of drugs on hyperactivity and aggression induced by raphe lesions in rats. Pharmacol. Biochem. Behav., 9, 821-826.

Yen, C.Y., Stanger, R.L. & Millman, N. (1959). Ataractic suppression of isolation-induced aggressive behaviour. Arch. int. de Pharmocodynamie et de Therapie, 123, 179-185.

Yodyingyuad, U., de la Riva, C., Abbott, D.H., Herbert, J. & Keverne, E.B. (1985). Relationship between dominance hierarchy, cerebrospinal fluid levels of amine transmitter metabolites (5-hydroxyindole acetic acid and homovanillic acid) and plasma cortisol in monkeys. Neurosci., 16, 851-858.

DIFFERENTIAL EFFECTS OF BENZODIAZEPINES AND 5-HT$_{1A}$ AGONISTS ON DEFENSIVE PATTERNS IN WILD RATTUS

D.C.Blanchard, K.Hori, R.J.Rodgers[1], C.A.Hendrie[1], R.J.Blanchard.
Bekesy Laboratory of Neurobiology, University of Hawaii, U.S.A., 96822, and
[1]Pharmacoethology Laboratory, School of Psychology, University of Bradford, U.K. BD7 1DP.

Over a century ago, Darwin (1872) provided a firm basis for considering the response of lower animals to predators and other environmental threats as an essential precursor to human fear and anxiety reactions. Surprisingly, however, defensive behaviours have received relatively little attention as natural animal models for the investigation of fear and anxiety. In this context, ethological studies have identified a number of specific behaviours shown by feral animals to threatening stimuli, descriptions that have since been confirmed and extended for several species under controlled laboratory conditions. We have recently developed a battery of tests designed to elicit a wide range of active and passive defensive activities in wild rats (Blanchard et al., 1986a,b). These tests measure flight, freezing, boxing (defensive upright), vocalization (defensive threat) and jump attack responses to a number of nonpainful threat stimuli, including human approach and contact, dorsal contact, stimulation of the vibrissae, and an anesthetized conspecific. Although all of these reactions may be elicited in laboratory rats by painful stimulation, and some components occur in lab rats to nonpainful threat stimuli, laboratory rats appear to have been selected by breeders for a lack of defensive threat and attack and possibly flight, such that wild rats may be a more sensitive and appropriate subject for the study of changes in the defense pattern following administration of pharmacological compounds (Blanchard & Blanchard, 1987).

The Defense Battery has now been used to assess the effects of 3 benzodiazepines and 2 5-HT$_{1A}$ agonists on the defensive repertoire of wild R. rattus.

Methods
Adult male and female wild R. rattus, trapped on Oahu, served as subjects. These subjects has been individually housed in the laboratory for a minimum of 30 days prior to testing. Naive animals (N=20, with approximately equal properties of males and females) were used for each compound and, within each study, treatments were administered in counterbalanced order (ITI= 4 days). Drugs used were chlordiazepoxide (5-20mg/kg), midazolam (1-10mg/kg), diazepam (1-5mg/kg), buspirone (5-20mg/kg), and gepirone (5-20mg/kg). Injections were given i.p. 30 min prior to testing.

The Defense Test Battery provides measures of flight and avoidance of a human experimenter in an oval runway approximately 10 m in circumference, permitting unlimited forward motion: Line crossings in the same situation are also measured prior to the appearance of the experimenter.

Following flight testing a barrier is placed in the runway. Approaches to the subject by the experimenter in the resultant straight alley, 5 m long, provide measures of freezing as a function of the distance between the subject and the experimenter, while defensive threat (vocalization) and defensive attack, and, flight from the experimenter often occur as this distance decreases to zero. The subject is then placed in a small compartment and reactions to brush stimulation of the vibrassae, to an anesthetized conspecific brought up to the subject by hand and to attempted pickup by the experimenter are recorded. Data were analyzed by ANOVA or non-parametric tests as appropriate for scale of measurement and distribution of scores. Because of the large number of tests reported here, details of statistical tests are not given, and significance levels will be reported only in the table.

Results

The three benzodiazepines produced extremely similar behavioural profiles and only the results for diazepam will be presented here. The two $5-HT_{1A}$ agonists were again relatively similar in terms of behavioural effects, but were less so than the benzodiazepines, and both buspirone and gepirone results for selected measures are given in Table 11.1.

Diazepam, buspirone and gepirone generally failed to significantly reduce flight and avoidance in the circular runway, or freezing in the straight alley.

Defensive threat, defensive upright, jump attacks and bites were common for wild rat controls as the experimenter approached closely in the straight alley, as was flight past the experimenter: Diazepam at all doses profoundly reduced each element of the defensive threat/attack pattern, in this situation, simultaneously increasing flight. Buspirone and gepirone decreased defensive upright in the same situation with gepirone also reducing biting, and the 20 mg/kg dose of buspirone decreased flight as well.

Table 11.1 Defensive behaviour in wild rats given diazepam, buspirone, and gepirone

Drugs	Diazepam				Buspirone				Gepirone			
dose (mg/kg)	0.0	1.0	2.5	5.0	0.0	5.0	10.0	20.0	0.0	5.0	10.0	20.0
Initial movement	75.0	34.0*	34.4*	13.7*	42.9	70.0	74.9	57.7	54.7	65.9	84.0	108.3*
Vibrissae brush												
Defensive upright	3.8	3.8	3.9	3.8	3.2	3.2	2.6	2.5*	3.3	2.7*	2.7*	2.7
Defensive threat	2.1	1.7	0.9*	0.7*	2.3	1.8	1.6	1.8	2.3	1.6*	1.4*	1.2*
Biting	0.7	1.2	1.2	1.0	0.5	0.4	0.9	0.4	0.6	0.4	0.0	0.0
Conspecific												
Defensive upright	4.0	4.0	3.9	4.0	3.4	3.4	2.5*	3.2	3.6	2.9	2.6*	2.5*
Defensive threat	2.4	2.5	2.6	2.2	3.3	2.8	2.6	3.0	2.5	1.8*	1.6*	1.6*
Biting	3.8	3.7	3.9	3.7	3.3	2.8	2.8*	2.5*	2.9	2.2	1.7*	1.6*
Experimenter rating	3.4	3.5	3.3	2.8*	3.2	2.7	2.6	2.6*	3.4	2.5	2.3*	2.2*
Reactions to experimenter in straight alley												
% Boxing	75.0	25.0*	15.0*	20.0*	55.0	40.0	15.0*	35.0	85.0	70.0	45.0*	20.0*
% Biting	55.0	20.0*	10.0*	15.0*	40.0	15.0	20.0	25.0	45.0	25.0	20.0	5.0*
% Vocalization	35.0	5.0*	0.0*	0.0*	40.0	25.0	30.0	30.0	35.0	25.0	10.0	10.0
% Jump attacks	30.0	5.0	0.0*	0.0*	30.0	5.0	15.0	10.0	15.0	15.0	5.0	0.0
% Flight	25.0	65.0*	85.0*	75.0*	40.0	30.0	15.0	10.0*	35.0	30.0	20.0	20.0

* $p<.05$

In contrast, when the vibrissae brush or the anesthetized conspecific were used as threatening stimuli, in situations where flight from the stimulus was not possible, diazepam failed to produce a consistent reduction in defensive threat and attack, affecting only defensive threat to vibrissae stimulation. However, gepirone reduced defensive upright and defensive threat to both these stimuli, as well as biting to the conspecific. The buspirone results showed lesser reductions in the same set of reactions, but these often failed to reach an acceptable level of significance. Experimenter ratings of subjects' defensive reactions to pickup were reduced for diazepam, buspirone and gepirone at the higher doses given, with gepirone showing the greaters apparent effect. The numerous additional measures of defense which were taken failed to show any systematic effect of any of these compounds.

However, one additional measure, not intended to reflect defense, did show profound and systematic changes. Initial movement patterns in the oval runway prior to appearance of the human experimenter were extremely consistent within compound classes and extremely dissimilar for the benzodiazepines as opposed to the $5-HT_{1A}$ agonists: Movements decreased reliably as a function of increasing dose levels for diazepam and, increased reliably for gepirone and buspirone. These changes did not appear to be due to sedation, as avoidance of the approaching experimenter and flight speed, both in the same situation, remained high for all groups.

Discussion

Both benzodiazepines and $5-HT_{1A}$ agonists attenuated some aspects of defensive threat/attack in wild rats, while not altering such defensive behaviours as avoidance, and the speed of flight. However, at the doses given, diazepam produced this effect only in a situation in which flight was possible, and in competition with the defensive threat/attack behaviours, while the $5-HT_{1A}$ agonists, especially gepirone, produced a broader reduction in defensive threat and attack in situations where flight was not possible.

The finding that all these compounds reduced ratings of the subject defensiveness to pick up suggests the great importance of defensive threat/attack/biting in determining such ratings. The sensitivity of the Defense Battery to similarities and dissimilarities in the behavioural effects of anxiolytics from different chemical classes strongly suggests its potential utility in further refining our understanding of the neurohumoral substrated of defense, and, hence, fear and anxiety.

Acknowledgements: This Research was supported by NIH Grant RR0361-01A1.

References

Blanchard, R.J. & Blanchard, D.C. (1987). An ethoexperimental approach to the study of fear. Psychol. Rec., **37**, 305-316.

Blanchard, R.J., Blanchard, D.C., Flannely, K.J. & Hori, K.. (1986a). Ethanol changes patterns of defensive behaviour in wild rats. Physiol. Behav., **38**, 645-650.

Blanchard, R.J., Flannely, K.J. & Blanchard, D.C. (1986b). Defensive behaviours of laboratory and wild Rattus norvegicus. J. Comp. Psychol., 100, 101–107.

Darwin, C. (1872). The Expression of the Emotions in Man and Animals. London: John Murray.

COMPETITION FOR SUCROSE-PELLETS IN TRIADS OF MALE, WISTAR RATS: THE EFFECTS OF EIGHT 5-HT-AGONISTS

C. Gentsch, M.Lichtsteiner, H.Feer.
Psychiatric University Clinic Basle Biochem. Laboratory, CH-4025 Basel, Switzerland.

Within groups of rats rank-specific behavioural patterns have been discerned by either recording offensive and defensive postures and/or observing each individual's performance for a rewarding stimulus of limited availability (e.g. food or fluid or a sexual partner). Having recently described a stable intragroup rank-order when assessing each individual's tendency to partake in competition for sucrose-pellets (Gentsch, Lichtsteiner & Feer, 1988), we asked whether the competition-rate of so-called high- or poor-performing rats could be affected by pharmacological interventions. Among others, serotonergic drugs influence the competition-rates for sucrose-pellets as follows: by inhibiting 5-HT-synthesis in poor-performing rats these animals temporarily overcome their characteristic abstention from competition and, after administration of quipazine, a 5-HT-agonist, the high-performing rats' competition-rate is temporarily attenuated (Gentsch et al., unpublished). Such findings which are corroborating previous observations (Kostowski, Plewako &Bidzingski, 1984), indicate that the competition-rate might be related to the activity of the 5-HT-system.

Advances in in-vitro binding studies have identified several 5-HT-receptor subtypes (5-HT_{1A}, 5-HT_{1B}, 5-HT_{1C} and 5-HT_2). Since quipazine, the previously used 5-HT-agonist, is of only limited specificity, we attempted to further elucidate whether any of these receptor-subtypes is primarily involved in regulating the high-performing rats' competition-rate.

Material and methods
Groups of 3 male, adult Wistar rats (local strain (Fuellinsdorf Albino)) were housed and tested in macrolon cages (42x26x17) with a L12/D12 cycle (lights on 12.00). Cages were equipped with a table-like cylinder onto which sucrose-pellets (Precision Food Pellets, formula F; 20 mg (Noyes Company, Lancaster)) could be delivered. Food (Nafag No 890) and tap water were available ad libitum.

Our test-procedure can briefly be summarized as follows (for a more extensive description we refer to our previous report (Gentsch, Lichtsteiner & Feer, 1985), having ascertained that all rats of a given triad instantaneously consumed all presented sucrose-pellets, 30 pellets were manually delivered per triad and day (always between 08.00 and 09.00) at approximately 2 min. intervals. By recording three behavioural items "nearby" (being close to the pellet-dispenser), "compete" (trying to snatch the pellet) and "winner" (ingesting the pellet) and differentially weighting these (0.33, 2.0 and 1.0 points, respectively) for each rat an overall score, representing its tendency to partake in competition, was formed. Selfevidently, with the 30 pellets per triad and day the individual's maximal score is 100 per day. With such a scoring system in most triads a high-, a medium- and a poor-performing rat was discernable.

In the present experiments drugs were administered to high-performing rats of stable groups, only six competition-trials were performed per week (the test-free day, which normally preceeded drug-treatment by at least 2 days, did, thereby, not appreciably influence the acquired levels of competition).

Prior to drug-administration all rats were accustomed to the injection-procedure (i.p.; saline). Drugs were always freshly dissolved in either saline or distilled water and given 60 min. prior to the start of the competition-trial. The two other rats of a triad had vehicle-injections.

By taking the mean pre-drug score (mean of 3 days) as 100%, the drug-effects were expressed as percentages. In order to compare the effects of the diverse treatments, for all drugs those molar concentrations inducing a 50% reduction were graphically estimated.

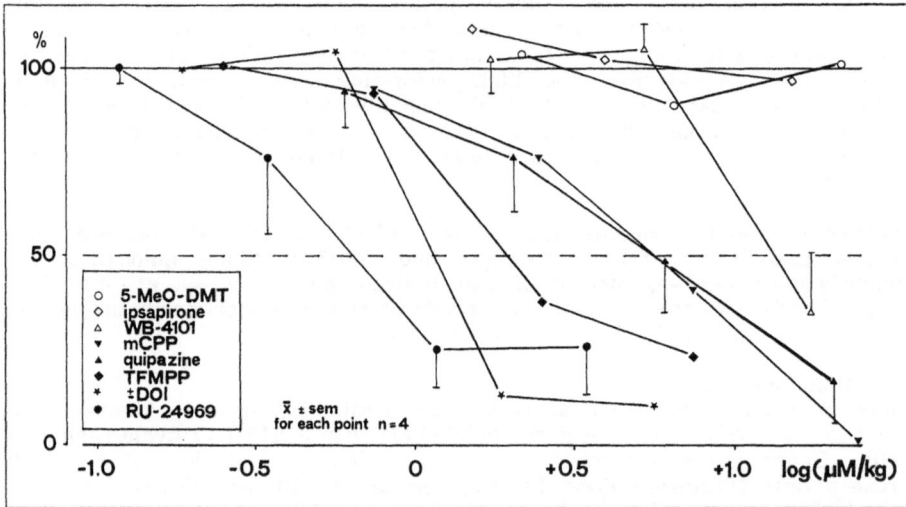

Fig. 12.1 Effects of 8 different 5-HT-agonists on the competition-rate in high-performing rats

Drug injections (i.p.) were given 60 min. prior to starting the competition-trial.

Each point represents the mean ± S.E.M. of 4 rats; for clarity, deviations are examplified for RU 24969, quipazine and WB-4101, only.
abscissa: drug-concentration (as log (µM/kg).
ordinate: overall score following drug-treatment, expressed as percentage of the mean pre-drug score (mean of the 3 preceeding days). 100% always corresponded to a score between 80 and (for details see Material and Methods).

Results
All groups of rats (N=4 per drug-concentration) had pre-drug mean scores between 80 and 90 and, these high levels were normally observable again on the second post-drug trial. 5-HT-agonists decreased (if at all) the scores of high-performing rats as indicated in figure 12.1.

For the drug-concentrations inducing a 50% reduction of the pre-drug performance-rate the following order of potency was obtained: RU24969 (0.6 µM/kg = 0.34 mg/kg) > DOI (1.0 µM/kg = 0.36 mg/kg) > TFMPP (2.0 µM/kg = 0.53 mg/kg) > quipazine (5.5 µM/kg = 1.81 mg/kg), mCPP (5.6 µM/kg = 1.51 mg/kg) > WB-4101 (14.9 µM/kg = 5.69 mg/kg) > Ipsapirone (>15.3 µM/kg = >6.7 mg/kg) 5-MeODMT (>21.8 µM/kg = >6.7 mg/kg). Linear regression analyses between these potencies (as log µM/kg) and the pIC_{50}-values for $5-HT_{1A}$-, $5-HT_{1B}$-, $5-HT_{1C}$- and $5-HT_2$-receptor subtypes, respectively (these values have been determined in in-vitro binding studies and put at our disposal by D.Hoyer (Sandoz Ltd., Basle) revealed the following non-significant correlation-coefficients: $5-HT_{1A}$ (r = + 0.533); $5-HT_{1B}$ (r = - 0.505); $5-HT_{1C}$ (r = - 0.424) and $5-HT_2$ (r = - 0.572). A significant correlation (r = + 0.768; 2p <0.05)) was found between (pIC_{50} ($5-HT_{1A}$)-pIC_{50} ($5-HT_{1B}$)) and the drugs' potencies to inhibit the high-performing rats' competition-rate.

Discussion
The present experiments revealed that some 5-HT-agonists (RU 24969 > DOI > TFMPP > quipazine, mCPP > WB-4101) temporarily attenuated the competition-rate in high-performing rats, whereas others (5-MeODMT and ipsapirone), at least in the concentrations used here, failed to have such an effect. Although one risks to underestimate a drug's full in-vivo action by just considering in-vitro pIC_{50}-values our findings would indicate that the behavioural inhibition in high-performing rats depends on the drug's action at the $5-HT_{1B}$- and is inverse to its specificity for the $5-HT_{1A}$-site.

Due to the limitation in space, we will concentrate for the rest of this discussion on the following aspects: Kennett, Dourish & Curzon (1987), as did other authors before, observed a clear anorexic activity of three $5-HT_{1B}$-agonists (RU 24969, TFMPP and mCPP). Such findings contrast the hyperphagia induced by $5-HT_{1A}$-agonists (e.g. ipsapirone (Wong & Reid, 1987)). At first glance our results seem to confirm such observations, since, here again, $5-HT_{1B}$-agonists reduced the consumption of sucrose-pellets, whereas the $5-HT_{1A}$-agonists ipsapirone and 5-MeODMT were ineffective (probably, our experimental set-up is less suitable to observe hyperphagia in high-performing rats due to the high, initial scores). However, with view to such data we re-examined the effects of RU 24969, TFMPP and quipazine as follows (unpublished data): when individually housed for the minimal time required, all our rats immediately start to consume sucrose-pellets (Gentsch, Lichtsteiner & Feer, 1988). We assessed the drug-effects in individually housed rats by injecting (i.p.; 60 min prior to deliver the pellets) RU 24969 (0.2 or 2.0 mg/kg), TFMPP (0.2 or 2.0 mg/kg) or quipazine (0.67 of 6.7 mg/kg). For the lower drug-concentrations (0.2, 0.2 and .67 mg/kg, respectively) none of the drugs affected the ingestion-rates and all rats immediately consumed all 15 sucrose-pellets, as did untreated controls. After 2.0, 2.0 or 6.7 mg/kg, respectively, the ingestion-rates tended to decrease in some cases (-33%, ±0%, -33%, respectively). However, all drug-concentrations were less effective as compared to their effects in the group-situation (RU 24969: -22% and -74%; TFMPP: -77% and -76%; quipazine: -24% and -83%, respectively).

Thus, the drug-induced inhibition, as observed in our high-performing rats with the group, seems not to be exclusively caused by the drug's anorexic activity. At least to some extent, it reflects the rat's attenuated tendency to partake in competition, too. With view to the differences between our own and Kennett's findings, two methodological differences should be considered. We always assess the ingestion-rate of highly palatable sucrose-pellets, whereas Kennett, Dourish & Curzon measured the intake of "normal food". Furthermore, our observations are carried out within the dark-phase, whereas Kennett's group tested in the light-phase (low food-intake). Our methodology (active animals and palatable food of limited availability) may render rats more eager and highly motivated to get the food.

In summary, within groups of rats, the high-performing animal's competition-rate is affected by 5-HT-agonist, depending on its 5-HT$_{1B}$-selectivity. Such findings corroborate previous observations that serotonergic drugs affect rank orders in rats, hinting at interindividual differences in 5-HT-mechanisms depending on (or causing) their performance in a competitive situation.

References

Gentsch, C., Lichtsteiner, M. & Feer, H. (1985). Competition for sucrose-pellets in triads of male Wistar rats: the individuals' performances are differing but stable. Behav. Brain Res. (In Press).

Kennett, G.A., Dourish, C.T. & Curzon, G. (1987). 5-HT$_{1B}$ agonists induce anorexia at a postsynaptic site. Eur. J. Pharmacol., **141**, 429-435.

Kostowski, W., Plewako, M. & Bidzinski, A. (1984). Brain serotonergic neurons: their role in a form of dominance-subordination in rats. Physiol. Behav., **33**, 365-371.

Wong, D.T. & Reid, L.R. (1987). Fenfluramine antagonizes the stimulation of food intake induced by the putative 5-hydroxytryptamine 1A agonist, ipsapirone, in non-fasted rats. J. Pharm. Pharmacol., **39**, 570-571.

DEPRESSION AND PSYCHOSIS

A.R. Cools
Psychoneuropharmacological Research–Unit, Dept. Pharmacology, University of Nijmegen, P.O. Box 9101, 6500 HB Nijmegen, The Netherlands

Today, it is generally accepted that serotonin anyhow plays a role in the pathophysiology of depression. Moreover, serotonin uptake inhibitors are known to be therapeutically effective as antidepressants. In contrast, there is no evidence that serotonin is directly involved in the aetiology of psychosis, and evidence that serotonergic agents have unequivocal antipsychotic effects, is also lacking. These facts underly the different strategies, assessed in the research on depression and serotonin and that of psychosis and serotonin, respectively. The most striking difference concerns the research on animal models. In the animal research on depression, attention is mainly focussed on animal models for depression (see contributions of Willner; Deakin; Kennet and Curzon; Martin et al.).

In contrast, the animal research on psychosis mainly deals with animal models for antipsychotics (see contribution of van der Heyden). A second difference concerns the attention given to the function of serotonin in the central nervous system as far as it concerns its role in processes giving rise to the behavioural disorders under discussion. Thus, there is a high amount of animal data allowing the creation of theoretical frameworks for the role of serotonin in processes underlying depression (see contributions of Willner and Deakin), whereas such data on the role of serotonin in processes underlying psychosis are very limited (see contribution of van der Heyden; of Geyer and of Barnes and his colleagues). A third difference between the strategies used in the research on depression and psychosis respectively concerns the attention given to the role of distinct subtypes of serotonin receptors. So, the research on depression and serotonin has already led to a hypothetical framework, in which the $5-HT_1$ and $5-HT_2$ receptors are postulated to play their own role in the pathophysiology and pharmacotherapy of depression (see contribution of Deakin), whereas the research on psychosis and serotonin is still limited to the question whether a particular subtype of serotonin receptors is mediating an antipsychotic effect.
Against this background, it is not amazing that extensive clinical studies on the efficacy of serotonergic agonists and/or antagonists in psychosis are not yet available. Nevertheless there are some promising clinical data opening the perspective that $5-HT_2$ antagonists such as ritanserin may be effective in the treatment of negative symptoms of schizophrenia (see contribution of van Meert and colleagues). Moreover, there are some intriguing animal data giving rise to the notion that $5-HT_3$ antagonists such as GR 38032F, ICS 205–930 and BRL 43694 as well as serotonergic agonists with a mixed profile such as befiperide, viz. a $5-HT_{1A}$ and $5-HT_2$ agonist, may have a certain therapeutic efficacy (see contribution of van der Heyden, and of Barnes and colleagues).

On the other hand, there are now at least three clinical studies indicating that $5-HT_2$ antagonists such as ritanserin may have a therapeutical value in the treatment of certain forms of depression (see contribution of van Meert and colleagues), whereas clinical studies on the putative antidepressant effects of

$5-HT_1$ agonists such as gepirone and buspirone are on their way (see contribution of Deakin).

In view of the above-mentioned considerations and findings, it becomes understandable why the first chapter of this section is devoted to the role of serotonin in the pathophysiology of serotonin in depression. In his chapter, Willner starts to review clinical data supporting the hypothesis that at least a serotonin dysfunctioning is involved in the pathophysiology of depression. Both clinical data and data collected in animal studies on serotonin, stress and social isolation bring him thereupon to the suggestion that a serotonin deficiency plays an important role in the pathogenesis of depression. Finally, he summarises evidence in favour of the fascinating theory that lack of adequate reinforcement due to a disturbed childhood social environment lowers the serotonin turnover with the consequence of losing control of impulsive behaviour. The latter, in turn, may produce maladaptive behaviour precipitating depression. Willner's approach provides a fresh look, challenging researchers to validate this testable hypothesis. The resulting nature of serotonin dysfunctioning in the brain of patients suffering from depression is discussed by Deakin in the next chapter.
By analising the function of distinct serotonergic fibre systems in regulating behaviour, by integrating available knowledge about the mechanism of antidepressants, and by providing new data about the function of $5-HT_1$ and $5-HT_2$ receptors, Deakin shows that the so-called serotonin excess theory of depression needs not to be in conflict with the so-called serotonin deficiency theory of depression. In fact, he develops the very attactive concept that excessive activitiy at the level of $5-HT_2$ receptors may give rise to excessive sensitivity to averise stimuli precipitating fear and anxiety which, in turn, may lead to depression, and that a deficient activity at the level of $5-HT_1$ receptors prevents both the display of adaptive-protective responses and the resilience to adversity with the consequence of learned helplessness and depression. In other words, Deakin elucidates how two seemingly opposite theories about the role of serotonin in depression can be reconciled. It is this concept that brings Deakin to the conclusion that both $5-HT_1$ agonists and $5-HT_2$ antagonists may have a therapeutic efficacy in patients suffering from anxiety and/or depression. Indeed, antidepressant effects of $5-HT_2$ antagonists such as ritanserin have already been seen in the clinic as mentioned by van Meert and his colleagues in their contribution. And, in line with Deakin's hypothesis, Kennet and Curzon illustrate in their contribution that the putatively antidepressant effects of $5-HT_1$ agonists such as buspirone, 8-OH-DPAT and ipsapirone are indeed predicted on the basis of the effects of these compounds in their animal model of depression, viz. lack of adaptation of behavioural and biochemical effects elicited by restraint stress in rats.

As mentioned above, the research on psychosis and serotonin is focussed on animal models for antipsychotics rather than on animal models for psychosis.
In the following chapter, van der Heyden discusses one of these animal models in detail, viz. the conditioned avoidance response in rats. First, he stresses that there is little evidence that serotonin is directly involved in schizophrenia. Then, he proceeds with presenting an overview of recently collected data about the ability of a great variety of serotonin agonists and antagonists to suppress the conditioned avoidance response.

IV: Depression and Psychosis   155

Van der Heyden shows that, apart from the 5-HT$_{1A}$ and 5-HT$_2$ agonist befiperide and the mixed 5-HT$_{1A/B}$ agonist eltoprazine, none of the tested 5-HT$_{1A}$, 5-HT$_2$ and 5-HT$_3$ agonists and/or antagonists are active in the conditioned avoidance paradigm. On the basis of the available data including the calculated correlations between behavioural activities and receptor affinities as revealed by binding studies, he thereupon reaches the conclusion that inhibition of conditioned avoidance response and hence possible antipsychotic action in man can be achieved through stimulation but not blockade of a particular subclass of serotonin receptors. The latter conclusion differs from that based on studies in which animal models of psychosis are used. In the startle habituation paradigm, viz. an animal model of schizophrenia with an impressive construct and face validity, 5-HT$_2$ antagonists such as ritanserin were found to be highly effective (see contribution of Geyer), suggesting that 5-HT$_2$ antagonists rather than 5-HT agonists may have an antipsychotic action. In another animal model mimicking a postulated behavioural correlate of psychosis, viz. the dopamine-induced hyperactivity elicited from the nucleus accumbens, 5-HT$_3$ antagonists such as GR 38032F, ICI 205-930 and BRL 43694 were found to be effective (see contribution of Barnes and his colleagues), suggesting again that 5-HT antagonists rather than 5-HT agonists may have an antipsychotic action. Recalling the concept of Deakin, the available data may be reconciled in case one assumes that, like in depression, there is a kind of "see-saw" between the function of distinct subtypes of serotonin receptors. From this point of view the available data open the perspectives that both a deficiency at the level 5-HT$_{1A}$ receptors and an excess at the level of 5-HT$_2$ and 5-HT$_3$ receptors can contribute to the pathophysiology schizophrenia. Anyhow, it is clear that we are just at the beginning of a promising exploration of the role of serotonin in the pathophysiology and pharmacotherapy of depression and schizophrenia.

The following chapters provide a firm foundation for setting up research devoted to answering challenging questions put forward by the various authors in their contributions.

TOWARDS A THEORY OF SEROTONERGIC DYSFUNCTION IN DEPRESSION

Paul Willner
Dept. of Psychology, City of London Polytechnic, Old Castle St., London E1 7NT, England

The minimal requirements for a theory relating brain dysfunction to psychopathology are that such a theory should adequately characterize the physiological disorder, and explain both its origin and its functional significance. In this paper, I have attempted to construct the outlines of a theory of serotonergic dysfunction in depression that meets these criteria.

Characterization
While it is widely accepted that serotonergic transmission is abnormal in at least a sub-group of depressed patients, the fundamental question of whether transmission through serotonergic synapses is reduced or increased remains largely unresolved. The following analysis distinguishes four categories of evidence bearing on this problem: Data that are largely irrelevant; data that may be argued either way; data that point to a serotonin deficiency, and data that clearly support the hyposerotonin hypothesis.

Irrelevant data
Numerous studies have measured aspects of peripheral 5-HT function in depression, including urinary concentrations of the 5-HT metabolite 5-HIAA, and various measures in blood platelets, including monoamine oxidase (MAO) activity, 5-HT uptake and ^{3}H-imipramine binding (see Willner, 1985a for review). However, there is no clear relationship between any peripheral measure and central 5-HT function. Hence, while platelets may prove to be a useful model for identifying biological markers of depression (Langer & Raisman, 1983; Stahl & Meltzer, 1978), peripheral studies are currently of no value for the assessment of central 5-HT dysfunction.

Much of the impetus for the hyperserotonin hypothesis of depression came from studies showing that some antidepressants have 5-HT antagonist properties in a supported animal model of depression: The suppression of operant behaviour by the 5-HT precursor 5-HTP (Aprison et al., 1982).

However, the actions of 5-HTP in this model are mediated peripherally: They are blocked by the decarboxylase inhibitor benserazide and the 5-HT antagonist BW501, neither of which cross the blood-brain barrier (Carter et al., 1978; Leander, 1986). While some antidepressants may also have central 5-HT antagonist properties, their primary effect, 5-HT uptake inhibition, plays a very minor role at peripheral 5-HT receptors. As with the platelet studies, therefore, evidence from the 5-HTP induced behavioural depression model does not bear directly on the question of how antidepressants alter central 5-HT function.

Ambiguous data: neurochemical changes in depression
Of far greater relevance to a dysfunction at central 5-HT synapses is the frequently reported reduction in cerebrospinal (CSF) concentrations of 5-HIAA. A high proportion – perhaps as much as 50% – of 5-HIAA measured in lumbar

CSF may arise from spinal sources (Banki & Molnar, 1981). Nevertheless, in animals, CFS values closely reflect changes in forebrain 5-HIAA levels (Mignot et al., 1985; Palfreyman et al., 1982), and in people, a strong correlation has been demonstrated between 5-HIAA levels measured post-mortem in lumber CSF and in cerebral cortex (Stanley et al., 1985). CSF 5-HIAA therefore provides a reasonably good estimate of forebrain 5-HT turnover.

Table 13.1: Studies of central 5-HT turnover in depression

Reduced	No change
Basal CSF 5-HIAA levels	
Agren, 1980b	Ashcroft et al., 1973
Asberg & Traskman, 1981	Banki & Arato, 1983
Ascroft & Sharman, 1960	Bowers, 1972
Ascroft et al., 1966	Bowers et al., 1969
Banki 1977a,b	Fotherby et al., 1963
Coppen et al., 1972b	Goodwin & Post, 1972
Dencker et al., 1966	Goodwin et al., 1973a
McLeod & McLeod, 1972	Jori et al., 1975
Mendels et al., 1972	Papeschi & McCLure, 1971
Traskman et al., 1981	Roos & Sjostrom, 1969
Van Praag & Korf 1971a	Sjostrom & Roos, 1972
Van Praag et al., 1970	Subramanyam, 1975
Van Praag et al., 1973	
Post-Probenecid 5-HIAA accumulation	
	Banki 1977b
Berger et al., 1980	
	Bowers, 1974a
Bowers, 1972	
Roos & Sjostrom, 1969	Goodwin et al., 1973a
Sjostrom, 1973	Jori et al., 1975
Sjostrom & Roos, 1972	
Van der Heyden et al., 1981	
Van Praag et al., 1970/1973	

See Willner (1985a) for references

Decreased CSF 5-HIAA in depressed patients, relative to age- and sex-matched controls, has been reported in around 50% of studies of basal 5-HIAA levels, and in a somewhat higher proportion of studies that have used probenecid pretreatment to block the transport of 5-HIAA out of the CSF. Table 13.1 summarizes many of the earlier studies; more recent studies continue to report either reduced levels of 5-HIAA in some patient groups (Asberg et al., 1984; Stokes et al., 1987; van Praag, 1982a), or no change (Gerner et al., 1984; Gjerris et al., 1987; Korf et al., 1983). Some studies have reported a bimodal distribution of 5-HIAA values (Asberg et al., 1976a); usually, however, distributions are unimodal, with depressed patients clustering towards the lower end of the normal range.

The interpretation of CSF 5-HIAA data is not entirely straightforward (Meltzer & Lowy, 1987), and there is also one report of increased 5-HIAA in depressed women, though not in men (Koslow et al., 1983). Nevertheless, there seem sufficient grounds here to conclude that forebrain 5-HT turnover is reduced in a substantial proportion of depressed patients. Considering that CSF 5-HIAA is influenced by, inter alia, age, sex, height, season and time of day; that spinal sources contribute to CSF 5-HIAA; and that only certain patients display the abnormality, it should not be surprising that differences between normals and depressed patients are somewhat unreliable. A post-mortem reduction of 5-HIAA in the frontal cortex of drug-free depressed patients has also been recently reported (Ferrier et al., 1986), though like the CSF literature, the post-mortem studies are also inconsistent (Cooper et al., 1986; Cross et al., 1983; Shaw et al., 1975), and methodologically fraught (Gottfries, 1980).

The inconsistencies in CSF studies largely disappear when attention is focussed on a specific subgroup of depressed patients: Those who have attempted suicide. Following the initial report of Asberg et al. (1976b), low CSF 5-HIAA levels in depressed suicide attempters have been confirmed in numerous studies, and in many countries (Table 13.2: Asberg et al., 1986b); Brown & Goodwin, 1986). For reasons at present obscure, low 5-HIAA may be less strongly related to suicidal behaviour in bipolar patients (Brown & Goodwin, 1986), but in unipolar patients the relationship between suicide and low 5-HIAA has held up in ten studies out of

Table 13.2 CSF 5-HIAA in (unipolar) depressed suicide attempters

Country	Low 5-HIAA?	Reference
Sweden	Yes	Asberg et al., 1976b; Traskman et al., 1981
	No	Vestergaard et al., 1978
	Yes	Agren, 1983
Netherlands	Yes	Van Praag, 1982a
Spain	Yes	Perez de los Cobos et al., 1984
Hungary	Yes	Banki t al., 1984
Britain	Yes	Montgomery & Montgomery, 1982
USA	Yes	Leckman et al., 1980
	Maybe	Roy-Byrne et al., 1983
India	Yes	Palannapian et al., 1983

See Asberg et al. (1986b), Brown & Goodwin (1986) for references.

twelve (Brown & Goodwin, 1986), with a tendency in the same direction in an eleventh (Roy-Byrne et al., 1983).

While the 5-HIAA data appear to provide prima facia evidence that serotonergic transmission in depression is decreased rather than increased, evidence also exists that a decrease in turnover may be offset by an increase in the sensitivity of postsynaptic 5-HT receptors. Post-mortem studies have reported a non-significant increase in the number of $5-HT_2$ receptors in depressed patients (Ferrier et al., 1986), and a significant increase in suicide victims (Stanley & Mann, 1983; Mann et al., 1986), though there are also negative reports, in both populations (Cooper et al., 1986; Crow et al., 1984).

There is also some functional evidence for 5-HT receptor supersensitivity in depression, in the form of an increased cortisol response to a 5-HTP challenge (Meltzer et al., 1984), though the 5-HT receptors mediating other hormonal responses do not appear to be supersensitive (see Meltzer & Lowy, 1987).

The observation of an inverse correlation between CSF 5-HIAA levels and the corticol response to 5-HTP (Koyama et al., 1987) suggests that if postsynaptic 5-HT receptors are indeed supersensitive in some depressed people, this effect if probably secondary to a reduction in 5-HT turnover. In this case, the net effect would probably still be a reduction in serotonergic transmission. However, this is far from certain, on the basis of the evidence reviewed.

Data pointing to decreased transmission: The effects of antidepressants
A similar uncertainty attends the interpretation of many of the effects of antidepressant drugs on central serotonergic systems. The classical picture, in which antidepressants enhance serotonergic transmission by inhibiting 5-HT uptake (Carlsson et al., 1969) has been challenged in recent years by the discovery that on chronic treatment, most antidepressant drugs 'downregulate' $5-HT_2$ receptors (Peroutka & Snyder, 1980; see Willner, 1985b for review). The net effect of these opposing changes is difficult to assess. Part of the problem is an uncertainty over the functional significance of a reduction in the number of $5-HT_2$ receptors: Receptor 'downregulation' is no guarantee of a decrease in the response to stimulation by receptor agonists. Behavioural responses to 5-HT receptor agonists following chronic antidepressant drug treatment are very variable. In general, receptors appear to be functionally subsensitive if animals are tested within a few hours of the final drug administration, but supersensitive at longer time intervals. There has been some discussion of which of these observations is the more relevant (Fuxe et al., 1983; Willner, 1985b), but the issue remains unresolved. It should also be noted that the majority of these studies have examined head twitching or the '5-HT syndrome', which reflect the activity of spinal rather than forebrain 5-HT systems (Jacobs & Klemfuss, 1975).

While some of the antidepressant data are difficult to interpret, two phenomena are not. Unlike antidepressant drugs, electroconvulsive shock (ECS) increases the number of $5-HT_2$ receptors, and also clearly increases behavioural responses to 5-HT agonists (Green et al., 1986). And in electrophysiological studies, both ECS and antidepressant drugs clearly and reliably increase transmission through forebrain 5-HT synapses. After chronic administration, tricyclics and various atypical antidepressants, including iprindole and mianserin, increase postsynaptic responses to iontophoretic application of 5-HT or the 5-HT agonists 5-MeODMT. The specific 5-HT uptake inhibitors do not sensitize postsynaptic receptors, but do desensitize presynaptic inhibitory autoreceptors, causing a gradual increase in the firing rate of 5-HT neurons and the release of 5-HT (reviewed by Willner, 1987). As a result of these diverse actions, drugs in both classes are found to increase the postsynaptic reponse to electrical stimulation of 5-HT pathways (Table 13.2: Blier & De Montigny, 1983; Chaput et al., 1986 Wang & Aghajanian, 1980). The response to electrical stimulation was also increased by the MAOIs amiflamine, clorgyline and phenelzine (Blier et al., 1986a,b), despite the fact that these drugs actually desensitized postsynaptic responses to 5-HT (Blier et al., 1986a,b; Olpe & Schellenberg, 1981). (These studies illustrates the importance of not over-interpreting the response to receptor agonists.)

Clear evidence for decreased transmission.
(i) Symptomatology
Perhaps the most striking evidence that depression is associated with serotonergic underactivity is the symptomatic similarity between the low 5-HIAA subgroup of depressed patients and animals bearing lesions to the ascending 5-HT pathways. 5-HT lesioned animals are typically hyperactive and aggressive (Albert & Walsh, 1982; Valzelli, 1984). Those depressed patients who have low CSF 5-HIAA levels are characterized by high levels of anxiety and agitation (Banki et al., 1981; Leckman et al., 1980; Traskman-Bendz et al., 1986) and by aggressive behaviour: Low 5-HIAA is found not only in suicide attemptors, (Asberg et al., 1986b; van Praag et al., 1986), but also in murderers (Linnoila et al., 1983) and other violent offenders (Brown & Goodwin, 1986). Excessive serotonergic transmission would have exactly the opposite effect of suppressing aggressive behaviour (Albert & Walsh, 1982; Valzelli, 1984).

(ii) Effects of acute 5-HT manipulations on mood
The effects of acute 5-HT manipulations on mood provide further support for the hypothesis that depression is associated with 5-HT deficiency rather that 5-HT excess. Two important studies have recently demonstrated that normal subjects fed a tryptophan-free diet experienced a transient lowering of mood at a time corresponding to the maximal depletion of central 5-HT; significantly, the mood change could not be over-ridden by instructing the subjects to attribute any unpleasant feelings to their amino acid meal, or by supplying a positive, supporting environment (Smith et al., 1987; Young et al., 1985). Conversely, a rapid antidepressant effect has recently been observed three hours following an acute dose of fenfluramine (Ward et al., 1985), and mood elevations in normal subjects have been reported following the administration of tryptophan (Charney et al., 1982) or 5-HTP (Puhringe et al., 1976; Trimble et al., 1975).

(iii) 5-HT precursors as antidepressants
5-HT precursors also possess some limited clinical antidepressant activity. Early studies using the 5-HT precursor tryptophan were equivocal (Cole et al., 1980; van Praag, 1982b). However, it is reasonably well established that tryptophan potentiates the antidepressant action of monoamine oxidase inhibitors (MAOIs), though not of tricyclic antidepressants (Gelenberg et al., 1982; van Praag, 1982n), and a recent large outpatient study found tryptophan alone to be superior to placebo and as effective as amitriptyline (Thompson et al., 1982). With very few exceptions, controlled studies have supported an antidepressant action of the

more proximate 5-HT precursor, 5-HTP, either alone (Angst et al., 1977; van Praag, 1982b) or in combination with an MAOI (Mendlewicz & Youdim, 1980), a tricyclic (van Praag et al., 1974) or tryptophan (Quadbeck et al., 1984). While it is not certain that 5-HT precursors act by enhancing central 5-HT transmission, a peripheral action is excluded by the use in some studies of a peripheral decarboxylase inhibitor (van Praag et al., 1974). A further possibility, that 5-HTP acts by impairing catecholaminergic transmission, is effectively ruled out by the observation that the antidepressant effect of 5-HTP was potentiated by tyrosine (van Praag, 1983a).

Nevertheless, it still remains to be demonstrated that the antidepressant activity of 5-HT precursors does result from an enhancement of the release of 5-HT. Furthermore, a substantial minority of patients treated with 5-HTP relapsed while still under treatment (van Praag, 1983a).

(iv) 5-HT uptake inhibitors as antidepressants
In contrast to the questionable efficacy of 5-HT precursors, the efficacy of selective 5-HT uptake inhibitors is now well established: A long list of 5-HT uptake inhibitors shown to be effective in controlled studies includes alaproclate, cianopramine, citalopram, femoxetine, fluoxetine, fluvoxamine, indalipine, paroxetine and zimelidine (Asberg et al., 1986a). Again, the problem lies in establishing that 5-HT uptake inhibition is actually the mechanism of action of these drugs. In fact, the majority of supposedly selective 5-HT uptake inhibitors are metabolized in vivo to agents that are relatively potent as noradrenaline (NA) uptake inhibitors. However, there are exceptions. Citalopram and fluoxetine, for example, are effective antidepressants (Asberg et al., 1986a) which selectively inhibit 5-HT uptake in vitro and also in vivo (Hytell et al., 1984; Mishra et al., 1979).

Unlike tricyclic and many other antidepressants, citalopram and fluoxetine do not appear to 'downregulate' $5-HT_2$ receptors either (Hytell et al., 1984; Peroutka & Snyder, 1980). As these drugs are not 5-HT antagonists, the clear implication that they enhance serotonergic transmission is marred only by the fact that on acute administration, 5-HT uptake inhibitors reduce the firing rate of 5-HT neurons (Quineaux et al., 1982). However, following chronic treatment, firing rate returns to normal as a result of autoreceptor desensitization (Chaput et al., 1986). An overall enhancement of serotonergic transmission is further indicated by the fact that the behavioural and endocrinological effects of 5-HT uptake inhibitors are in general similar to those of 5-HT precursors and receptor agonists (eg. Meltzer et al., 1981; Ogren et al., 1982; Stark et al., 1985).

(v) Reversal of antidepressant action
Finally, it can be demonstrated that in some circumstances, an increase in serotonergic activity mediates the therapeutic actions of antidepressants. A small minority of animal models of depression display symptoms (hyperactivity and aggression) characteristic of 5-HT depleted animals (Willner, 1984). Of these, the most widely studied is the olfactory bulbectomized (OB) rat (Cairncross et al., 1979; Jesberger & Richardson, 1986). In addition to their behavioural deficits, these animals show a number of other abnormalities charactistic of depression, including elevated levels of circulating corticosteroids (Cairncross et al., 1979; Jesberger & Richardson, 1986) and a

reduction in platelet ^{3}H-5HT uptake (Leonard, 1987). The behaviour of OB animals is normalized by tricyclic or atypical antidepressants, but not by drugs of other classes (Cairncross et al., 1979; Jesberger & Richardson, 1986). The observation that acute treatment with 5-HT uptake inhibitors is sufficient to normalize behaviour in OB animals, in contrast to the usual requirement for chronic treatment with most other antidepressants, gives an indication that the syndrome results from a reduced level of 5-HT function (Broekkamp et al., 1980; Joly & Sanger, 1986).

The implication that antidepressants act in this model by potentiating serotonergic transmission is confirmed by the demonstration that the therapeutic effects of subchronic treatment with imipramine of mianserin were abolished by acute administration of the 5-HT receptor antagonist metergoline (Broekkamp et al., 1980).

In fact, this technique has also been used clinically: The antidepressant effects of imipramine or the MAOI tranylcypromine were abolished, in a small number of patients, by acute administration of the 5-HT synthesis inhibitor para-chlorophenylalanine (PCPA) (Shopsin et al., 1975, 1976). If the present analysis of the case for a serotonergic hypofunction in depression remains unconvincing, the issue could very readily be resolved conclusively by the repetition of these studies using a 'cleaner' 5-HT antagonist.

Origin
If there is a serotonergic deficiency in depression, what is its origin? The animal literature provides two potential answers to this question, which will now be evaluated.

Stress
Exposure to severe stress is known to cause an increase in 5-HT turnover (eg. Weiss et al., 1981), but if stress is prolonged this leads to a depletion of brain 5-HT and turnover is consequently reduced (Anisman & Zacharko, 1982). A particularly interesting study demonstrated that forebrain 5-HIAA levels were decreased in female rats that failed to adapt to repeated restraint stress, while in male rats, successful adaptation was accompanied by an increase in the responsiveness of postsynaptic 5-HT receptors (Kennett et al., 1986).

Because stress has long been implicated in the etiology of depression (Anisman & Zacharko, 1982; Willner, 1985), these observations clearly provide a potential approach to understanding the origin of a 5-HT deficiency. Unfortunately, however, they are not easily reconciled with the clinical data. There are three problems. The first is the lack of symptomatic correspondence between stressed animals and the 'low 5-HT' group of depressed patients. As noted above, low 5-HT turnover in depressed patients is associated with increases in activity and aggression, but stressed animals show substantial reductions in locomotor activity (Anisman & Zacharko, 1982; Weiss et al., 1981) and are also hypoaggressive (Maier et al., 1972).

The second difficulty for the view that stress is responsible for low CSF 5-HIAA in depression comes from an important clinical study, which demonstrated that CSF 5-HIAA was normal in depressed patients who had experienced a recent stressful life event, but low in patients who had not experienced a recent life event (Roy et al., 1986). These observations suggest strongly that stress may actually raise 5-HT turnover in depressed patients.

The third problem is that serotonergic hypofunction appears to be a trait marker of depressive individuals, rather than a marker of the depressed state. If CSF levels of 5-HIAA are reduced in depression, they remain abnormally low after recovery in a substantial proportion of patients (Dencker et al., 1966; Goodwin et al., 1973; Mendels et al., 1972; Sjostrom & Roos, 1972; Traskman-Bendz, 1983; Traskman-Bendz et al., 1984; van Praag & de Haan, 1979).

Furthermore, patients with abnormally low CSF 5-HIAA levels are found to have a higher frequency of depressive episodes (van Praag & de Haan, 1979), as well as a family history of depression (Sedvall et al., 1980; van Praag & de Haan, 1979; Traskman-Bendz, 1983). Finally, with only a single exception (Ashcroft et al., 1966), if low CSF 5-HIAA was observed in depressed patients, the same abnormality was also present in mania (eg. Dencker et al., 1966; Mendels et al., 1972; Sjostrom & Roos, 1972). CSF 5-HIAA therefore differs, in its lack of relationship to the state of depression, from catecholamine markers such as CSF HVA or urinary MHPG, which are typically low in depression but high in the manic phase (see Willner 1985).

Social isolation
A second behavioural manipulation known to reduce 5-HT turnover is social isolation (Garattini et al., 1967; Segal et al., 1973; Yanai & Sze, 1983). Prolonged social isolation causes hyperactivity (Garzon & del Rio, 1981) and aggression (Garattini et al., 1967); Valzelli, 1984), both of these phenomena being reversible by antidepressant drugs (Garzon & del Rio, 1981; Sofia, 1969). Unlike stress, therefore, social isolation gives rise to symptoms reminiscent of those seen following the destruction of forebrain 5-HT pathways, and compatible with the symptoms of the 'low 5-HT' group of depressed patients. Furthermore, the neurochemical changes consequent on social isolation are long lasting, unlike the neurochemical sequelae of stress, which normalize spontaneously within a few days (Anisman & Zacharko, 1982; Willner, 1985).

Social isolation has not itself been implicated in the etiology of depression, except in the special case of anaclitic depression in infants (Schulterbrandt & Raskin, 1977). However, it might not be too far fetched to suppose that social isolation in animals may model the failure of socialization that can result from adverse childhood circumstances that predispose to depression, such as family discord or the loss of a loved one in childhood (Rutter & Madge, 1976; Brown & Harris, 1978; Lloyd, 1980). In fact, low CSF 5-HIAA levels are associated with low scores on socialization scales (Traskman-Bendz et al., 1986), and with a history of behavioural problems in childhood (Traskman-Bendz, 1983; Brown et al., 1986).

Poor socialization leads, in adult life, to inferior social skills and an absence of good social support. Both of these factors increase the risk of depression (Aneshensel & Stone, 1982; Brown & Harris, 1978; Lewinsohn, 1974); they also tend to be self-perpetuating, since people with poor social skills act in ways that alienate friends and drive away potential supporters (Lewinsohn, 1974).

The notion that isolation-induced 5-HT depletion may serve as an animal model for some of the social origins of depression receives support from studies of social cooperation in pairs of rats (Schuster et al., 1982). Cooperative behaviour was disrupted by single housing, or by the destruction of forebrain 5-HT pathways, but was restored by administration of the 5-HT agonist fluprazine; the demonstration of a direct correlation between 5-HT function and performance is remarkable, given the inverse relationship in other operant tasks (Berger & Schuster, 1987). As performance in the cooperation task depends upon the perception and utilization of cues emanating from the partner (Schuster et al., 1982), these results suggest that the crucial role of 5-HT may be to increase sensitivity to social cues.

Functional significance

If, as suggested, 5-HT deficiency is a biological correlate of poor socialization, what is its relationship to depression? Faulty socialization gives rise not only to depression, but also to a range of personality and other psychiatric disorders (Robins, 1966; Vaillant, 1975). However, 5-HT hypofunction is not specific to depression either. A shift of emphasis from depression to suicide followed the demonstration of a far stronger relationship of low 5-HIAA with suicidal behaviour than with depression; subsequently, the observation of low CSF 5-HIAA in murderers and other violent offenders led to a further widening of focus from suicide to aggression (see above). Of particular importance for this reappraisal are studies demonstrating low CSF 5-HIAA in non-depressed suicidal (Traskman et al., 1981; van Praag, 1983b) or violent (Brown & Goodwin, 1986; Linnoila et al., 1983) subjects. Evidence is now mounting that low 5-HT turnover is not directly related to aggression as such, but rather, to a lack of control over impulsive behaviour.

Table 13.3 Effects of chronic antidepressant treatment on postsynaptic responses at 5-HT synapses [a]

Drug	Response to 5-HT	Response to Raphe stimulation	Ref[b]
Tricyclics (imipramine, DMI)	Increased	Increased	1
5-HT uptake inhibitors (zimilidine, citalopram)	No change	Increased	2,3
MAO-A inhibitors (amiflamino, clorgyline)	Decreased	Increased	4,5

a) Only includes studies that tested with Raphe stimulation.

b) 1. Wang & Aghajanian (1980); 2. Blier & de Montigny (1983); 3. Chaput et al. (1986); 4,5. Blier et al. (1986a,b).

Two studies have found low CSF 5-HIAA to be associated with impulsive, but not with premediated acts of violence (Lidberg et al., 1985; Linnoila et al., 1983). There appears also to be a significant negative correlation between 5-HIAA and impulsivity, as measured by personality scaling techniques, in patient groups, suicide attempters and normal volunteers (Traskman-Bendz et al., 1986). Unlike suicide and aggression, a failure to exercise control over behaviour may be assumed to be a direct consequence of a failure of socialization (Arieti, 1967).

Low CSF 5-HIAA has also been reported in two other disorders characterized by poor impulse control, alcoholism and bulimia. In alcoholics, CSF 5-HIAA was normal whilst intoxicated, but decreased following a period of drying out (Ballenger et al., 1979). Similarly, post-probenecid 5-HIAA accumulation was low in weight-recovered anorectic patients who had a history of binge eating; a link between 5-HIAA and impulsivity is strongly supported by the finding that 5-HT turnover was normal in non-bulimic patients (Kaye et al., 1984). It will be of great interest to know whether 5-HT turnover is low in other disorders of impulse control, such as obsessive compulsive disorders or obesity, both of which may be successfully treated with 5-HT uptake inhibitors (Insel & Zohar, 1987; Stewart et al., 1986).

If low 5-HT turnover is reversible for a spectrum of psychiatric disorders, what determines which of them is manifest in a particular patients at a particular patient at a particular time? Although we are at present unable to supply an adequate answer to this questions, it is not too difficult to see what shape the eventual answer will take. Impulsive behaviours are, by definition, poorly thought out, and while they may lead to short-term gratification, they are very likely to be maladaptive in the long term, leading repeatedly to failure, demoralization, loss of self esteem and feelings of worthlessness. These consequences would usually precipitate a depressive episode (Becker, 1979). However, in response to repeated social failures, the outcome might be either a depression characterized by intense social dependence, or sociopathy, depending on the extent to which other people are a major source of gratification (see Beck, 1983). Similarly, situations that might provoke guilt in individuals with a strong conscience might lead instead to psychopathic behaviour if the conscience is weak (Klein, 1974; Prosen et al., 1983). The important point to note is that the explanation of what happens to people who are over-impulsive will largely be found outside of neuropharmacology, in an analysis of the social and psychological consequences of their behaviour.

Towards a theory
We are now in a position to construct a tentative theory of serotonergic dysfunction in depression. The central assumption is that the activity of forebrain 5-HT systems is controlled, inter alia, by social reinforcement. 5-HT turnover therefore decreases if the childhood social environment fails to provide adequate reinforcement. Such children do not learn self-control, and develop into impulsive, poorly socialized adults; they become depressed if the consequences of their impulsive behaviour damage their self-esteem. 5-HT turnover remains low because their social skills are poor and their social support inadequate. Antidepressants act in part by enhancing serotonergic transmission, so increasing sensitivity to social cues and mimicking an increase in social reinforcement. This theory gives rise to a large number of testable predictions,

Table 13.4 A theory serotonergic dysfunction in depression[a]
----
1. 5-HT turnover decreases if the childhood social environment fails to provide adequate reinforcement.

   <u>Predictions</u>: 1. Low CSF 5-HIAA in isolated or withdrawn children
   2. Low 5-HIAA related to childhood loss, unless compensated

2. Inadequately reinforced children do not learn self-control; they develop into impulsive poorly socialized adults

   <u>Predictions</u>: 1. Low CSF 5-HIAA in poorly socialized adults (*) and children
   2. Low 5-HIAA in disorders of impulse control (*)
   3. 5-HT uptake inhibitors effective in disorders of control (*)

3. Impulsive behaviour causes depression by damaging self-esteem

   <u>Predictions</u>: 1. Depression in impulsive people preceded by perceived failure
   2. Other forms of psychiatric disorder seen in pathologically impulsive individuals with effective coping strategies

4. 5-HT turnover remains low because social skills are poor and social support is weak

   <u>Predictions</u>: 1. Low CSF 5-HIAA related to poor social skills and absence of social support
   2. Low 5-HIAA remains low after recovery from depression unless social situation improves
   3. Elevation of low 5-HIAA by successfull social skills training

5. Antidepressants act in part by enhancing 5-HT transmission

   <u>Predictions</u>: 1. Reversal of clinical antidepressant action in low 5-HIAA sub-group by acute administration of 5-HT antagonists (*)
   2. Reversal of antidepressant action by 5-HT antagonists in animal models based on 5-HT lesion (*) or social isolation

6. Enhancement of 5-HT transmission enhances sensitivity to social cues

   <u>Predictions</u>: 1. Facilitation of social skills training by 5-HT uptake inhibitors.
   2. Better antidepressant response to 5-HT uptake inhibitors if socially unskilled
   3. Facilitation of performance in animal models of social cooperation by 5-HT agonists (*) and uptake inhibitors and by chronic antidepressant treatment.
----
a) Starred predictions have been partially confirmed (see text).

some of which are outlined in Table 13.4.

It is recognized that even if the predictions were fully confirmed, this account of serotonergic dysfunction in depression would remain incomplete in a number of respects: It includes no analysis, for example, of the underlying anatomy. The present approach does, however, place the problem of the role of 5-HT in depression into a broad psychobiological context; it is presented here in that spirit.

References

Albert, D.J. & Walsh, M.L. (1982). The inhibitory modulation of agonistic behaviour in the rat brain: a review. Neurosci. Biobehav. Rev., 6, 125-143.

Angst, J., Woggon, B. & Schoepf, H. (1977). The treatment of depression with 1-5-hydroxytryptophan versus imipramine. Arch. Psychiatr. Nervenkr., 224, 175-186.

Aneshensel, C.S. & Stone, J.D. (1982). Stress and depression: A test of the buffering model of social support. Arch. Gen. Psychiatr., 39, 1392-1396.

Anisman, H. & Zacharko, R.M. (1982). Depression: the predisposing influence of stress. Behav. Brain Sci., 5, 89-137.

Aprison, M.H., Hintgen, J.M. & Nagayama, H. (1982). Testing a new theory of depression with an animal model: Neurochem.-behav. evidence for postsynaptic serotonergic receptor involvement. In: New Vistas in Depression. (Eds. Langer, S.Z., Takahashi, R., Segawa, T. & Briley, M.). pp. 171-178. New York: Perganmon.

Arieti, S. (1978). The Intrapsychic Self. Basic Books, New York.

Asberg, M., Thoren, P., Traskman, L. (1976a). 'Serotonin depression": A biochemical subgroup within the depressive disorders? Sci. 191, 478-480.

Asberg, M., Traskman, L. & Thoren, P. (1976b). 5-HIAA in the cerebropsinal fluid: A biochemical suicide predictor? Arch. Gen. Psychiatr., 33, 1193-1197.

Asberg, M., Bertilson, L., Martensson, B., Scalia-Tomba, G.-P., Thoren, P. & Traskman-Bendz, L. (1984). CFS monoamine metabolites in melancholia. Acta Psychiatr. Scand., 69, 201-219.

Asberg, M., Eriksson, B., Martensson, B., Traskman-Bendz, L. & Wagner, A. (1986a). Therapeutic effects of serotonin uptake inhibitors in depression. J. Clin. Psychiatr., 47, (4, Suppl.), 23-35.

Asberg, M., Nordstrom, P. & Traskman-Bendz, L. (1986b). Cerebrospinal fluid studies in suicide: An overview. Ann. N.Y. Acad. Sci., 487, 243-255.

Ascroft, G.W., Crawford, T.B.B., Eccleston, D., Sharman, D.F., McDougall, E.J., Stanton, J.B. & Binns, J.K. (1966). 5-Hydroxyindole compounds in the cerebrospinal fluid of patients with psychiatric or neurological disease. Lancet, 2, 1049-1052.

Ballenger, J.C., Goodwin, F.K., Major, L.F. & Brown, G.C. (1979). Alcohol and central serotonin metabolism in man. Arch. Gen. Psychiatr., 36, 224-227.

Banki, C.M. & Molnar, G. (1981). Cerebrospinal fluid 5-hydroxyindoleacetic acid as an index of central serotonergic processes. Psychiatr. Res., 5, 23-32.

Banki, C.M., Molnar, G. & Vojnik, M. (1981). Cerebrospinal fluid amine metabolites, tryptophan and clinical parameters in depression: II. Psychopathological symptoms. J. Affect. Disord., 3, 91-99.

Beck, A.T. (1983). Cognitive therapy of depression: New perspectives. In: Treatment of Depression: Old controversies and new approaches. (Eds. Clayton, P.J. & Barrett, J.E.), pp. 265-290. New York: Raven.

Becker, J. (1979). Vulnerable self-esteem as a predisposing factor in depressive disorders. In: The psychobiology of the depressive disorders: Implications for the effects of stress. (Ed. Depue, R.A.), pp. 317-334. New York: Academic.

Berger, B.D. & Schuster, R. (1987). Pharmacological effects of social cooperation. In: Ethopharmacology of agonistic behaviour in animals and humans. (Eds. Olivier, B., Mos, J. & Brain, P.). Dordrecht, Martinus Nijhoff. (In Press.)

Blier, P. & De Montigny, C. (1983). Electrophysiological invesatigations on the effects of repeated zimelidine administration on serotonergic neurotransmission. J. Neurosci., 3, 1270-1278.

Blier, P., De Montigny, C. & Azzaro, A.J. (1986a). Effect of repeated amiflamine administration on serotonergic and noradrenergic transmission: Electrophysiological studies in the rat central nervous system. Naunyn-Schmiedebergs Arch. Pharmacol., 334, 253-260.

Blier, P., De Montigny, C. & Azzaro, A.J. (1986b). Modification of serotonergic and noradrenergic neurotransmissions by repeated administration of monoamine-oxidase inhibitors: Electrophysiological studies in the rat central nervous system. J. Pharmacol. Exp. Ther., 237, 987-994.

Broekkamp, C.L., Garrigou, D. & Lloyd, K.G. (1980). Serotonin-mimetic and antidepressant drugs on passive avoidance learning by olfactory bulbectimized rats. Pharmacol. Biochem. Behav., 13, 643-646.

Brown, G.L. & Goodwin, F.K. (1986). Cerebropsinal fluid correlates of suicide attempts and aggression. Ann. N.Y. Acad. Sci., 487, 175-188.

Brown, G.L., Kline, W.J., Goyer, P.F., Minichiello, M.D., Kreusi, M.J.P. & Goodwin, F.K. (1986). Relationship of childhood characteristics to cerebrospinal fluid 5-hy-droxyindoleacetic acid in aggressive adults. In: Biological Psychiatry. (Eds. Shagass, C., Josiassen, R.C., Bridger, W.H., Weiss, K.J., Stoff, D. & Simpson, G.M.), pp. 177-179. New York: Elsevier 1985, 7.

Brown, G.W. & Harris, T. (1978). Social origins of depression. London: Tavisotock.

Cairncross, K.D., Cox, B., Forster, C. & Wren, A.F. (1979). Olfactory projection systems, drugs and behaviour: a review. Psychoneuroendocrinology, 4, 253-272.

Carlsson, A., Corrodi, H., Fuxe, K. & Hokfelt, T. (1969). Effect of antidepressant drugs on the depletion of intraneuronal brain 5-hydroxytryptamine stores caused by 4-methyl-ethyl-meta-tyramine. Eur. J. Pharmacol., 5, 357-366.

Carter, R.B., Dykstra, L.A., Leander, J.D. & Appel, J.B. (1978). Role of peripheral mechanisms in the behavioural effects of 5-hydroxytryptophan. Pharmacol. Biochem. Behav., 9, 249-253.

Chaput, Y., De Montigny, C. & Blier, P. (1986). Effects of a selective 5-HT reuptake blocker, citalopram, on the sensitivity of 5-HT autoreceptors: Electrophysiological studies in the rat brain. Naunyn-Schmiedebergs Arch. Pharmacol., 333, 342-348.

Charney, D.S., Heninger, G.R., Reinhard, J.F., Sternberg, D. & Hafstad, K. (1982). The effect of intravenous L-tryptophan on prolactin and growth hormone and mood in healthy subjects. Psychopharmacology, 78, 217-222.

Cole, J.O., Hartmann, E. & Brigham, P. (1980). L-tryptophan: Clinical studies. McLean Hosp. J., 5, 37-71.

Cooper, S.J. Owen, F., Chambers, D.R., Crow, T.J., Johnston, J. & Poulter, M. (1986). Post-mortem neurochemical findings in suicide victims and depression: A study of the serotonergic system and imipramine binding in suicide victims. In: The biology of depression. (Ed. Deakin, J.F.W.), Royal college of psychiatrists, pp. 53-71. London: Gaskell Press.

Cross, A.J., Crow, T.J., Jnston, J.A., Perry, E.K., Perry, R.H., Blessed, G. & Tomlinson, B.E. (1983). Monoamine metabolism in senile dementia of Alzheimer type. J. Neurol. Sci., 60, 383-392.

Crow, T.J., Cross, A.J., Cooper, S.J., Deakin, J.F.W., Ferrier, I.N., Johnson, J.A., Joseph, M.H., Owen, F., Poulter, M., Lofthouse, R., Corsellis, J.A.N., Chambers, D.R., Blessed, G., Perry, E.K., Perry, R.H. & Tomlinson, B.E. (1984). Neurotransmitter receptors and monoamine metabolites in the brians of patients with Alzheimer-type dementia and depression, and suicides. Neuropharmacology, 23, 1561-1569.

Dencker, S.J., Malm, V., Roos, B.-E. & Werdinius, B. (1966). Acid monoamine metabolites of cerebrospinal fluid in mental depression and mania. J. Neurochem., 13, 1545-1548.

Ferrier, I.N., McKeith, I.G., Cross, A.J., Perry, E.K., Candy, J.M. & Perry, R.H. (1986). Postmortem neurochemical studies in depression. Ann. N.Y. Acad. Sci., 487, 128-142.

Fuxe, K., Ogren, S.-O., Agnati, L.F., Benfenati, F., Cavicchioli, L., Fredholm,

B., Andersson, K., Farabegoli, C. & Eneroth, P. (1983). Regional variations in 5-HT receptor populations and in ^{3}H-imipramine binding sites in their responses to chronic antidepressant treatment. In: Frontiers in Neuropsychiatric Research. (Eds. Usdin, E., Goldstein, M., Friedhoff A.J. & Georgotas, A.), pp. 33-54. London: MacMillan.

Garattini, S., Giacalone, E. & Valzelli, L. (1967). Isolation, aggressiveness and brain 5-hydroxytryptamine turnover. J. Pharm. Pharmacol., **19**, 338-339.

Garzon, J. & del Rio, J. (1981). Hyperactivity induced in rats by long-term isolation: further studies on a new animal model for the detection of antidepressants. Eur. J. Pharmacol., **74**, 287-294.

Gelenberg, A.J., Gibson, C.J. & Wojcik, J.D. (1982). Neurotransmitter precursors for the treatment of depression. Psychopharmacol. Bull, **18**, 7-18.

Gerner R.H., Fairbanks, L., Anderson, G.M., Young, J.G., Scheinin, M., Linnoila, M., Hare, T.A., Shaywitz, B.A. & Cohen, D.J. (1984). CSF neurochemistry in depressed, manic, and schizophrenic patients compared with that of normal controls. Am. J. Psychiatr., **141**, 1533-1540.

Gjerris, A., Sorensen, A.S., Rafaelen. O.J., Werdelin, L., Alling, C. & Linnoila, M. (1987). 5-HT and 5-HIAA in cerebrospinal fluid in depression. J. Affect. Disord., **12**, 13-22.

Goodwin, E.K., Olivier, J., Goodwin, F.K., Chase, T.N. & Post, R.M. (1973). Effect of probenecid on free 3-methoxy-4-hydroxyphenylethylene glycol (MHPG) and its sulfate in human cerebrospinal fluid. Neuropharmacology, **12**, 391-396.

Gottfries, C.G. (1980). Human brain levels of monoamines and their metabolites: Postmortem investigations. Acta Psychiatr. Scand., **61** (Suppl. 280), 49-61.

Green, A.R., Heal, D.J. & Goodwin, G.M. (1986). The effects of electroconvulsive therapy and antidepressant drugs on monoamine receptors in rodent brain - similarities and differences. In: Antidepressants and receptor function (CIBA Found. Symp. 123). pp. 246-267. Chichester: Wiley.

Hytell, J., Overo, K.F. & Arnt, J. (1984). Biochemical effects and drug levels in rats after long-term treatment with the specific 5-HT-uptake inhibitor, citalopram. Psychopharmacology, **83**, 20-27.

Insel, T.R. & Zohar, J. (1987). Psychopharmacologic approaches to obsessive-compulsive disorder. In: Psychopharmacology: The Third Generation of Progress. (Ed. Meltzer, H.Y.), pp. 1205-1210. New York: Raven.

Jacobs, B.L. & Klemfuss, H. (1975). Brainstem and spinal cord mediation of a serotonergic behavioural syndrome. Brain Res., **100**, 450-457.

Jesberger, J.A., & Richardson, J.S. (1986). Effects of antidepressant drugs on the behaviour of olfactory bulbectomized and sham operated rats. Behav. Neurosci.,

100, 256–274.

Joly, D. & Sanger, D.J. (1986). The effects of fluoxetine and zimelidine on the behaviour of olfacotry bulbectomized rats. Pharmacol. Biochem. Behav., 24, 199–204.

Kaye, W.H., Ebert, M.H., Gwirtsman, H.E. & Weiss, S.R. (1984). Differences in brain serotonergic metabolism between nonbulimic and bulimic patients with anorexia nervosa. Am. J. Psychiatr., 141, 1598–1601.

Kennett, G., Chaouloff, F., Marcou, M. & Curzon, G. (1986). Female rats are more vulnerable that males in an animal model of depression: The possible role of serotonin. Brain Res., 382, 416–421.

Korf, J., Burg, W. van den. & Hoofdakker, R.H. van den. (1983). Acid metabolites and precursor amino acids of 5-hydroxytryptamine and dopamine in affective and other psychiatric disorders. Psychiatr. Clin., 16, 1–16.

Koslow, S.H., Maas, J.W., Bowden, C.L., Davis, J.M., Hanin, I. & Javaid, J. (1983). CSF and urinary biogenic amines and metabolites in depression and mania. Arch. Gen. Psychiatr., 40, 999–1010.

Koyama, T., Lowy, M.T. & Meltzer, H.Y. (1987). 5-hydroxytryptophan–induced cortisol response and CSF 5-HIAA in depressed patients. Am. J. Psychiatr., 144, 334–337.

Langer, S.Z. & Raisman, R. (1983). Binding of [^{3}H] imipramine and [^{3}H] desipramine as biochemical tools for studies in depression. Neuropharmacology, 22, 407–413.

Leander, J.D. (1986). Peripheral action of serotonin as a model of depression. Biol. Psychiatr., 21, 842–844.

Leckman, J.F., Charney, D.S., Nelson. J.C., Heninger, G.R. & Bowers, M.B. (1980). CSF tryptophan, 5-HIAA and HVA in 132 psychiatric patients characterized by diagnosis and clinical state. In: Recent Advances in Neuropsychopharmacology. (Eds. Angrist, B. Burrows, G.D., Lader, M., Lingjaerde, O., Sedvall, G. & Wheatley, D.), pp. 289–297. New York: Pergamon.

Leonard, B.E. (1987). The olfactory bulbectomized rat: A test for selecting antidepressants or an animal model of depression? J. Psychopharmacol. ,1 (1, Suppl.) 7.

Lewinsohn, P.M. (1974). A behavioural approach to depression. In: The psychology of depression: contemporary theory and research. (Eds. Friedman, R.J. & Katz, M.M.), pp. 157–185. New York: Winston/Wiley.

Lidberg, L., Tuck, J.R., Asberg, M., Scalia–Tomba, G.-P. & Bertilsson, L. (1985). Homicide, suicide and CSF 5-HIAA. Acta Psychiatr. Scand., 71, 230–236.

Linnoila, M., Virkkunen, M., Scheinin, M., Nuutila, A., Rimon, R. & Goodwin,

F.K. (1983). Low cerebrospinal fluid 5-hydroxyindoleacetic acid concentration differentiates impulsive from non-impulsive violent behaviour. Life Sci., 33, 2609-2614.

Lloyd, C. (1980). Life events and depressive disorders reviewed. I. Events as predisposing factors. Arch. Gen. Psychiatr., 37, 541-548.

Maier, S.F., Anderson, C. & Lieberman, D.A. (1972). Influence of control of shock on subsequent shock-elicited aggression. J. Comp. Physiol. Psychol., 81, 94-100.

Mann, J.J., Stanley, M., McBride, A. & McEwen, B.S. (1986). Increased serotonin$_2$ and beta-adrenergic receptor binding in the frontal cortices of suicide victims. Arch. Gen. Psychiatr., 43, 954-959.

Mendels, J., Frazer, A., Fitzgerald, R.G., Ramsey, T.A. & Stokes, J.W. (1972). Biogenic amine metabolites in cerebrospinal fluid of depressed an manic patients. Science, 175, 1380-1382.

Mendlewicz, J. & Youdim, M.B.H. (1980). Antidepressant potentiation of 5-hydroxytryptophan by 1-deprenyl in affective illness. J. Affecxt. Disord., 2, 137-146.

Meltzer, H.Y. & Lowy, M.T. (1987). The serotonin hypothesis of depression. In: Psychopharmacology: The third generation of progress. (Ed. Meltzer, H.Y.) pp. 513-526. New York: Raven.

Meltzer, H.Y., Simonovic, M., Sturgeon, R.D. & Fang, V.S. (1981). Effect of antidepressants, lithium and electroconvulsive shock treatment on rat serum prolactin. Acta Psychiatr. Scand., 63 (Suppl. 290), 100-121.

Meltzer, H.Y., Uberkoman-Wiita, B., Robertson, A., Tricou, B.J., Lowy, M. & Perline, R. Effect of 5-hydroxytryptophan on serum cortisol levels in the major affective disorders. I. Enhanced response in depression and mania. Arch. Gen. Psychiatr., 41, 366-374.

Mignot, E., Serrano, A., Laude, D., Elghozi, J.-L., Dedek, J. & Scatton. B. (1985). Measurement of 5-HIAA levels in ventricular CSF (by LCEC) and in striatum (by in vivo voltammetry) during pharmacological modifications of serotonin metabolism in the rat. J. Neural Trans., 62, 117-124.

Mishra, R., Janowsky, A. & Sulser, F. (1979). Subsensitivity of the NE receptor coupled adenylate cyclase system in rat brain. Effects of nisoextine of fluoxetine. Eur. J. Pharmacol., 60, 379-382.

Ogren, S.-O, Fuxe, K., Archer, T., Johansson, G. & Holm, A.C. (1982). Behavioral and biochemical studies on the effects of acute and chronic administration of antidepressant drugs on central serotonergic receptor mechanisms. In: New vistas in depression. (Eds. Langer, S.Z., Takahashi, R., Segawa, T. & Briley, M.), pp. 11-19. New York: Pergamon.

Olpe, H.R. & Schellenberg, A. (1981). The sensitivity of cortical neurons to serotonin: Effect of chronic treatment with antidepressants, serotonin–uptake inhibitors and monoamine–oxidase–blocking drugs. J. Neural Trans., 51, 233–244.

Palfreyman, M.G., Huot, S. & Wagner, J. (1982). Value of monoamine metabolite determinations in CSF as an index of their concentrations in rat brain following various pharmacological manipulations. J. Pharmacol. Meth., 8, 183–196.

Peroutka, S.J. & Snyder, S.H. (1980). Long-term antidepressant treatment decreases spiroperidol-labelled serotonin receptor binding. Sciences, 210, 88–90.

Puhringe, W., Wirz-Justice, A., Graw, P. Lacoste, V. & Gastpar, M. (1976). Intravenous L-5--hydroxytryptophan in normal subjects: An interdisciplinary Precursor preloading study. I. Implications of reproducible mood elevation. Pharmakopsychiatria, 9, 260–268.

Quadbeck, H., Lehman, E. & Tegeler, J. (1984). Comparison of the antidepressant action of tryptophan, tryptophan-5-hydroxytrptophan combination and nomifensine. Neuropsychobiology, 11, 111–115.

Quineaux, N.B. Scuvee-Moreau, J. & Dresse, A. (1982). Inhibition of in vitro and ex vivo uptake of noradrenaline and 5-hydroxytryptamine by five antidepressants: correlation with reduction of firing rate of central monoaminergic neurons. Naunyn Schmiedebergs Arch. Pharmacol., 319, 66–70.

Robins, L.N. (1966). Deviant Children Grow Up. Baltimore: Williams & Wilkens.

Roy, A., Pickar, D., Linnoila, M., Doran, A. & Paul, S.M. (1986). Cerebrospinal fluid monoamine metabolite levels and the dexamethasone suppression test in depression. Arch. Gen. Psychiatr., 43, 356–360.

Roy-Byrne, P., Post, R.M., Rubinow, D.R., Linnoila, M., Savard, R. & Davis, D. (1983). CSF 5-HIAA and personal and family history of suicide in affectively ill patients: A negative study, Psychiatr. Res., 10, 263–274.

Rutter, M. & Madge, N. (1976). Cycles of Disadvantage. London: Heinemann.

Schulterbrandt, R.G. & Raskin, A. (ed). Depression in Childhood: Diagnosis, Treatment and Conceptual Models. New York: Raven.

Schuster, R.H., Rachlin, H., Rom., M. & Berger, D. (1982). An animal model of dyadic social interaction: Influence of isolation, competition, and shock-induced aggression. Aggress. Behav., 8, 116–121.

Sedvall, F., Fyro, B., Gullberg, B., Nyback, H., Wiesel, F.-A. & Wode-Helgodt, B. (1980). Relationships in healthy volunteers between concentrations of monoamine metabolites in cerebrospinal fluid and family history of psychiatric morbidity. Br. J. Psychiatr., 136, 366–374.

Segal, D.S., Knapp, S., Kuczenski, R. & Mandell, A.J. (1973). The effects of environmental isolation on behaviour and regional brain tyrosine hydroxylase and tryptophan hydroxylase activities. Behav. Biol., **8**, 47–53.

Shaw, D.M., Camps, F.E. & Eccleston, E.G. (1975). 5–Hydroxytryptamine in the hind–brains of depressive suicides. Brit. J. Psychiatr., **1113**, 1407–1411.

Shopsin, B., Friedman, E., Goldstein, M. & Gershon, S. (1975). The use of synthesis inhibitors in defining a role for biogenic amines during imipramine treatment in depressed patients. Psychopharmacol. Commun., **1**, 239–249.

Shopsin, B., Friedman, E. & Gershon, S. (1976). Parachlorophenylalanine reversal of tranylcypromine effects in depressed patients. Arch. Gen. Psychiatr., **33**, 811–819.

Sjostrom, R. & Roos, B.–E. (1972). 5–Hydroxyindoleacetic acid and homovanillic acid in cerebrospinal fluid in manic depressive psychosis. Eur. J. Clin. Pharmacol., **4**, 170–176.

Smith, S.E., Pihl, R.O., Young, S.N. & Ervin, F.R. (1987). A test of possible cognitive and environmental influences on the mood lowering effect of tryptophan depletion in normal males. Psychopharmacology, **91**, 451–457.

Sofia, R.D. (1969). Effects of centrally active drugs on experimentally–induced aggression in rodents. Life Sci., **8**, 705–716.

Stahl, S.M. & Meltzer, H.Y. (1978). A kinetic and pharmacological analysis of 5–hydroxytryptamine transport by human platelets and platelet storage granules: Comparison with central serotonergic neurons. J. Pharmacol. Exp. Ther., **205**, 118–132.

Stanley, M. & Mann, J.J. (1983). Increased serotonin–2 binding sites in frontal cortex of suicide victims. Lancet, **1**, 214–216.

Stanley, M., Traskman–Bendz, L. & Dorovini–Zis, K. (1985). Correlations between aminergic metabolites simultaneously obtained from human CSF and brain. Life Sci., **37**, 1279–1286.

Stark, P. & Hardison, D. (1985). A review of multicenter controlled studies of fluoxetine vs. imipramine in outpatients with major depressive disorder. J. Clin. Psychiatr., **46** (3, Sec. 2), 53–58.

Stark, P., Fuller, R.W. & Wong, D.T. (1985). The pharmacological profile of fluoxetine. J. Clin. Psychiatr., **46** (3, Sec. 2), 7–13.

Stewart, I.C., Donald, P. & Munro, J.F. (1986). Neuropharmacological treatment of obesity. In: Disorders of eating behaviour: a psychoendocrine approach (Advances in the Biosciences, Vol. 60). (Eds. Ferrari, E. & Brambilla, F.), pp. 295–303. New York: Pergamon.

Stokes, P.E., Maas, J.W., Davis, J.M., Koslow, S.H., Caspar, R.C. & Stoll, P.M.

(1987). Biogenic amine and metabolite levels in depressed patients with high versus normal hypothalamic–pituitary–adrenocortical activity (1987). Am. J. Psychiatr., 144, 868–872.

Thompson, J., Rankin, H., Ashcroft, G.W., Yates, C.M., McQueen, J.K. & Cummings, S.W. (1982). The treatment of depression in general practice: A comparison of L–tryptophan, amitriptyline, and a combination of L–tryptophan and amitriptyline with placebo. Psychol. Med., 12, 741–751.

Traskman, L., Asberg, M., Bertillson, L. & Sjostrand, L. (1981). Monoamine metabolites in CSF and suicidal behaviour. Arch. Gen. Psychiatr., 38, 631–636.

Traskman–Bendz, L. (1983). CSF 5–HIAA and family history of psychiatric disorders. Am. J. Psychiatr., 140, 1257.

Traskman–Bendz, L., Asberg, M., Bertilsson, P. & Thoren, P. (1984). CSF monoamine metabolites of depressed patients during illness and after recovery. Acta. Psychiatr. Scand., 69, 333–342.

Traskman–Bendz, L., Asberg, M. & Schalling, D. (1986). Serotonergic function and suicide behaviour in personality disorders. Ann. N.Y. Acad. Sci., 487, 168–174.

Trimble, M.R., Chadwick, D., Reynolds, E.H. & Marsden, C.D. Arch. Gen. Psychiatr., 32, 178–183.

Valzelli, L. (1984). Reflections on experimental and human pathology of aggression. Prog. Neuropsychopharmacol. Biol. Psychiatr., 8, 311–325.

Van Praag, H.M. (1982a). Depression, suicide and the metabolism of serotonin in the brain. J. Affect. Disord., 4, 275–290.

Van Praag, H.M. (1982b). Serotonin precursors in the treatment of depression. In: Serotonin in Biological Psychiatry. (Eds. Ho, B.T., Schoolar, J.C. & Usdin, E.),pp. 259–286. New York: Raven.

Van Praag, H.M. (1983a). In search of the mode of action of antidepressants: 5–HTP/tyrosine mixtures in depression. Neuropharmacology, 22, 433–440.

Van Praag, H.M. (1983b). CSF 5–HIAA and suicide in non–depressed schizophrenics. Lancet, 2, 977–978.

Van Praag, H.M. & de Haan, S. (1979). Central serotonin metabolism and frequency of depression. Psychiatr. Res., 1, 219–224.

Van Praag, H.M., Van den Burg, M., Bos, E.R.H. & Dols, L.C.W. (1974). 5–hydroxytryptophan in combination with clomipramine in "therapy-resistant" depression. Psychopharmacology, 38, 267–269.

Van Praag, H.M., Plutchik, R. & Conte, H. (1986). The serotonin hypothesis of (auto)aggression. Ann. N.Y. Acad. Sci., 487, 150–167.

Wang, R.Y. & Aghajanian, G.K. (1980). Enhanced sensitivity of amygdaloid neurons to serotonin and norepinephrine after chronic antidepressant treatment. Commun. Psychopharmacol., 4, 83-90.

Ward, N.G., Ang, J. & Pavinich, G. (1985). A comparison of the acute effects of dextroamphetamine and fenfluramine in depression. Biol. Psychiatr., 20, 1090-1097.

Weiss, J.M., Goodman, P.A., Losito, B.G., Corrigan, S., Charry, J.M. & Bailey, W.H. (1981). Behavioral depression produced by an uncontrollable stressor: relationship to norepinephrine, dopamine, and serotonin levels in various regions of rat brain. Brain Res., 3, 167-205.

Willner, P. (1984). The validity of animal models of depression. Psychopharmacology, 83, 1-16.

Willner, P. (1985a). Depression: a psychobiological synthesis. New York: Wiley.

Willner, P. (1985b). Antidepressants and serotonergic transmission: An integrative review. Psychopharmacology, 85, 387-404.

Willner, P. (1987). Sensitization to the actions of antidepressant drugs. In: Tolerance and sensitization to psychoactive drugs. (Eds. Emmett-Oglesby, M.W. & Goudie, A.J.), New Jersey: Humana Press, Clifton. (In Press.

Yanai, J. & Sze, P.Y. (1983) Isolation reduces midbrain tryptophan hydroxylase activity in mice. Psychopharmacology, 80, 284-285.

Young, S.N., Smith, S.E., Pihl, R.O. & Ervin, F.R. (1985). Tryptophan depletion causes a rapid lowering of mood in normal males. Psychopharmacology, 87, 173-177.

5-HT RECEPTOR SUBTYPES IN DEPRESSION

J.F.W. Deakin
University Department of Psychiatry, University Hospital of South Manchester, West Didsbury, Manchester M20 8LR. United Kingdom

Introduction
The idea that 5-HT neurones mediate the effects of punishment on behaviour was first suggested by Wise, Berger and Stein (1970) and fostered by subsequent reports that 5-HT depletion with PCPA or treatment with non-selective 5-HT receptor antagonists releases punished responding in the conflict paradigm (Geller & Blum, 1970; Graeff, 1974). Tye, Everritt & Iverson, (1977) showed that 5,7-dihydroxytryptamine (5,7-DHT) lesions of ascending 5-HT pathways in rats released bar pressing for food reward from inhibition by contingent foot shock. This was perhaps the clearest demonstration that ascending 5-HT pathways are involved in shock-induced suppression of behaviour. One criticism of 5-HT punishment theories is that they are vaguely formulated and do not take into account the various behavioural effects of punishment nor do they ascribe differential behavioural functions to the different 5-HT projections.

I have suggested elsewhere (Deakin, 1983) that the operation of the 5-HT punishment system may be seen at its simplest level in the spinal cord. It is known that spinal 5-HT projections are activated by noxious stimuli and it has been argued on the basis of electrophysiological experiments that spinal 5-HT projections have the function of extracting and transmitting pain messages (Le Bars, Dickenson & Besson, 1979). Ascending 5-HT projections from the dorsal and median raphe nucleus have the related function of mediating behavioural and motivational states arising from noxious stimulation or the threat of noxious stimulation. 5-HT terminals arising from the dorsal raphe nucleus are distributed to areas which receive a dopaminergic innervation – corpus striatum, accumbens and amygdala. It was suggested (Deakin, 1983) that dorsal raphe nucleus projections function as a negative incentive or 'stop' system with the function of guiding the organism away from conditioned fear stimuli. There is increasing evidence that dopaminergic projections may have the opposed behavioural function of mediating the attractive and energising behavioural effects of positive incentive stimuli. The median raphe nucleus projects to the hippocampus whose major catecholamine innervation is noradrenergic rather than dopaminergic.

Postsynaptic $5-HT_{1A}$ receptors show a striking degree of localization to the hippocampus (eg Glaser and Traber, 1985) and they may transmit the effects of 5-HT released from median raphe nucleus projections to hippocampus. The normal behavioural funtions of this system are obscure. In recent years it has been suggested that 5-HT systems are more concerned with suppressing motor outputs or restraining impulsive behaviour than in the mediation of central motive states such as fear of helplessness (Soubrie, 1986). It is hard to discriminate between 5-HT punishment theories and 5-HT-response inhibition theories on the basis of animal behavioural experiments since animal mental states cannot be directly accessed.

I will argue in this chapter that studies with human subjects suggest that 5-HT systems are involved in the mediation or modulation of aversive central motive states and that an account of the role of 5-HT in depression and anxiety in terms of 5-HT punishment theories is possible and as compatable with the evidence as its competitors.

5-HT$_2$ receptors, fear and anxiety

One piece of evidence that 5-HT is concerned with aversive central motive states is the probable efficacy of selective 5-HT$_2$ antagonists such as ritanserin in the treatment of anxiety. Ceulemans et al, (1985) reported a dose related anxiolytic effect of ritanserin that was comparable with lorazepam and significantly superior to placebo (fig. 14.1). It is difficult to see how 5-HT-response inhibition theories would account for this. However, the somewhat tortuous argument might be advanced that ritanserin disinhibited the patients behaviour so that they exposed themselves to anxiety-inducing stimuli and thereby became desensitized. The more straight forward explanation would

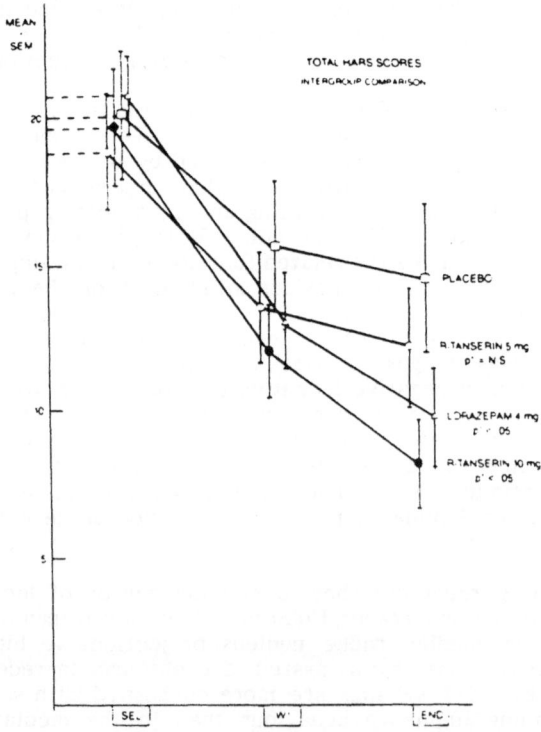

Fig. 14.1 Dose-related anxiolytic effect of the 5-HT$_2$ antagonist, ritanserin (Ceulemans et al., 1985).

be that 5-HT mediates or potentiates anxiety, and that antagonists reduce it. Recent evidence from our laboratory is compatable with this suggestion.

As part of a large clinical trial of ritanserin in neurotic out-patients, we have been monitoring autonomic indices of anxiety in trial patients. Patients sit in a sound attenuated room and listen to a series of randomly spaced tones of neutral intensity. Half way through the session they are subjected to an aversive stimulus of 100 Db of white noise lasting one second.

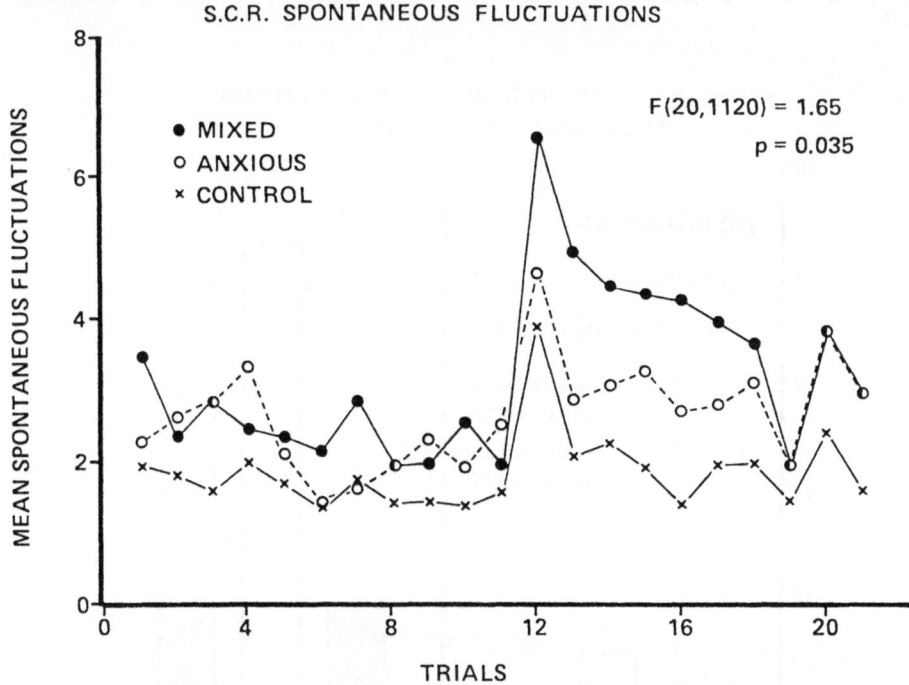

Fig. 14.2 Autonomic lability in neurotic patients and controls. Vertical axis = No. of spontaneous fluctuation in skin conductance per minute; horizontal axis = presentations of neutral tones through headphones. Prior to tone 12, subjections experienced a loud, aversive white noise (100 Db, 1 sec). Mixed = anxious and depressed patients (N = 10); anxious - patients with anxiety, without depression (N = 20); controls = age and sex matched controls (N = 30). The loud noise evokes greater autonomic lability in the patients compared with controls.

It can been seen from fig. 14.2 that this is followed by an increase in spontaneous fluctuations of skin conductance. This is an index of activation of central mechanisms controlling autonomic (sweat gland) activity. Patients with anxiety (n=20) and a mixed anxiety and depression (n=10) were compared with an age and sex matched control group of 30 subjects. The patients showed a greater degree and duration of autonomic activation following the aversive stimulus than did the controls. We do not yet know the effects of ritanserin in the patients. However, in a separate experiment 10 normal subjects pretreated with ritanserin (5 mg, p.o.) showed a markedly attenuated autonomic response to the aversive stimulus compared to non-drugged (n=10) and placebo treated (n=10) normal subjects (fig. 14.3).

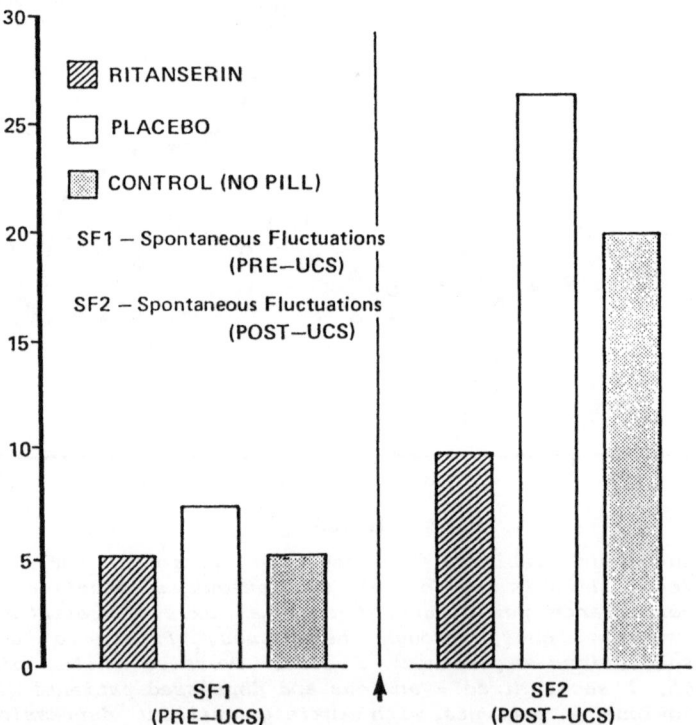

Fig. 14.3 Effects of ritanserin (5 mg) pretreatment upon autonomic lability in healthy controls (N = 10 per group). Vertical axis = number of spontaneous fluctuations in skin conductance prior to the loud noise (USC) and after it. Ritanserin had no effect in basal conditions but strongly antagonized the increased autonomic lability following the white noise.

This result suggests that ritanserin reduces the limbic-autonomic component of anxiety, the component which is exaggerated in anxious – depressed patients. Again it is difficult to see how this effect of ritanserin can be explained in terms of 5-HT-response suppression theories.

$5-HT_3$ receptor antagonists have been developed in recent years and animal studies suggest they have anxiolytic effects (Brittain et al, 1987; Jones et al., 1988). The role of $5-HT_3$ receptors in the psychology of punishment is as yet unclear and there are no data available from studies of anxious humans $5-HT_2$.

Receptor Down-Regulation and the Anxiolytic Effect of Antidepressants
There is a good deal of evidence that tricyclic antidepressants are effective in the treatment of anxiety. Two large trials demonstrated that a tricyclic antidepressant was superior to a benzodiazepine and to placebo in the treatment of anxiety symptoms (Johnstone et al, 1980; Kahn, 1986; see also Ancill, 1984) in neurotic out patients. Generally speaking, therapeutic effects were small. The beneficial effects of the antidepressants could not be explained by the presence of patients with a panic disorder in the trials.

In view of the probable anxiolytic effects of $5-HT_2$ receptor antagonists, it is tempting to suggest that the shared ability of tricyclic antidepressants to induce acute or delayed decreases in $5-HT_2$ receptor function (Peroutka and Snyder, 1980) is more related to the anxiolytic effects of antidepressants than to their antidepressant actions. The fact that ECT treated rats show increases in $5-HT_2$ receptor numbers and function (Vetulani, Lebrecht and Pilc, 1981) reinforces the view that $5-HT_2$ receptor down regulation may not be directly related to antidepressant effects of drugs. ECT is effective in treating severe depressive illnesses but therapeutic effects of the seizures are largely confined to patients with psychotic depressive illnesses (Clinical Research Centre 1984). There are other reasons to think that depressive psychosis is distinct from other types of depression. It is a defendable point of view that ECT is as much an antipsychotic treatment as an antidepressant since it is also effective in mania and schizophrenia (Taylor and Fleminger, 1980). Furthermore, there is evidence that the presence of marked anxiety symptoms may be a contra-indication to ECT (Carney, Roth & Garside, 1965). Perhaps this is because ECT increases $5-HT_2$ receptor numbers in contrast to antidepressant drugs (effective anxiolytics) which decrease $5-HT_2$ receptor function.

It seems clear that nonpsychotic depression and anxiety symptoms do not have a markedly differential response to drug treatment. These findings emphasize the close relationship between anxiety and depression. Other clinical findings also emphasize this close relationship.

The neurobiology of anxiety and depression
Goldberg et al (1987) rated symptoms of anxiety and depression in a very large number of patients attending their general practitioner with psychological symptoms. The statistical technique of latent trait analysis was used to reduce the variation in symptom patterns to a minimum number of artificial, statistically generated variables. It was found that two dimensions accounted for 80% of the variance in symptom pattern. One dimension correlated most highly with symptoms of anxiety and the other most highly with symptoms of

depression. However the two dimensions were highly correlated (r=0.7) so that anxiety symptoms were quite good measures of the depression dimension as well as of the anxiety dimension. The reverse was also true; symptoms of depression correlated significantly with the anxiety dimension.

The reason that two distinct dimensions emerged may be that they have distinct psychobiological substrates. It has been suggested that pathological anxiety results from an excessive sensitivity to aversive stimuli and an over-ready acquisition of conditioned fear responses (eg Gray, 1971). Such processes would be mediated by excessive function in 5-HT systems according to 5-HT-punishment theories. In contrast, depression has been conceptualized as a lost efficacy of rewards such that behaviour is no longer controlled by positive reinforcers (eg Akiskal, 1976). This idea lies at the heart of the DSM III definition of depressed mood – "a pervasive loss of interest and pleasure". Catecholamine neurones have been implicated in reward processes in animal behavioural studies. Thus deficient catecholamine function could underlie the depressives insensitivity to rewards.

The reason that anxiety and depression are correlated dimensions of psychological disturbance may be that reward and punishment processes, and 5-HT and catecholamine function are reciprocally related. Depression could thus result from a primary deficiency in brain reinforcement systems (catecholamines ?) or as a secondary consequence of excessive inhibition from overactive 5-HT punishment systems. According to this view, excessive 5-HT neurotransmission could be associated with depression as well as with anxiety. Indeed, symptoms of anxiety are often prodromal to the development of depression.

5-HT function in depressed patients

Recent post-mortem brain studies in patients with chronic depression or who have died by suicide have reported increased numbers of 5-HT receptor binding sites in frontal cortex and this would be compatible with 5-HT excess theories of depression (Stanley & Mann, 1983). However, this finding has not been replicated in other studies (Cooper et al.,1986). Nevertheless, it is intriguing that many antidepressant drugs are $5-HT_2$ receptor antagonists (eg. amitriptyline, trazodone, mianserin, clomipramine) and all established antidepressants induce a delayed down-regulation of $5-HT_2$ receptor numbers (Peroutka and Snyder, 1980). Thus, antidepressant drugs do have actions which would counteract the increased number of $5-HT_2$ receptors found in some post-mortem brain studies.

A difficulty for 5-HT excess theories is that they stand in opposition to the longer tradition that a deficiency of 5-HT neurotransmission is associated with depression. Many studies have reported reduced levels of csf 5HIAA in depressed patients compared with controls (eg Asberg et al., 1976; Goodwin et al., 1973). However the magnitude of the difference is small (about 20%) and there are several negative studies. One recent study found no difference in 5HIAA concentrations in depressed patients compared to controls but csf concentrations of 5-HT itself were increased in depressed patients (Gjerris et al.,1987). Post-mortem brain studies of 5-HT and its metabolites are frequently adduced in support of 5-HT deficiency theories but again the evidence is in fact contradictory and there are many negative studies (see Cooper et al.,1986). A recent development is that tritiated imipramine binding sites, a marker for 5-HT terminals, are reportedly reduced in some post mortem brain studies (Stanley, Virgilio & Gershon, 1982; Crow et al. ,1984;Perry et al., 1983). Again there are two negative reports and even one report of increased imipramine binding (Cooper et al.,1986; Meyerson et al., 1982).

The equivocal nature of the results of biological studies in depressed patients is highlighted by a post-mortem brain studie carried out by my colleagues and I at the Clinical Research Centre (Owen et al. 1983). Brains of patients who committed suicide were grouped according to whether there was evidence for clear suicidal intent or preceeding clinical depression. It can be seen from table 14.1 that clinically subdividing the patients does not reveal signification differences between patients and controls in concentration of 5HIAA or in 5-HT$_2$ receptor binding sites (^{3}H-Ketanserin).

Table 14.1 5-HT receptors and 5HIAA in frontal cortex.
Samples from controls and suicides.

GROUP	^{3}H-Ketanserin	^{3}H-5-HT	^{3}H-Imipramine	5HIAA
Controls	87.0 ± 22.6 (n = 20)	82.1 ± 19.5 (n = 18)	205.0 ± 34.0 (n = 20)	190.9 ± 154.9 (n = 18)
Def. suicid.	74.1 ± 38.9 (n = 9)	65.9 ± 28.7 (n = 7)	184.6 ± 58.2 (n = 9)	136.9 ± 69.7 (n = 9)
Poss. suicid.	89.2 ± 35.7 (n = 8)	75.0 ± 17.6 (n = 8)	184.1 ± 56.2 (n = 8)	143.0 ± 84.1 (n = 8)
Def. poss.	81.2 ± 37.1 (n = 17)	70.7 ± 23.0 (n = 15)	186.1 ± 54.1 (n = 17)	139.8 ± 94.4 (n = 17)
Depressives	65.1 ± 32.5 (n = 11)	65.9 ± 17.6 (n = 10)	182.4 ± 54.7 (n = 11)	145.2 ± 65.6 (n = 11)

(from Owens et al., 1983)

There is a trend for reductions in 5HIAA in the depressed patients but it does not reach significance – the variation in 5HIAA concentrations in the controls is notable. The equivocal results of biological studies in depression are in contrast to those in Alzheimer's disease where no abnormalities were predicted but where post mortem brain studies have revealed gross deficits in biochemical markers of pre and post synaptic 5-HT function (Crow et al., 1984).

ANTIDEPRESSANTS AND 5-HT FUNCTION

Are 5-HT reuptake blockers effective antidepressants?

5-HT deficiency theories led to the development of drugs to counteract the supposed deficiency in 5-HT neurotransmission by selectively inhibiting 5-HT reuptake. It is critically important for theories of 5-HT in depression to know whether such drugs are effective antidepressants. This question is surprisingly difficult to answer. In a third of the placebo-controlled clinical trials of tricyclics, no significant differences between drug and placebo emerged in the treatment of depressive illnesses (Morris & Beck 1974). A single trial is not sufficient to establish that a new drug is an effective antidepressant. New antidepressants are often compared with established 'reference' antidepressants without a placebo treated group. The almost invariable result is that symptoms of depression improve equally on the two drugs. This does not mean that the new drug is effective; a placebo treated group might have improved as much as the two drug treated groups. This point is illustrated by a recent trial of the 5-HT reuptake blocker fluvoxamine in depressed out-patients (Norton et al, 1984). Fluvoxamine and imipramine treated patients improved equally. However, despite the large number of patients evaluated, there were no significant differences between drug-treated patients and those receiving placebo. Had a placebo treated group not been included the authors might have been tempted to reach the conclusion that fluvoxamine was an effective antidepressant and as effective as imipramine. Other placebo-controlled trials do suggest that the 5-HT reuptake blockers fluoxetine and fluvoxamine do have antidepressant activity (eg Amin et al., 1984). However, there is evidence, particularly from self-rating scales, that they may be less effective than older, non-selective antidepressants or that their clinical spectrum of activity differs (see Young, Coleman and Lader, 1987). The probable antidepressant efficacy of fluoxetine and fluvoxamine does not rule decisively in favour of 5-HT deficiency theories because both drugs have delayed actions on noradrenergic neurotransmission which could account for their antidepressant activity (Claassen 1983; Racagni and Bradford, 1984). The critical drug is citalopram because it is clear that this drug does not induce down regulation of 5-HT_2 receptors nor does it affect noradrenergic neurotransmission. The Danish University Antidepressant Group (1986) compared citalopram with clomipramine in a large number of depressed patients and obtained an almost unique result; patients treated with citalopram improved less than those treated with clomipramine. The question therefore arises as to whether citalopram has any antidepressant activity; only placebo-controlled trials can establish this.

In the absence of a published placebo controlled trial, the results of a small study carried out by Dr. Pennell and myself in Manchester are presented in table 14.2. Randomization of patients did not work out well and only 8 received citalopram compared to 13 who received placebo. However, the interpretability of this trial is enhanced by the fact that both groups started off with similar Hamilton depression rating scores and by the fact that no patients dropped out of the trial during the drug or placebo treatment phase. It can be seen from improvement scores that many patients on placebo showed minimal improvements and only three showed a significant resolution of their symptoms whereas 5 of the 8 patients treated with citalopram showed significant improvements in their Hamilton depression rating scores. The differences in improvement scores

Table 14.2

	Citalopram (8)	Placebo (13)
Initial hamd	21.5	23.2
Improvement	-1	-3
	4	-1
	6	0
		1
	12	1
	12	1
	13	4
	13	4
	13	6
		7
		11
		14
		18

Effects of 4 weeks treatment with citalopram or matching placebo on Hamilton depression ratings in patients suffering from major depressive disorder.

between the two groups is statistically significant. If further large-scale placebo-controlled trials of citalopram reveal it to have antidepressant properties this will have to be accommodated by any 5-HT theory of depression. Such a finding would pose difficulties for 5-HT excess theories of depression.

Evidence that antidepressants enhance 5HT function

While receptor binding studies and behavioural studies suggest that chronic antidepressant treatment decreases 5-HT neurotransmission, electrophysiological studies suggest the opposite. DeMontigny and Aghajanian (1984) showed that iontophoresis of 5-HT into the vicinity of forebrain units caused a decrease in firing rates. In animals chronically treated with antidepressants, forebrain units became more sensitive to the depressant effects of 5-HT. In other words, chronic antidepressant therapy was associated with an enhanced neuronal responsiveness of 5 HT. The 5-HT receptor subtype involved is not clear. It is clear that $5-HT_2$ antagonists do not block the electrophysiological effects of 5-HT, suggesting that they are mediated by some kind of $5-HT_1$ receptor mediated neurotransmission. However, there is another possibility. Lakoski and Aghajanian (1985) found that the $5-HT_2$ antagonist, ketanserin, far from abolishing the depressant effects of 5-HT on neuronal firing, enhanced 5-HT's effect. They suggest that removing a $5-HT_2$ receptor mediated component reveals a purer and more potent non $5-HT_2$ - depressant effect of 5-HT. This raises the possibility that antidepressant drug-treatment enhances neuronal responsiveness to 5-HT by down-regulation of $5-HT_2$ receptor numbers rather than directly enhancing $5-HT_1$ mediated

Table 14.3 Conflicting views on 5-HT and depression

5-HT excess	5-HT deficiency
$5\text{-}HT_2$ receptors p.m.	low csf 5HIAA
	low imipramine binding
Acute antidepressants	
$5\text{-}HT_2$ antagonism:	5-HT reuptake blockers:
Amitriptyline	Citalopram
Mianserin	Fluoxetine
Trazodone	Fluvoxamine
Ceronic antidepressants	
$5\text{-}HT_2$ receptor binding	5-HT effects on
	neuronal firing
	(....)

neurotransmission.

Summary of conflicting 5–HT excess vs 5–HT deficiency theories

It can be seen from table 14.3 and from the foregoing discussion that the results of biochemical investigations in patients are equivocal.

The effects of antidepressant drugs on 5–HT neurotransmission are more certain but produce opposing strands of evidence. Considering only the positive biological studies in patients, there is evidence for a deficiency of presynaptic 5–HT neuronal function and increased post-synaptic $5\text{-}HT_2$ receptor function in depressed patients. It has been suggested that these two abnormalities are causally related. According to one view (fig. 14.4) deficient pre-synaptic 5–HT neuronal function causes a compensatory upregulation of post-synaptic $5\text{-}HT_2$ receptors. However, this is not a satisfying explanation of the findings because several studies in animals failed to find that destruction of 5–HT neurones causes an increase in $5\text{-}HT_2$ receptor numbers (Leysen et al., 1983; Quick & Azmitea, 1983). Furthermore, treatment with selective $5\text{-}HT_2$ antagonists does not increase $5\text{-}HT_2$ receptor binding sites. On the contrary, decreases occur (Leysen et al., 1986). This is in contrast to other monoamine receptors which proliferate when afferent monoamine neurones are destroyed or blocked with antagonists. Other studies suggest that lesions which induce 5–HT depletions of more than 70% do induce an increase in $5\text{-}HT_2$ receptor binding sites (Heal et al., 1985). However, no biological study in depression reveals deficits in 5–HT neuronal function of this magnitude.

Another explanation suggests the primary abnormality is an increase in the

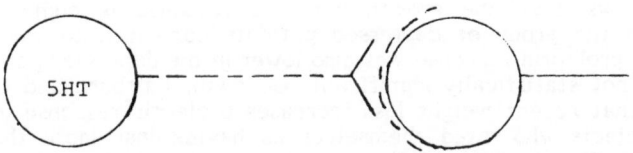

1. Primary reduction in 5-HT release
   Secondary increase in receptors.

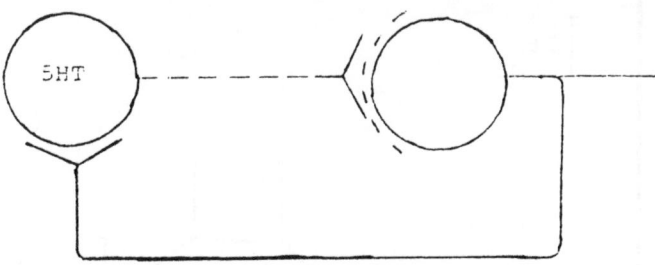

2. Primary increase in receptors
   Secondary inhibition BRT release

Fig. 14.4 Resolutions of conflicting evidence

number of 5-HT$_2$ receptor binding sites in depression and that this causes a negative feedback inhibition of presynaptic 5-HT neuronal functioning (fig. 14.4).

Again this is an unsatisfactory resolution because there is no evidence that 5-HT$_2$ receptors inhibit or influence presynaptic 5-HT neuronal function. My colleague Dr. Pennell and I attempted to discriminate between 5-HT excess and deficiency theories using a neuroendocrine strategy divised by Charney et al. (1982). We predicted that, according to 5-HT excess theories, depressed patients should have exaggerated pituitary hormonal responses to intravenous infusions of 5-HT precursor tryptophan. In contrast the deficiency theory predicted blunted responses to tryptophan.

Hormonal responses to tryptophan in depression
Depressed patients and controls were starved overnight. A cannula was inserted and patients remained supine and awake for the two and a half hour test. Baseline blood samples were taken at time 0 and after 60 minutes. They then received an intravenous infusion of L-tryptophan (LTP; 100 mg/kg). Growth hormone and prolactin were assayed in serial blood samples following the infusion. LTP induced no changes in cortisol secretion.

Fig. 14.5 shows that the growth hormone response is highly significantly attenuated in the group of depressed patients compared to the response in controls. The prolactin response was also lower in the depressed patients, but the difference is not statistically significant. Goodwin, Fairburn and Cowan (1987) have shown that recent weight loss increases prolactin response to LTP. When depressed subjects who rated themselves as having lost more than 3 kg are removed from the present analysis, then depressed patients without weight loss shows significantly attenuated responses.

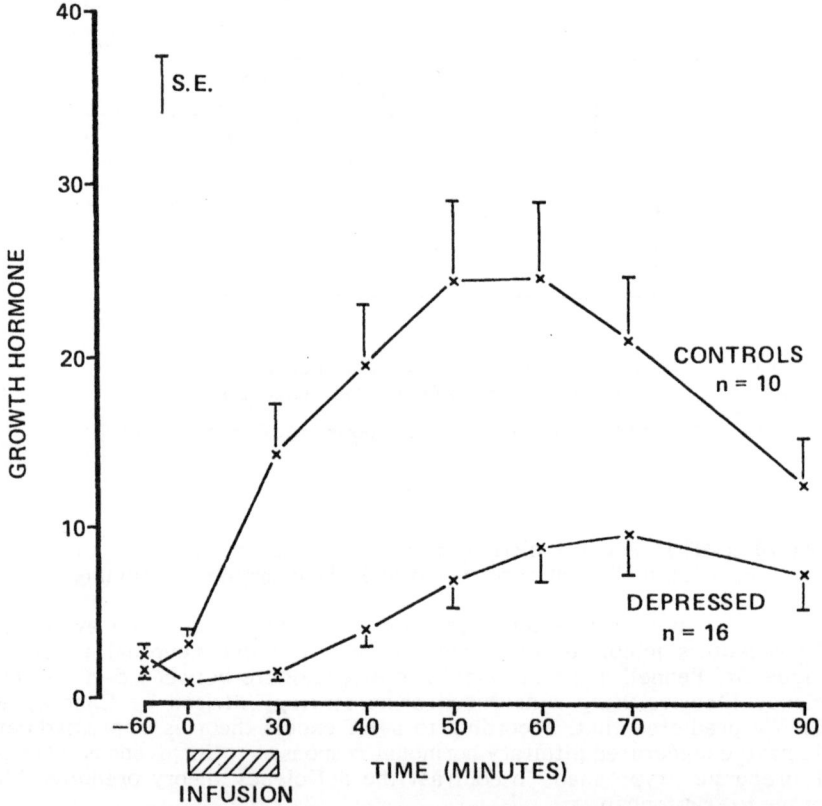

Fig. 14.5 Mean serium Growth Hormone concentrations at different times before and after infusions of LTP (100 mg/kg, i.v.). Vertical bars - sem. Depressives were drug free and met DSM III criteria for major depressive disorder.

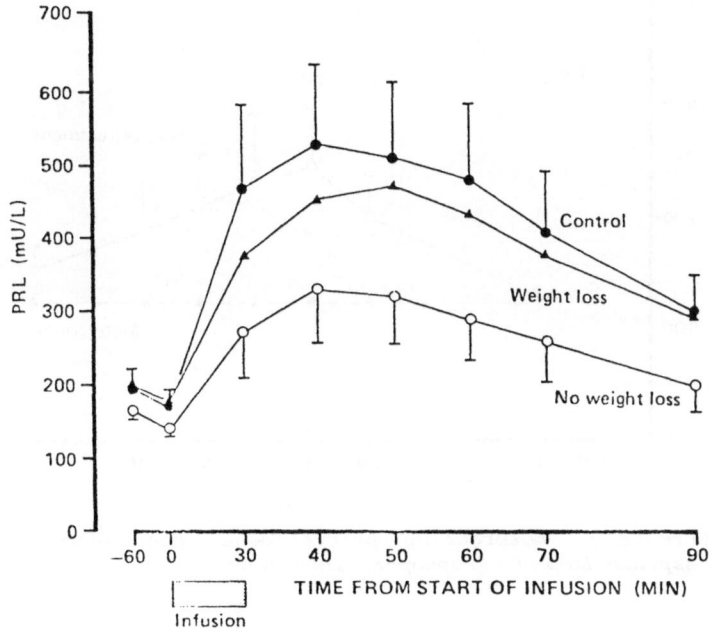

Fig. 14.6 Mean serium prolactin responses before and after 100 mg/kg LTP infusion. Controls (N = 10). Depressives devided into those without weight loss as rated on Beck Depression Inventory (N = 13) and those with significant weight loss (N = 3).

The results shown in fig. 14.6 are very similar to those obtained by Cowan & Charig (1987).

Blunted prolactin responses to LTP were also reported by Heninger et al. (1984) and following fenfluramine by Siever et al. (1984). These results appear to corroborate 5-HT deficiency theories and to refute 5-HT excess theories of depression. However, before this interpretations can be made it is necessary to demonstrate that the hormonal responses are indeed mediated by 5-HT receptors and not by some nonspecific action of LTP. Anderson & Cowen (1986) have reported that pretreating normal subjects with clomipramine potentiates prolactin and growth hormone responses suggesting the involvement of 5-HT. We have attempted to determine which 5-HT receptors are involved by using antagonists. Neither metergoline, the nonspecific $5-HT_1$ and $5-HT_2$ antagonists, nor ritanserin, a selective $5-HT_2$ antagonist, prevented the increase in growth hormone which follows LTP infusion. These results suggest that $5-HT_1$ and $5-HT_2$ receptors do not mediate growth hormone responses to LTP. The possibility remains that they may be mediated by $5-HT_3$ receptors. Fig. 14.7 shows that metergoline completely abolished prolactin responses to LTP in 4 healthy volunteers, a result confirmed by McCance et al. (1988).

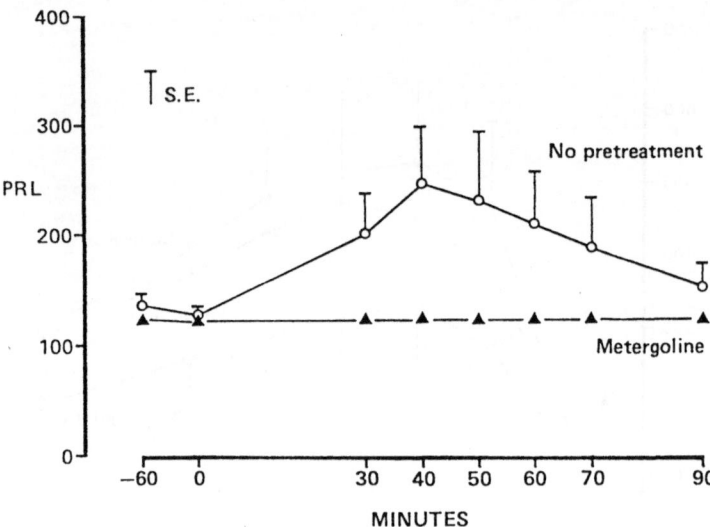

Fig. 14.7 Effects of metergoline pretreatment upon serum prolactin response to i.v. tryptophan infusions.

In contrast pretreatment with ritanserin, in two subjects, did not. This suggests that 5-HT$_1$ receptors may mediate prolactin responses to LTP. Interestingly, ritanserin, far from antagonising prolactin responses to LTP, caused a doubling and this result has been confirmed by Charig et al. (1987). This effect of ritanserin suggests that 5-HT$_2$ receptors normally oppose the rise in prolactin following LTP infusions which is mediated by 5-HT$_1$ receptors (fig. 14.8).

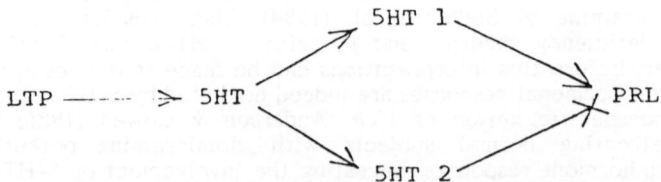

Ritanserin blocks 5-HT$_2$ receptors, removing an inhibitory component of 5-HT upon PRL secretion thus potentiating PRL responses to LTP.

Fig. 14.8 5-HT receptor subtypes involved in PRL responses to LTP.

Opposed function of $5-HT_1$ and $5-HT_2$ receptors

There are 3 other reports that $5-HT_2$ antagonists potentiate a $5-HT_1$ receptor mediated response (table 14.4). The ability of ketanserin to potentiate the neuronal effects of iontophorefically applied 5-HT have already been mentioned.

Table 14.4 $5-HT_2$ antagonists enhance $5-HT_1$ effects on:

Prolactin secretion	(Charig et al., 1987)
Neuronal firing	(Dakoski et al., 1985)
RU 24696 locomotion	(Goodwin and Green, 1985)
VIP inhibition c AMP	(Weis et al., 1986)

RU 24969 is a $5-HT_1$ receptor agonist which evokes hyperactivity in mice which is potentiated by pretreatment with ritanserin. The addition of vasoactive intestinal polypeptide (VIP) to hippocampal homogenates evokes a rise in cyclic AMP. 5-HT inhibits the response to VIP and this action of 5-HT is potentiated by ketanserin.

There are a number of other circumstances where $5-HT_1$ and $5-HT_2$ receptors appear to mediate opposed effects. These are summarized in table 14.5. Other examples are described elsewhere in this volume.

Table 14.5 Opposed effects of $5-HT_1$ and $5-HT_2$ receptors on:

Sleep	(Solomon et al., 1988)
Sexual behaviour	(Wilson and Hunter, 1985; Gorsalka, Thisvol)
Temperature regulation	(Gudelsky et al., 1986)
5-HT syndrome	(Harrison Read, 1983)
Morphine analgesia	(Gonzalez and Stolz, 1985)
Suprahyoid muscle twitching	(Gardner et al., 1987)

In some of the examples listed, it is not clear whether apparently opposed $5-HT_1$ and $5-HT_2$ receptor effects are a consequence of a presynaptic influence of $5-HT_1$ active drugs upon 5-HT release and of the post-synaptic localization of $5-HT_2$ effects (morphine, anti-nociception, thermoregulation; table 14.5). In other cases the opposed functions are clearly both post-synaptic (eg. sexual behaviour; Mendelson & Gorzalka, this volume). It is probably not a general rule that $5-HT_1$ and $5-HT_2$ receptors oppose each other since they are differentially localized. For example, high concentrations of both receptor subtypes are found in frontal cortex but in the hippocampus $5-HT_1$ receptors predominate very considerably over $5-HT_2$.

5-HT receptors and neuroendocrinology in depression

$5-HT_1$ and $5-HT_2$ receptors have opposed effects in mediating prolactin responses to tryptophan. This allows two explanations of the blunted responses seen in depressives (fig. 14.9).

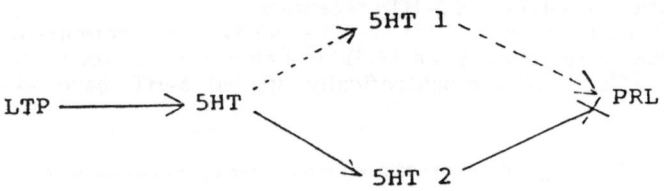

1. Decreased 5-HT$_1$ function

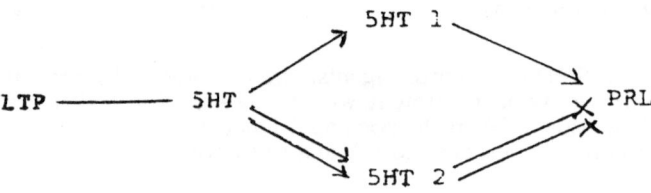

2. Excessive 5-HT$_2$ function

Fig. 14.9 Two mechanisms for blunted PRL responses to LTP in depression.

Deficient functioning in pathways which involve 5-HT$_1$ receptors is one way in which the blunted responses could come about. Another possibility is an excessive inhibitory influence on prolactin secretion mediated by 5-HT$_2$ receptors. It is not possible on the basis of our results to state whether the abnormality lies at the level of the receptor. The abnormality could occur in systems afferent to or beyond a synapse involving 5-HT$_1$ or 5-HT$_2$ receptors.
It is not clear whether such a scheme could explain reports of exaggerated cortisol responses to 5-Hydroxytryptophan (5-HTP) in depressives (Meltzer et al., 1984). If cortisol responses to 5-HTP are mediated by 5-HT$_2$ receptors, enhanced responses in depressive would be compatible with excessive functioning in systems involving 5-HT$_2$ receptors. However, there is no evidence that this is the case. Disturbed feedback control of cortisol secretion is another possible explanation of exaggerated cortisol responses to 5-HTP in depression. For the moment therefore, these results are not germane to the issue of the role of 5-HT receptor subtypes in depression.

5-HT receptor subtypes and depression
If blunted prolactin responses to LTP indicate an imbalance in the relative functioning of 5-HT$_1$ and 5-HT$_2$ receptors as indicated in fig. 14.9, perhaps the same is true of depression itself. This would go some way to reconciling the conflicting lines of evidence on the role of 5-HT and depression summarized in table 14.3. It can be seen that evidence for 5-HT excess theories applies entirely to the 5-HT$_2$ receptor. Perhaps evidence for deficiency theories applies to

$5-HT_1$ receptors. A small reduction in 5-HT release playing upon excessively sensitive $5-HT_2$ systems would lead to a relative deficit in $5-HT_1$ neurotransmission.

5-HT receptor subtypes and the mechanism of action of antidepressant drugs

If the 5-HT receptor inbalance theory of depression is correct then antidepressant drugs would need to cause an increase in $5-HT_1$ receptor function relative to $5-HT_2$ to restore the imbalance. This prediction is compatible with much of the evidence and is summarized in table 14.6.

Table 14.6 Antidepressants and systems involving 5-HT receptor subtypes

	$5-HT_2$	$5-HT_1$
Depression	↑	↓

Amitriptyline
  reuptake blockade
  $5-HT_2$ antagonist
	↓	↑

Trazodone
  $5-HT_2$ antagonist
  McPP metabolite
	↓	↑

Mianserin	↓	↑
Buspirone	↓	↑
Gepirone	↓	↑

Amitriptyline, clomipramine and some other tricyclics are effective 5-HT re-uptake blockers. This immediate action and the delayed desensitization of autoreceptors combine to increase synaptic 5-HT availability. Several 5-HT re-uptake blocking tricyclics are also effective postsynaptic $5-HT_2$ receptorblockers leading to a relative increase in $5-HT_1$ neurotransmission over $5-HT_2$. Trazodone is an effective antidepressant (Al-Yassiri, Ankier & Bridges, 1979) whose mode of action is obscure. Its main pharmacological action is $5-HT_2$ receptor antagonism but its metabolite, mCPP is a $5-HT_1$ agonist (Fuller et al., 1981). Thus trazodone may restore $5-HT_1$ neurotransmission relative to $5-HT_2$. Mianserin is a $5-HT_2$ receptor antagonist and this action alone would tend to restore the proposed inbalance between $5-HT_1$ and $5-HT_2$ receptor functioning in depression.

The ability of estabished antidepressant drugs to increase neuronal responsivity to iontophoretically applied 5-HT has been discussed above. This appears to reflect an enhancement of $5HT_1$ and $5-HT_2$ receptor functioning in depression. The ability of established antidepressant drugs to increase neuronal responsivity to iontophoretically applied 5-HT has been discussed above. This appears to reflect an enhancement of $5-HT_1$ receptor functioning compared to $5-HT_2$ thus neutralising the proposed imbalance in depression. The probable clinical efficacy of citalopram poses a difficulty for the scheme in table 14.6. However, it has been pointed out that pure 5-HT re-uptake blockers may be less effective antidepressants than older, non-selective drugs. One reason may be that they do not effect catecholamine neurotransmission but another may be that they do not have differential effects on the functioning of 5-HT receptor subtypes. The recent results of Soubrie (this volume) seem relevant to this point. They report that high doses of 5-HT re-uptake blockers are ineffective in reversing learned helplessness, an animal model of depression. However low doses are effective and higher doses become effective if a $5-HT_2$ antagonist is concurrently administered.

It would be interesting but probably too laborious to find out whether $5-HT_2$ antagonists potentiate the antidepressant effects of 5-HT re-uptake blockers. It will also be interesting to know whether $5-HT_1$ receptor agonists are effective antidepressants. Buspirone is marketed as an anxiolytic but the clinical trial of Goldberg and Finnerty (1979) also suggests that buspirone is more effective than diazepam or placebo in treating symptoms of depression in neurotic outpatients. The closely related drug, gepirone, is currently being evaluated for antidepressant activity. If these agents prove to be effective antidepressants this will be compatible with the scheme in table 14.6. Not only are they $5-HT_1$ receptor agonists but they also induce delayed decreases in $5-HT_2$ binding sites, just as most other antidepressants (New et al, 1985; DeSouza et al., 1986; Blackshear et al., 1986). Thus $5-HT_1$ agonists have a dual and differential effect on $5-HT_1$ receptor subtype functioning which would reverse the proposed imbalance in depression.

It was suggested earlier that the ability of antidepressants to decrease $5-HT_2$ receptor functioning may be more directly related to their anxiolytic actions than to antidepressant actions. It may be that putative anxiolytic actions of $5-HT_1$ receptor agonists in humans are related to their ability to down-regulate $5-HT_2$ binding sites on chronic administration.

Some studies appear incompatible with the idea that chronic administration of antidepressants enhance $5-HT_1$ mediated neurotransmission. Goodwin, DeSouza and Green (1987) reported that rats chronically treated with antidepressants are less sensitive to the stereotopy-inducing effects of 8OHDPAT, a $5-HT_1$ agonist. They suggested that chronic antidepressants reduce $5-HT_1$ receptor function in animals. However this result needs to be set against the electrophysiological evidence (see above) and a behavioural study (Charanjit et al., 1987) suggestive of enhanced $5-HT_1$ receptor function following antidepressant treatment. Furthermore, it is not known whether antidepressant treatments change $5-HT_1$ receptor binding sites in rat brain. Behavioural evidence for reduced $5-HT_1$ receptor function following antidepressant drug treatment may simply indicate that antipressants drugs are working through systems involving $5-HT_1$ receptors and that diminished functioning is an adaptive response to this rather than the mechanism by which the drugs work.

The proposed relative excess of $5-HT_2$ neurotransmission in depression and the ability of antidepressant drugs to counteract this can be seen as compatible with 5-HT punishment theories. However, the postulated deficit in $5-HT_1$ neurotransmission and the evidence that antidepressant drugs may reverse this cannot readily be accounted for by current formulations of 5-HT punishment theories.

Behavioural functions of $5-HT_1$ receptors
It is suggested that $5-HT_1$ receptors may mediate adaptive-protective responses to aversive stimulation. Kennett et al. (1985) immobilised rats by taping them to a wire grid for two hours. When placed in a open field 24 hours later the stressed animals showed less locomotion and defecated more than non-stressed controls. However, after seven daily immobilisations, no differences were observed in open field behaviour between the chronically stressed animals and unstressed controls. This tolerance or adaptation to chronic stress was associated with the emergence of enhanced $5-HT_1$ mediated behavioural responses to the agonist 5-methoxy-N,N-dimethyl tryptamine. Interestingly, female rats did not become tolerant to the stress and did not show enhanced $5-HT_1$ mediated behaviours. This may be a model for the greater susceptibility of women to depression. The model is discussed more fully in Chapter 16.

The behavioural syndrome of learned helplessness may result from the failure of adaptive-protective responses to stress. The adaptive mechanisms which operate in the Kennet et al. paradigm may be insufficient to cope with the severity and duration of aversive stimuli which are necessary to induce learned helplessness. If $5-HT_1$ receptors mediate tolerance and adaptation to aversive situations then a loss of $5-HT_1$ neurotransmission would be associated with learned helplessness and drugs which restore or enhance $5-HT_1$ receptor function should reverse learned helplessness. Sherman and Petty (1980, 1982) reported that learned helplessness in rats was associated with reduced levels of frontal cortical 5-HIAA and that infusions of 5-HT into the frontal cortex reversed the behavioural deficits. Exactly as predicted by this theory, Soubrie et al. report that the $5-HT_1$ agonist 8-OH-DPAT reverses learned helplessness (chapter 29). Furthermore, they demonstrated that the ability of 8-OH-DPAT to reverse learned helplessness was not attenuated by prior destruction of brain 5-HT neurones with the neurotoxin 5,7-DHT. Thus 8-OH-DPAT reversed learned helplessness by a direct post-synaptic action which was independent of its effects on presynaptic neurones.

It is intriguing that highest concentrations of $5-HT_1$ binding sites in brain are found in the hippocampus. The hippocampus also has high concentration of glucocorticoid receptors. There is evidence that CNS glucocorticoid receptors are intimately concerned with behavioural adaptation to stress. The hippocampus is selectively innervated by the median raphe nucleus. Perhaps median raphe nucleus projections of hippocampal $5-HT_1$ receptors mediate protective adaptive responses to aversive stimuli and when this system fails, learned helplessness in animals and depression in humans is the result.

Summary
It is proposed that 5-HT punishment theories remain viable and relevant to the

Table 14.7 Systems involving

	5-HT$_2$ receptors	5-HT$_1$ receptors
Function	Punishment Fear	Resilience, tolerance to aversion
Change in depression	↑	↓
Symptoms	Anxiety	Learned helplessness Depression Suicide
Antidepressants	↓	↑

aetiology of anxiety and depression. However, such theories need to take account of the differential effects of aversive stimuli on behaviour, the different 5-HT projections and of different 5-HT receptor subtypes.

It is suggested that 5-HT$_2$ receptors, perhaps associated with dorsal raphe nucleus projections to frontal cortex, amygdala and other areas, mediate conventionally studied effects of aversive stimuli on behaviour. In animals, these include behavioural inhibition, conditioned fear responses, and guidance away from fear-associated stimuli. These effects of aversive stimuli are studied in the conflict paradigm, avoidance behaviour and the open field test among others. Selective 5-HT$_2$ antagonists show some efficacy in these models (Colpaert et al., 1985). 5-HT$_1$ agonists do not have consistant anxiolytic effects in animal models. Perhaps they need to be administered repeatedly to induce 5-HT$_2$ receptor down-regulation before consistant anxiolytic effects will be seen.

In humans, excessive 5-HT$_2$ neurotransmission may be associated with fear and anxiety due to an excessive sensitivity to aversive stimuli. Secondary inhibition of positive reinforcement mechanism may give rise to symptoms of depression. Antidepressant drugs may exert their anxiolytic effects by their ability to decrease 5-HT$_2$ neurotransmission, and as a secondary consequence, symptoms of depression are improved.

It is suggested that aversive stimuli also activate 5-HT$_1$ receptor mechanism, perhaps associated with median raphe nucleus projections to the hippocampus, which set in train adaptive-protective responses and mediate a resilience to adversity. These 5-HT$_1$ protective mechanism may fail due to constitutional vulnerability or in the face of overwhelming environmental stressors resulting in learned helplessness and depression. It is argued that antidepressants restore 5-HT$_1$ neurotransmission and that this contributes directly to their therapeutic effects.

References

Akiskal, H.S. & McKinney Jr., W.T. (1975). Overview of recent research in depression: integration of ten conceptual models into a comprehensive clinical frame. Arch. Gen. Psychiat., 32, 285-303.

Al-Yassiri, M.M., Ankier, S.I. & Bridges, P.K. (1979). Trazodone - a new Antidepressant. Life Sci., 28, 2449-2458.

Amin, M.M., Ananth, J.V., Coleman, B.S., Darcourt, G., Farkas, T., Goldstein, B., Lapierre, Y.D., Paykel, E. & Wakelin, J.S. (1984). Fluvoxamine: antidepressant effects confirmed in a placebo-controlled international study. Clin. Neuropharmacol., 7, 580-581.

Ancill, R.J., Poyser, J., Davey, A. & Kennerson, A. (1984). Management of mixed affective symptoms in primary care: a critical experiment. Acta Psychiatr. Scand., 70, 463-469.

Anderson, I.M. & Cowen, P.J. (1986). Clomipramine enhances prolactin and growth hormone responses to l-tryptophan. Psychopharmacol., 89, 131-133.

Asberg, M., Thonen, P., Traskman, L., Bertilsson, L., Ringberger, V. (1976). 'Serotonin depression' a biochemical subgroup within the Affective Disorders. Sci., 191, 478-480.

Blackshear, M.A., Martin, L.L., Sanders-Bush, E. (1986). Adaptive changes in the $5HT_2$ finding site after chronic administration of agonists and antagonists. Neuropharmacol., 25, 1267-1271.

Brittain, R.T., Butler, A. & Coates, I.H. et al. (1987). GR 380327, a novel selective $5HT_3$ receptor antagonist. Br. J. Pharmacol., 90, 87P.

Carney, M.W.P., Roth, M. & Garside, R.F. (1965). The diagnosis of depressive syndromes and the prediction of ECT response. Br. J. Psychiat., 111, 659-674.

Ceuleman, D.L.S., Hoppenbrouwers, M.L.J.A., Gelders, Y.G. & Reyntjens, A.J.M. (1985). The influence of ritanserin, a serotonin antagonist, in anxiety disorders: a double blind placebo-controlled study versus lorazepam. Pharmacol. Psychiat., 18, 303-305.

Charanjit, S., Cohen, R.M., Hill, J.L., Murphy, D.L. & Zohar, J. (1987). Long-term imipramine treatment enhances locomotor and food intake suppressant effects of m-chlorophenylpiperazine in rats. Br. J. Pharmacol., 91, 747-752.

Charig, E.M., Anderson, I.M., Robinson, J.M., Nutt, D.J. & Cowen, P.J. (1987). L-tryptophan and prolactin release: evidence for interaction between $5-HT_1$ and $5HT_2$ receptors. Human Psychopharmacol., 1, 93-97.

Charney, D.S., Heninger, G.R., Renhard, J.F., Sternberg, D.E. & Hafstead, E.M. (1982). Effect of Intravenous L-Tryptophan on Prolactin and Growth hormone and mood in Healthy Subjects. Psychopharmacol., 77, 217-222.

Claassen, V. (1983). Review of the animal pharmacology and pharmacokinetics of Fluvoxamine. Br. J. Clin. Pharmac., 15, 349S-355S.

Clinical Research Centre, Division of Psychiatry (1984). The Northwick Park ECT Trial. Predictors of Response to real and simulated ECT. Br. J. Psychiat., 144, 227-237.

Colpaert, F.C., Meert, T.F., Niemegeers, C.J.E. & Janssen, P.A.J. (1985). Behavioural and 5HT antagonist effects of ritanserin: a pure and selective antagonist of LSD discrimination in rat. Psychopharmacol., 16, 45-54.

Cooper, S.J., Owen, F., Chambers, D.R., Crow, T.J., Johnson, J.A. & Poulter, M. (1986). Post-mortem neurochemical findings in suicide and depression: a study of the serotonergic system and imipramine binding in suicide victims. In: The Biology of Depression. (Ed. Deakin, J.F.W.), pp. 1-25. London: Gaskell.

Cowen, P.J. & Charig, E.M. (1987). Neuroendocrine responses to tryptophan in major depression. Arch. Gen. Psychiat., 44, 958-966.

Crow, T.J., Cross, A.J., Cooper, S.J. et al., (1984). Neurotransmitter receptors and monoamine metabolites in the brains of patients with Alzheimer-type dementia and depression and suidices. Neuropharmacol., 23, 1561-1569.

Danish University Antidepressant Group (1986), Citalopram: Clinical Effect Profile in Comparison with Clomipramine. A Controlled Multicenter Study. Psychopharmacol., 90, 131-138.

De Montigny, C. & Aghajanian, G.K. (1984). Tricyclic antidepressants: long-term treatment increases responsivity of rat forebrain neurones to serotonin. Sci., 202, 1303-1306.

DeSouza, R.J., Goodwin, G.M., Green, A.R. & Heal, D.J. (1986). Effect of chronic treatment with $5HT_1$ agonist (8-OH-DPAT and RU 24969) and antagonist (ipsapirone) drugs on the behavioural responses of mice to 5HT1 and 5HT2 agonists. Br. J. Pharmacol., 89, 377-384.

Deakin, J.F.W. (1983). Roles of serotonin in escape, avoidance and other behaviours. Psychopharmacol., Cooper, S.J. OUP. 149-193.

Fuller, R.W., Snoddy, H.D., Mason, N.R. & Owen, J.E. (1981). Disposition and pharmacological effects of m-chlorophenylpiperazine in rats. Neuropharmacol., 20, 155-162.

Gardner, C.R., Guy, A.P & James, V. (1987). Opposite effects of $5HT_1$ and $5HT_2$ agonists on suprahyoid muscle twitching. Br. J. Pharmacol., 91, 471P.

Geller, N.E. & Blum, K. (1970). The effects of 5-HTP on parachlorophenylalanine (P-CPA) attenuation of conflict behaviour. Eur. J. Pharmacol., 9, 319-324.

Gjerris, A., Sorensen, A.S., Rafaelsen, O.J., Wederlin, L., Alling, G. & Linnoila, M. (1987). 5HT and 5HTIAA in cerebrospinal fluid in depression. J. Aff. Disorder, 12, 13–22.

Glaser, T & Traber, J. (1985). Binding of the putative anxiolytic TVX Q 7821 to hippocampal 5-hydroxytryptamine (5HT) recognition sites. Arch. Pharmacol., 29, 211–215.

Goldberg, D.P., Bridges, K., Duncan-Jones, P. & Grayson, D. (1987). Dimensions of neuroses seen in primary care settings. Psycholog. Med., 17, 461–470.

Goldberg, H.L. & Finnerty, R.J. (1979). The comparative efficacy of buspirone and diazepam in the treatment of anxiety. Am. J. Psychiat., 136, 1184–1187.

Gonzalez, J.P. & Stolz, J.F. (1985). Functional antagonism between subtypes of 5-hydroxy-tryptamine receptors on morphine antinociception. Br. J. Pharmacol., 85, 248P.

Goodwin, F.K., Post, R.M., Dunnmer, D.L. & Fordon, E.K. (1973). Cerebrospinal fluid amine metabolites in affective illness: the probenecid technique. Am. J. Psychiat., 130, 73–79.

Goodwin, G.M., DeSouza, R.J. & Green, A.R. (1987). Attenuation by electroconvulsive shock and antidepressant drugs of $5HT_{1A}$ receptor-mediated hypothermia and serotonin syndrome produced by 8-OH-DPAT in the rat. Psychopharmacol., 91, 500–505.

Goodwin, G.M. & Green, A.R. (1985). A behavioural and biochemical study in mice and rats of putative selective agonists and antagonists for $5HT_1$ and $5HT_2$ receptors. Br. J. Pharmacol., 84, 743–753.

Goodwin, G.M., Fairbairn, C.G & Cowen, P.J. (1987). The effects of dieting and weight loss on neuroendocrine responses to L-tryptophan, clonidine and apomorphine in volunteers: important implications for neuroendocrine investigation in depression. Arch. Gen. Psychiat., 44, 952–957.

Graeff, F.G. (1974). Tryptamine antagonists and punished behaviour. J. Pharmacol. Exp. Ther., 189, 344–350.

Gray, G. (1971). The psychology of fear and stress. London: Weidenfeld and Nicolson.

Gudelsky, G.A., König, J.I. & Meltzer, H.Y. (1986). Thermoregulatory responses to serotonin (5-HT) receptor stimulation in the rat: evidence for opposing roles of $5-HT_2$ and $5-HT_{1A}$ receptors. Neuropharmacol., 25, 1307–1313.

Harrison-Read, P.E. (1983). "Wet dog shake" behaviour in rats may reflect functionally-opposed indoleaminergic systems involving different 5-HT receptors. Brit. J. Pharmacol., 78, 92P.

Heal, D.J., Philpot, K., Molyneux, S.G. & Metz, A. (1983). Intracerebroventricular administration of 5,7-dihydroxy-tryptamine to mice increases both head-twitch response and the number of cortical 5-HT$_2$ receptors. Neuropharmacol., 24, 1201-1205.

Heninger, G.R., Charney, D.S. & Sternberg, D.E. (1984). Serotonergic function in depression. Arch. Gen. Psychiat., 41, 398-402.

Johnstone, E.C., Cunningham Owens, D.G., Frith, C.D., McPherson, K., Dowie, C., Riley, G. & Gold, A. (1980). Neurotic illness and its response to anxiolytic and antidepressant treatment. Psychol. Med., 10, 321-328.

Jones, B.J., Costall, B., Domeney, A.M., Kelly, M.E., Naylor, R.J., Oakley, N.R. & Tyers, M.B. (1988). The potential anxiolytic activity of GR 38032, a 5-HT$_3$ receptor agonist. Brit. J. Pharmacol., 93, 985-993.

Kahn, R.J., McNair, M., Lipman, S.R., Covi, L., Rickels, K., Downing, R., Fisher, S. & Frankenthaler, L.M. (1986). Imipramine and chlordiazepoxide in depressive and anxiety disorders. Arch. Gen. Psychiat., 43, 79-84.

Kennett, G.A., Dickinson, S. & Curzon, G. (1985). Enhancement of some 5-HT-dependent behavioural responses following repeated immobilization in rats. Brain Res., 330, 253-263.

Lakoski, J.M. & Aghajanian, G.K. (1985). Effects of ketanserin on neuronal responses to serotonin in the prefrontal cortex, lateral geniculate and dorsal raphe nucleus. Neuropsychopharmacol., 24, 265-273.

Le Bars, D., Dickenson, A.M. & Besson, J.M. (1979). Diffuse noxious inhibitory controls (DNIC) II. Lack of effect on non-convergent neurones, supraspinal involvement and theoretical implications. Pain, 6, 305-327.

Leysen, J.E., Van Gompel, P., Gommeren, W., Woestenborghs, R. & Janssen, P.A.J. (1986). Down regulation of serotonin-S2 receptor sites in rat brain by chronic treatment with the serotonin-S2 antagonists: ritanserin and setoperone. Psychopharmacol., 88, 434-444.

Leysen, J.E., Van Gompel, P., Verweimp, M. & Niemegeers, C.J.E. (1983). Role and localization of serotonin 2 (S2)-receptor binding sites: effects of neuronal lesions. In: Molecular Pharmacology to Behaviour. (Eds. Mandel, P. & Defeudis, F.V.), pp. 373-383. New York: Raven Press.

McCance, S.L., Cowen, P.J., Waller, H. & Grahame-Smith, D.G. (1988). The effect of metergoline on endocrine responses to L-Tryptophan. J. Psychopharmacol. 2, 90-94.

Meltzer, H.Y., Perline, R., Tricou, B.J., Lowy, M. & Robertson, A. (1984). Effect of 5-hydroxy-tryptophan on serum cortisol levels in major affective disorders: 1. enhanced response in depression and mania. Arch. Gen. Psychiat., 41, 366-374.

Meyerson, L.R., Wennogle, L.P., Abel, M.S., et al., (1982). Human brain receptor alterations in suicide victims. Pharmacol. Biochem. Behav., 17, 159-163.

Morris, J.B. & Beck, A.T. (1974). The efficacy of antidepressant drugs. Arch. Gen. Psych., 30, 667–674.

New, J.S., Eison, M.S., Riblett, C.A., Lobeck, W.G., Yevich, J.P. & Temple, D.L. (1985). BMY 13803: a novel anxioselective agent with potential antidepressant activity. 189 ACS Meeting Abstract: 89.

Norton, K.R.W., Sireling, L.I., Bhat, A.V., Rao, B. & Paykel, E.S. (1984). A double-blind comparison of fluvoxamine, imipramine and placebo in depressed patients. J. Affect. Disord., 7, 297–308.

Owen, F., Cross, A.J., Crow, T.J., et al., (1983). Brain 5-HT_2 receptors and suicide. Lancet II, 1256.

Peroutka, S.J. & Snyder, S.H. (1980). Long-term antidepressant treatment decreases spiroperidol-labelled serotonin receptor binding. Sci., 210, 88–90.

Perry, E.K., Marshall, E.F., Blessed, G., et al., (1983). Decreased imipramine binding in the brains of patients with depressive illness. Br. J. Psychiat., 1412, 188–192.

Petty, F. & Sherman, A.D. (1983). Learned helplessness induction decreases in vivo cortical serotonin release. Pharmacol. Biochem. Behav., 18, 649–650.

Quik, M. & Azmitia, E. (1983). Selective destruction of the serotonergic fibers of the fornix-fimbria and cingulum bundle increases 5-HT_1 but not 5-HT_2 receptors in rat midbrain. Eur. J. Pharmacol., 90, 377–384.

Racagni, G. & Bradford, D. (1984). Biochemical and behavioural changes in chronic fluvoxamine administration. Proc. 14th CINP Congr., Florence, Italy.

Sherman, A.D. & Petty, F. (1980). Neurochemical basis of the action of antidepressants on learned helplessness. Behav. Neurol. Biol., 30, 119–134.

Sherman, A.D. & Petty, F. (1982). Additivity of neurochemical changes in learned helplessness and imipramine. Behav. Neurol. Biol., 35, 344–353.

Siever, L.J., Murphy, D.L., Slater, S., De la Vega, E. & Lipper, S. (1984). Plasma prolactin changes following fenfluramine in depressed patients compared to controls: an evaluation of central serotonergic responsivity in depression. Life Sci., 34, 1029–1039.

Solomon, R.A., Sharpley, A.L. & Cowen, P.J. (1987). 5-HT antagonists and slow wave sleep. Br. J. Clin. Pharmacol. 25, 125–126P.

Soubrie, P. (1986). Reconciling the role of central serotonin neurones in human and animal behaviour. Behav. Brain Sci. 9, 319–364.

Stanley, M. & Mann, J.J. (1983). Increased serotonin-2 binding sites in frontal cortex of suicide victims. Lancet I, 214–216.

Taylor, P. & Fleminger, J.J. (1980). ECT for schizophrenia. The Lancet II, 1380-1383.

Tye, N.C., Everitt, B.J. & Iversen, S.D. (1977). 5-hydroxy-tryptamine and punishment. Nature, **268**, 741-743.

Vetulani, J., Lebrecht, U. & Pilc., A. (1981). Enhancement of responsiveness of the central serotonergic system and serotonin-2 receptor density in rat frontal cortex by electroconvulsive treatment. Eur. J. Pharmacol., **76**, 81-85.

Weiss, S., Sebben, M., Kemp., D.E. & Bockaert, J. (1986). Serotonin 5-HT_1 receptors mediate inhibition of cyclic amp production in neurons. Eur. J. Pharmacol., **120**, 277-280.

Wilson, C.A. & Hunter, A.J. (1985). Progesterone stimulates sexual behaviour in female rats by increasing 5-HT activity on 5-HT_2 receptors. Brain Res., **333**, 223-229.

Wise, C.D., Berger, B.D. & Stein, L. (1970). Serotonin: a possible mediator of behavioural suppression induced by anxiety. Dis. Nerv. Syst., GWAN. Suppl., **31**, 34-37.

Young, J.P.R., Coleman, A. & Lader, M.H. (1987). A controlled comparison of fluoxetine and amitriptyline in depressed out-patients. Brit. J. Psychiat., **151**, 337-340.

MODULATION OF THE 5-HT SYSTEM AND ANTIPSYCHOTIC ACTIVITY

Jan A.M. van der Heyden,
Department of Pharmacology, Duphar B.V., P.O.Box 2, 1380 AA Weesp, The Netherlands

Introduction

The hypothesis on the involvement of 5-HT in schizophrenia is not new, but dates from the beginning of the 50's. In 1947 Stoll reported the similarity between psychosis induced by LSD and that occurring in schizophrenia. Several years later Gaddum et al. (1953) described the 5-HT antagonistic properties of LSD which was followed shortly thereafter by a publication by Woolley and Shaw (1954) ending as follows: "In summary, the suggestions we wish to make are the following: (1) serotonin probably plays a role in maintaining normal mental processes; (2) metabolically induced deficiency of serotonin may contribute to the production of some mental disorders; (3) serotonin or a long-acting derivative of it may prove capable of alleviating disorders similar to schizophrenia". This is truly a remarkable conclusion for a time when 5-HT was described as a hormone and antagonists were called anti-metabolites.

Table 15.1 Neurotransmitters involved in schizophrenia

neurotransmitter	psychotogen	drug	mechanism
dopamine	amphetamine	neuroleptics*	D_2 antagonist
		SCH23390	D_1 antagonist
		3-PPP	presynaptic agonist
serotonin	LSD, mescaline	neuroleptics	$5-HT_2$ antagonist
		befiperide	$5-HT$ agonist
		GR38032F	$5-HT_3$ antagonist
noradrenaline	yohimbine?	prazosine	α_1 antagonist
		clonidine	α_2 agonist
		propranolol	β antagonist
GABA	?	benzodiazepines	indirect agonist
		baclofen	agonist
opiates	morphine, PCP	naloxone	antagonist
		γ-endorphine	agonist
		rimcazole	sigma-antagonist
peptides	?	CCK-analogue	agonist
		TRH-analogue	agonist

*: these are the only drugs mentioned in this table of which the clinical efficacy has been well established.
?: unknown (uncertain)

Presently, many transmitters are proposed to be involved in schizophrenia (see Table 15.1).

The most widely accepted dopamine (DA) theory of schizophrenia states that an overactivity of the DA system is the underlying cause of this disease. This theory receives support from the finding that amphetamine, a drug enhancing DA functioning, mimics psychosis in normal humans (Connell, 1958; Angrist and Gershon, 1970; Griffith et al., 1972) and from the fact that all clinically effective antipsychotic drugs are neuroleptics which all block DA receptors (Carlsson and Lindqvist, 1963; van Kammen, 1979; Niemegeers and Janssen, 1979). More recently developed drugs diminish DA functioning by interaction with a subclass of DA receptors like the D1 specific antagonist SCH 23390 (Ioro et al., 1983) and presynaptic receptor agonist 3-PPP (Hjorth et al., 1981).

As mentioned above, 5-HT has been considered to be involved in schizophrenia for a long time. Many drugs affecting the 5-HT system induce psychosis in man (Stoll, 1947; Shulgin, 1978; Hollister, 1978). The antipsychotic activity of neuroleptics can be interpreted as supportive evidence for the involvement of 5-HT since most of these drugs potently block the 5-HT$_2$ receptor as well as the D$_2$ receptor (see Table 15.4). Recently, two non-dopaminolytic compounds affecting the 5-HT system (Befiperide and GR 38032F) have been proposed as potential antipsychotic drugs (Van der Heyden et al., 1986, 1987; Costall et al., 1987).

An altered noradrenergic transmission is also often regarded to be involved in schizophrenia (Hornykiewicz, 1982; Rodnight, 1983; Van Kammen & Antelman, 1984). Yohimbine can be regarded as a psychotogen and drugs diminishing the noradrenergic functioning either pre- or postsynaptically have been reported to ameliorate schizophrenic symptoms (Atsmon & Blum, 1978; Freedman et al., 1982). Increasing inhibitory processes in the brain through enhancement of GABA transmission either indirectly with benzodiazepines or directly with GABA agonists has been reported to result in beneficial effects in schizophrenic patients (Donaldson et al., 1983; Linnoila et al., 1976; Simpson et al., 1976). The putative involvement of the opiate system in schizophrenia is based upon the activity of psychotogens like morphine and PCP and the claimed therapeutic efficacy by some investigators of the opiate antagonists naloxone and rimcazole and the opiate agonist γ-endorphin (Mueser & Dysken, 1983; Van Ree & De Wied, 1981; Davidson et al., 1982; Chouinard & Ammable, 1984). Finally the use of analogues of peptides like cholecystokinin (CCK) and thyrotropin-releasing hormone (TRH) in treatment of schizophrenia has been reported (Wilson et al., 1973; Moroji et al., 1982).

Several approaches can be distinguished which are directed towards the foundation of a 5-HT hypothesis of schizophrenia.
1) Measurement of the central 5-HT activity as reflected by the concentration of 5-HT or its metabolite 5-HIAA in body fluids of schizophrenic patients.
    The results obtained are inconsistent and do not allow general conclusions about an abnormal activity of the 5-HT system in schizophrenic patients (Wyatt et al., 1973; Matthysse & Sugarman, 1978; Curzon et al., 1980; Potkin et al., 1983)
2) Measurement of the formation of methylated indolamines to detect any

abnormal methylation processes in schizophrenic patients resulting in the formation of endogenous psychotogens possibly interacting with 5-HT receptors. The results obtained sofar remain equivocal (Gillin et al., 1976; Murray et al., 1979; Baldessarini et al., 1979).
3) Establish the relation between drugs inducing psychosis in man and their mechanism of action. Several drugs stimulating the 5-HT system are well known hallucinogens, Glennon et al. (1984) found a significant correlation between 5-HT$_2$ binding affinity and hallucinogenic potency in man for a number of phenalkylamine derivatives. Another approach is of course to establish the therapeutic value of drugs affecting the 5-HT system. At this moment, however, no drugs are yet available that selectively affect the 5-HT system and have beneficial effects in the treatment of schizophrenia.
4) Since 5-HT is intimately involved in sleep mechanisms, much effort has been spent on establishing a relation between disturbance of sleep patterns and schizophrenia (Feinberg et al., 1969; Steru et al., 1969; Zarcone et al., 1969). These results do not allow any general conclusions to be drawn.

The above mentioned approaches have not yet demonstrated an unequivocal relation between 5-HT activity and schizophrenia. The major difference with the well established DA hypothesis of schizophrenia, however, is the lack of selective serotonergic drugs with proven clinical efficacy in the treatment of this disease. Demonstration of the clinical efficacy of drugs like befiperide, which stimulates 5-HT receptors without affecting DA receptors (Van der Heyden et al., 1986, 1987) may result in additional support for a role of 5-HT in schizophrenia.

When developing new drugs for the treatment of schizophrenia, several approaches can be followed. The classic profile of an antipsychotic drug is of course one that includes D_2 receptor blockade. Such an activity can be detected in animal models ranging from in vitro binding techniques to behavioural models. A compound possessing such an activity is very likely to be an effective antipsychotic agent. Blockade of DA receptors, however, not only predicts antipsychotic activity but is also held responsible for the induction of unwanted extrapyramidal (motor) side effects (Honma and Fukushima, 1976; Ljungberg and Ungerstedt, 1978; Tarsy, 1983). Thus, besides a high likelihood of therapeutic efficacy, one can also expect a high incidence of the troublesome extrapyramidal side-effects including tardive dyskinesia. One approach to circumvent these problems may be the development of drugs modulating DA activity through some other mechanism than D_2 receptor blockade. The animal models used range from in vitro functional tests (e.g. release of ^{3}H-DA), to behavioural models. Presynaptic DA agonists (e.g. 3-PPP, Hjorth et al., 1981) have been developed on this basis. Theoretically any drug diminishing DA activity may have antipsychotic activity.

A third and speculative approach is to develop a drug possessing antipsychotic activity without affecting DA neurotransmission at all. Obviously, evidence for a possible therapeutic efficacy can only be provided by behavioural models that have a good predictive value for clinical efficacy (i.e. good correlation with activity in man). The conditioned avoidance behaviour models may qualify as such a test. The first description of such a procedure dates from 1932 when Warner described the "association span of the white rat". After the accidental discovery of chlorpromazine as an effective treatment for schizophrenia in the early 50's, Cook et al. (1955) were the first ones to describe the inhibitory action of chlorpromazine on conditioned avoidance. Since then the conditioned avoidance response (CAR) has often been employed in the preclinical evaluation of antipsychotic drugs. A good correlation has been shown to exist between the clinical efficacy of these drugs and their ability to inhibit CAR behaviour (Kuribara & Tadokoro, 1980; Arnt, 1982).

All clinically used antipsychotics are active in reducing the conditioned avoidance response (CAR) in a selective manner. That is, doses which selectively impair the CAR do so without a significant change in escape responses (ER) (Herz, 1960; Davidson & Weidley, 1976). The conditioned avoidance response is maintained by relatively weak conditioned stimuli (CS) which acquire their properties by being paired temporally with some unconditioned stimulus (UCS). The escape response, on the other hand, has a stronger response strength, by virtue of its being maintained directly by the UCS (shock, for example). It has been proposed that this selectivity is due to the differential response strength (Grilly et al., 1984; Bignami, 1978), suggesting that these drugs may have their effects by filtering out irrelevant or weak stimuli, but leaving behaviour maintained by more salient stimuli intact. In this respect a remark from Weil-Malherbe (1978) about the similarity of the influence of hallucinogenic drugs and acute schizophrenia is appropriate: "The flooding of consciousness with an excess of sensory stimuli may, in both states, lead to a jamming of the 'filter' which separates the relevant from the irrelevant and the real from the unreal".

Though the value of inhibition of CAR behaviour has only been shown for clinically effective antipsychotics which block D_2 receptors it can be hypothesised that this D_2 receptor blockade is not a prerequisite for inhibition of CAR and hence antipsychotic activity. We therefore tested a variety of serotonergic drugs in a CAR procedure described earlier (Van der Heyden & Bradford, 1985). Since the differential response strengths of the avoidance and escape responses greatly determines the selectivity of drug effects in the CAR procedure it is of great importance to carefully choose the appropriate intensity of the conditioned and unconditioned stimulus, maintaining the avoidance and escape response respectively. Thus it has been shown that various centrally acting drugs besides the neuroleptics, including sedatives, anxiolytics and analgetics can selectively inhibit CAR (Bignami, 1978; Houser, 1978, Cornfeldt et al., 1978; Sanger, 1985; Maffi, 1959). The selectivity of some of these drugs, however, has been shown not to be present at low shock intensities, maintaining the escape response (Grilly et al., 1984).

The CAR paradigm we employed was a simple procedure which can be acquired within a single 15-trial session (i.e. avoidance response rate >80%). In order to make the acquisition of the response as simple as possible for the animals, a species specific defence reaction (i.e. running) and a clearly discriminated compound stimulus (being placed in the experimental chamber + light) were employed. The conditioned stimulus (CS) for the one-way conditioned avoidance behaviour was a light provided by a 12W light bulb in the shock compartment, that was activated immediately upon placing the animal in this compartment. Ten seconds after the onset of the CS an electric shock serving as the unconditioned stimulus (UCS) was applied for 10 sec to the grid floor.

Each session consisted of 15 trials with an intertrial interval of 60 seconds. An avoidance response was recorded if a rat avoided the UCS by moving to the other (non-shock) compartment within 10 sec after the onset of the CS. If it failed to avoid, the rat could still escape the UCS by moving to the other compartment during the ensuing 10 sec period (escape response). If it failed to respond to the UCS a failure was recorded. The effect of varying the strength of the UCS (shock) on acquisition of the appropriate response within a single 15-trial session is shown in Figure 15.1.

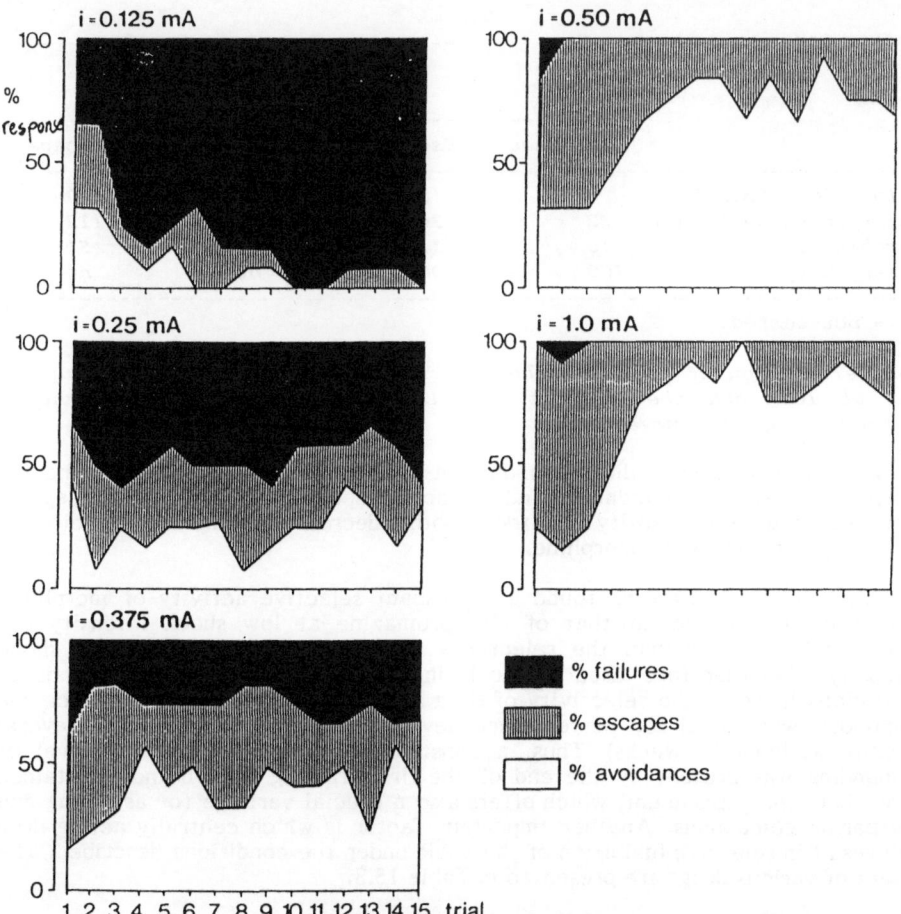

Figure 15.1: *Effect of varying shock intensity on the acquisition of avoidance behaviour. The data shown are the mean percentual responses per trial of groups of 12 animals.*

Shock levels of 0.5 mA and higher met the criterium of reaching at least 80% avoidance responses. Training and testing the animals either at a 0.5 mA or 1.0 mA shock intensity did not affect the disruption of avoidance behaviour by haloperidol and chlorpromazine but did decrease their effect on escape responses (i.e. increased selectivity, see Table 15.2).

Table 15.2 Effect of shock intensity on selectivity of drug effect

Drug (route)	Shock intensity 0.5 mA		1.0 mA	
	Inhibition (mg/kg) ED_{50} disruption of			
	Avoidance	Escape	Avoidance	Escape
haloperidol (p.o.)	0.9	8.2	0.9	30
chlorpromazine (p.o.)	22	34	27	112
morphine (i.p.)	10	11	32	51
diazepam (p.o.)	> 100	> 100	nt	nt

nt = not tested.

For calculation of the ED_{50} the response of vehicle treated group is set at 100% and the ED_{50} is calculated by linear interpolation of the dose-response curve.

At low shock intensity, diazepam did not affect avoidance behaviour whereas morphine disrupted avoidance and escape responses to the same degree. Increasing the UCS intensity resulted in both a decreased disruption of avoidance and escape behaviour for morphine.

At this high shock level we found a significant selective activity of morphine, that was comparable to that of chlorpromazine at low shock intensity but substantially smaller than the selectivity of chlorpromazine at similar shock intensity. We therefore used a shock intensity of 0.5 mA in order not to artificially increase the selectivity of drug effects. We also found that once the behaviour was acquired, performance levels remained constant over several months (at least 17 weeks). Thus, a constant and reliable base line level of responding was present by the end of the first training session and maintained throughout the experiment, which offers also a crucial variable for assessing and comparing compounds. Another important factor is which centrally acting drug will result in selective inhibition of the CAR under the conditions described. The effect of various drugs are presented in Table 15.3.

All DA antagonists selectively inhibit CAR (for details see Table 15.4) whereas DA agonists like apomorphine and amphetamine facilitate avoidance behaviour.

The results obtained with serotonergic drugs varies depending upon the nature of the receptor interaction (for details see below and Tables 15.6 and 15.7). Of the adrenergic drugs tested only clonidine showed a highly selective activity. This is

Table 15.3 Effects of centrally acting drugs in the CAR procedure

Drug Mechanism + drug used	route	avoidance inhibition ED_{50} or (%) inhibition at highest dose (mg/kg)	escape inhibition ED_{50}
Dopamine			
antagonists		active	less active
agonists		facilitation *	-
Serotonin			
antagonists		inactive	-
agonists		active/inactive	active/inactive
Noradrenaline			
yohimbine	p.o.	>30 (31%)	>30
propranolol	p.o.	>50 (21%)	>50
clonidine	p.o.	1.2	>10
guanabenz	p.o.	>42	>42
guanfacine	p.o.	>30	>30
Monoamine uptake inhibition			
amitriptyline	p.o.	>100 (30%)	>100
clovoxamine	p.o.	>100	>100
fluvoxamine	p.o.	>100	>100
Opiate			
morphine	i.p.	9.2	10.4
naloxone	i.p.	>10	>10
Acetylcholine			
atropine	p.o.	>180 (38%)	>180
Benzodiazepine			
Ro-15-1788	p.o.	>100	>100
chlordiazepoxide	p.o.	>120 (24%)	>120
diazepam	p.o.	>100	>100
Barbiturate			
butobarbital	p.o.	120	135
phenobarbital	p.o.	>200	>200
Anticonvulsant			
carbamazepine	p.o.	>100	>100

* indicated by a decreased response latency.
ED_{50}'s were calculated as described in Table 15.2.

not necessarily a false positive activity since some inconsistent data on clonidine have been reported about its efficacy in the treatment of schizophrenia (Freedman et al., 1982).

Surprisingly, two other α_2 adrenoceptor agonists tested, guanfacine and guanabenz were inactive in the CAR. Of the various other drugs tested none showed a selective inhibitory effect on CAR. Only morphine and butobarbital simultaneously affected both avoidance and escape behaviour at high doses. Excluding the serotonergic drugs, which will be discussed later, these data strongly support the specificity of this CAR procedure for clinically effective antipsychotic drugs.

Another way for drugs not blocking DA receptors to predict antipsychotic activity may be the inhibition of DA mediated behaviour. For the serotonin system both inhibitory and stimulatory influences on DA mediated behaviour have been reported, depending upon the nature of the behaviour studied. Thus increased DA functioning can result in hyperlocomotion, stereotypy, yawning and hypothermia. All these behaviours can be blocked by DA antagonists (i.e. neuroleptics) and by increased activity of the 5-HT system except for stereotyped behaviour which is enhanced by an increased 5-HT activity (Grabowska et al, 1973, 1974; Costall & Naylor, 1974, 1978; Kostowski, 1975; Dray et al., 1978; Menon & Vivovia, 1981; Carter & Pycock, 1978, 1981; Ortman et al., 1982). We used the inhibition of apomorphine-induced climbing of mice to detect such an activity in the serotonergic drugs tested. Briefly, the climbing intensity of mice after administration of 1 mg/kg s.c. of apomorphine was determined on a 2-point intensity scale. Drugs were administered either 30 or 60 min before apomorphine (i.p. or p.o. respectively). In order to characterise the interaction of the drug tested with the DA and/or 5-HT receptors we measured the affinity for the D_2 and $5-HT_{1A}$, $5-HT_{1B}$, $5-HT_{1C}$ and $5-HT_2$ receptors according to well described methods (Gozlan et al., 1983; Sills et al., 1984; Creese et al., 1977; Creese & Snyder, 1978). The affinity for all receptors was measured in rat brain tissue except for the $5-HT_{1C}$ assay, which uses pig choroid plexus and [^{3}H]-5-HT as a ligand.

We measured the activity of several clinically effective neuroleptics, belonging to various different structure classes in the CAR procedure, the apomorphine-induced climbing and their affinity for the D_2 and $5-HT_2$ receptor (Table 16.4). These results were compared with each other and with the average clinical dose.

The receptor binding data in Table 15.4 show that, with the exception of molindone and sulpiride, all neuroleptics block both D_2 and $5-HT_2$ receptors to a similar extent. This fits in both the DA- and 5-HT hypothesis of schizophrenia. Except for sulpiride, all neuroleptics selectively inhibit CAR, that is, escape responses are disrupted at higher doses than needed for inhibition of avoidance responses. The lowest selectivity was found for chlorpromazine whereas haloperidol showed the highest selectivity. All drugs tested also potently inhibited the apomorphine-induced climbing behaviour in mice. The ED_{50} values obtained in this model were always well below those obtained for inhibition of avoidance behaviour. This does not necessarily point to an uncoupling of DA blockade and CAR inhibition but is most likely explained by a

Table 15.4 Activity of clinically effective neuroleptics

Drug	clinical dose (mg/day)*	affinity D_2 (pK_i)	affinity $5-HT_2$ (pK_i)	inhibition avoidance (ED_{50}, mg/kg PO)	inhibition escape (ED_{50}, mg/kg)	inhibition climbing (ED_{50}, mg/kg)
loxapine	15 - 60	8.1	8.1	0.4	1.2	0.02
haloperidol	3 - 20	8.9	7.3	0.9	8.2	0.3
butaclamol	5 - 50	9.1	7.4	1.9	3.8	0.6
fluphenazine	1 - 10	8.5	7.9	2.4	6.0	0.5
trifluperazine	6 - 30	8.2	7.4	4.8	7.7	1.3
prochloroperazine	50 - 250	8.3	7.5	5.8	9.9	2.0
molindone	15 - 60	8.8	5.7	10	60	2.0
clopipazan	5 - 40	8.1	8.8	12	>50	NT
chlorpromazine	150 - 600	8.1	8.2	22	34	7.4
clozapine	100 - 600	7.2	7.9	26	70	13.0
chlorprothixene	100 - 400	8.4	8.2	32	130	3.1
thioridazine	150 - 600	8.0	8.0	100	>120	6.0
sulpiride	200 -1200	7.4	4.6	>220	>220	200

* data taken from Van Praag (1978) and Psychotropics (1986)

Compounds are ordered in according to decreasing activity in inhibiting avoidance response. All drugs were administered orally 60 min before testing. Climbing test refers to antagonism of apomorphine-induced climbing in mice, inhibition of avoidance and escape was measured in the CAR procedure in rats as described in the text. Affinity for D_2 and $5-HT_2$ receptors was determined by displacement of ^3H-spiroperidol from rat striatal and cortical brain tissue, respectively. NT = not tested.

difference in sensitivity of both models at the given experimental conditions.

The rank order correlation coefficient between clinical dose and inhibition of avoidance response was 0.79 (n=13, p=0.001) confirming the value of this test for predicting clinical efficacy, in accordance with previously reported data by Kuribara & Todakoro (1980) and Arnt (1982).

An even higher correlation coefficient was found for the inhibition of apomorphine-induced climbing behaviour and the clinical dose (r=0.87, n=12, p<0.001). A weaker correlation coefficient of 0.61 (n=13, p=0.014) was found for the affinity for the D_2 receptor and clinical dose, which may be attributed to an in vitro-vivo difference and/or to the structural diversity of the compounds tested. No correlation between affinity for $5-HT_2$ receptors and clinical dose was found. As can be expected for DA antagonists, the inhibition of avoidance behaviour correlated well with inhibition of climbing behaviour (r=0.94). A weak but significant correlation was also found for affinity for D_2 receptors and inhibition of avoidance and climbing behaviour (r=0.57 and 0.65, respectively). Again no relation between $5-HT_2$ affinity and behavioural activity was found.

Though the relation between avoidance inhibition and clinical potency may be the result of the DA antagonistic activity of the drugs tested it cannot be

Table 15.5 Correlation coefficients for clinically effective neuroleptics

	Spearman correlation coefficients			
	inhibition avoidance	inhibition climbing	affinity D_2 receptor	affinity 5-HT_2 receptor
clinical dose	0.79 (0.001)	0.87 (<0.001)	0.61 (0.014)	-0.09 (0.39)
inhibition avoidance		0.94 (<0.001)	0.57 (0.022)	-0.15 (0.31)
inhibition climbing			0.65 (0.011)	-0.04 (0.45)

Data are the rank order correlation coefficients of the data shown in Table 15.4. The values in brackets indicate the significance two-sided (p-values) of the calculated correlation coefficient.

excluded that inhibition of CAR can also predict antipsychotic activity in man for drugs not blocking the DA receptors.

On basis of this assumption we tested a variety of serotonergic drugs in the CAR procedure and compared this activity with their ability to inhibit apomorphine-induced climbing and their affinity for D_2 and various 5-HT receptors.

The results obtained with the various 5-HT antagonists tested in the CAR and apomorphine-induced climbing procedure are described in Table 15.6. On basis of the receptor binding data presented here and literature data the selectivity of the drugs for the 5-HT receptors and D_2 receptor was determined. Of the non-selective 5-HT antagonists tested both methiotepine and metergoline but not methysergide inhibited avoidance behaviour. The former two compounds, however, potently block DA receptors (D_2 affinity metergoline see Spano et al., 1978) as confirmed by their inhibition of apomorphine-induced climbing behaviour which can explain their activity in the CAR procedure. None of the 5-HT_2/5-HT_{1C} selective antagonists attenuated avoidance behaviour but did show inhibitory activity in the climbing test. The 5-HT_{1A} selective antagonist propranolol was not active in either model. The 5-HT_3 antagonists tested (Fozard, 1984; Bittain et al., 1987) did not affect apomorphine climbing behaviour. MDL 72222 inhibited avoidance behaviour, but at slightly higher doses (factor 1.3) escape responses were also inhibited, indicating a non-specific inhibition of CAR. GR38032F was not active in the CAR.

Since only very few of the compounds tested were active in the CAR procedure, calculation of rank order correlation coefficients of CAR inhibition with

Table 15.6 Activity of 5-HT antagonists

Drug	D_2	Affinity(pK_i)				inhibition avoidance (ED_{50}, mg/kg)	inhibition escape (ED_{50}, mg/kg)	inhibition climbing (ED_{50}, mg/kg)
		$5-HT_{1A}$	$5-HT_{1B}$	$5-HT_{1C}$	$5-HT_2$			
methiotepine	9.3	7.6	7.3	8.6	9.0	1	2.4	0.2
metergoline	NT	8.4	8.2	9.0	8.6	20	>20	12
methysergide	6.8	7.7	7.3	8.3	7.8	>20	>20	NT
ketanserine	6.3	<6.2	<6.2	7.4	8.7	>10	>10	4
ritanserine	7.9	6.1	6.2	9.3	8.5	>20	>20	54
mianserine	6.0	6.1	6.5	8.5	8.0	>20	>20	28
propranolol	NT	6.7	6.2	6.0	NT	>20	>20	>20
MDL 72222	<5.3	<5.2	<4.5	<5.3	<5.5	16	21	>10
GR 38032F	<5.5	<5.2	<4.5	5.2	<5.2	>10	>10	>10

NT: not tested. For the behavioural experiments all drugs were administered intraperitoneally 30 min before testing. Data obtained as described for Table 15.4. Affinity for $5-HT_{1A}$ and $5-HT_{1B}$ receptor measured by displacement of $[^3H]$-8-OHDPAT and $[^3H]$-5-HT in the presence of 8-OHDPAT in the whole rat brain respectively. The affinity for the $5-HT_{1C}$ receptor was measured by displacement of $[^3H]$-5-HT in the pig choroid plexus.

receptor affinity data does not make sense. Though only based on six pairs of data we calculated a rank order correlation coefficient between inhibition of apomorphine-induced climbing and affinity for the $5-HT_2$ and D_2 receptor of 0.94 and 0.82 (p=0.003 and 0.022), respectively. No significant (at 5% level) correlation with any other receptor affinity was found. On basis of these data no important role for selective 5-HT antagonists (i.e. those not concomitantly blocking DA receptors) in the development of new antipsychotic drugs can be predicted. The results obtained with several both direct and indirect 5-HT agonist are presented in Table 15.7.

The most selective $5-HT_{1A}$ agonists tested, 8-OH-DPAT and ipsapirone did not affect avoidance behaviour but both inhibited apomorphine-induced climbing at high doses. The activity of buspirone in the CAR occurs at doses strongly blocking the DA receptors and is therefore probably not due to its $5-HT_{1A}$ agonistic properties. The same holds true for its activity in the climbing model. The mixed $5-HT_{1A/1B}$ agonist RU 24969 is not active in either behavioural model. The mixed $5-HT_{1A/1C}$ agonist 5-MeO-DMT inhibits avoidance behaviour but also attenuates escape responses at doses only a factor 1.4 higher. No effect on climbing behaviour was observed. The mixed $5-HT_{1B/1C}$ agonists TFMPP and quipazine both attenuated avoidance behaviour. The effect of quipazine, however, was a non-selective activity, avoidance and escape response inhibition occurred at virtually similar doses. TFMPP but not quipazine inhibited climbing behaviour. The difference between these drugs may be due to the lower affinity of quipazine for 5-HT receptors than TFMPP or a less selective influence on the 5-HT receptors. The mixed $5-HT_{1A/B}$ agonist eltoprazine (DU 28853), (Olivier et al., 1987) selectively inhibits CAR and attenuates apomorphine-induced behaviour. Befiperide, a drug primarily affecting the $5-HT_{1A}$ and $5-HT_2$ receptors also inhibits CAR and climbing behaviour. The very broadly active $5-HT_1$ and $5-HT_2$ agonist d-LSD was not effective in suppressing avoidance behaviour.

Table 15.7 Activity of 5-HT agonists

Drug	D_2	5-HT$_{1A}$	Affinity(pK_i) 5-HT$_{1B}$	5-HT$_{1C}$	5-HT$_2$	inhibition avoidance (ED_{50}, mg/kg)	inhibition escape (ED_{50}, mg/kg)	inhibition climbing (ED_{50}, mg/kg)
TFMPP	6.1	6.7	7.5	7.9	6.1	2.2	6.6	10
befiperide	6.7	7.7	6.1	6.7	7.4	4.2	9.2	21
buspirone	7.4	7.8	5.2	6.1	6.0	8.2	>15	8
eltoprazine	6.0	7.4	7.2	7.1	5.8	9.4	>14	13
5-MeO-DMT	<5.5	8.2	7.3	7.6	5.6	16.5	23	>10
quipazine	<5.4	5.6	6.6	7.1	6.0	19	23	>10
8-OHDPAT	5.7	8.6	5.8	5.8	5.2	>5	>5	4
ipsapirone	6.4	8.3	5.6	4.9	5.6	>10	>10	23
RU 24969	5.9	8.1	8.1	7.3	5.8	>5	>5	>5
LSD	7.0	8.6	7.8	8.4	9.1	>0.4	>0.4	NT
fenfluramine*	<5.4	5.7	5.6	6.1	<5.2	7.4	>30	>10
5-HTP*,+	NT	5.6	5.2	<5.3	4.5	37	63	50
fluvoxamine*	<5.5	<5.2	5.1	<5.0	5.9	>20	>20	>10

NT: not tested, *: indirectly acting agonists, + pretreatment with benserazide and pargyline 20 and 25 mg/kg ip, respectively 30 min before 5-HTP.
All drugs were administered intraperitoneally 30 min before testing. Data were obtained as described in Table 15.6.

Fenfluramine, which most likely stimulates all 5-HT receptors through release of the transmitter itself, selectively inhibits avoidance behaviour but does not affect apomorphine-induced climbing. The 5-HT precursor 5-HTP attenuated avoidance behaviour and also inhibited climbing. No such effects were found for the selective 5-HT reuptake blocker fluvoxamine. The results described in Table 15.7 for the directly acting 5-HT agonists were used for calculation of the rank order correlation coefficient between the various activations.

As shown in Table 15.8 the only significant correlation found was a reciprocal relationship between the affinity for 5-HT$_{1A}$ receptors and inhibition of avoidance behaviour. Such a result should, however, be cautiously interpreted since it is based upon a limited amount of data on not highly-selective drugs.

Both the calculations of correlation between behavioural activities and receptor affinities for the 5-HT agonists and antagonists do not take into account the possible vitro-vivo differences and the possibility that more than one type of 5-HT receptor is involved in the activity of a drug. Therefore, the lack of a simple and direct relationship between a receptor affinity and behavioural activity is not surprising.

The data shown seem to justify the conclusion that inhibition of conditioned avoidance behaviour and hence possible antipsychotic activity in man can be achieved through stimulation but not by blockade of 5-HT receptors. The 5-HT$_{1A}$ receptor does not seem to be involved but further work is needed to establish the importance of the other 5-HT$_1$ receptor sub-types and 5-HT$_2$ receptor.

A serotonergic drug that has recently been developed as a putative antipsychotic

Table 15.8 Correlation coefficients for 5-HT agonists

	Spearman correlation coefficients	
	inhibition avoidance	inhibition climbing
inhibition climbing	0.31 (0.21)	
affinity $5\text{-}HT_{1A}$	-0.70 (0.01)	0.14 (0.36)
affinity $5\text{-}HT_{1B}$	-0.19 (0.37)	0.47 (0.10)
affinity $5\text{-}HT_{1C}$	0.24 (0.27)	0.28 (0.25)
affinity $5\text{-}HT_2$	0.44 (0.10)	0.02 (0.48)
affinity D_2	0.23 (0.26)	0.48 (0.10)

Data are the rank order correlation coefficients of the data shown in table 15.7.
The values in brackets refer to the significance two-sided (p-values) of the calculated correlation coefficient.

drug on basis of inhibition of conditioned avoidance behaviour is befiperide (see fig. 15.2). The activity of this drug is not due to blockade of DA receptors as established both in vitro and in vivo (Van der Heyden et al., 1986; Schipper et al., 1986). The most likely mechanism of action is an activation of 5-HT receptors (Van der Heyden et al., 1987). The advantages of befiperide over the presently used neuroleptics are obvious: since it does not block the DA receptors the emergence of the most troublesome extrapyramidal side effects, including tardive dyskinesia is highly unlikely. Clinical studies with this drug will show whether the conditioned avoidance response procedure can be used as a predictor for antipsychotic activity in man and will also provide some direct information about the use of serotonin agonists in the treatment of schizophrenia.

Acknowledgements: The author wishes to thank Marijke Mulder for preparing the manuscript.

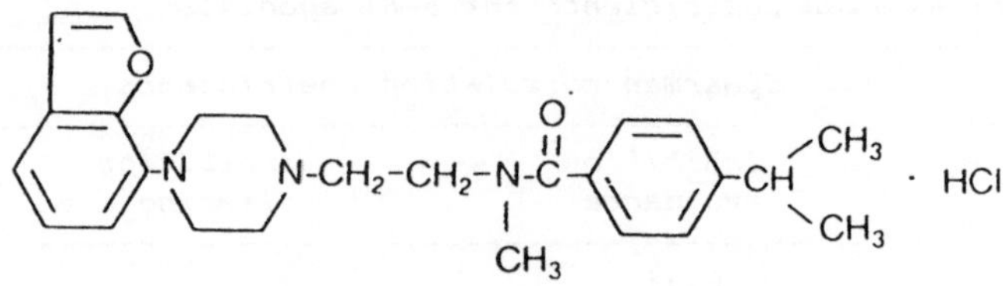

218 Van der Heyden

Befiperide hydrochloride

Figure 15.2: Structure of putative antipsychotic befiperide

References

Angrist, B.M., Gershon, S. (1970). The phenomenology of experimentally induced amphetamine psychosis – preliminary observations. Biol Psychiatry, **2**, 95–107.

Atsmon, A. & Blum, I. (1978). The discovery. In: propranolol and schizophrenia. (Eds. Roberts, E. & Amacher, P.) pp 5–38. New York, Alan R. Liss.

Baldessarini, R.J., Stramentinolo, G. & Lipinski, J.F. (1979). Methylation hypothesis. Arch. Gen. Psychiatry, **36**, 303–307.

Bignami, G. (1978). Effects of neuroleptics, ethanol, hypnotic–sedatives, tranquilizers, narcotics and minor stimulants in aversive paradigms. In: Psychopharmacology of Aversively Motivated Behaviour, (Eds. Anisman, H., Bignami, G.) pp. 385–408, New York: Plenum Press.

Brittain, R.T., Butler, A., Coates, I.H., Fortune, D.H., Hagan, R., Hill, J.M., Humber, D.C., Humphrey, P.P.A., Hunter, D.C., Ireland, S.J., Jack, D., Jordau, C.C., Oxford, A., & Tyers, M.B. (1987). GR 38032 F, a novel selective 5–HT$_3$ receptor antagonist. Brit. J. Pahramcol., **90**, 87.

Carlsson, A., Lindqvist, M. (1963). Effect of chlorpromazine and haloperidol on formation of 3–methoxy tyramine and normetanephrine. Acta Pharmac. Tox., **40**, 140–144.

Carter, C.J. & Pycock, C.J. (1978). Differential effects of central serotonin manipulation on hyperactive and stereotyped behaviour. Life Sci., **23**, 953–960.

Carter, C.J. & Pycock, C.J. (1981). The role of 5–hydroxytryptamine in dopamine–dependent stereotyped behaviour. Neuropharmacol, **20**, 261–265.

Chouinard, G. & Ammable, L. (1984). An early phase II clinical trial of BW 234U in the treatment of acute schizophrenia in newly admitted patients. Psychopharmacology, **84**, 282–284.

Connell, P.H. (1958). Amphetamine psychosis. Maudsley monograph No. 5. Chapman Hall London.

Cook, L., Weidley, E.F., Morris, R.W. & Matis, P.A. (1955). Neuropharmacological; and behavioural effects of chlorpromazine (Thorazine hydrochloride). J. Pharmacol. Exp. Ther., 113, 11-12.

Cornfeldt, M.L., Fiedling, S., Kruse, H., Billey-Nichuck, J., DobsonC. & Wilker, J. (1978). Clonidine inhibition of amphetamine induced rat circling and other psychopharmacological effects. Pharmacologist, 20, 162.

Costall, B., Domeney, A.M., Kelley, M.E., Naylor, R.J. & Tyers, M.B. (1987). The antipsychotic potential of GR 38032F, a selective antagonist of $5-HT_3$ receptors in the central nervous system. Brit. J. Pharmacol., 90, 89P.

Costall, B. & Naylor, R.J. (1974). Stereotyped and circling behaviour induced by dopaminergic agonists after lesions of the midbrain raphe nuclei. Eur. J. Pharmacol., 29, 206-222.

Costall, B. & Naylor, R.J. (1978). Neuroleptic interaction with the serotoninergic-dopaminergic mechanisms in the nucleus accumbens. J.Pharm. Pharmacol., 30, 257-259.

Creese, I., Schneider, R. & Snyder, S.H. (1977). [^{3}H] spiroperidol labels dopamine receptors in pituitary and brain. Eur. J. Pharmacol., 46, 377-381.

Creese, I. & Snyder, S.H. (1978). [^{3}H] spiroperidol labels serotonin receptors in rat cerebral cortex and hippocampus. Eur. J. Pharmacol., 49, 201-202.

Curzon, G., Kantamaneni, B.D., Van Voxel, P., Gillman, P.K., Partlett, J.R. & Bridges, P.V. (1980). Substances related to 5-hydroxytryptamine in plasma and in lumbar and ventricular fluids of psychiatric patients. Acta Psychiatr. Scand. (Suppl.), 280, 3-19.

Davidson, J., Miller, R., Wingfield, M., Zung, W. & Dren, A.T. (1982). The first clinical study of BW234U in schizophrenia. Psychopharmacol. Bull., 18, 173-176.

Davidson, A.B., & Weidley, E. (1976). Differential effects of neuroleptic and other psychotropic agents on acquisition of avoidance in rats. Life Sci., 18, 1279-1284.

Donaldson, S.R., Belenberg, A.J. & Baldessarini, R.J. (1983). The pharmacological treatment of schizophrenia: A progress report. Schizophr. Bull., 12, 1-28.

Dray, A., Davies, J., Oakley, M.R., Tongroach, P. & Vellucci, S. (1978). The dorsal and medial raphe projections to the substantia nigra in the rat: electrophysiological, biochemical and behavioural observations. Brain Res., 151, 431-442.

Feinberg, I., Brain, M., Koresko, R.L. & Gottlieb, F. (1969). Stage 4 sleep in schizophrenia. Arch. Gen. Psychiat., 21, 262-266.

Fozard, J.R. (1984). MDL 72222, a potent and highly selective antagonist at neuronal 5-hydroxytryptramine receptors. M.S. Arch. Pharmacol., 326, 36-44.

Freedman, R., Kirch, R., Bell, J., Adler, L.E., Pecevich, M., Pachtman, E. & Renver, P. (1982). Clonidine treatment of schizophrenia: Double-blind comparison to placebo and neuroleptic drugs. Acta Psychiatr. Scand., 65, 35-45.

Gaddum. J.H. (1953). Antagonism between lysergide acid diethylamine and 5-hydroxytryptamine. J. Physiol., 121. 15P.

Gillin, J.C., Kaplan, J., Stillman, R. & Wyatt, R.J. (1976). The psychodelic model of schizophrenia - the case of N,N-dimethyltryptamine. Am. J. Psychiatry, 133, 203-208.

Glennon, R.A.., Titeler, M. & McKenney, J.D. (1984). Evidence for $5-HT_2$ involvement in the mechanism of action of hallucinogenic agents. Life Sci., 35, 2505-2511.

Gozlan, H., El Mestikawy, S., Pichat, L., Glowinsky, J. & Hamon, M. (1983). Identification of presynaptic serotonin autoreceptors using a new ligand: ^{3}H-PAT. Nature, 305, 140-142.

Grabowska M. (1974). Influence of midbrain raphe lesion on some pharmacological and biochemical effects of apomorphine in rats. Psychopharmacol., 39, 316-322.

Grabowska M., Antkiewicz, L., May, J. & Michaluk, J. (1973). Apomorphine and central serotonin neurons. Pol. J. Pharmacol. Pharm., 25, 29-39.

Griffith, J.D., Cavanaugh, J., Held, J. & Oates, J.A. (1972). Dextroamphetamine: Evaluation of psychotomimetic properties in man. Arch. Gen. Psychiatry, 26, 97-100.

Grilly, D.M., Johnson, S.K., Minardo, R., Jacoby, D. & LaRiccia J. (1984). How do tranquilizing agents selectively inhibit conditioned avoidance responding? Psychopharm., 84, 262-267.

Herz, A. (1960). Drugs and the conditioned avoidance response. In: Int. Rev. of Neurobiology. (Eds. Pfeiffer, C.C. & Smythies, J.R.) Vol. II, pp 229-277, New York, Academic Press.

Hjorth, S., Car lsson, A., Wikström, H., Lindberg, P., Sanchez, D., Hacksell, U., Arvidsson, L.E., Svensson, U. & Nilsson, J.L.G. (1981). 3-PPP, a new centrally acting DA-receptor agonist with selectivety for autoreceptors. Life Sci., 28, 1225-1238.

Hollister, L.E. (1978). Psychotomimetic drugs in man. In: Handbook of Psychopharmacology. (Eds. Iversen, L.L., Iversen, S.D. & Snyder, S.H.) Vol 11. Stimulants, pp. 389-424, New York and London: Plenum Press.

Honma, T. & Fukushima, H. (1976). Correlation between catalepsy and dopamine decrease in the rat striatum induced by neuroleptics. Neuropharmacology, 15, 601-607.

Hornykiewicz, O. (1982). Brain catecholamines in schizophrenia - a good case for noradrenaline. Nature, 299, 484-486.

Houser, V.P. (1978). The effect of drugs on behaviour controlled by aversive stimuli. In: Contemporary Research in Behavioural Pharmacology, (Eds. Blackman, D.E. & Sanger, D.J.) pp 69-157, New York: Plenum Press.

Ioro, L.C., Barnett, A., Leitz, F.H., Houser, V.P. & Korduba, C.A. (1983). SCH 23390, a potential benzazepine antipsychotic with unique interactions on dopaminergic systems. J. Pharmacol. Exp. Ther., 226, 462-468.

Kostowski, W. (1975). Interactions between serotonergic and catecholaminergic systems in the brain. Pol. J. Pharmacol. Pharm. (suppl.), 27, 15-24.

Kuribara, H. & Tadokoro, S. (1980). Correlation between antiavoidance activities of antipsychotic drugs in rats and daily clinical doses. Pharmacol. Biochem. & Behavior., 14, 181-192.

Linn oila, M., Viukari, M. & Hietala, O. (1976). Effect of sodium valproate on tardive dyskinesia. Brit. J. Psychiatry, 129, 114-119.

Ljungberg, T. & Ungerstedt, U. (1978). Classification of neurolpetic drugs according to their ability to inhibit apomorphine-induced locomotion and gnawing: evidence for two different mechanisms of action. Psychopharmacol., 56, 239-247.

Maffii, G. (1959). The secondary conditioned response of rats and the effects of some psychopharmacological agents. J. Pharm. and Pharmac., 11, 129-139.

Matthysse, A. & Sugarman, J. (1978). Neurotransmitter theories of schizophrenia. In: Handbook of psychopharmacology, Vol. 10, Neuroleptics and schizophrenia. (Eds. Iversen, L.L., Iversen, S.D. & Snyder, S.H.) pp 221-240, New York, Plenum Press.

Menon, M.K. & Vivovia, C.A. (1981). Modification of apomorphine hypothermia by drugs affecting brain 5-hydroxytryptamine function. Eur. J. Pharmacol., 76, 223-227.

Moroji, T., Watamabe, N., Aaki, N. & Itoms, S. (1982). Antipsychotic effects of caerulein, a decapeptide chemically related to cholecystokinin octapeptide, on schizophrenia. Int. Pharmacopsychiatry, 17, 255-273.

Mueser, K.T. & Dysken, M.W. (1983). Narcotic antagonists in schizophrenia: a methodological review. Schizophrenia Bull., 9, 213-225.

Murray, M.R., Oon, M.C.H., Rodnight, R., Birley, J.L.T. & Smith, A. (1979). Increased excretion of dimethyltryptamine and certain features of psychosis.

Arch. Gen. Psychiatry., 36, 644–649.

Niemegeers, C.J.E., Janssen P.A.J. (1979). A systematic study of the poharmacological activities of dopamine antagonists. Life Sci., 24, 2201–2216.

Olivier, B., Mos, J., Van der Heyden, J., Schipper, J., Tulp, M., Berkelmans, B. & Bevan, P. (1987). Serotonergic modulation of agonistic behaviour. In: Ethopharmacology of agonistic behaviour in animals and humans. (Eds. Olivier B., Mos, J. & Brain P.F.) pp 162–186, Dordrecht, Martinus Nijhoff.

Ortmann, R., Bischoff, S., Radeke, E., Buech, O. & Delini–Stula, A. (1982). Correlations between different measures of antiserotonin activity of drugs: study with neuroleptics and serotonin receptor blockers. N.S. Arch. Pharmacol., 321, 265–270.

Potkin, S.G., Weinberger, D.R., Linnoila, M. & Wyatt, R.J. (1983). Low CSF 5-hydroxyindoleacetic acid in schizophrenic patients with enlarged cerebral ventricles. Am. J. Psychiatr., 140, 21–25.

Psychotropics (1986). Amsterdam, Lundbeck, Nederland.

Rodnight, R. (1983) Schizophrenia: some current neurochemical apporaches. J. Neurochem., 41, 12–21.

Sanger, D.J. (1985) The effects of clozapine on shuttle–box avoidance responding in rats: comparisons with haloperidol and chlordiazepoxide. Pharmacol. Biochem. Behav. 23, 231–236.

Schipper, J., Van der Heyden, J.A.M., Kruse, C.G. & Tulp, M.Th.M. (1986). Befiperide, a putative antipsychotic, blocks apomorphine induced stereotypy without blocking dopamine receptors. Proc. of 16th Soc. Neurosci. meeting, abstract 131.2.

Shulgin, A.T. (1978). Psychotomimetic drugs: structure–activity relationships. In: Handbook of Psychopharmacology. Vol. 11. Stimulants. (Eds. Iversen, L.L., Iversen, S.D. & Snyder, S.H.) pp 243–333, New York and London: Plenum Press.

Sills, M.A., Wolfe, B.B. &Frazer, A. (1984). Determination of selective and non–selective compounds for the $5-HT_{1A}$ and $5-HT_{1B}$ receptor subtypes in rat frontal cortex. J. Pharmacol. Exp. Ther., 231, 480–487.

Simpson, G.M., Branchey, M.H. & Shrivastava, R.M. (1976). Baclofen in schizophrenia. Lancet, 1, 967–968.

Spano, P.F., Biggio, G., Casu, M., Gessa, G.L., Bareggi, S.R., Govini, S. & Trabucchi, M. (1978). Interaction of metergoline with striataldopamine system. Life Sci., 23, 2383–2392.

Steru M., Fram, D.H., Wyatt, R., Grinspoon, L. & Turnsky, B. (1969). All–night sleep studies of acute schizophrenics. Arch. Gen. Psychiat., 20, 470–477.

Stoll, W.A. (1947). Lysergsäure–diäthylamid, ein phantasticum aus der Mutterkorngruppe. Schweiz. Arch. Neurol. Psychiat., **60**, 279.

Tarsy, D. (1983). Neuroleptic–induced extrapyramidal reactions: classification, description, and diagnosis. Clin. Neuropharmacol., **6**, S9–S26.

Van der Heyden, J.A.M. & Bradford, L.D. (1985). Selectivity of the conditioned avoidance response procedure. Proceedings IVth World Congress of Biological Psychiatry, abstract 131.1

Van der Heyden, J.A.M., Schipper J. & Tulp, M.Th.M. (1987). Befiperide, serotonergic properties of a new non–dopaminergic putative antipsychotic. Proc. 17th meeting. Soc. for Neurosci., abstract 128.4.

Van Kammen, D.P. (1979). The dopamine hypothesis of schizophrenia revisited. Psychoneuroendocrinology, **4**, 37–46.

Van Kammen, D.P. & Antelman, S. (1984). Impaired noradrenergic transmission in schizophrenia? Life Sci., **34**, 1403–1413.

Van Praag, H.M. (1978). Psychotropic Durgs. Asse, van Gorkum.

Van Ree, J.M. & De Wied, D. (1981). Endorphines in schizophrenia. Neuropharmacology, **20**, 1271–1277.

Warner, L.H. (1932). The association spar of the white rat. J. Genet. Psychol., **41**, 57.

Weil–Malherbe, H. (1978). Serotonin and schizophrenia. In: Serotonin in health and disease, Vol. III: The central nervous system. (Ed. Essman, W.B.) pp 231–291, New York and London, Spectrum Publications.

Wilson, I.C., Lara, P.P. & Prange, A.J. Jr. (1973). Thyrotropin–releasing hormone in schizophrenia. Lancet, ii, 43–44.

Woolley, D.W. & Shaw, E. (1954). A biochemical and pharmacological suggestion about certain mental disorders. Proc. Natn. Acad. Sci., U.S.A., **40**, 228–231.

Wyatt, R.J., Vaughan, T., Kaplan, J., Galantar, M. & Green, R. (1973). 5–hydroxytryptophan and chronic schizophrenia – A preliminary study. In: serotonin and behaviour. (Eds. Barchas, J. & Usdin, E.) pp. 487–498, New York and London, Academic Press.

Zarcone, V., Gulevich, G., Pivik, T., Azumi, K. & Dement, W. (1969). REM deprivation and schizophrenia. Biol. Psychiatry., **1**, 179–184.

MECHANISM OF ACTION OF 8–OH–DPAT ON A RAT MODEL FOR HUMAN DEPRESSION

G.A. Kennett, G. Curzon.
Department of Neurochemistry, Institute of Neurology, Queen Square, London. WC1N 3BG United Kingdom

We have developed a depression model with various similarities to the human illness, based on the behavioural effects of 2 h restraint stress in the rat, i.e. (a) increased plasma corticosterone during stress, (b) decreased activity 24 h later in an open field, and (c) hypophagia. On repeating the stress each day, adaptation occurs i.e. behaviour becomes normal (Kennett et al., 1985a). Failure to adapt is the depression model. Adaptation was associated with increased postsynaptic 5–HT function as 24 h after repeated stress, components of the 5–HT behavioural syndrome induced by the 5–HT agonist 5–methoxy–N,N–dimethyltryptamine (5–MeODMT) were increased. The hypophagia is not due to stress–induced ulceration (Donohoe et al., 1987) and is of interest as anorexia nervosa may be a manifestation of depression (Brambilla et al., 1985).

Although plasma corticosterone rises in stress and may have acute anxiolytic effects, repeated corticosterone injection was maladaptive, decreasing open field activity, food intake and the response to 5–MeODMT (Dickinson et al., 1985). Conversely, the corticosterone synthesis inhibitor metyrapone accelerated adaptation to repeated stress and increased the response to 5–MeODMT (Kennett et al., 1985b).

Stressed female rats had defective adaptation and unaltered responses to 5–MeODMT (Kennett et al., 1986). Plasma corticosterone rose higher than in males. If it was decreased to male values with metyrapone then adaptation and responses to 5–MeODMT altered to the male pattern (Haleem et al., unpublished).

The model implies that high corticoid responses, low 5–HT functional activity at certain sites and female sex oppose adaptation and predispose to depression. Much evidence indicates associations between these factors and depression.
Furthermore, they are also associated with vulnerability to affective illness (Coryell and Zimmerman, 1987; Sedvall et al., 1980).

The association between adaptation and increased responses to 5–MeODMT suggested that $5-HT_{1A}$ agonists might enhance adaptation as some of these responses (forepaw treading and tremor) are mediated by $5-HT_{1A}$ receptors (Tricklebank et al., 1985). Therefore the effects of 8–OH–DPAT and other $5-HT_{1A}$ agonists on the model were investigated.

Method
Male Sprague–Dawley rats (200–250 g) were restrained by taping to wire grids for 2 h commencing between 14.00 and 15.00 h. The rats were then released, placed in their home cages and returned to a holding room. Drug treatments were as described below. The next day, between 10.30 and 12.30 h, each rat was placed in the centre of an open field in a quiet room under normal white light. Squares entered with all four paws and faecal pellets dropped were scored over

the first 5 min.

Results and discussion
Effects of 5-HT_{1A} agonists and other drugs on behavioural deficits after restraint stress

8-OH-DPAT given to rats 2 h after release from restraint, dose dependently opposed the locomotor deficit apparent on placement in an open field on the next day (table 16.1). Two other 5-HT_{1A} agonists, buspirone and ipsapirone, had similar effects. The drugs to some degree opposed the hypophagia due to the stress (Dourish et al., 1987) and also opposed the increased defaecation in the open field (Kennett et al., 1987a). Single treatments as above with the benzodiazepine anxiolytics chlordiazepoxide (5.0, 10.0 mg/kg i.p.) or diazepam (5 mg/kg i.p.) or with the antidepressant desipramine (5 mg/kg i.p.) were ineffective; pretreatment for 14 days with desipramine (5 mg/kg i.p.) or another antidepressant sertraline (5 mg/kg s.c.) significantly reversed the locomotor deficit but not the increased defaecation.

The above antidepressant-like effects of 5-HT_{1A} agonists seemed unlikely to be due to direct postsynaptic or presynaptic activation as their known behavioural consequences are transient (Dourish et al., 1985) but the antidepressant-like effects are evident a day later. Another possibility was that the relatively large doses of 8-OH-DPAT required acted by desensitising 5-HT_{1A} somatodendritic autoreceptors. As these receptors mediate the hyperphagic effect of 8-OH-DPAT (Hutson et al., 1986) this response was used to test for receptor desensitisation as follows.

Table 16.1 Reversal by 5-HT_{1A} agonists of open-field locomotor deficit following restraint stress.

Drug	No of squares crossed	
	Unstressed	Stressed
Saline	90 ± 6	21 ± 3**
8-OH-DPAT 60 µg/kg	92 ± 9	17 ± 5**
8-OH-DPAT 250 µg/kg	88 ± 7	44 ± 7**+
8-OH-DPAT 1000 µg/kg	82 ± 7	50 ± 6**++
Saline	81 ± 5	21 ± 4
Ipsapirone 5 mg/kg	73 ± 7	27 ± 6**
Ipsapirone 10 mg/kg	77 ± 7	55 ± 8++
Ipsapirone 4 mg/kg	86 ± 6	52 ± 8*++

See "Methods" for details. Drugs were given 2 h after release from 2 h stress. Values are mean ± SEM, n=10/group. Significant differences were determined by 2-tailed Mann-Whitney U test following a significant ANOVA. Significant differences from respective unstressed controls *$P<0.05$, **$P<0.01$, from stressed saline treated groups +$P<0.05$, ++$P<0.01$.

Attenuation of the hyperphagic response to 8-OH-DPAT by pretreatment with 5-HT$_{1A}$ agonists

8-OH-DPAT (1.0 mg/kg s.c.) caused marked hyperphagia in rats. This or similar treatments with buspirone (4 mg/kg s.c.) or ipsapirone (10 mg/kg s.c.) significantly attenuated the hyperphagic response to 8-OH-DPAT (1.0 mg/kg or 60 µg/kg s.c.) on the next day (Kennett et al., 1987b). 8-OH-DPAT (1.0 mg/kg s.c.) did not affect food intake after saline injection on the next day.

The attenuated hyperphagic response to 8-OH-DPAT was still apparent on the fifth day after the 8-OH-DPAT (1.0 mg/kg s.c.) or buspirone (4 mg/kg s.c.) pretreatments but not 10 days after 8-OH-DPAT pretreatment.

Attenuation of the effect of 8-OH-DPAT on 5-HT metabolism by 8-OH-DPAT pretreatment

Pretreatment with 8-OH-DPAT also reduced the effect of subsequent 8-OH-DPAT injection on 5-HT metabolism. Thus, rats killed 30 min after giving 8-OH-DPAT (60 µg/kg s.c.) had a significant 33% decrease of 5-HIAA in the raphe but a non-significant increase of 8% if given 8-OH-DPAT (1 mg/kg s.c.) 21 h previously.

Lack of effect of pretreatment with 8-OH-DPAT on postsynaptic response to 8-OH-DPAT

The above results show that pretreatment with a large dose of 8-OH-DPAT opposes behavioural and neurochemical effects associated with activation of presynaptic 5-HT$_{1A}$ receptors. However, effects on postsynaptic 5-HT receptors as indicated by the 5-HT syndrome were unaltered.

This finding implies that the presynaptic effects were specific and not merely a result of activation of 8-OH-DPAT metabolism by its previous injection.

Mechanism of antidepressant-like action of 8-OH-DPAT

The results suggest that the rapid antidepressant-like action of 5-HT$_{1A}$ agonists is mediated by their ability to desensitise 5-HT$_{1A}$ presynaptic receptors. This is consistent with reports that some chronic antidepressant treatments have similar effects as indicated by the electrophysiological effects of infusing 5-HT onto the raphe cell bodies (Blier and DeMontigny, 1985). The findings are also consistent with preliminary clinical evidence that the 5-HT$_{1A}$ agonists buspirone (Schweitzer et al., 1986) and gepirone (Amsterdam et al., 1987) have antidepressant activity.

Desensitisation of presynaptic 5-HT$_{1A}$ receptors appears to impair feedback control of 5-HT release at terminals as raphe stimulation-induced 5-HT metabolism is enhanced in some brain regions 24 h after giving 8-OH-DPAT (1.0 mg/kg) (Kennett and Curzon, unpublished).

References

Amsterdam, J.D., Berwish N., Potter, L. & Rickels, K. (1987). Open trial of gepirone in the treatment of major depressive disorder. Current Therap. Res., 41. 185–193.

Blier, P. & De Montigny, C. (1985). Serotonergic but not noradrenergic neurons in rat central nervous system adapt to long term treatment with monoamine oxidase inhibitors. Neuroscience, 16, 949–955.

Brambilla, F., Cavagnini, F., Invitti, C., Poterzio, F., Lampertilo, M., Sali, L., Maggioni, M., Candolfi, F., Panerai, A.E. & Muller, E.E. (1985). Neuroendrocine and psychopathological measures in anorexia nervosa: resemblances to primary affective disorders. Psychiat. Res., 16, 165–176.

Coryell, W. & Zimmerman, M. (1987). HPA-axis abnormalities in psychiatrically well controls. Psychiat. Res., 20, 265–273.

Dickinson, S.L., Kennett, G.A. & Curzon, G. (1985). Reduced 5-hydroxytryptamine-dependent behaviour in rats following chronic corticosterone treatment. Brain Res., 345, 10–18.

Donohoe, T.P., Kennett, G.A. & Curzon, G. (1987). Immobilisation stress-induced anorexia is not due to gastric ulceration. Life Sci., 40, 467–472.

Dourish, C.T., Hutson, P.H. & Curzon, G. (1985). Low doses of the putative serotonin agonist 8-hydroxy-2-(di-n-propylamino) tetralin (8–OH–DPAT) elicit feeding in the rat. Psychopharmacology, 86, 197–204.

Dourish, C.T., Kennett, G.A. & Curzon, G. (1987). The $5-HT_{1A}$ agonists 8-OH-DPAT buspirone and ipsapirone attenuate stress-induced anorexia in rats. J. Psychopharmacol., 1, 23–30.

Hutson, P.H., Dourish, C.T. & Curzon, G. (1986). Neurochemical and behavioural evidence for mediation of the hyperphagic action of 8-OH-DPAT by 5-HT cell body autoreceptors. Eur.J. Pharmacol., 129, 347–352.

Kennett, G.A., Dickinson, S.L., & Curzon, G. (1985a). Enhancement of some 5-HT-dependent behavioural responses following repeated immobilisation in rats. Brain Res., 330, 253–263.

Kennett, G.A., Dickinson, S.L. & Curzon, G. (1985b). Central serotonergic responses and behavioural adaptation to repeated immobilisation: the effect of the corticosterone synthesis inhibitor metyrapone. Eur.J. Pharmacol., 119, 143–274.

Kennett, G.A., Chaouloff, F., Marcou, M. & Curzon, G. (1986). Female rats are more vulnerable than males in an animal model of depression: the possible role of serotonin. Brain Res., 382, 416–421.

Kennett, G.A., Marcou, M., Dourish, C.T. & Curzon, G. (1987b). Single administration of $5-HT_{1A}$ decreases presynaptic but not postsynaptic receptor-mediated responses: releationship to antidepressant-like action. Eur.J. Pharmacol., 138, 53–60.

Schweitzer, E.E., Amsterdam, J.D., Rickels, K., Kaplan, M. & Droba, M. (1986). Open trial of buspirone in the treatment of major depressive disorder.

Psychopharmacol. Bull., 22, 183-186.

Sedvall, G., Fyro, B., Gallberg, B., Nyback, H., Wiesel, F.A., & Wodehelgodt, B. (1980). Relationships in healthy volunteers between concentrations of monoamine metabolites in cerebrospinal fluid and family history of psychiatric morbidity. Brit. J. Psychiat., 136, 365-374.

Tricklebank, M.D., Forler, C. & Fozard, J.R. (1985). Subtypes of the 5-HT receptor mediating the behavioural response to 5-methoxy-N,N-dimethyl-tryptamine in the rat. Eur.J. Pharmacol., 117, 15-24.

REVERSAL OF HELPLESS BEHAVIOUR IN RATS BY SEROTONIN UPTAKE INHIBITORS

P.Martin, A.M.Laporte, P.Soubrie*, S.El Mestikawy°, M.Hamon°.
Departement de Pharmacologie, INSERM U 302* et U 288°, Faculte de Medicine, Hopital Pitie-Salpetriere, 91 Bld. de l'Hopital, 75013 Paris, France

Although serotonergic systems are thought to be involved in the mechanisms of action of antidepressants in both humans and animals (Sherman and Petty, 1980), there is no evidence to suggest that serotonin uptake blockers are efficacious in animal models of depression much as the forced swimming test or the learned helplessness paradigm.

The present study was undertaken to test the effects of serotonin uptake inhibitors citalopram, fluvoxamine, indalpine and zimelidine, on the learned helplessness paradigm, an animal model of depression that is highly sensitive to imipramine-like drugs (Kametani et al., 1983; Sherman and Petty, 1982).

Rats were first (day 1) exposed to 60 scrambled, randomized inescapable shocks (0.8 mA, 15 duration, every min ± 15 s). On day 3, avoidance training was initiated in automated two-way shuttle-box, animals being subjected to 30 avoidance trials (intertrial intervals being 30 s). Avoidance sessions were performed for 3 consecutive days in the morning and the number of escape failures, referred to as absence of crossing response during shock (0.8 mA, 35 s duration) delivery was recorded. Drugs were injected i.p. on 5 consecutive days, i.e. 6h after inescapable shock pretreatment on day 1 and then, twice per day in the morning, 30 min after each shuttle-box session and at 18.00 h (Martin et al., 1986).

Our results showed that, as compared with saline-treated rats, citalopram (1 mg/kg/day), fluvoxamine (4 mg/kg/day), indalpine (1, 2, 4 mg/kg/day) and zimelidine (1, 2 mg/kg/day) reversed significantly escape deficits seen at the third shuttle-box session (Figure 17.1). At higher doses, however, all the compounds tested. (citalopram (2 mg/kg/day), fluvoxamine (8 mg/kg/day), indalpine (8, 16 mg/kg/day) and zimelidine (4, 8 mg/kg/day)), lost their ability to reverse escape failures (Figure 17.1).

In order to assess the involvement of serotonergic neurons in the antidepressant effects of 5-HT uptake inhibitors, the effects exerted by indalpine, in 5,7-dihydroxytryptamine (5,7-DHT) -lesioned rats and in sham animals were compared.

After desimipramine (25 mg/kg i.p.) - pretreatment, anesthetized rats were either sham operated or infused with 5,7-DHT (5 µg of free base in 0.6 µl saline containing 0.02% ascorbic acid) into the midbrain raphe area. Three weeks later, animals were exposed to learned helplessness training and shuttle-box sessions as described previously. After behavioural testing, animals were sacrificed and [^{3}H]-5-HT uptake was assessed in the hippocampus, cerebral cortex and striatum. Our results showed that damage to serotonergic neurons associated with a 65% reduction in [^{3}H]-5-HT uptake markedly reduced, as

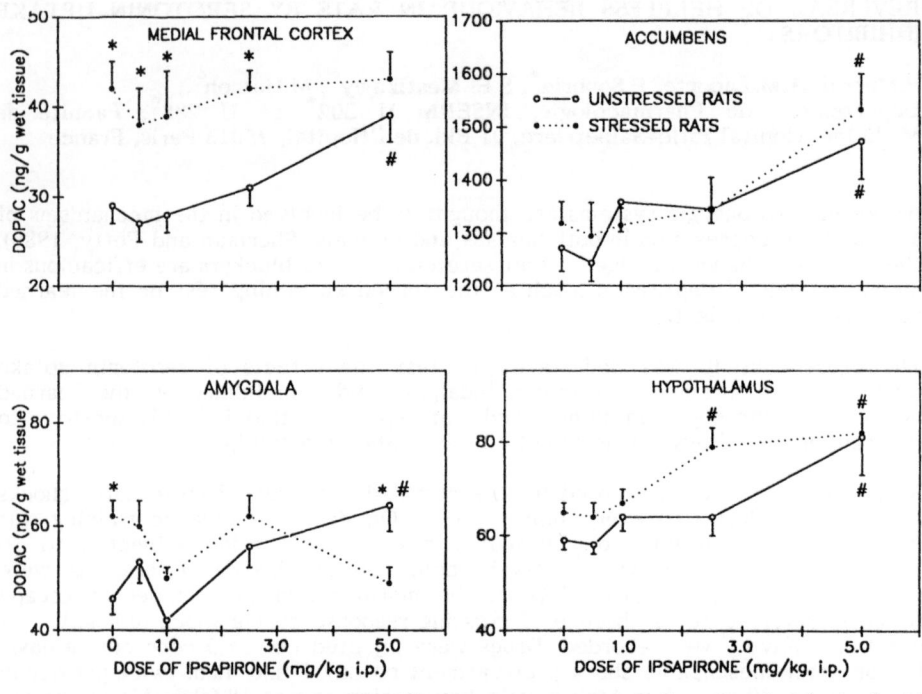

Figure 17.1 Effect of ipsapirone on regional forebrain dopac concentrations.

The data represent mean ± S.E.M., n=6-16.
* Significant difference from the corresponding non-stressed rats, p<0.05.
# Significant difference from the corresponding vehicle-injected rats, p<0.05.

compared to sham rats, the ability of indalpine (1 mg/kg/day) to reverse escape failures.

The number of escape failures observed at the third shuttle-box session was: saline-treated (sham or 5,7-DHT): 17.2 ± 2.3, sham + indalphine 7.2 ± 1.5, lesioned rats treated 18.6 ± 2.7.

In conclusion, these data support the involvement of serotonergic neurons in the learned helplessness syndrome. They suggest that a moderate stimulation of 5-HT transmission (perhaps in relation to some receptor and/or regional specificity) is required to reverse helpless behaviour. These results are in keeping with data obtained by Mc Elroy et al. (1982) showing that a marked enhancement of 5-HT transmission impaired the avoidance acquisition in a two-way avoidance paradigm.

References

Sherman A.D., Petty F. Neurochemical basis of the action of antidepressants on learned helplessness. Behav. Neur. Biol. 1980, **30**, 119-134.

Kametani H., Nomuras, Shimizu J. The reversal effect of antidepressants on the escape deficit induced by inescapable shock in rats. Psychopharmacology, 1983, **80**, 206-208.

Sherman A.D., Petty F. Additivity of neurochemical changes in learned helplessness and imipramine. Behav. Neur. Biol., 1982, **35**, 344-353.

Martin P, Soubrie P, Simon P. Shuttle-box deficits induced by inescapable shocks in rats reversal by the beta-adrenoreceptor stimulants clenbuterol and salbutamol. Pharm. Bioch. Behav., 1986, **24**, 177-181.

MacElroy J.F., Dupont A.F., Feldmann R.S. The effects of fenfluramine and fluoxetine on the acquisition of a conditioned avoidance response in rats. Psychopharmacology, 1982, **77**, 356-359.

RITANSERIN (R 55 667), AN ORIGINAL THYMOSTHENIC

T.F.Meert, C.J.E.Niemegeers, Y.G.Gelders, P.A.J.Janssen.
Janssen Research Foundation, B-2340 Beerse, Belgium

Ritanserin or R 55 667 (Figure 18.1) has been described to be a potent serotonin-$5HT_2$ antagonist, different from classical serotonin antagonists in that it selectively binds to serotonin-$5HT_2$ but not to serotonin-$5HT_1$ receptor sites (Leysen et al., 1985). Ritanserin is also a pure LSD antagonist in contrast to other serotonin antagonists such as cyproheptadine, metergoline and methysergide which act as mixed agonist/antagonists in the LSD-saline discrimination procedure (Colpaert et al., 1982). Clinically, ritanserin was found to improve the quality of sleep (Idzikowski et al., 1986) and to possess mood elevating properties (Reyntjens et al., 1986; Hoppenbrouwers et al., 1986).

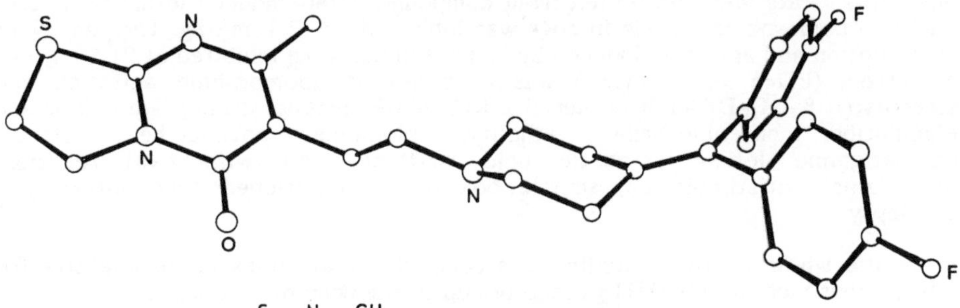

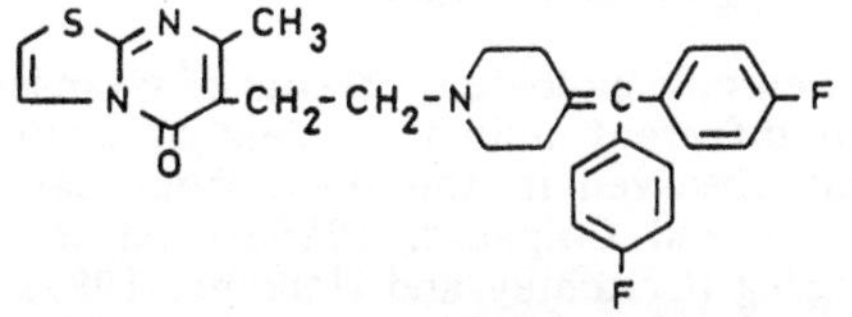

Figure 18.1 Chemical and tridimensional structure of ritanserin.

General pharmacology
Ritanserin has been described as a potent, selective and long acting serotonin-$5HT_2$ antagonist. In in vitro binding studies the K_i-value of 0.28 nM indicated a high affinity of ritanserin for serotonin-$5HT_2$ receptors (Leysen et al., 1985; Leysen, 1987). Histamine-H_1 (15 nM), dopamine-D_2 (22 nM), α_1-adrenergic (35 nM) and α_2-adrenergic (60 nM) receptors were occupied at much higher concentrations. In a variety of other binding assays ritanserin displayed no activity (>1,000 nM in $5-HT_{1A}$, $5-HT_{1B}$, ß-adrenergic, cholinergic- muscarinic, µ-opiate and benzodiazepine binding assays). In vitro occupation of $5HT_2$ receptors occurred at low doses of ritanserin and was even more selective for these receptors than was suggested by

the in vitro affinities (Leysen et al., 1985).

Antagonism of endogenous serotonin and of various serotonergic stimuli was generally most pronounced 2 h after the administration of ritanserin (Awouters et al., 1987).

The most sensitive expression of peripheral $5HT_2$ antagonism was reversal of tryptamine–induced lesions were inhibited at 0.028 mg/kg. In other tests potent central $5HT_2$ antagonism was manifest: Inhibition of tryptamine–induced clonic seizures of the forepaws (0.037 mg/kg) and coarse body tremors (0.11 mg/kg), inhibition of 5–HTP (0.074 mg/kg) and mescaline–induced (0.085 mg/kg) head twitches. The discriminative stimulant action of LSD was blocked at 11.6 mg/kg; no generalisation with LSD was observed at doses up to 40.0 mg/kg. No evidence of partial agonist activity was found.

Ritanserin was further studied in a large number of tests related to different mediators. Rats were protected from compound 48/80–induced lethality at 2.35 mg/kg. Apomorphine emesis in dogs was inhibited at 12.1 mg/kg. Oxygen hyper consumption and agitation induced by amphetamine were inhibited at 21.5 mg/kg. No effect (ED_{50} >40.0 mg/kg) was observed on apomorphine agitation and stereotypy, 8–OH–DPAT hypothermia, RU 24969 circling, citalopram licking and ejaculation, norepinephrine lethality, clonidine antidiarrheal effect, physostigmine lethality, acetic acid writhing, tail withdrawal reaction, intracranial selfstimulation, shuttle box dogs, conditioned food intake and catalepsy.

From the whole of these studies it is concluded that ritanserin is selective for interactions mediated by $5HT_2$ receptors up to the dose of 10.0 mg/kg.

Psychopharmacology and neurophysiological effects of ritanserin
Ritanserin was tested in different animal models of anxiety. Disinhibition of exploratory behaviour was observed in the open field between 0.04 and 10.0 mg/kg (Meert, 1986; Meert and Colpaert, 1986a) and in the plus–maze test between 0.0025 and 2.5 mg/kg (Critchley and Handley, 1987). Almost no activity was found in a lick suppression test (Colpaert et al., 1985) and the shock probe conflict procedure (Meert and Colpaert, 1986b), two classical conflict tests using electric shock as an inhibitory stimulus.

Ritanserin was devoid of benzodiazepine or opiate–like abuse potential and in contrast to the benzodiazepines, ritanserin induced no state dependency or motor incoordination nor was there any interaction with alcohol.

Neurophysiologically, ritanserin was found to increase slow wave sleep without interfering with sleep onset and REM sleep (Dugovic and Wauquier, 1987; Wauquier and Dugovic, 1987).

Clinical findings with ritanserin
Ritanserin was described to improve the quality of sleep and to increase slow wave sleep (Idzikowski et al., 1986), effects which might help to restore the energetic functions during the night (Janssen, 1987). Ritanserin was furthermore tested in different psychiatric and neurological disorders. Beneficial effects

were observed in general anxiety disorders (Bressa et al., 1987), panic disorders, neurotic depression and dysthymic disorders (Reyntjens et al., 1986); also in negative symptoms of schizophrenia and EPS especially when given in addition to conventional neuroleptic treatment (Reyntjens et al., 1986; Hoppenbrouwers et al., 1986) and in tension headache and in parkinson tremor (Maertnes de Noordhout and Delwaide, 1986; Hildebrand and Delecluse, 1987). In the more than 2000 patients treated with doses of 5 to 20 mg daily no important side-effects have been reported. Globally, ritanserin was clinically characterized as an original thymosthenic (Hoppenbrouwers et al., 1986; Reyntjens et al., 1986).

Conclusion

Ritanserin can be described as a pure, selective and long acting $5HT_2$ antagonist having pure but rather weak LSD antagonistic properties. Ritanserin was found to possess disinhibitory effects different from the benzodiazepines and to increase slow wave sleep without having any effect on REM sleep or sleep onset.

Clinically, ritanserin is an original and well tolerated thymosthenic agent which is especially active in anxiety disorders and neurotic depression, which improves the quality of sleep and which on top of administration of neuroleptics reduces negative symptoms in schizophrenia as well as EPS.

References

Awouters, F., Niemegeers, C.J.E., Megens, A.A.H.P., Meert, T.F. & Janssen, P.A.J. (1987). The pharmacological profile of ritanserin, a very specific central serotonin–$5HT_2$ antagonist. J. Pharmac. Exp. Ther., (submitted).

Bressa, G.M., Marini, S. & Gregori, S. (1987). Serotonin–$5HT_2$ receptor blockade and generalized anxiety disorders. A double-blind study on ritanserin and lorazepam. Int. J. Clin. Pharm. Rev., 7, 111-119.

Colpaert, F.C., Meert, T.F., Niemegeers, C.J.E. & Janssen, P.A.J. (1985). behavioural and 5-HT antagonist effects of ritanserin. A pure and selective antagonist of LSD discrimination in rat. Psychopharmacol., 86, 45-54.

Colpaert, F.C., Niemegeers, C.J.E. & Janssen, P.A.J. (1982). A drug discrimination analysis of lysergic acid diethylamine (LSD): in vivo agonist and of pirenperone, a LSD-antagonist. J. Pharmac. Exp. Ther., 221, 206-214

Critchley, M.A.E. & Handley, S.L. (1987). Effects in the X-maze anxiety model of agents acting at $5-HT_1$ and $5-HT_2$ receptors. Psychopharmacology, in press.

Dugovic, C. & Wauquier, A. (1987). $5-HT_2$ receptors could be primarily involved in the regulation of slow wave sleep in the rat. Eur. J. Pharmac., 137, 145-146.

Hildebrand, J. & Delecluse, F. (1987). Effects of ritanserin, a selective serotonin–$5HT_2$ antagonist on parkinson rest tremor. Curr. Ther. Res., 41, 298-300.

Hoppenbwouwers, M.L., Gelders, Y. & Van den Bussche, G. (1986). Ritanserin (R 55 667) an original thymostenic. Boll. Chim. Farm., **125**, 1369–1479.

Idzikowski, C., Mills, F.J. & Glennard, R. (1986). 5-Hydroxytryptamine-2 antagonist increases human slow wave sleep. Brain Res., **378**, 164–168.

Janssen, P.A.J. (1987). Does ritanserin, a potent serotonin–$5HT_2$ antagonist, restore energetic functions during the night. R. Soc. Med., **80**, 409–413.

Leysen, J.E. (1987). The use of 5-HT receptor agonists and antagonists for the characterization of their respective receptor sites. In Neuromethods, Neuropharmacology II: "Drugs as tools in neurotransmitter research". (Eds. Boulton, A.A., Baker, G.B. & Juorio, A.U. Clifton) Humans Press, in press.

Leysen, J.E., Gommeren, W., Van Gompel, P., Wynants, J., Janssen, P.F.M. & Laduron, P.M. (1985). Receptor binding properties in vitro and in vivo of ritanserin. A very potent and long acting $5HT_2$ antagonist. Molec. Pharmac., **27**, 600–611.

Maertens de Noordhout, A. & Delwaide, P.J. (1986). Open pilot trial of ritanserin in parkinsonism. Clin. Neuropharmacol., **9**, 480–484.

Meert, T.F. (1986). A comparative study of the effects of ritanserin (R 55 667) and chlordiazepoxide on rat open field behaviour. Drug Dev. Res., **8**, 197–204.

Meert, T.F. & Colpaert, F.C. (1986a). Effects of $5HT_2$-antagonists in two conflict procedures that involve exploratory behavio. Psychopharmacology, **89**, S23.

Meert, T.F. & Colpaert, F.C. (1986b). The scock probe conflict procedure. A new assay responsive to benzodiazepines, barbiturates and related compounds. Psychopharmacology, **88**, 445–450.

Reyntjens, A., Gelders, Y.G., Hoppenbrouwers, M.L.J.A. & Van den Bussche, G. (1986). Thymostenic effects of ritanserin (R 55 667), a centrally acting serotonin–$5HT_2$ receptor blocker. Drug Dev. Res., **8**, 205–211.

Wauquier, A. & Dugovic, G. (1987). Slow wave sleep enhancing effects by the specific serotnin–$5HT_2$ antagonist ritanserin. Clin. Neurol. Neurosurg., **89**, 112.

THE ABILITIES OF 5-HT$_3$ ANTAGONISTS TO INHIBIT THE HYPERREACTIVITY CAUSED BY DOPAMINE INFUSION INTO THE NUCLEUS ACCUMBENS OF THE RAT

N.M.Barnes, B.Costall, A.M.Domeney, M.E.Kelly, R.J.Naylor, M.B.Tyers*
Postgraduate School of Studies in Pharmacology, University of Bradford, Bradford, BD7 1DP and * Neuropharmacology Department, Glaxo Group Research Ltd., Ware, SG12 ODJ, United Kingdom

The selective 5-HT$_3$ receptor antagonist, GR38032F, has been shown to inhibit the hyperactivity caused by dopamine infused persistently into the rat mesolimbic nucleus accumbens (Costall et al., 1987). Since such an inhibition is only effected by neuroleptic agents and by lithium, it was suggested that GR38032F may represent the first of a new class of antipsychotic agents. In the present studies we extend this observation to two further selective 5-HT$_3$ receptor antagonists, ICS 205-930 and BRL 43694 (Richardson et al., 1985; Fake et al., 1987).

Methods
Female Sprague-Dawley (Bradford strain) rats were anaesthetised with chloral hydrate (200 mg/kg -1 s.c.) and placed in a Kopf stereotaxic frame. Chronically indwelling guide cannulae (constructed of stainless steel tubing 0.65 mm diameter held bilaterally in Perspex holders) were implanted using standard stereotaxic techniques to terminate 3.5 mm above the centre of the nucleus accumbens (Ant. 9.4, Vert. 0.0, Lat ±1.6). The guides were kept patent during a 14 day recovery period using stainless steel stylets, 0.3 mm diameter, which extended 0.5 mm beyond the guide tips. After the 14 days animals were anaesthetised with fluothane for the subcutaneous implantation into the scapala region of two Alzet osmotic minipumps attached via polythene tubing to stainless steel injections units (0.3 mm diameter with a 0.65 mm diameter cuff) which were made to fit permanently into the previously implanted guides in place of the stylets, but terminated bilaterally at the centre of the nucleus accumbens. The pumps had been previously filled with a dopamine solution or its solvent and the entire injection unit primed overnight at 37°C (see Costall et al., 1983).

The pumps delivered dopamine (25 µg 24h^{-1}) or its solvent at a constant rate of 0.47 µl h^{-1} from the time of implantation and, although the pumps were designed to deliver solution for 14 days, removal on day 13 precluded any "fall-off" effect.

Behavioural experiments were conducted between 07 h 30 min and 10 h 30 min in a quiet room maintained at 22 ± 2°C. Rats were taken from the holding room and allowed 1h to adapt to the new environment. Locomotor activity was assessed in individual, screened Perspex cages (25 x 15 x 15 cm high) (banked in groups of 30) each fitted with one photocell unit along the longer axis 3.5 cm from the side; this position hads been found to minimise spurious activity counts due to, for example, preening and head movements when the animal is stationary.

Interruptions of the light beam were recorded every 5 min. At this time animals

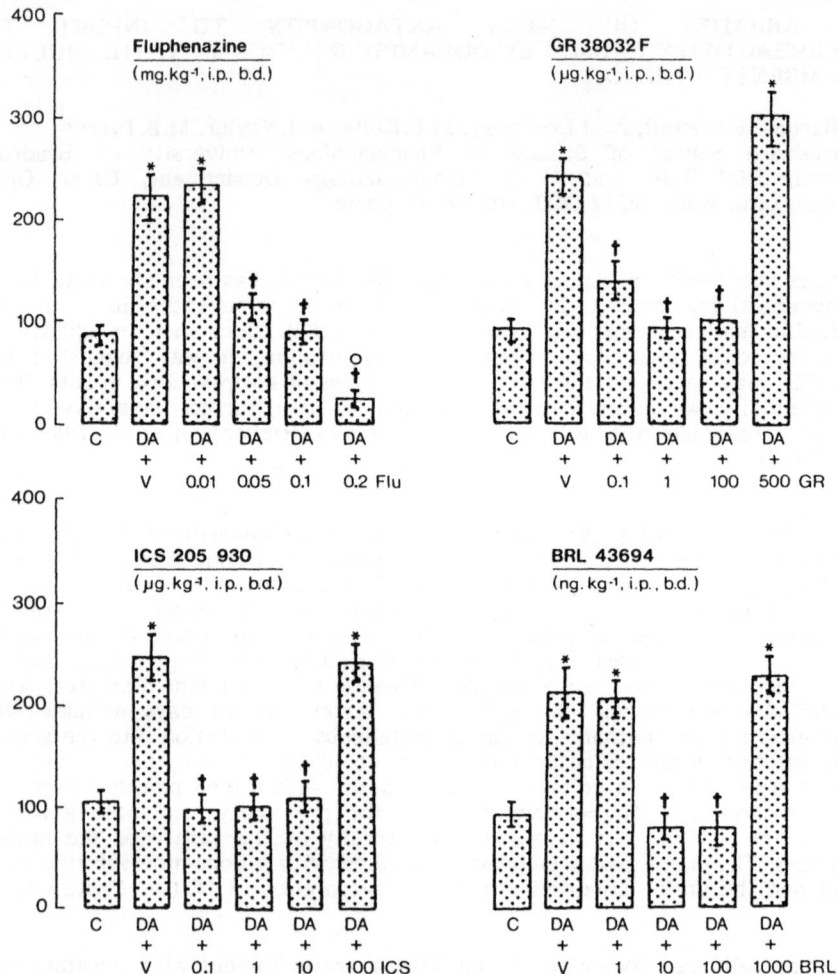

Figure 19.1 Maximum locomotor activity response measured during the time of the 2nd peak of hyperactivity caused by dopamine infusion into rat nucleus accumbens (day 9-12). C indicates the response of animals receiving intra-accumbens vehicle. Other animals received intra-accumbens dopamine combined with twice daily (b.d.) injections of vehicle, fluphenazine, GR38032F, ICS 205-930 or BRL 43694 (doses and units indicated). n=6. S.E.M.s given. Significant elevation of locomotor activity decreases as °P<0.001. Antagonism of the dopamine-induced hyperactivity is indicated as +p<0.001.

were also observed for the presence of any non-specific change in locomotor activity, e.g. sedation, prostration, stereotyped movements, that could interfere with the recording of locomotor activity. Infusion sites were analysed histologically on completion of the experiments. In two animals the infusion site was found to be just posterior to the nucleus accumbens, in the area of the caudate-putamen. Data from these animals was excluded. Results were analysed using two-way analysis of variance (repeated measure analysis) followed by Dunnett's t test.

GR38032F (1, 2, 3, 9-tetrahydro-9-methyl-3[2-methyl-1H-imidazol-1-yl) methyl]-4H-carboazol-4-one, HCl $2H_2O$), BRL 43694 (endo-N-(9-methyl-indazole-3-carboxamide) (Chemistry Research Department, Glaxo Group Research Ltd., Ware), ICS 205-930 ([3α-tropanyl]-1H-indole-3-carboxylic acid ester) (Sandoz) and fluphenazine .HCl (Squibb) were prepared in saline. Dopamine HCl (Koch Light) was prepared in a N_2 bubbled solution of 0.1% sodium metabisulphite.

Results
Dopamine infused into the rat nucleus accumbens caused biphasic hyperactivity responding with peaks occurring on days 3-5 and 9-12 (in the order of 250 counts 50 min^{-1} compared to vehicle control values which were in the range 80-95 counts 60 min^{-1}). This hyperactivity was antagonised by fluphenazine (0.05-0.2 mg/kg i.p. b.d.), GR38032F (0.1-100 µg/kg i.p. b.d.), ICS 205-930 (0.1-10 µg/kg i.p. b.d.) and by BRL 43694 (10-100ng/kg i.p. b.d.). As the dose of fluphenazine increased locomotor activity of GR38032F, ICS 205-930 and BRL 43694 were increased the antagonism of the dopamine hypearactivity observed during dopamine infusion. 500 µg/kg GR38032F inhibited the first peak of hyperactivity, but not the second. In contrast 100 µg/kg ICS 205-930 and 0.1µg/kg BRL 43694 failed to inhibited either peak of hyperactivity, indeed, in some animals the enhanced locomotor responding even persisted during the normal 'trough' between two hyperactivity peaks.

Discussion
Three selective 5-HT_3 receptor antagonists, GR38032F, ICS 205-930 and BRL 43694, were shown to inhibit a raised mesolimbic dopamine function. That this inhibition may be due to an action on dopamine turnover in limbic terminal areas has been suggested by Hagan et al. (1987) who showed that GR38032F inhibits both the hyperactivity and dopamine metabolism induced by stimulation of the VTA. In the accumbens experiments described here, for all compounds there would appear to be a loss of selectivity of action at high dosage. Other studies have shown that the ability to antagonise a dopamine hyperactivity is not linked with other typical neuroleptic effects. Thus, GR38032F, ICS 205-930 and BRL 43694 all failed to induce catalepsy or to antagonise apomorphine or amphetamine-induced stereotyped behaviour (at doses up to and including 1 mg/kg), failed to initiate circling on unilateral injection into the striatum (doses up to 10 µg), either spontaneously or following the peripheral administration of apomorphine to reveal hemispheric asymmetries, and failed to displace ^3H-spiperone in radioligand binding assays. These data would indicate that the 5-HT_3 receptor antagonists may exert antipsychotic activity with an absence of the side effects normally associated with neuroleptic therapy.

References

Costall, B., Domeney, A.M. & Naylor, R.J. (1983). A comparison of the behavioural consequences of chronic stimulation of dopamine receptors in the nucleus accumbens of rat brain effected by a continuous infusion or by single daily injections. Naunyn-Schmiedeberg's Arch. Pharmacol., 324, 27–33.

Costall, B., Domeney, A.M., Naylor, R.J. & Tyers, M.B. (1987). Effects of the 5-HT_3 receptor antagonist, GR38032F, on raised dopaminergic activity in the mesolimbic system of the rat and marmoset brain. Br. J. Pharmac., in press.

Fake, C.S., King, F.D. & Sanger, G.J. (1987). BRL 43694: A potent and novel 5-HT_3 receptor antagonist. Br. J. Pharmac., 91, 335P.

Richardson, B.P., Engel, G., Donatsch, P. & Stadler, P.A. (1985). Identification of serotonin M-receptor subtypes and their specific blockade by a new class of drugs. Nature, 316, 126–131.

Hagan, R.M., Butler, A., Hill, J.M., Jordan, C.C., Ireland, S.J. & Tyers, M.B. (1987). Effect of the 5-HT_3 receptor antagonist, GR38032F, on responses to injection of a neurokinin agonist into the ventral tegmental area of the rat brain. Eur. J. Pharmac., 138, 303–305.

5-HT$_2$ ANTAGONISTS INCREASE TACTILE STARTLE HABITUATION IN AN ANIMAL MODEL OF A HABITUATION DEFICIT IN SCHIZOPHRENIA

Mark A. Geyer
Department of Psychiatry, University of California at San Diego, La Jolla, California, U.S.A.

Studies of the habituation of startle responses to exteroceptive stimuli provide unique opportunities for cross-species explorations into attentional deficits in schizophrenia. We previously reported that LSD impairs response habituation in a tactile startle paradigm in rats (Geyer et al., 1987; Braff & Geyer, 1980). We have also described a similar impairment of acoustic startle habituation in schizophrenic patients (Geyer & Braff, 1982). Hence, the study of startle habituation may have implications for the neurobiological substrates of the deficits in habituation observed in schizophrenia.

Binding sites for serotonin exist in at least two forms: 5-HT$_1$ sites, labeled by [^{3}H]-5-HT and thought to mediate the inhibitory electrophysiological effects of serotonin in brain; and 5-HT$_2$ sites, labeled by [^{3}H]-spiperone and thought to mediate the excitatory effects of serotonin (Peroutka & Snyder, 1979). The present studies examined the effects of 5-HT$_2$ antagonists on tactile startle habituation. Since serotonergic systems have been implicated in both startle reactivity and habituation, the test paradigm was designed to enable the detection of increases or decreases in either the level of behavioural reactivity or the rate of startle habituation.

Methods
Male Sprague-Dawley rats (250-300 g) from Simonson Laboratories were used, each being tested only once. A microcomputer system was used to control stimulus presentations and monitor startle responses from two cylindrical stabilimeters, as previously described (Geyer & Braff, 1987). Each test session consisted of 201 presentations of 25 psi air-puff (tactile) stimuli directed at the animal's back and presented at 15 sec intervals. Animals were placed in the stabilimeter 30 or 60 min after drug injection, the first trial beginning 5 min later.

The 5-HT$_2$ antagonists cyproheptadine (Sigma), ketanserin, ritanserin (both Janssen Pharmaceuticals), and cinanserin (E.R. Squibb and Sons), were administered subcutaneously at the salt doses shown in Table 20.1.

For each experiment, a mixed-design analysis of variance (ANOVA) was done with the drug treatment as a between subjects factor and blocks of 10 trials as a within subjects factor. The first and last 10-trial blocks were also analyzed in separate one-way ANOVAs.

Results
The 5-HT$_2$ antagonists cyptoheptadine increased the rate of habituation of tactile startle at all doses tested. The initial level of responding following cyproheptadine administration, as measured by the first 10-trial block, was not significantly different from controls. A mixed-design ANOVA revealed a

Table 20.1 Effects of serotonin-2 antagonists on startle habituation*

compound dose	first block	ANOVA	last block	ANOVA	drug-by-trial interaction
Cyproheptadine					
0.25 n=12	103 ±7.1	n.s.	65 ±8.2	$F(1,22)=6.99$ $p<0.05$	$F(19,418)=1.54$ $p=0.069$
0.50 n=11	104 ±6.8	n.s.	54 ±12.0	$F(1,21)=11.7$ $p<0.005$	$F(19,399)=4.01$ $p<0.0001$
1.00 n=11	109 ±7.0	n.s.	63 ±9.1	$F(1,21)=7.32$ $p<0.05$	$F(19,399)=2.85$ $p<0.0001$
2.00 n=12	110 ±6.0	n.s.	76 ±7.2	$F(1,22)=3.54$ $p=0.0731$	$F(19,418)=3.44$ $p<0.0001$
1.00 n=10 (65 min pre-inj)	96 ±8.1	n.s.	59 ±4.1	$F(1,19)=9.30$ $p<0.01$	$F(19,316)=4.48$ $p<0.0001$
Ketanserin					
0.5 n=22	98 ±4.1	n.s.	70 ±7.2	$F(1,43)=7.9$ $p<0.01$	n.s.
1.0 n=12	101 ±4.7	n.s.	55 ±6.8	$F(1,33)=14.7$ $p<0.001$	$F(19,627)=3.1$ $p<0.0001$
2.5 n=12	96 ±4.8	n.s.	62 ±6.7	$F(1,33)=10.4$ $p<0.01$	n.s.
Ritanserin					
0.4 n=12	101 ±3.5	n.s.	67 ±8.5	$F(1,21)=11.4$ $p<0.005$	$F(19,399)=4.6$ $p<0.0001$
1.0 n=9	108 ±3.8	n.s.	70 ±10.1	$F(1,18)=7.9$ $p<0.05$	$F(19,342)=4.2$ $p<0.0001$
Cinanserin					
2.0 n=11	97 ±5.3	n.s.	78 ±12.9	n.s.	n.s.
5.0 n=12	94 ±5.6	n.s.	90 ±9.3	n.s.	$F(19,418)=2.12$ $p<0.005$
12.5 n=12	90 ±5.2	n.s.	70 ±10.7	$F(1,22)=3.01$ $p=0.096$	$F(19,418)=1.48$ $p=0.09$

* Startle response amplitudes are presented as the mean (±S.E.M.) percentage of control values for both the first and last (twentieth) 10-trial blocks. F and p values (n.s. means p>0.1) are also given from the one-way ANOVA on each of these measures. The drug-by-trial interaction was derived from a two-way mixed-design ANOVA with drug and blocks of 10 trials as factors.

significant drug-by-trial interaction for all but the low dose group, due to the more rapid response decrement exhibited by the drug-treated animals (Table 20.1). To determine whether this increase in startle habituation was a reflection of a delayed onset of a drug effect which lowers reactivity, animals given 1.0 mg/kg cyproheptadine were tested at 65 rather than 35 min after injection. With either injection interval, drug-treated animals were similar to controls during the first 10 trials, and subsequently showed an increased rate of habituation (Table 20.1).

In addition to its high affinity for $5-HT_2$ binding sites (ki=6.5nM) cyproheptadine also acts as a weak antagonist at $5-HT_1$ sites (ki=700nM). Hence, three other $5-HT_2$ antagonists, ketanserin, ritanserin, and cinanserin, which have less activity at $5-HT_1$ sites (Janssen, 1982; Leysen et al., 1985) were also tested. All three doses of ketanserin increased habituation. That is, initial levels of responding were unaffected while final values were reduced (Table 20.1). Ritanserin has similar effects, as summarized in Table 20.1. While the magnitude of the effect of ritanserin appeared to be less than that of ketanserin, the overall ANOVA and specific comparisons of each dose individually produced significant drug-by-trials interactions [$F(38,551)=3.3$ $p<0.0001$; Table 20.1), reflecting the acceleration of startle habituation. Cinanserin, while inactive at 2.0 mg/kg, increased habituation and produced a significant drug-by-trials interaction at 5.0 mg/kg.

Discussion

The results of these studies support the hypothesis that the impairment of startle habituation produced by LSD (Geyer et al., 1978; Braff & Geyer, 1980) is mediated by its agonist action of $5-HT_2$ receptors. The $5-HT_2$ antagonists cyproheptadine, ketanserin, ritanserin, and cinanserin all significantly increased the rate of habituation of tactile startle. It appears that $5-HT_1$ receptors had little involvement in this phenomenon, since cinanserin and ketanserin have no appreciable $5-HT_1$ activity (Janssen, 1982; Leysen et al., 1985), yet each increased the rate of starle habituation. In conclusion, the pattern of results obtained with the post-synaptic serotonergic antagonists is consistent with the suggestion (Jacobs, 1984) that hallucinogenic drugs exert many of their behavioural effects via actions as $5-HT_2$ agonists. Ketanserin and related $5-HT_2$ antagonists significantly accelerated startle habituation while having no effect on initial levels of reactivity. That is, the $5-HT_2$ blockers produced effects opposite to those produced by LSD in this paradigm (Geyer et al., 1978), suggesting that LSD may be exerting its effects as a $5-HT_2$ agonist. The effects of the $5-HT_2$ antagonists reported here are consistent with our report that the serotonin synthesis inhibitor, parachlorophenylalanine accelerates tactile startle habituation (Geyer & Braff, 1987). Taken together, these findings further corroborate the hypothesis that the serotonergic system is critically involved in the expression of startle habituation and that this involvement is mediated via $5-HT_2$ receptors. It remains to be determined whether the deficit in startle habituation exhibited by schizophrenics is related to abnormalities of 5-HT systems and if $5-HT_2$ antagonists would reverse the deficit.

Acknowledgments: This work was supported by grants from NIDA ((DA02925) and NIMH (MH00188).

References

Braff, D.L. & Geyer, M.A. (1980). Acute and chronic LSD effects on rat startle: Data supporting an LSD-rat model of schizophrenia. Biol. Psychiatry, 15, 909-916.

Davis, M. (1980). Neurochemical modulation of sensory-motor reactivity: acoustic and tactile startle reflexes. Neurosci. Biobehav. Rev., 4, 241-263.

Geyer, M.A. & Braff, D.L. (1982). Habituation of the blink reflex in normals and schizophrenic patients. Psychophysiology, 19, 1-6.

Geyer, M.A. & Braff, D.L. (1987). Alterations of startle responding in schizophrenia and related animal models: Toward understanding the neurobiology of schizophrenia. Schizophr. Bull., in press.

Geyer, M.A. , Peterson, L., Rose, G., Horwitt, D., Light, R., Adams, L., Zook, L., Hawkins, R. & Mandell, A.J. (1978). Effects of LSD and mescaline derived hallucinogens on sensory integrative function: Tactile startle. J. Pharmacol. Exper. Therap., 207, 837-847.

Jacobs, B.L. (1984). Postsynaptic serotonergic action of hallucinogens. In: Hallucinogens: Neurochemical, behavioural and clinical perspectives. (Ed. Jacobs B.L.) Raven Press, New York, pp 188.

Janssen, P.A. (1982). The pharmacology of specific, pure and potent serotonin $5-HT_2$ or S2 antagonists. In: Advances in Pharmacology and Therapeutics II. (Eds. Yoshida, H., Hagihara, Y., Ebashi, S.) vol. 4, Oxford and New York, Pergamon Press, pp. 21-33.

Leysen, J.E., Gommeren, W., Van Gompel, P., Wynants, J., Janssen, P.F.M. & Laduron, P.M. (1985). Receptor-binding properties in vitro and in vivo of ritanserin: A very potent and long acting serotonin-S2 antagonist. Mol. Pharmacol., 27: 600-611.

Peroutka, S.J. & Snyder, S.H. (1979). Multiple serotonin receptors: Differential binding of [^{3}H]-5-hydroxytryptamine, [^{3}H]-lysergic acid diethylamine and [^{3}H]-spiroperidol. Mol. Pharmacol., 16, 687-699.

FEEDING BEHAVIOUR

A.R.Cools
Psychoneuropharmacological Research-Unit, Dept. Pharmacology, University of Nijmegen, P.O.Box 9101, 6500 HB Nijmegen, The Netherlands

For the past thirty years basic research on the peripheral and central control of feeding behaviour has evolved in the notion that food intake cannot anymore be conceptualized in terms of a simple "depletion-repletion" model, in which the hypothalamus with its so-called satiety and appetitive centres is the major focus. The same holds true for the traditional view that only noradrenaline and dopamine modulate food intake. Evidence accumulated in the past decades emphasizes the important role of serotonin in the peripheral and central control of food intake in its broadest sense.

Stimuli from the food promote feeding by activating the brain. This information enters the central nervous system which is hierarchically organized, and is used to modulate outputs of increasing orders of complexity, from brain stem to cortex. At the lowest level in the hierarchy of the neuronal and hormonal organization of feeding behaviour, food intake satifies physiological needs that results from deprivation and stimulate hunger, viz. it serves to produce an output that modulates peripheral metabolism. At the highest level in the hierarchy, however, food intake not only eliminates the disquieting and unpleasant sensations that accompany hunger, but also satifies the quieting and pleasant sensations that accompany appetite. Given the non-linear hierarchical nature of these levels in the organization of feeding behaviour, it is evident that even drug-induced changes in the periphery can have far-reaching consequences for emotional and cognitive factors that modulate food intake. Thus, drug-induced changes in gastric emptying, energetic balances and hormonal activity which directly modulate the neuronal activity at lower levels in the cerebral organization of feeding behaviour, can indirectly alter the neuronal activity at higher levels in the organization and, accordingly alter complex features of this behaviour such as those responsible for craving for a particular nutrient. In other words, an extremely careful analysis of drug-induced changes in food intake is required in order to delineate which level in the hierarchy is affected by the treatment under study.

The chapter on the involvement of serotonergic mechanisms in human feeding, by Goodall and Silverstone, provides an overview of the available approaches to attack these problems in human studies. This chapter on the role of serotonin in food intake outlines the effects of drugs which indirectly enhance the serotonergic transmission by precursor, re-uptake inhibitor and releasing agents, at different levels in the hierarchical organization of feeding behaviour. Particular attention is paid to the ability of serotonergic agents to affect the higher levels, viz. levels modulating appetitive and nutrient selection, the latter issue being far from settled. Overall, this chapter provides direct evidence that serotonin has anyhow an inhibitory role in the control of feeding in healthy subjects and patients suffering from eating disorders.

The chapter on serotonin and feeding, by Samanin, constitutes an overview of our current understanding of serotonin and of its receptors in the central and peripheral control of ingestive behaviour in animals, especially rats. Following a brief review of evidence delineating the attenuating role of serotonergic fibres arising in the median raphe nuclei, this chapter outlines which hierarchical levels in the organization of feeding behaviour are affected by fenfluramine, viz. a drug which promotes the release of serotonin and inhibits its re-uptake. Next, this chapter describes recent evidence indicating that activation of $5-HT_{1A}$ receptors located on cell bodies in the median raphe nuclei promotes feeding, whereas activation of $5-HT_{1B}$ receptors attenuates feeding. In contrast to the absence of any important role of $5-HT_2$ receptors in the central control of feeding behaviour, it appears likely that $5-HT_2$ receptors may affect gastric emptying and, accordingly, modulate food intake. The final part of this chapter deals with data suggesting that serotonin in the nucleus accumbens can attenuate stress-induced overeating by decreasing the responsiveness to food-related or external cues.

Taking together both chapters, there is ample evidence that peripheral and central serotonin modulates food intake at distinct levels in the hierarchical organization of feeding behaviour, opening thereby perspectives for the therapeutic efficacy of serotonergic agents in feeding disorders. To which degree serotonin forms part of a feedback loop, whereby the consumption of carbohydrate diminishes the tendency to consume further carbohydrate, is still open for discussion: apart from the original studies, more recent human and animal studies (Goodall and Silverstone, chapter 21; Lawton and Blundell, chapter 26) failed to find direct evidence in favour of this hypothesis.

The above-mentioned role of $5-HT_{1A}$ and $5-HT_{1B}$ receptors in the control of food intake is more extensively discussed by Hutson et al. (chapter 23), Leander (chapter 24), Montgomery et al. (chapter 25) and Gentsch et al (chapter 12). Apart from the study reported in the latter chapter which used a slightly different methodology for the drug-induced effects, all remaining studies underline the opposing role of $5-HT_{1A}$ and $5-HT_{1B}$ receptors. The contribution of Hutson et al. (chapter 23), in addition, reveals that these receptors of which activation results in a suppression of food intake, are postsynaptically located, and belong to $5-HT_{1B}$ and $5-HT_{1C}$ subtypes. The latter finding together with the fact that the $5-HT_{1A}$ receptors of which activation promotes feeding, are located on the cell bodies of the serotonergic fibres arising in the median raphe nuclei emphasizes the important role of the median raphe system in the control of feeding. Still, the limited availability of serotonergic agonists of distinct subtypes of serotonergic receptors together with the actual lack of selective antagonists of these subtypes postpone definite conclusions about the differential role of $5-HT_{1A}$, $5-HT_{1B}$, $5-HT_{1C}$ and $5-HT_2$ receptors in this respect. Accordingly, it remains open for discussion whether patients suffering from distinct eating disorders will differentially benefit from serotonergic agents that selectively interacts with one of the mentioned subtypes of serotonin receptors.

PHARMACOLOGICAL EVIDENCE FOR THE INVOLVEMENT OF SEROTONERGIC MECHANISMS IN HUMAN FEEDING

Elisabeth Goodall, Trevor Silverstone
Academic Unit of Human Psychopharmacology, Medical College of St.Bartholomew's Hospital, London, U.K.

Introduction
The role of serotonin (5-HT) in human feeding can be explored using drugs which, in animals, act on central 5-HT neurotransmission. Such drugs can either enhance transmission by providing precursor loading (tryptophan), by increasing release from presynaptic neurones (fenfluramine) (Garattini and Samanin, 1976) or by blocking its reuptake (fluoxetine) (Goudie, Thornton & Wheeler, 1976), or reduce transmission by blocking post-synaptic 5-HT receptors (metergoline, methysergide) (Blundell and Latham, 1980; Garattini, 1986).

Evidence for serotonergic involvement in human feeding has come from studies in normal volunteers, obese patients, patients with eating disorders and in those with neuroleptic-induced obesity. Whereas with animals, one can only observe what they do, not what they feel, in human subjects, by asking carefully framed questions, and using visual analogue scales (VAS), one can find out what feelings of hunger, appetite and satiety the subjects is currently experiencing and to what degree. These measures are both reliable and sensitive; there is a positive and significant correlation between VAS hunger ratings taken just prior to a meal and the amount of food subsequently eaten at the meal (Goodall, Trenchard & Silverstone, 1987).

It is important, though, that these subjective experiences are carefully defined beforehand. The term 'hunger' as applied to humans can be too limited to describe those physiological sensations which accompany significant food deprivation; in this sense, hunger is entirely physiologically determined. The term 'appetite' on the other hand, can more appropriately refer to a person's overall desire to eat however it may be brought about (Silverstone, 1982).

While physiologically determined hunger can obviously influence appetite, the desire to eat at any given time (i.e. appetite) can be partly independent of hunger, with cognitive and emotional factors playing a role. In the pharmacologcal study of human feeding such cognitive and emotional factors playing a role. In the pharmacological study of human feeding such cognitive and emotional factors may blur the response to drug manipulation. Furthermore, the qualitative aspects of food, involving the sight, smell and taste act as rewards to reinforce the initiation and maintenance of the eating response. These aspects need to be borne in mind when evaluating the pharmacological evidence concerning the role of neurotransmitters in such a complex activity as human feeding.

Serotonergic control of nutrient selection
In addition to maintaining energy balance, eating also involves the choosing of food stuffs which will ensure a balanced diet, and there is currently much discussion as to whether the regulation of nutrient intake, at least in the

short-term, is under serotoninergic control (Fernstrom, 1987). Tryptophan hydroxylase, the enzyme that converts tryptophan to 5-HT, is usually not fully saturated and thus the rate of 5-HT synthesis depends on the availability of tryptophan. In animal studies Wurtman and his group have shown that the level of tryptophan in the brain is dependent on the ratio of tryptophan to other long chain neutral amino-acids (LNAAs) in the blood (Wurtman, 1983). Following a high protein meal this ratio is low and the other LNAAs are actively transported across the blood brain barrier at the expense of tryptophan (TRP). Following a meal high in carboxydrate, in the presence of insulin (released in response to carbohydrate intake) the TRP/LNAA ratio is high and transport to tryptophan into the brain is facilitated (Fernstrom and Wurtman, 1971, 1972). The synthesis of 5-HT in the brain is thus enhanced. From this Wurtman and Wurtman postulate that choice of food intake is influenced by the level (or functional availability) of 5-HT in the brain and sought to show in rats that a drug, such as fenfluramine, which increased serotonin transmission would selectively lower carbohydrate intake (Wurtman & Wurtman, 1979; Wurtman, 1983).

Similarly, in obese human subjects, they found that dextrofenfluramine-enhanced 5-HT neurotransmission selectively decreased intake of carbohydrate but not protein snacks in those described specifically as having a need for carbohydrate at certain times (Wurtman & Wurtman, 1984; Wurtman et al., 1985). Such a hypothetical extension of the biochemical data concerning the relationship between nutrients consumed, the brain tryptophan level and rate of 5-HT synthesis which includes a behavioural feedback loop to control subsequent carbohydrate appetite has recently been disputed by Fernstrom (1987). While similar information on the neurotransmitter control of food preferences can be obtained in humans it needs to be emphasised that the hedonic aspects of food may play a larger role in determinating preferences, and that responses may be continued by past experiences to a greater extent than in animals.

Drugs enhancing 5-HT neurotransmission
1. Tryptophan
A 2g dose of L-TRP reduced total food intake by 13% compared to placebo 45 minutes after administration in 17 lean young men screened to include only those who did not show any tendency to restrained eating. This substantial reduction was replicated in a further study involving 15 subjects with two and three grams TRP but not 1 g (Hrboticky, Leiter & Anderson, 1985). A buffet meal consisting of a variety of meats, cheeses bread rolls and cookies was provided. In both studies the carbohydrate items (rolls and cookies) were significantly reduced but relative macronutrient intake was not affected apart from a marginally significant increase in the protein / carbohydrate ratio in the first study. The authors consider that changes in macronutrient selection might have been more apparent had there been more variety as subject preferences may have influenced the behavioural expression of physiological regulation.

TRP at all doses significantly raised the plasma TRP/LNAA ratio in both studies 30 minutes after administration. Hroticky et al. (1985) discussed the possibility of TRP exerting a peripheral influence on food intake as the TRP/LNAA ratio was raised as much by 1g as by the anorexogenic doses. In women, who were uncontrolled for the menstrual cycle tryptophan failed to modify energy intake but did reduce carbohydrate intake expressed as a percentage of total intake

(Leiter, Hrboticky & Anderson, 1987).

The authors found that the behavioural effects of tryptophan were influenced by the menstrual cycle and by the degree of restraint shown by women subjects. Energy and carbohydrate intake were found to be correlated with plasma TRP in the follicular phase, but not in the luteal phase, when women following placebo had eaten 20% more calories, carbohydrates and desserts than in the follicular phase. Tryptophan did not suppress carbohydrate consumption in restrained eaters. Hill (1986) found TRP to have a subtle effect on CHO ingestion. TRP, embedded in a chocolate bar, suppressed subsequent selection of carbohydrate, but only following a high protein meal. As the authors remark (Blundell and Hill 1987) the data do suggest a serotoninergic influence on nutrient selection but one which cannot be revealed with crude experimental procedures.

Patients
In a single case study (Cole and Lapierre, 1986), 1g tryptophan, given for six weeks to a non-depressed bulimic woman led to marked reduction in urge to binge within one week. Bingeing had ceased after three weeks but normal calorie intake was maintained. Discontinuation and reinstitution were paralleled by increase and decrease in bingeing. Psychotherapy was given concurrently until the drug was no longer needed. Although an isolated case must be viewed with caution, it would appear that increased availability of 5-HT had a facilitating effect on satiety until changes in attitudes and lifestyle could maintain normal satiety. This is an example of the interaction of pharmacological and psychological factors in the control of eating behaviour.

A further indication that 5-HT is involved in human feeding and that 5-HT dysfunction may be implicated in the etiology of eating disorders comes from a study of the prolactin response to TRP in dieting volunteers (Goodwin, Fairburn and Cowen, 1987). A marked increase in the prolactin response to L-TRP was found in normal weight women volunteers undertaking a 3 week period of dieting but not in male volunteers indicating that dieting may selectively alter neurotransmitter function in women.

2. Fenfluramine
Normal subjects
The following three studies illustrate different experimental approaches to the measurement of anorectic activity of dextrofenfluramine in normal human subjects.

i) In a study comparing the anorectic activity of the dextrorotatory isomer d-fenfluramine (d-FF) with the racemic mixture (dl-FF), which has been used as an anorectic for many years, 30, 40, and 60 mg d-FF and 30 and 60 mg dl-FF were administered four hours before presentation of food to 16 subjects (12 female and 4 male) (Silverstone, Smith and Richards, 1987). The meal consisted of a tray containing preweighed portions of a large variety of foods. Food remaining after the 30 minute meal was reweighed to ascertain the amount consumed. Food intake was significantly reduced ($p<0.01$) after all doses apart from 30 mg dl-FF with d-FF having twice the anorectic potency of dl-FF. Hunger ratings fell and reached significance after four hours. d-FF appeared to be more extensively metabolised to nor-fenfluramine than dl-FF but differences in plasma levels were not consistent with differences in the anorectic potency. The emerge of side effects did appear to be related to the plasma level of FF as the number of subjects reporting side effects over an 8 hour period following 30 mg d-FF were half that after 60 mg dl-FF.

ii) Hill and Blundell (1986) designed a model system to investigate the effects of acute and subchronic doses of d-FF over a period of five days on several aspects of food intake. They gave a standardised test meal on day 1 to measure the effect of d-FF on total food intake and designed a meal to test the drug's effect on nutrient intake on day 5. In addition, diet diaries were kept by the subjects over the five days so that the parttern of food intake throughout the day could be examined. They used a method, of temporal tracking of subjective sensations by VAS and questionaires throughout the test meals in order to quantify temporal changes related to elements in the meal (Hill, Magson and Blundell, 1984). d-FF significantly reduced total food intake (calculated from both the test meals and diet diaries) compared to placebo on each of the five days, and reduced the number of medium-sized eating episodes (400-3199 kJ).

Energy intake was plotted hourly from the diaries and it was found that the energy deficit caused by d-FF occurred during the three hour period following drug administration. Subjective motivation to eat before and during a meal was markedly affected by d-FF. Temporal tracking of hunger, appetite (desire to eat), prospective consumption and fullness on day 1 showed that subjects were less hungry, more full and wanting to eat less before and during the meal following d-FF. During the meal they also recorded significantly less desire to eat. After the meal d-FF prevented the recovery of desire to eat, hunger and prospective consumption. Interestingly, pre-meal desire to eat was the only subjective feeling altered by d-FF on day 5, although a significant 18% reduction in food intake was seen. The authors considered that their results implied a facilitation and intensification of satiety by d-FF. Hill and Blundell (1986) did not find any effect on nutrient selection apart from a reduced selection of high protein items from a preference checklist on day 1 and the prevention of the restoration of checking high carbohydrate items 2 hours after a meal. No differences in actual food intake were found.

iii) Somewhat in contrast to the above approaches is the technique which allows the auromatic recording of food intake over several hours (Silverstone, Fincham and Brydon, 1980).

Although it cannot provide fine detail of eating behaviour the automated food dispenser (AFD) can measure temporal aspects of eating such as latency to start eating, timing and duration of bouts of eating and the development of satiation from study of cumulative intake curves. The AFD contains four refrigerated channels and is connected to a pen recorder so that every time a subject takes an aliquot of food the action if recorded. Using this apparatus, containing four snacks of varying carbohydrate content and sweetness and recording food intake over two hours, Goodall and Silverstone (1987) found d-FF to have a strongly anorectic effect in healthy male subjects. d-FF did not have a selective effect on snacks of varying carbohydrate content.

However when the effect of d-FF was examined separately for non-sweet and sweet tasting foods it was found that the anorectic action was much greater for the non-sweet food; it had no effect on the intake of sweet food until the second hour and then although it was reduced, the different from placebo did not reach statistical significance. Among the possible explanations for this differential effect is that the hedonic property of sweetness may have attenuated the anorectic effects of the drug. The reduced motivation to eat found by Hill and Blundell (1986) might have been counteracted in the case of sweet food (but not non-sweet food) by the greater reinforcing character of sweetness. Subjective feelings of hunger and satiety (fullness) were recorded hourly using VAS. Hunger was significantly reduced 4 hours after drug administration. It is interesting to note that although d-FF would appear to affect satiety when the subjective dimension of satiation is explored post-prandially using VAS worded 'fullness' no differential effect was found in this study in that of Hill and Blundell (1986).

Patients
Two groups of patients with refractory obesity took part in a placebo controlled double blind trial of d-FF (Finer et al., 1985). Patients attending either their General Practitioner (GP) or a hospital obesity clinic received either 15 mg d-FF twice daily or placebo for three months. Seventy-six patients completed the trial. Weight loss of those on d-FF in the GP group became significant after six weeks, reaching 5.3 kg after 12 weeks; the weight loss of 3.9 kg on placebo never became significant. In the hospital group there was a highly significant difference in total weight loss between those on d-FF and those on placebo. Sensi, Della Loggia, Del Ponte and Guagnano (1985) found that after 3 months treatment with fenfluramine (60 mg once daily or 20 mg twice daily) given in conjunction with a hypocaloric diet obese patients had lost significantly more weight than those on diet alone. In the following 3 months all groups continued to lose weight so that after 6 months there was no significant difference in weihgt loss between the groups but body weight within each group, including placebo, was significantly different from baseline. These results indicate that although fenfluramine can be effective in the short term non-drug related behavioural changes can influence weight loss in the longer term.

Weight gain is a serious problem in patients receiving long-term neuroleptic medication (Silverstone, Smith and Goodall, 1988) and is thought to be due to the noradrenergic and serotonergic receptor blocking activity that these drugs have in addition to dopamine receptor blockade. A double blind trial of d-FF was therefore undertaken in thirty-three obese patients with neuroleptic induced obesity the majority of whom has schizophrenia (Goodall et al. submitted for publication). Sixteen patients completed the 12 week trial. Weight loss was 5.4 kg for those receiving d-FF and 2.8 kg for those on placebo. The rate of weight loss was significantly greater for those on d-FF. Weight loss correlated with plasma d-FF and with d-FF and the metabolite nor-dFF combined thus giving further evidence for the role of 5-HT in human eating.

Robinson, Checkley and Russell (1985) gave single double blind doses 60 mg dl-FF to 15 young women with bulimia nervosa. A large quantity of food which the subjects usually binged upon was presented for half an hour two hours after drug administration. The amount of food eaten following placebo ranged from 90-976 kcals. FF significantly reduced the calories consumed compared to placebo, with the quantity eaten being inversely correlated with plasma fenfluramine level. Fenfluramine significantly decreased the number of subjects showing bulimic symptoms of bingeing or vomiting on the ward. Patients rated their VAS hunger differently from matched controls. d-FF restored the normal fall in hunger ratings after food which was seen in the controls but not in the patients when they were on placebo.

The evidence from these clinical studies of fenfluramine support the theory that 5-HT has a central role in the behavioural control of eating. However, Robinson has suggested that the drug-induced reduction in food intake may be partially the result of reduced gastric emptying and consequent gastric distention (Robinson, Moran and McHugh, 1986). In monkeys sucrose intake was strongly related to the rate of gastric emptying which was reduced by dl-FF (Robinson et al., 1986). Likewise, in humans fenfluramine can reduce gastric emptying (Horowitz et al., 1985). Using a gamma camera in 8 obese subjects they found that FF significantly reduced the linear gastric emptying rate of solid food (ground beef with labelled chicken liver) but had no effect on liquid (dextrose) comsumed at the same time. The dose of FF in mg/kg was lower than that given by Robinson et al. (1987).

3. 5-HT reuptake inhibitors
Fluoxetine (FXT) is a selective reuptake inhibitor with minimal effect on noradrenalin and dopamine uptake mechanisms (Lemberger et al., 1978, 1982). Ferguson (1986) has shown that significantly greater weight loss was found in obese non-depressed outpatients after three weeks administration of an average dose of 65 mg FXT compared to placebo. After 8 weeks the average weight loss of the 32 out of 50 patients who completed the trial of FXT was 10.3 lbs, while 30 patients on placebo had lost 3.5 lbd. Ferguson attempted to test the hypothesis that weight loss had occurred through modification of carbohydrate craving. Patients rated desire for carbohydrate on a scale from 0-10 at the start of the trial. Those who 'craved carbohydrate' (rating 6-10) lost more on FXT than those who were in the middle range or were indifferent to carbohydrate but the effect did not reach significance. Had he monitored desire for carbohydrate throughout the trial a more meaningful analysis of the effect of FXT on carbohydrate craving and weight loss might have been made. We have examined in more detail the appetite suppressant effects of FXT in non-obese volunteers. (McGuirk and Silverstone in preparation). Eleven males received FXT 60 mg daily or matching placebo for 14 days, each 14 day period being followed by a 28 day washout. The order was random and the trial was double-blind. Food intake of a cafetaria type meal from the AFD and VAS for appetite and weight were recorded weekly in each trial period. Food intake was significantly reduced by FXT compared to placebo on certain days and time points. Subjects felt less hungry on FXT but these differences did not reach significance. Subjects lost weight in response to treatment for 14 days.

Drugs reducing 5-HT neurotransmission
5-HT receptor antagonists
Much of the evidence that part of fenfluramine's action is via central mechanisms comes from the use of relatively specific 5-HT receptor blockers. In rats the effect of d-FF is blocked by the 5-HT antagonist metergoline (MTG) which crosses the blood brain barrier but not by xylamidine which does not (Borsini et al., 1982). MTG is known to have central activity in human subjects

(Delitala et al., 1977). We therefore examined the effect of MTG on d–FF induced anorexia in human volunteer subjects (Goodall and Silverstone, 1987). MTG attenuated the anorectic action of d–FF on total food intake and on the intake of non–sweet foods while having no effect on these foods when given alone. In contrast MTG significantly increased the intake of sweet foods compared to placebo from 30–80 minutes during the two hours that the food was available. d–FF and MTG interacted together giving an intake of sweet food that was no different from placebo. These results are consistent with those in animals and add support to the view that 5–HT mechanisms are involved in human eating. The increase in eating which we have found following MTG adds support to Wurtman's theory that 'carbohydrate craving' is due to reduced functional availability of 5–HT.

Conclusions

In conclusion the evidence presented suggests that 5–HT mechanisms are involved in human feeding. While little evidence has been found in support of the view that 5–HT effects nutrient selection, there is some evidence to suggest that the nature of the food itself can modulate the pharmacological manipulation of human feeding.

References

Blundell, J.E. & Hill, A.J. (1987). Nutrition, serotonin and appetite: case study in the evolution of a scientific idea. Appetite, **8**, 183–194.

Blundell, J.E. & Latham, C.J. (1980). Characterisation of adjustments to the structure of feeding behaviour following pharmacological treatment: effects of amphetamine and fenfluramine and the antagonism produced by pimozide and methergoline. Pharmacol. Biochem. & Behav., **12**, 717–722.

Borsini, F., Bendotti, C., Aleotti, A., Samanin, R. & Garattini, S. (1982). d–fenfluramine and d–norfenfluramine reduce food intake by acting on different serotonin mechanism in the rat brain. Pharmacol. Res. Commun., **14**, 671–678.

Cole, W. & Lapierre, Y.D. (1986). The use of tryptophan in normal–weight bulimia. Can. J. of Psych., **31**, 755–756.

Delitala, G., Masala, A., Alagna, S., Devilla, L. & Rovasio, P.P. (1977). Inhibition of prolactin release by metergoline administration in man. Biomed., **27**, 31–33.

Ferguson, J.M. (1986). Fluoxetine induced weight loss in humans. In: Advances in the Biosciences, Vol. 60, pp 313–318, Oxford, Pergamon Press.

Fernstrom, J.D. (1987). Food induced changes in brain serotonin synthesis: is there a relationship to appetite for specific macronutrients? Appetite, **8**, 163–182.

Fernstrom, J.D. & Wurtman, R.J. (1971). Brain serotonin content: increase following ingestion of carbohydrate diet. Sci., **174**, 1023–1025.

Fernstrom, J.D. & Wurtman, R.J. (1972). Brain serotonin content: physiological regulation by plasma neutral amino acids. Sci., 178, 414-416.

Finer, N., Craddock, D., Lavielle, R. & Keen, H. (1985). Dextrofenfluramine in the treatment of refractory obesity. Current Ther. Res., 38, 847-854.

Garattini, S. (1976). Effects of d-fenfluramine on eating disorders. In: Advances in the Biosciences, Vol. 60, pp 327-341, Oxford, Pergamon Press.

Garattini, S. & Samanin, R. (1976). Anorectic drugs and brain neurotransmitters. In: Appetite & Food Intake. (Ed. Silverstone, T.) pp 83-108, Berlin, Dahlem Konferenzen.

Goodall, E. & Silverstone, T. (1987). The effect of the 5-HT releasing drug fenfluramine and the receptor blocker, metergoline, on food intake in human subjects. Annals of the New York Acad. of Sci., 499, 321-323.

Goodall, E., Trenchard, E. & Silverstone, T. (1987). Receptor blocking drugs and amphetamine anorexia in human subjects. Psychopharmacol., 92, 484-494.

Goodwin, G.M., Fairburn, C.G. & Cowen, P.J. (1987). Dieting changes serotonergic function in women, not men: implications for the aethiology of anorexia nervose? Psych. Med., 17, 839-842.

Goudie, A.J., Thornton, E.W. & Wheeler, T.J. Effects of Lilly 110140, a specific inhibitor of 5-hydroxytryptamine uptake, on food intake and on 5-hydroxytryptophan-induced anorexia. Evidence for serotonergic inhibition of feeding. J. of Pharm. & Pharmacol., 28, 318-320.

Hill, A.J. (1986). Investigation of some short-term influences on hunger, satiety and food consumption in man. Unpublished PhD Thesis for the University of Leeds.

Hill, A.J. & Blundell, J.E. (1986). Model system for investigating the action of anorectic drugs: effect of d-fenfluramine on food intake, nutrient selection, food preference, meal patterns, hunger and satiety in healthy human subjects. In: Disorders of Eating Behaviour. (Eds. Ferrari, E. & Brambilla, F.). A Psycho Neuroendocrine Approach, pp 377-389, Oxford, Pergamon Press.

Hill, A.J., Magson, L.D. & Blundell, J.E. (1984). Hunger and palatability: tracking ratings of subjective experience before, during and after the consumption of preferred and less preferred food. Appetite, 5, 361-371.

Hrboticky, N., Leiter, L.A., Anderson, H.A. (1985). Effects of L-tryptophan on short term food intake in lean men. Nutrition Res., 5, 595-607.

Horowitz, M., Collins, P.J., Tuckwell, V., Vernon-Roberts, J. & Shearman, D.J.C. (1985). Fenfluramine delays gastric emptying of solid food. Brit. J. of Clin. Pharmacol., 19, 849-851.

Leiter, L.A., Hrboticky, N. & Anderson, G.H. (1987). Effects of L-tryptophan on food intake and selection in lean men and women. Ann. of New York Acad. of

Sci., **499**, 327-328.

Lemberger et al. (1987). Fluoxetine, a selective serotonin inhibitor. Clin. Pharmacol & Therap., **23**, 421-429.

Robinson, P.H. (1987). Fenfluramine and gastric emptying. Am. J. of Physiol., **252**, R433.

Robinson, P.H., Checkley, S.A. & Russell G.F.M. (1985). Suppression of eating by fenfluramine in patients with bulimia nervosa. Brit. J. of Psych., **146**, 169-176.

Robinson, P.H., Moran, T.H. & McHugh, P.R. (1986). Inhibition of gastric emptying and feeding by fenfluramine. Am. J. of Physiol., **250**, R764-R769.

Silverstone, T. (1982). Preface. In: Drugs and Appetite. (Ed. Silverstone, T.) pp vii-xi, London, Academic Press.

Silverstone, T., Fincham, J. & Brydon, J. (1980). A new technique for the continuous measurement of food intake in man. Am. J. of Clin. Nutr., **3**, 1852-1855.

Silverstone, T., Smith, G. & Goodall, E. (1987). The prevalence of obesity in patients receiving depot antipsychotic. Brit. J. of Psych. (in press).

Silverstone, T., Smith, G. & Richards, R. (1987). A comparative evaluation of dextro-fenfluramine and dl-fenfluramine on hunger, food intake, psychomotor function and side effects in normal human subjects. In: Body weight control. (Eds. Bender, A.E. & Brookes, L.J.) pp 240-246, London, Churchill Livingstone.

Sensi, S., Della Loggia, F., Del Ponte, A. & Guagnano, M.T. (1985). Long-term treatment with fenfluramine in obese subjects. Int. J. of Clin. and Pharmacol. Res., **5**, 247-253.

Wurtman, R.J. (1983). Behavioural effects of nutrient. Lancet, May 21, 1145-1147.

Wurtman, J.J. & Wurtman, R.J. (1979). Drugs that enhance central serotoninergic transmission diminish elective carbohydrate consumption by rats. Life Sci., **24**, 895-904.

Wurtman, J.J. & Wurtman, R.J. (1984). d-fenfluramine selectively decreases carbohydrate but not protein intake in obese subjects. Int. J. of Obes., **8**, (suppl. 1), 79-84.

Wurtman, J.J., Wurtman, R.J., Mark, S., Tsay, R., Gilbert, W. & Growdon, J. (1985). Int. J. of Eating Disorders, **4**, 89-99.

SEROTONIN AND FEEDING

Rosario Samanin
Instituto di Richerche Farmacologische "Mario Negri", Via Eritrea 62, 20157 Milan, Italy

Introduction
Recent gains in the knowledge of the pharmacology of serotonin (5-HT), together with increasingly sophisticated descriptions and quantitation of feeding behaviour, have consolidated the notion that 5-HT in the brain has an inhibitory influence on feeding. Recent studies have also suggested a role for 5-HT in peripheral tissue.

Although the exact role of 5-HT in physiological mechanisms controlling hunger and satiety is not clear, important aspects of feeding have been found to be controlled by changes in 5-HT activity which have little or no effect on other behaviours. Significant progress has been made in investigating the particular mechanisms by which 5-HT reduces food intake, including the role of 5-HT in feeding disorders usually associated with stress and emotional behaviour.

This chapter summarizes several findings which, in the author's opinion, are of particular interest for these issues.

Evidence that 5-HT in the brain acts to inhibit feeding: the utility of studies with fenfluramine
An agent which has contributed considerably to the hypothesis of a role of brain 5-HT in feeding control is fenfluramine, a drug which releases 5-HT from nerve terminals and inhibits its reuptake (Garattini et al., 1975), and is used as an anorextic agent in humans (Pinder et al., 1986a).

The first evidence that fenfluramine uses brain 5-HT to cause anorexia came from the finding that electrolytic lesions of the nucleus raphe medianus, where part of the 5-HT neurons innervating the forebrain originate, blocked fenfluramine lowering food intake of food-deprived rats (Samanin et al., 1972). Shortly after, studies with intracerebral injections of a neurotoxin for 5-HT neurons confirmed these findings (Clineschmidt, 1973).

The robustness of the phenomenon has been questioned by some authors who found that electrolytic lesions and intracerebral injections of 5-HT neurotoxins failed to modify fenfluramine's anorexia (Sugrue, Goodlet & McIndewar, 1975; Carey, 1976; Hoebel et al., 1978). The possible reasons for these discrepancies have recently been discussed by Samanin (1983) and Blundell (1984). Here only three major points will be discussed. The extent and location of the brain lesion is crucial. It was confirmed by Fuxe et al. (1975) and subsequently by Davies et al. (1983) that the integrity of the nucleus raphe medianus and closely surrounding tissue is important for the anorectic activity of fenfluramine in starved rats. It could be significant that lesions in this nucleus, but not in other 5-HT containing nuclei, were found to cause a body weight gain in rats (Geyer et al., 1976).

Moreover, hyperphagia and body weight gain were found in rats in which injection of a 5-HT neurotoxin in the ventromedial hypothalamus had depleted 5-HT in the hypothalamus, septum and hippocampus (Waldbilling, Bartness & Stanley, 1981). Three areas receiving substantial 5-HT innervation from the nucleus raphe medianus (Fuxe et al., 1975; Azmitia, 1978).

The time between treatment and testing may be critical too, particularly in studies with 5-HT neurotoxins. In free feeding rats we recently found that food intake was increased or not affected depending on whether it was measured three days or more that ten days after an intracerebroventricular injection of 5,7-dihydrotryptamine (Bendotti & Samanin, 1986). That adaptive changes after 5-HT neurotoxin administration may be important in interpreting behavioural modifications was also suggested by Berge et al. (1986) who found that nociceptive thresholds and morphine analgesia were reduced during the first post-injection days after 5-HT neurotoxins, but returned to control values during the second week, with a mechanism involving functional adaptation of 5-HT neurotransmission.

An addition complication is that in animals treated with fenfluramine D and L stereoisomers and their respective nor-metabolites form: These have different biochemical and functional effects. D-fenfluramine (DF) and D-norfenfluramine (DNF) have more potent effects than L-isomers on 5-HT mechanisms and food intake (Garattini et al., 1979) while L-fenfluramine and its metabolite preferentially affect brain catecholamines (Invernizzi et al., 1986), and at relatively high doses may depress food intake by a non 5-HT mechanism (Garattini et al., 1986).

To overcome the difficulty of controlling the variables introduced by brain lesions and fenfluramine racemate, we studied the effect of DF and DNF in animals treated with mergoline, a potent central 5-HT antagonist, and xylamidine, a 5-HT antagonist that very poorly penetrates the brain. The anorectic activity of 2.5 mg/kg DF and 1.25 mg/kg DNF in food-deprived rats was completely prevented by mergoline but was not affected by a dose of xylamidine well above the level reported to have peripheral antiserotoninergic effect. The same dose of xylamidine was found to block the effect on food intake of 2 mg/kg 5-HT (Borsini et al., 1982). These findings are summarized in Fig. 22.1. These studies, which have been confirmed recently for fenfluramine (Fletcher & Burton, 1986a) and DF (Carruba et al., 1986), clearly support the hypothesis that DF reduces food intake of rats by interacting with brain 5-HT.

The effects of fenfluramine and DF on food intake cannot be attributed to interference with sensorimotor performances or malaise (Blundell, 1984; Rowland &Carlton, 1986). Moreover, fenfluramine and DF cause anorexia in humans (Silverstone & Goodall, 1986a) at doses with no major adverse reactions. The effect is antagonized by 5-HT antagonists suggesting a role of 5-HT in DF's effects on human feeding (Silverstone & Goodall, 1986b).

What aspects of feeding are regulated by brain 5-HT?

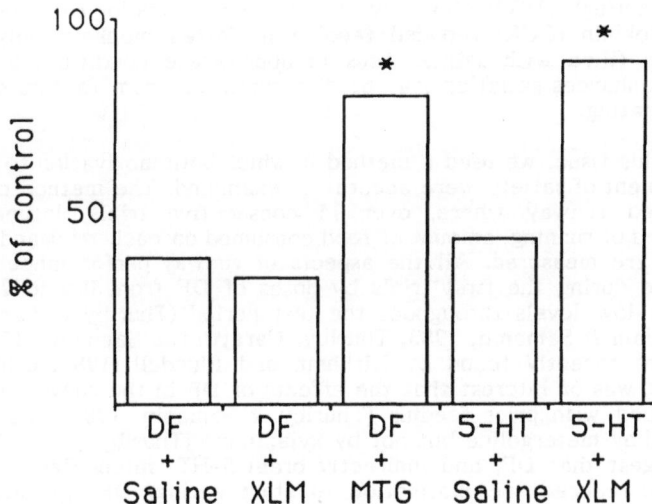

Fig. 22.1 Effect of metergoline (MTG, 1 mg/kg i.p., 3 h before) and xylamidine (XLM, 2 mg/kg p.o., 1 h before) on reduction of food intake of deprived rats (trained over two weeks to take their daily food during 4h) by d-fenfluramine (DF, 2.5 mg/kg i.p.) and serotonin (5-HT, 2 mg/kg i.p.). Mean 1 h food intake. None of the antagonists significantly modified food intake.
*$p<0.01$ vs saline (Tukey's test).

Hunger vs satiety
In a series of studies in which changes in the behavioural structure of feeding were monitored in food-deprived and free-feeding rats, Blundell and his colleagues (Blundell, Latham, & Leshem, 1976; Blundell & Latham 1978; Blundell et al., 1979) found that fenfluramine slowed the rate of eating and reduced meal size with no effect on meal-to-meal intervals. They interpreted this as an intensification of the process of prandial satiation. Since fenfluramine, unlike d-amphetamine, had no effect on the start of eating, Blundell and colleagues suggested that the drug had no effect on hunger defined as the process that stimulates the onset of eating.

The distinction between reduced hunger and increased satiation may not be simple when artificial replenishment or pharmacological activation of a satiety factor is done before a meal. In fact, an artificially-induced satiety state, like that induced by food ingestion, is expected to reduce the interest in starting eating (Wise & Raptis, 1985; Thurlby & Samanin, 1981). However, there is evidence that in consumatory component of feeding may be affected before the appetitive response by pre-feeding (Wise & Raptis, 1985).

Moran and McHugh (1982) elegantly showed that "physiological" concentrations of cholecystokinin (CCK) reduced feeding in fasted monkeys only when the stomach was filled with saline. Thus in appropriate conditions a factor that specifically enhances satiation may be differentiated from factors that control the start of eating.

To address this issue, we used a method in which both motivation to start eating and development of satiety were accurately examined. The method consists of a food-rewarded runway where, over 15 consecutive trials, latency to start running, speed of running, amount of food consumed on each trial and cumulative food intake are measured. All the aspects of runway performance and feeding were reduced during the first trials by doses of DF from 0.6 to 2 mg/kg and remained at low levels throughout the test period (Thurlby & Samanin, 1981; Thurlby, Grimm & Samanin, 1983; Thurlby, Garattini & Samanin, 1985). Similar findings were recently found by Kirkham and Blundell (1986) using a similar procedure. It was of interest that the effects of DF in the runway were similar to those found with prior feeding (Thurlby & Samanin, 1981). Moreover, they were blocked by metergoline but not by xylamidine (Thurlby et al., 1985). These findings suggest that DF, and indirectly brain 5-HT, intensifies the process of satiation and reduced the motivation to start eating. DF was also found to reduce the number of food reinforcements of a food maze (Rech, Borsini & Samanin, 1984).

Studies using meal pattern analysis have not confirmed that fenfluramine specifically intensified satiation (Grinker et al., 1980; Burton, Cooper & Popplewell, 1981). In experiments on conditioned satiation, Booth, Gibson and Backer (1986) recently reached the conclusion that reduced motivation or ability to eat from the start of the meal largely contributes to fenfluramine's effect on meal size. In this study Booth suggested that fenfluramine may affect ingestive movements possibly by enhancing an inhibitory influence of 5-HT on the swallowing reflex (Kessler & Jean, 1985). This is unlikely in view of the runway finding that fenfluramine affected motivation to eat in a condition where no actual food ingestion was required by rats (Thurlby & Samanin, 1981). In humans fenfluramine appears to delay the onset of feeding (Kyriakides & Silverstone, 1979) and reduce hunger ratings (Silverstone & Goodall, 1986a). In view of the evidence that brain 5-HT can cause satiety (Papadokos, Knudson & Leibowitz, 1986), it is likely that DF, through its action on brain 5-HT, induces a satiety sate which, like food ingestion, reduces the interest to initiate a meal.

Rewarding effect of food
One important aspect of DF's action is its ability to reduce palatability-induced ingestion in free-feeding animals. The orosensory characteristics of food play an important role in motivated ingestive behaviour and may be the main determinant of feeding in situations where subjects can choose from various nutritional sources (Cytawa & Trojniar, 1978; Davis & Levin, 1977; Samanin & Garattini, 1984). Moreover, the incentive value of the food stimulus is considered a major determinant of meal maintenance and may contribute to the start of intake (Le Magnen, 1983; Halmi et al., 1987; Mook et al., 1980).

The intake of sucrose solutions by sated rats has been proposed as a useful model to study the rewarding effects of food (Waldbilling & O'Callaghan, 1980) and we have used this method to characterize better the effect of DF on feeding. Doses of DF equal to those lowering food intake in deprived rats strongly reduced consumption of a sucrose solution in sated rats and the effect was blocked by metergoline but not xylamidine (Borsini, Bendotti & Samanin, 1985). In agreement with these findings, fluoxetine, a selective inhibitor of 5-HT uptake, was reported to suppress the ingestion of saccharin solutions at doses lower than those reducing food intake in deprived rats (Leander, 1987).

The findings with DF and fluoxetine suggest that brain 5-HT is involved in the mechanisms that control the rewarding effect of food. In view of the close relationship between rewarding effects of food and meal size (Le Magnen, 1983; Halmi et al., 1987), it is possible that the satiating-effects of DF is, at least partly, due to its ability to reduce the reward impact of food stimuli.

Nutrient selection
In experiments in which food deprived rats had access to two diets containing different proportions of carbohydrate and protein, Wurtman and Wurtman (1977) found that fenfluramine selectively suppressed carbohydrate intake. Based on previous findings that carbohydrate ingestion in fasted rats raises brain tryptophan (TRP) levels and 5-HT synthesis (Fernstrom & Wurtman, 1971) the attractive proposal was made of a reciprocal influence between carbohydrate intake and brain 5-HT function (Wurtman & Wurtman, 1979; Wurtman, 1985). The essential characteristics of the model are the following: carbohydrate ingesation, by increasing insulin release, raises the ratio between TRP and other large neutral aminoacids (LNAA) with consequent increases in brain TRP levels and 5-HT synthesis. This results in increased 5-HT transmission which reduces the appetite for carbohydrates. This then lowers the TRP/LNAA ratio and brain TRP levels. 5-HT synthesis falls and the appetite for carbohydrates returns to close the behavioural loop.

This hypothesis has been challenged by several authors whose main objections are set out in a series of articles published in the June 1987 issue of Appetite. Only the main points of criticism will be briefly summarized here. Various authors agree that the results by Wurtman and his collaborators are uninterpretable in terms of macronutrient selection since the sensory characteristics of the diet could have effects on the experimental subjects which are confounded with the nutrient composition (Blundell & Hill, 1987; Booth, 1987; Fernstrom, 1987; Peters and Harper, 1987).

In animals which could select each macronutrient (including fat) independently, the selective effect of fenfluramine has not been confirmed (Orthen-Gambill & Kanarek, 1975). Other studies in which rats were offered dietary choices found no selective action (Blundell, 1983; Booth et al., 1986; Curtis-Prior & Prouteau, 1983; McArthur & Blundell, 1983; Rowland & Carlton, 1986).

In apparent agreement with his model Wurtman et al., found that fenfluramine suppressed high-carbohydrate snacking in a selected group of obese patients (Wurtman et al., 1981; Wurtman & Wurtman, 1984). A major objection to this study is that in patients choosing snacks containing high carbohydrate levels the reduction by fenfluramine would necessarily appear carbohydrate-selective (Fernstrom, 1987, Rowland & Carlton, 1986). Fernstrom in noting that there is no known requirement for carbohydrates in humans, calls for precision in the definition of "carbohydrate cravers" who may be people simply interested in appealing foods. In view of the inhibitory influence of brain 5-HT on the rewarding effect of food, fenfluramine's effects on snacking may reflect a reduction in the incentive value of palatable food. Experiments on macronutrient intake of normal weight patients during maeltimes (Blundell, 1984) revealed no major selective action of fenfluramine.

Another point of criticism raised by Fernstrom (1987), Leathwood (1987) and Peters and Harper (1987) is that, although in well-defined conditions such as a single meal in fasted rats, it is possible to show a relation between carbohydrate or protein intake and brain TRP levels and 5-HT synthesis, this cannot be extrapolated to conditions such as free feeding. There is in fact evidence that changes in the proportions of carbohydrate and proteins consumed by free-feeding rats have no effect on brain TRP of 5-HT (Leathwood, 1987). It also may be questioned whether an increase in 5-HT synthesis actually results in increased 5-HT transmission. Although it has been recently proved that exogenous administered TRP may increase the release of 5-HT, particularly in activated neurons (De Simoni et al., 1987), a dose of TRP (100 mg/kg) which raised the brain 5-HT concentrations by 50% was reported to have no effect on food intake or selection (Peters, Bellissimo & Harper, 1984).

Although one may agree that no convincing evidence has been provided on a role of brain 5-HT in the control of macronutrient intake, there is probably room for further studies to separate mechanisms selectively controlling the intake of one macronutrient unambiguously from other confounding variables. These studies will certainly enhance our knowledge of the mechanisms controlling nutrient intake in mammals and may also help in deciding whether the serotonergic hypothesis of nutrient selection should be abandoned or modified.

Which 5-HT receptor subtypes mediate the effects on feeding?
Samanin et al. (1977a) reported that quipazine, an agent which directly stimulates 5-HT receptors, markedly reduced food intake in rats at doses causing no apparent change in behaviour. The effect was completely prevented by metergoline but not by various procedures known to interfere with catecholamines (Samanin et al., 1977b). m-Chlorphenylpiperazine (m-CPP), a potent displacer of ^{3}H-5-HT binding to brain membranes with little effect on catecholamine mechanisms (Samanin et al., 1980; Invernizzi et al., 1981) also caused a dose-related reduction of food intake in rats and this effect was reduced by pretreatment with metergoline (Samanin et al., 1979).

M-CPP also inhibited the motivation to start eating and the consumatory response in a food-rewarded runway (Thurlby & Samanin, 1981). These findings suggested that agents which directly stimulate 5-HT receptors cause marked anorexia in laboratory animals (Samanin, Mennini & Garattini, 1980).

Radioligand studies have suggested the existence of two subtypes of central 5-HT receptors named $5-HT_1$ and $5-HT_2$ (Peroutka & Snyder, 1979) and at least three subtypes of $5-HT_1$ sites, named $5-HT_{1A}$ and $5-HT_{1B}$ and $5-HT_{1C}$, have been proposed recently (Pedigo, Yamamura & Nelson, 1981; Pazos, Hoyer & Palacios, 1984). While the studies with quipazine and m-CPP

Table 22.1 Effect of ritanserin and metergoline on cortical 5-HT$_2$ receptors and anorectic activity of d-fenfluramine (DF) in food deprived rats

Treatment	mg/kg i.p.	food intake saline	(g/rat/1 h) DF	% reduction of DF effect (a)
Control	-	8.7 ± 1.4	0.8 ± 0.2 **	-
Ritanserin	0.5	10.5 ± 0.9	3.1 ± 1.0	6
Ritanserin	1.0	8.6 ± 0.6	3.0 ± 0.6	29
Ritanserin	2.0	7.5 ± 0.6	3.2 ± 0.3	45
Control	-	11.5 ± 0.8	2.0 ± 0.5 **	-
Metergoline	0.125	12.3 ± 1.0	6.0 ± 0.8	34
Metergoline	0.5	12.4 ± 0.2	9.7 ± 1.1	72
Metergoline	1.0	11.3 ± 1.1	11.7 ± 0.9	104

^{3}H-spiperone binding (5-HT$_2$ receptors) in vivo was measured in frontal cortex as previously described by Barone et al. (1985). ED$_{50}$ (mg/kg i.p.) for occupancy of 5-HT$_2$ receptors were 0.4 ± 0.1 for ritanserin and >0.1 for metergoline.
Food intake was measured in rats trained over two weeks to take their daily food in 4 h. Each value is the mean ± S.E. of at least rats.
Metergoline and ritanserin were injected 2.5 h and 30 min before DF (2.5 mg/kg i.p.). DF was administered 30 min before access to food. F interaction was 3.72 (df 3/32) p<0.02 for ritanserin treatment (the data were transformed for homogeneity in log (x+1) and 11.7 (df 5/46) p<0.01 for metergoline treatment (factorial analysis, ANOVA).
**p<0.01 vs control (Tukey's test).
(a) The % reduction was derived from the formula:

[Control - DF] - [antagonist - (antagonist + DF)]
─── X 100
                [Control - DF]

prove that direct activation of postsynaptic 5-HT receptors reduces food intake, they do not clarify which receptor is involved as they have poor selectivity for 5-HT receptor subtypes. Metergoline aslo lacks selectivity on 5-HT receptors. The recent availability of a very selective 5-HT$_2$ receptor antagonist, ritanserin (Leysen et al., 1985), provides a means of obtaining information on whether 5-HT$_1$ or 5-HT$_2$ receptors are involved in feeding control.

Since DF increases the availability of 5-HT at post-synaptic sites innervated by 5-HT nerve terminals, we reasoned that if activation of 5-HT$_2$ receptors was important for the anorectic activity, doses of ritanserin that occupied the central 5-HT$_2$ receptors should antagonize the effect of DF. We therefore compared the effects of different doses of ritanserin and metergoline on DF anorexia with their ability to occupy central 5-HT$_2$ receptors in vivo. As shown in Table 22.1, metergoline dose-dependently reduced the effect of DF, causing complete antagonism at 1 mg/kg, a dose which occupied less than 50 % of cortical 5-HT$_2$ receptors. Ritanserin, instead, at a dose (0.5 mg/kg) ensuring more than 50% occupancy of 5-HT$_2$ receptors had no effect or only partially prevented the anorexia produced by DF at doses (1-2 mg/kg) causing maximal occupation of 5-HT$_2$ receptors. The fact however that 1 mg/kg ritanserin, but not metergoline, significantly reduced the hypothalamic and cortical levels of d-norfenfluramine (data not shown), a metabolite which is more potent than the parent compound in reducing food intake (Garattini et al., 1979), strongly that this effect may account for the activity of 1 and 2 mg/kg ritanserin. One mg/kg ritanserin was recently reported to have no effect on the reduction of food intake caused by centrally administered 5-HT (Massi & Marini, 1987).

These results suggest that 5-HT receptors other than 5-HT$_2$ mediate the anorectic activity of DF and argue against central 5-HT$_2$ receptors having any important role in the regulation of feeding.

But which 5-HT$_1$ receptors subtype (5-HT$_{1A}$, 5-HT$_{1B}$ or even 5-HT$_{1C}$) does play a major role in feeding control? To address this issue we recently studied the effect on food intake of 8-hydroxy-2-(di-n-propylamino) tetralin (8-OH-DPAT), a selective 5-HT$_{1A}$ receptor agonist (Middlemiss & Fozard, 1983) and 5-methoxy-3-(1,2,3,6-tetrahydro-4-piridinyl)1H indole succinate (RU 24969), a selective 5-HT$_1$ receptor agonist with similar affinity for 5-HT$_{1A}$ and 5-HT$_{1B}$ binding sites in the rat brain (Hamon et al., 1986). The compounds were studied in free-feeding and food-deprived rats and 5,7-dihydroxytryptamine or metergoline used to antagonize their effect on food intake. At doses from 0.125 to 0.5 mg/kg 8-OH-DPAT significantly increased food intake by free-feeding rats while 0.5 mg/kg reduced eating by food-deprived rats (Bendotti & Samanin, 1987). The increase in eating was prevented by intraventricular injection of the 5-HT neurotoxin (Bendotti & Samanin, 1986) suggesting that 5-HT neurons were involved. The median raphe nuclei are particularly rich in 5-HT$_{1A}$ binding sites (Verge et al., 1985) while presynaptic receptors in nerve terminals in rodents are mainly of the 5-HT$_{1B}$ type (Engel et al., 1986). 8-OH-DPAT was therefore injected into the nuclei dorsalis and medianus raphe to study its effect on food intake. Microinjections of 0.5 and 1 µg of 8-OH-DPAT in both nuclei significantly increased eating by free-feeding rats, suggesting that activiation of autoreceptors located in serotonergic cell bodies in midbrain raphe stimulates feeding. The decrease of food intake observed with 0.6 mg/kg in food-deprived rats was associated with marked flat body posture, extension of the forepaws and hindlimb abduction which obviously interfered with eating. Moreover, it was not modified by metergoline treatment. These findings are summarized in fig. 22.2.

It has been argued that 8-OH-DPAT causes animals to eat foods by inducing gnawing since a reduction rather than an increase in glucose consumption was found with 60-120 µg/kg in non-deprived rats (Fletcher, 1987). However, no gnawing response was noted in animals in which overeating was induced by

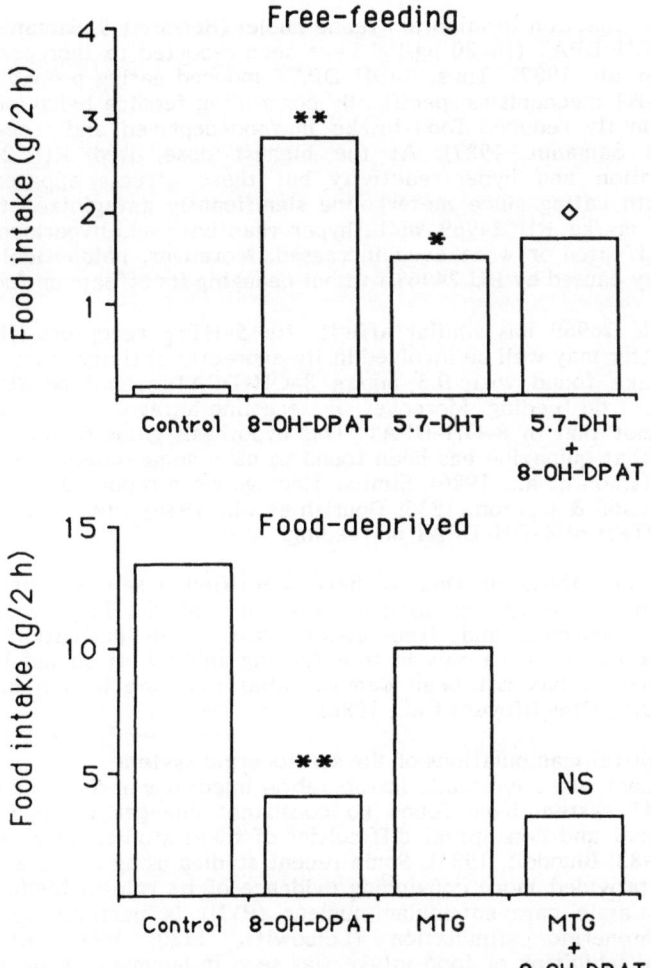

Fig. 22.2 Effect of 0.5 mg/kg s.c. 8-OH-DPAT on eating of free-feeding and food-deprived rats. 5,7-DHT 150 µg/20 µl was injected i.c.v. 3 days before testing. Metergoline (MTG, 2 mg/kg i.p.) was injected 3 h before testing. The bars represent the mean of 5-7 rats.

◊ p<0.01 (ANOVA 2 x 2); *p<0.05; **p<0.01 vs controls (Tukey's test); NS=not significant p<0.05 (ANOVA 2 x 2).

8-OH-DPAT injection in midbrain raphe nuclei (Bendotti & Samanin, 1986). Low doses of 8-OH-DPAT (10-30 µg/kg) have been reported to increase milk intake (Coughlan et al., 1987). Thus, 8-OH-DPAT induced eating probably reflects an effect on 5-HT mechanisms specifically controlling feeding behaviour. RU 24969 dose-dependently reduced food-intake in food-deprived and free-feeding rats (Bendotti & Samanin, 1987). At the highest dose used RU 24969 caused hyperlocomotion and hyper-reactivity but these effects apparently did not interfere with eating since metergoline significantly antaginized the anorectic effect of 5 mg/kg RU 24969 while hyper-reactivity and hyperlocomotion were either not affected or were even increased. Moreover, haloperidol blocked the hypermotility caused by RU 24969 without changing its effects on food intake.

Although RU 24969 has similar affinity for $5-HT_{1B}$ receptors (Hamon et al., 1986) the latter may well be involved in its anorectic activity since the decrease of food intake found with 0.5 mg/kg 8-OH-DPAT cannot be attributed to a specific effect on feeding. Moreover, metergoline antagonized the effect of RU 24969, but not that of 8-OH-DPAT. This hypothesis gains further support from the finding that quipazine has been found to have some selectively for $5-HT_{1B}$ receptors (Hamon et al., 1986). Similar findings were reported by Dourish et al. (Dourish, Hutson & Curzon, 1985; Dourish et al., 1986) who first described the enhancing effect of 8-OH-DPAT on feeding.

Our studies and those of Dourish have identified a selective involvement of $5-HT_1$ receptors in feeding control; activation of $5-HT_{1B}$ receptors causes anorexia in deprived and free-feeding rats while activation of $5-HT_{1A}$ receptors increases eating only in free-feeding animals. It should be mentioned however, that it has not been demonstrated that the human brain contains $5-HT_{1B}$ binding sites (Hoyer et al., 1986).

Effect of central manipulations of the serotonergic system
Various authors who have made intracerebral injections of 5-HT or lesions of the central 5-HT system have found no consistent changes in food intake. The methodological and conceptual difficulties of these studies have been discussed (Samanin, 1983; Blundell, 1984). Some recent studies using cerebral injections of 5-HT have provided more convincing evidence of its role in feeding regulation. The hypothalamic paraventricular nucleus (PVN) is particularly sensitive to direct serotonergic stimulation (Leibowitz, 1980; Weiss et al., 1986). Dose-related inhibition of food intake was seen in hungry rats on administering into the PVN doses of 5-HT (1-10 µg) that had no apparent effect on sensorimotor performance. 5-HT (threshold dose 100 ng) had a greater effect on eating stimulated by noradrenaline injected into the PVN of satiated rats.

The nucleus accumbens was recently founds to be sensitive to the inhibitory effect of 5-HT on a particular type of overeating associated with behavioural activation, i.e. eating caused by muscimol injected in the nucleus raphe dorsalis (Bendotti & Samanin, 1986). Interestingly, doses of 5-HT (2.2-8.8 µg) which caused dose-related inhibition of muscimol-induced eating had no effect on food intake of deprived rats, suggesting that the brain's regional sensitivity to the anorectic effect of 5-HT may depend on the animals' nutritional state and/or arousal level.

If brain 5-HT acts by inhibiting feeding, one would expect reduced 5-HT transmission to enhance food intake. As mentioned above, evidence of this has recently been provided by the finding that direct activation by a $5-HT_{1A}$ agonist of autoreceptors located on serotonergic cell bodies in the midbrain raphe nuclei -- a condition which reduced 5-HT transmission -- significantly increased eating by satiated rats (Bendotti & Samanin, 1986; Dourish et al., 1986).

Role of peripheral 5-HT
Recent studies indicate that peripherally administered 5-HT reduces food intake without interfering with sensorimotor performance or causing malaise (Fletcher & Burton, 1984, 1985; Pollock & Rowland, 1981; Montgomery, Fletcher & Burton, 1986). $5-HT_2$ receptors may be involved in this effect since ritanserin, a selective antagonist of these receptors, prevented the effect of 5-HT on food intake (Massi & Marini, 1987). Using the technique of microstructural analysis in food-deprived rats Fletcher and Burton (1986b) found that 1-4 mg/kg 5-HT reduced bout size and duration with no effect on other parameters such as bout frequency, mean eating rate and median inter-pellet interval. In the authors' opinion these data suggested but did not prove that 5-HT in the periphery was involved in the control of satiety. Peripherally administered 5-HT delays gastric emptying (Fletcher & Burton, 1985), an effect which according to Moran and McHugh (1982) may reduce meal size.

However, the effects on gastric emptying reversed anorexia but not the slowing of gastric emptying induced by 5-HT (Fletcher & Burton, 1985). It has also been suggested that fenfluramine reduces meal size (Robinson, Moran & McHugh, 1986) and lengthens the postmeal interval relative to meal size in free-feeding rats by inhibiting gastric emptying (Davies et al., 1983). Although no causal relationship has yet been clearly established between inhibition of gastric emptying and reduction of food intake, it is likely that fenfluramine and 5-HT, by slowing the rate of gastric emptying, may affect some aspects of feeding not easily revealed by measuring food intake in deprived rats. It has also been suggested that peripherally administered 5-HT reduces the incentive value of food-related stimuli (Montgomery & Burton, 1986).

5-HT and eating disorders
In some individuals chronic stress may cause overeating and obesity associated with anxiety and depressive reactions (Morley, Levine & Rowland, 1983). Examples of the relationship between stress and feeding in humans are the so called "reactive obesity" and "night eating syndrome" described respectively by Bruch (1973) and Stunkard (1955).

Ingestive behaviour can be induced in animals by various conditions which have in common the ability to provoke non-specific motivational excitement (Antelman & Caggiula, 1977; Robbins & Koob, 1980; Mittleman & Valenstein, 1981). There is evidence that brain dopamine is involved in stress-induced ingestive behaviour (Antelman & Caggiula, 1977; Mittleman & Valenstein, 1985), and the mesolimbic system in particular plays an important role in eating induced by electrical stimulation of the lateral hypothalamus (Mittleman & Valenstein, 1981) and in "displacement drinking" (Robbins & Koob, 1980). Selective involvement of this system has also been found for eating and behavioural excitement caused by muscimol injected into the nucleus raphe dorsalis, where most 5-HT neurons

Table 22.2 Effect of 5-HT injected bilaterally into the nucleus accumbens or into caudate putamen on eating induced by muscimol injection in the dorsal raphe or by starvation

Treatment	dose µg/2µl	eating (g/30 min) muscimol	induced by starvation
Caudate putamen			
Vehicle	-	5.5 ± 1.1	3.5 ± 2.4
5-HT	2.2	5.1 ± 1.0	4.4 ± 0.8
5-HT	4.4	5.7 ± 0.9	4.3 ± 0.6
5-HT	8.8	6.2 ± 0.7	-
Nucleus accumbens			
Vehicle	-	4.2 ± 0.6	5.7 ± 0.9
5-HT	2.2	4.6 ± 1.3	5.6 ± 0.5
5-HT	4.4	2.0 ± 0.7*	4.5 ± 0.5
5-HT	8.8	0.5 ± 0.3**	5.0 ± 1.0

Values are mean ± S.E. of 5-8 rats.
*$p<0.05$; **$p<0.01$ vs vehicle (Dunnett's test).

innervating the forebrain originate (Bendotti et al., 1986a).

Drugs enhancing 5-HT transmission are particularly effective in reducing eating caused by tail-pinch (Antelman et al., 1979) and muscimol injection (Borsini et al., 1983) and - as shown in Table 22.2 - doses of 5-HT which had no effect on feeding by deprived rats inhibited muscimol-induced eating when injected in the nucleus accumbens, a terminal area of the mesolimbic dopamine system (Bendotti, Garattini & Samanin, 1986b). Other findings indicate that 5-HT inhibits dopamine-dependent behaviours mediated in the nucleus accumbens (Costall et al., 1976; Kruszewksa, Romandini & Samanin, 1986).

Dopamine in the nucleus accumbens is closely involved in mediating reinforcement and may facilitate feeding (Wise, 1981; Taylor & Robbins, 1986; Evans & Vaccarino, 1986) and impairment of central dopamine mechanisms reduced the reward quality of food (Wise et al., 1978).

There is increasing evidence that the mesolimbic system plays an important role in antidepressant activity (Spyraki & Fibiger, 1981; Cervo & Samanin, 1987); Maj et al., 1987) and we have recently found that stimulation of 5-HT autoreceptors in the nucleus raphe dorsalis causes overeating and antidepressant-like effects which may involve activation of the mesolimbic dopamine system (Bendotti & Samanin, 1986; Cervo & Samanin, 1987) confirming a close relationship between feeding and affective behaviour (Katz et al., 1984; Rothenberg, 1986; Halmi et al., 1987). These findings support the suggestion that overeating is a defensive response to attenuate the affective reactions consequent to behavioural arousal (Bruch, 1973). Zimelidine, a selective 5-HT uptake blocker which shows antidepressant and anorectic activity in humans, was recently reported to inhibit the reinforcing effects of d-amphetamine, which are mediated by the mesolimbic dopamine system (Kruszewska et al., 1986). Attenuation of the overactivity of the mesolimbic dopamine system may therefore be one mechanism by which drugs enhancing 5-HT transmission reduce overeating associated with emotional stress. It has been suggested that some obese persons may be hyper-responsive to internal and external signals (Rodin, 1976) and central dopamine neurons are involved in the response to sensory stimulation (Keller, Stricker & Zigmond, 1983; Louilot, Le Moal & Simon, 1986); thus agents enhancing 5-HT transmission may reduce overeating by some obese individuals by attenuating their responsiveness to food-related or external cues.

Fenfluramine has been seen to reduce overeating associated with emotional disorders such as that observed in bulimia nervosa (Robinson, Checkley & Russell, 1985). It is possible that the ability to attenuate the overactivity of the dopaminergic system through an action on 5-HT is involved in the effect of fenfluramine in this form of "self-administrative" behaviour. Neuropeptide Y (NPY), a potent endogenous orexigenic proposed to play a role in the pathogenesis of bulimia (Morley, 1987), may also be involved since d-fenfluramine was recently reported to be particularly effective in reducing overeating caused by NPY injection in the hypothalamic paraventricular nucleus (Bendotti, Garattini & Samanin, 1987).

Conclusions

Recent studies with d-fenfluramine, a selective releaser of 5-HT, suggest that both initiation and termination of a meal and the rewarding effect of food are affected by changes in central 5-HT transmission. It is instead questionable whether agents increasing 5-HT transmission cause any selective suppression of carbohydrate intake.

Anorexia caused by brain 5-HT in rats is mediated by $5-HT_{1B}$ receptors while peripheral 5-HT seems to use $5-HT_2$ receptors to reduce food intake. Activation of $5-HT_{1A}$ receptors results in overeating by satiated rats, probably by reducing 5-HT transmission through stimulation of autoreceptors located in the midbrain raphe nuclei. Finally, it is suggested that attenuation of central dopamine activity, particularly in the mesolimbic system, is one mechanism by which agents enhancing 5-HT transmission may control overeating associated with emotional stress.

References

Antelman, S.A., Caggiula, A.R., Eichler, A.J. & Lucik, R.R. (1979). The importance of stress in assessing the effects of anorectic drugs. Current Medicinal Research Opinion, 6, 73-82.

Antelman, S.A. & Caggiula (1977). Tail of stress-related behaviour: A neuropharmacological analysis. In: Animal models in psychiatry and neurology. (Eds. Hanin, I. & Usdin, E.) pp. 227-245. Oxford: Pergamon Press.

Azmitia, E.C. (1978). The serotonin-producing neurons of the midbrain median and dorsal raphe nuclei. In: Handboek of psychopharmacology, Vol. 9. (Eds. Iversen, L.L., Iversen, S.D. & Snyder, S.H.). pp 233-304, New York: Plenum Press.

Barone, D., Luzzani, F., Assandri, A., Galliani, G., Mennini, T. & Garattini, S. (1985). In vivo stereospecific [^{3}H]spiperone binding in rat brain: Characteristics, regional distribution, kinetics and pharmacological properties. European Journal Pharmacology, 116, 63-74.

Bendotti, C., Beretta, C., Invernizzi, R. & Samanin, R. (1986a). Selective involvement of dopamien in the nucleus accumbens in the feeding response elicited by scimol injection in the nucleus raphe dorsalis of sated rats. Pharmacology Biochemistry Behavior, 24, 1189-1193.

Bendotti, C., Garattini, S. & Samanin, R. (1986b). Hyperphagia caused by muscimol injection in the nucleus raphe dorsalis of rats: its control by 5-hydroxytryptamine in the nucleus accumbens. Journal Pharmacy Pharmacology, 38, 541-543.

Bendotti, C., Garattini, S. & Samanin, R. (1987). Eating caused by neuropeptide-Y injection in the paraventricular hypothalamus: response to (+)-fenfluramine and (+)-amphetamine in rats. Journal Pharmacy Pharmacology, 39, 900-903.

Bendotti, C. & Samanin, R. (1986). 8-Hydroxy-2(di-n-propylamino)tetralin (8-OH-DPAT) elicits eating in free-feeding rats by acting on central serotonin neurons. European Journal Pharmacology, 121, 147-150.

Bendotti, C. & Samanin, R. (1987). The role of putative 5-HT$_{1A}$ and 5-HT$_{1B}$ receptors in the control of feeding rats. Life Sciences, 41, 635-642.

Berge, O-G., Fasmer, O.B., Flatmark, T. & Hole, K. (1983). Time course of changes in nociception after 5,6-dihydroxytryptamine lesions of descending 5-HT pathways. Pharmacology Biochemistry Behavior, 18, 637-643

Blundell, J.E. (1983). Problems and processes underlying the control of food selection and nutrient intake. In: Nutrition and the brain. (Eds. Wurtman, R.J. & Wurtman, J.J.). Vol. 6, pp 163-221, New York, Raven Press.

Blundell, J.E. (1984). Serotonin and appetite. Neuropharmacology, 23, 1537-1551.

Blundell, J.E. & Latham, C.J. (1978). Pharmacological manipulation of feeding behaviour. Possible influences of serotonin and dopamine on food intake. In: Central mechanisms of anorectic drugs. (Eds. Garattini, S. & Samanin, R.) pp 83-109, New York, Raven Press.

Blundell, J.E., Latham, C.J. & Lesham, M.B. (1976). Differences between the anorexic actions of amphetamine and fenfluramine. Possible effects on hunger and satiety. Journal Pharmacy Pharmacology, 28, 471-477.

Blundell, J.E., Latham, C.J., Moniz, E., Mc.Arthur, R.A. & Rogers, P.J. (1979). Structural analysis of the actions of amphetamine and fenfluramine on food intake and feeding behaviour in animals and in man. Current Medical Research Opnion, 6 suppl., 1, 34-54.

Booth, D.A. (1987). Central dietary "feedback onto nutrient selection". Not even a scientific hypothesis. Appetite, 8, 195-201.

Booth, D.A., Gibson, E.L. & Backer, B.J. (1986). Gastromotor mechanism of fenfluramine anorexia. Appetite, 7 suppl., 57-69.

Borsini, F., Bendotti, C., Aleotti, A., Samanin, R. & Garattini, S. (1982). d-Fenfluramine and d-norfenfluramine reduce food intake by acting on different serotonin mechanisms in the rat brain. Pharmacological Research Communications, 14, 671-678.

Borsini, F., Bendotti, C., Przewlocka, B. & Samanin, R. (1983). Monoamine involvement in the overeating caused by muscimol injection in the rat nucleus raphe dorsalis and the effects of d-fenfluramine and d-amphetamine. European Journal Pharmacology, 94, 109-115.

Borsini, F., Bendotti, C. & Samanin, R. (1985). Salbutamol, d-amphetamine and d-fenfluramine reduce sucrose intake in freely fed rats by acting different neurochemical mechanisms. International Journal Obesity, 9, 277-283.

Bruch, H. (1973). Eating disorders. New York: Basic Books.

Burton, M.J., Cooper, S.J. & Popplewell, D.A. (1981). The effect of fenfluramine on the microstructure of feeding and drinking in the rat. British Journal Pharmacology, 72, 621-633.

Carey, R.J. (1976). Effects of selective forebrain depletions of norepinephrine and serotonin on the activity and food intake effects of amphetamine and fenfluramine. Pharmacology Biochemistry Behavior, 5, 519-523.

Carruba, M.O., Mantegazza, P., Memo, M., Missale, C., Pizzi, M. & Spano, P.F. (1986). Peripheral and central mechanisms of action of serotoninergic anorectic drugs. Appetite, 7 suppl., 105-113.

Cervo, L. & Samanin, R. (1987). Evidence that dopamine mechanisms in the nucleus accumbens are selectively involved in the effect of desipramine in the forced swimming test. Neuropharmacology, 26, 1469-1472.

Clineschmidt, B.V. (1973). 5,6-Dihydroxytryptamine. Suppression of the anorexigenic action of fenfluramine. Eurpean Journal Pharmacology, 24, 405-409.

Costall, B., Naylor, R.J. Marsden, C.D. & Pycock, C.J. (1976). Serotoninergic modulation of the dopamine response from the nucleus accumbens. Journal Pharmacy Pharmacology, 28, 523-526.

Coughlan, J., Dourish, C.T. Gilbert, F. & Iversen, S.D. (1987). The 5-HT$_{1A}$ agonist 8-OH-DPAT increases sweetened milk consumption in rats. British Journal Pharmacology, 91 suppl., 423P.

Curtis-Prior, P.B. & Prouteau, M. (1983). Qualitative and quantitative effects of fenfluramine and tiflorex on food consumption in trained rats offered dietary choices. International Journal Obesity, **7**, 575–581.

Cytawa, J. & Trojniar, W. (1978). Hedonesthesia: The nervous process determining motivated ingestive behaviour. Acta Neurobiologica Exeprimentalis, **38**, 139–151.

Davies, R.F., Rossi, J. III, Panksepp, J., Bean, N.J. & Zolovick, A.J. (1983). Fenfluramine anorexia. A peripheral locus of action. Physiology Behavior, **30**, 723–730.

Davis, J.D. & Levin, M.W. (1977). A model for the control of ingestion. Psychological Reviews, **84**, 379–412.

De Simoni, M.G., Sokola, A., Fodritto, F., Dal Tose, G. & Algeri, S. (1987). Functional meaning of tryptophan-induced increase of 5-HT metabolism as clarified by in vivo voltammetry. Brain Research, **411**, 89–94.

Dourish, C.T., Hutson, P.H. & Curzon, G. (1985). Low doses of the putative serotonin agonist 8-hydroxy-2-(di-n-propylamino) tetralin (8-OH-DPAT) elicit feeding in the rat. Psychopharmacology, **86**, 197–204.

Dourish, C.T., Hutson, P.H., Kennett, G.A. & Curzon, G. (1986). 8-OH-DPAT-induced hyperphagia. Its neural basis and possible therapeutic relevance. Appetite, 7 suppl., 124–140.

Engel, G., Gothert, M., Hoyer, D., Schlicker, E. & Hillenbrand, K. (1986). Identity of inhibitory presynaptic 5-hydroxytryptamine (5-HT) autoreceptors in the rat brain cortex with 5-HT_{1B} binding sites. Naunyn-Schmiedebergs Archives Pharmacology, **332**, 1–7.

Evans, K.R. & Vaccarino, F.J. (1986). Intra-nucleus acumbens amphetamine. Dose-dependent effects on food intake. Pharmacology Biochemistry Behavior, **25**, 1149–1151.

Fernstrom, J.D. (1987). Food-intake changes in brain serotonin synthesis:. Is there a relationship to appetite for specific macronutrients? Appetite, **8**, 163–182.

Fernstrom, J.D. & Wurtman, R.J. (1971). Brain serotonin content. Increasing following ingestion of carbohydrate diet. Science, **174**, 1023–1025.

Fletcher, P.J. (1987). 8-OH-DPAT elicits gnawing, and eating of solid but not liquid foods. Psychopharmacology, **92**, 192–195.

Fletcher, P.J. & Burton, M.J. (1984). Effects of manipulations of peripheral serotonin on feeding and drinking in the rat. Pharmacology Biochemistry Behavior, **20**, 835–840.

Fletcher, P.J. & Burton, M.J. (1985). The anorectic action of peripherally administered 5-HT is enhanced by vagotomy. Physiology Behavior, **34**, 861–866.

Fletcher, P.J. & Burton, M.J. (1986a). Dissociation of the anorectic actions of 5-HTP and fenfluramine. Psychopharmacology, 89, 216-220.

Fletcher, P.J. & Burton, M.J. (1986b). Microstructural analysis of the anorectic action of peripherally administered 5-HT. Pharmacology Biochemistry Behavior, 24, 1133-1136.

Fuxe, K., Farnebo, L.O. Hamberger, B. & Ogres, S.O. (1975). On the in vivo and in vitro activity of fenfluramine and its derivatives on central monoamine neurons, especially 5-hydroxytryptamine neurons, and their relation to the anorectic action of fenfluramine. Postgraduate Medical Journal, 51, suppl. 35-45.

Garattini, S., Buczko, W., Jori, A. & Samanin, R. (1975). The mechanism of action of fenfluramine. Postgraduate Medical Journal, 51 suppl. 1, 27-35.

Garattini, S., Caccia, S., Mennini, T., Samanin, R., Consolo, S. & Ladinsky, H. (1979). Biochemical pharmacology of the anorectig drug fenfluramine. A review. Current Medical Research Opnion, 6, suppl. 1, 15-27.

Garattini, S., Mennini, T., Bendotti, C., Invernizzi, R. & Samanin, R. (1986). Neurochemical mechanism of action of drugs which modify feeding via the serotoninergic system. Appetite, 7 suppl., 15-38.

Geyer, M.A., Puerto, A., Dawsey, W.J., Knapp, S., Bullard, W.P. & Mandell, A.J. (1976). Histological and enzymatic studies of the mesolimbic and mesostriatal serotonergic pathways. Brain Research, 106, 241-256.

Grinker, J.A., Drewnowski, A., Enns, M. & Kissileff, H. (1980). Effects of d-amphetamine and fenfluramine on feeding patterns and activity of obese and lean Zuckar rats. Pharmacology Biochemistry Behavior, 12, 265-275.

Halmi, K.A., Ackerman, S., Gibbs, J. & Smith, G. (1987). Basic biological overview of the eating disorders. In: Psychopharmacology. The third generation of progress, (Ed. Meltzer, H.Y.) pp 1255-1266, New York, Raven Press.

Hamon, M., Cossery, J-M., Spampinato, U. & Gozlan, H. (1986). Are there selective ligands for $5-HT_{1A}$ and $5-HT_{1B}$ receptor binding sites in brain? Trends Pharmacological Sciences, 7, 336-338.

Hoebel, B.G., Zemlan, F.P., Trulson, M.E., MacKenzie, R.G., DuCret, R.P. & Norelli, C. (1978). Differential effects of p-chlorophenylalanine and 5,7-dihydrotryptamine on feeding in rats. Annals New York Academy Sciences, 305, 590-594.

Hoyer, D., Pazos, A., Probst, A. & Palacios, J.M. (1986). Serotonin receptors in the human brain. I. Characterization and autoradiographic localization of $5-HT_{1A}$ recognition sites. Apparent absence of $5-HT_{1B}$ recognition sites. Brain Research, 376, 85-96.

Invernizzi, R., Beretta, C., Farattini, S. & Samanin, R. (1986). D- and L-isomers of fenfluramine differ markedly in their interaction with brain serotonin and catecholamines in the rat. European Journal Pharmacology, 120, 9-15.

Invernizzi, R., Cotecchia, S., De Blasi, A., Mennini, T., Pataccini, R. & Samanin, R. (1981). Effects of m-chlorophenylpiperazine on receptor binding and brain metabolism of monoamines in rats. Neurochemistry International, 3, 239-244.

Katz, J.L., Kupferberg, A., Pollack, C.P., Walsh, B.T., Zumoff, B. & Weiner, H. (1984). Is there a relationship between eating disorder and affective disorder? New evidence from sleep recordings. American Journal Psychiatry, 141, 753-759.

Keller, R.W. Jr., Stricker, E.M. & Zigmond, M.J. (1983). Environmental stimuli but not homeostatic challenges produce apparent increases in dopaminergic activity in the striatum. An analysis by in vivo voltammetry. Brain Research, 279, 159-170.

Kessler, J.P. & Jean, A. (1985). Inhibition of the swallowing reflex by local application of serotonergic agents into the nucleus of the solitary tract. European Journal Pharmacology, 118, 77-85.

Kirkham, T.C. & Blundell, J.E. (1986). Effect of naloxone and naltrexone on the development of satiation measured in the runway. Comparisons with d—amphetamine and d-fenfluramine. Pharmacology Biochemistry Behavior, 25, 123-128.

Kruszewska, A., Romandini, S. & Samanin, R. (1986). Different effects of zimelidine on the reinforcing properties of d-amphetamine and morphine on conditioned place preference in rats. European Journal Pharmacology, 125, 283-286.

Kyriakides, M. & Silverstone, T. (1979). Comparison of the effects of d-amphetamine and fenfluramine on hunger and food intake in man. Neuropharmacology, 18, 1007-118.

Leander, J.D. (1987). Fluoxetine suppresses palatability-induced ingestion. Psychopharmacology, 91, 285-287.

Leathwood, P. (1987). Food-composition, changes in brain serotonin synthesis and appetite for protein and carbohydrate. Appetite, 8, 202-205.

Leibowitz, S.F. (1980). Neurochemical systems of the hypothalamus. Control of feeding and drinking behaviour and water-electrolyte excretion. In: Handbook of the hypothalamus. (Eds. Morgane, P.J. & Panksepp, J.) Vol. VI, part A. Behavioral studies of the hypothalamus, pp. 299-437. New York, Marcel Dekker.

Le Magnen, K. (1983). Body energy balance and food intake: A neuroendocrine regulatory mechanism. Physiological Reviews, 63, 314-386.

Leysen, J.E., Gommeren, W., Van Gompel, P., Wynants, J., Janssen, P.F.M. & Laduron, P.M. (1985). Receptor-binding properties in vitro and in vivo of ritanserin: a very potent and long acting serotonin S$_2$ antagonist. Molecular Pharmacology, 27, 600–611.

Louilot, A., Le Moal, M. & Simon, H. (1986). Differential reactivity of dopaminergic neurons in the nucleus accumbens in response to different behavioural situations. An in vivo voltammetric study in free moving rats. Brain Research, 397, 395–400.

Maj, J., Wedzony, K. & Klimek, V. (1987). Desimipramine given repeatedly enhances behavioural effects of dopamine and d-amphetamine injected into the nucleus accumbens. European Journal Pharmacology, 140, 179–185.

Massi, M. & Marini, S. (1987). Effect of the 5-HT$_2$ antagonist ritanserin on food intake and on 5-HT-induced anorexia in the rat. Pharmacology Biochemistry Behavior, 26, 333–340.

McArthur, R.A. & Blundell, J.E. (1983). Protein and carbohydrate self-selection. Modification of the effects of fenfluramine and amphetamine by age and feeding regimen. Appetite, 4, 113–124.

Middlemiss, D.N. & Fozard, J.R. (1983). 8-Hydroxy-2-(di-n-propylamino) tetralin discriminates between subtypes of the 5-HT$_1$ recognition site. European Journal Pharmacology, 90, 151–153.

Mittleman, G. & Valenstein, E.S. (1981). Strain differences in eating and drinking evoked by electrical stimulation of the hypothalamus. Physiology Behavior, 26, 371–378.

Mittleman, G. & Valenstein, E.S. (1985). Individual differences in non-regulatory ingestive behaviour and catecholamine system. Brain Research, 348, 112–117.

Montgomery, A.M.J. & Burton, M.J. (1986). Effects of peripheral 5-HT on consumption of flavoured solutions. Psychopharmacology, 88, 262–266.

Montgomery, A.M.J., Fletcher, P.J. & Burton, M.J. (1986). Behavioural and pharmacological investigations of 5-HT hypophagia and hyperdipsia. Pharmacology Biochemistry Behavior, 25, 23–28.

Mook, D.G., Bryner, C.A., Rainey, L.D. & Wall, C.L. (1980). Release of feeding by the sweet taste in rats. Oropharyngeal satiety. Appetite, 1, 299–315.

Moran, T.H. & McHugh, P.R. (1982). Cholecystokinin suppresses food intake by inhibiting gastric emptying. American Journal Physiology, 242, R491–R497.

Morley, J.E. (1987). Behavioral pharmacology for eating and drinking. In: Psychopharmacology: The third generation of progress. (Ed. Meltzer, H.Y.) pp. 1267–1271, New York, Raven Press.

Morley, J.E., Levine, A.S. & Rowland, N.E. (1983). Stress induced eating. Life Sciences, 32, 2169–2182.

Orthen-Gambill, N. & Kanarek, R.B. (1982). Differential effects of amphetamine and fenfluramine on dietary self-selection in rats. Pharmacology Biochemistry Behavior, 16, 303-309.

Pazos, A., Hoyer, D. & Palacios, J.M. (1984). The binding of serotonergic ligands to the porcine choroid plexus. Characterization of a new type of serotonin recognition site. European Journal Pharmacology, 106, 539-546.

Pedigo, N.W., Yamamura, H.I. & Nelson, D.L. (1981). Discrimination of multiple [^{3}H]5-hydroxytryptamine binding sites by the neuroleptic spiperone in rat brain. Journal Neurochemistry, 36, 220-226.

Peroutka, S.J. & Snyder, S.H. (1979). Multiple serotonin receptors. Differential binding of [^{3}H]5-hydroxytryptamine, [^{3}H]lysergide acid diethylamide and [^{3}H]spiroperidol. Molecular Pharmacology, 16, 687-699.

Peters, J.C., Bellissimo, D.B. & Harper, A.E. (1984). l-Tryptophan injection fails to alter nutrient selection by rats. Physiology Behavior, 32, 253-259.

Peters, J.C. & Harper, A.E. (1987). A skeptical view of the role of central serotonin in the selection and intake of protein. Appetite, 8, 206-210.

Pinder, R.M., Brogden, R.N., Sawyer, P.R., Speight, T.M. & Avery, G.S. (1975). Fenfluramine. A review of its pharmacological properties and therapeutic efficacy in obesity. Drugs. 10, 241-323.

Pollock, J.D. & Rowland, N. (1981). Peripherally administered serotonin decreases food intake in rats. Pharmacology Biochemistry Behavior, 15, 179-183.

Rech, R.H., Borsini, F. & Samanin, R. (1984). Effects of d-amphetamine and d-fenfluramine on performance of rats in a food maze. Pharmacology Biochemistry Behavior, 20, 489-493.

Robbins, T.W. & Koob, G.F. (1980). Selective disruption of displacement behaviour by lesions of the mesolimbic dopamine system. Nature, 285, 409-412.

Robinson, P.H., Checkley, S.A. & Russell, G.F.M. (1985). Suppression of eating by fenfluramine in patients with bulimia nervosa. British Journal Psychiatry, 146, 169-176.

Robinson, P.H., Moran, T.H. & McHugh, P.R. (1986). Inhibition of gastric emptying and feeding by fenfluramine. American Journal Physiology, 250, R764-R769.

Rodin, J. (1976). The role of perception of internal and external signals on the regulation of feeding in overweight and nonobese individuals. In: Appetite and food intake. (Eds. Silverstone, T.) pp 265-283. Berlin, Dahlem Konferenzen.

Romandini, S., Merlo Pich, E., Esposito, E., Kruszweska, A.Z. & Samanin, R.

(1986). The effect of intracerebroventricular 5,7-dihydroxytryptamine on morphine analgesia is time-dependent. Life Sciences, 38, 869-875.

Rothenberg, A. (1986). Eating disorder as a modern obessive-compulsive syndrome. Psychiatry, 49, 45-53.

Rowland, N.E. & Carlton, J. (1986). Neurobiology of an anorectic drug. Fenfluramine. Progress Neurobiology, 27, 13-62.

Samanin, R. (1983). Drugs affecting serotonin and feeding. In: Biochemical pharmacology of obesity. (Ed. Curtis-Prior, P.B.) pp 339-356, Amsterdam, Elsevier.

Samanin, R., Bendotti, C., Candalaresi, G. & Garattini. S. (1977b). Specificity of serotoninergic involvement in the decrease of food intake induced by quipazine in the rat. Life Sciences, 21, 1259-1266.

Samanin, R., Bendotti, C., Miranda, F. & Garattini. S. (1977a). Decrease of food intake by quipazine in the rat. Relation to serotoninergic receptor stimulation. Journal Pharmacy Pharmacology, 29, 53-54.

Samanin, R., Caccia, S., Bendotti, C., Borsini, F., Borsini, E., Invernizzi, R., Pataccini, R. & Mennini, T. (1980). Further studies on the mechanism of serotonin-dependent anorexia in rats. Psychopharmacology, 68, 99-104.

Samanin, R., & Garattini. S. (1984). Mechanisms of anorexia. In: Advances human psychopharmacology. (Eds. Burrows, G.D. & Werry, J.S.). Vol. 3 (pp.357-397). Greenwich, JAI Press.

Samanin, R., Ghezzi, D., Valzelli, & Garattini. S. (1972). The effects of selective lesioning of brain serotonin or catecholamine containing neurons on the anorectic activity of fenfluramine and amphetamine. European Journal Pharmacology, 19, 318-322.

Samanin, R., Mennini, T., Ferraris, A., Bendotti, C., Borsini, F. & Garattini. S. (1979). m-Chlorphenylpiperazine. A central serotonin agonist causing powerful anorexia in rats. Naunyn-Schmiedeberg'sa Archives Pharmacology, 308, 159-163.

Samanin, R., Mennini, T., & Garattini. S. (1980). Evidence that it is possible to cause anorexia by increasing release and/or directly stimulating postsynaptic serotonin receptors in the brain. Progress Neuro-Psychopharmacology, 4, 363-369.

Silverstone, T. & Goodall, E. (1986a). Serotoninergic mechanisms in human feeding. The pharmacological evidence. Appetite, 7 suppl., 85-97.

Silverstone, T. & Goodall, E. (1986b). Recent studies on the clincial pharmacology of anorectic drugs. In 5th International Congress on Obesity, Jerusalem, Israel, September 14-19, 1986. Abstracts book p 47.

Spyraki, C. & Fibiger, H.C. (1981). Behavioural evidence for supersensisivity of postsynaptic dopamine receptors in the mesolimbic system after chronic administration of desipramine. European Journal Pharmacology, 74, 195-206.

Stunkard, A.J. (1955). The night-eating syndrome. A pattern of food intake among certain obese patients. American Journal Medicine, 19, 8-85.

Sugrue, M.F., Goodlet, I. & McIndewar, I. (1985). Failure of depletion of rat brain 5-hydroxytryptamine to alter fenfluramine-induced anorexia. Journal Pharmacy Pharmacology, 27, 950-953.

Taylor, J.R. & Robbins, T.W. (1986). 6-hydroxytryptamine lesions of the nucleus accumbens, but not of the caudate nucleus, attenuate enhanced responding with reward-related stimuli produced by intra-accumbens d-amphetamine. Psychopharmacology, 90, 390-397.

Thurly, P.L., Garattini, S. & Samanin, R. (1985). Effects of serotonin antagonists on the performance of a simple food acquisition task in rats treated with fenfluramine isomers. Pharmacological Research Communications, 17, 1129-1138.

Thurly, P.L., Grimm, V.E. & Samanin, R. (1983). Feeding and satiation observed in the runway. The effects of d-amphetamine and d-fenfluramine compared. Pharmacology Biochemistry Behavior, 18, 841-846.

Thurly, P.L., & Samanin, R. (1981). Effects of anorectic drugs and prior feeding on food-rewarded runway behaviour. Pharmacology Biochemistry Behavior, 14, 799-804.

Verge, D., Daval, G., Patey, A., Gozlan, E., El Mestikawy, S. & Hamon, M. (1985). Presynaptic 5-HT autoreceptors on serotonergic cell bodies and/or dendrites but not terminals are of the 5-HT_{1A} subtype. European Journal Pharmacology, 113, 463-464.

Waldbilling, R.J., Bartness, T.J. & Stanley, B.G. (1981). Increased food intake, body weihgt, and adiposity in rats after regional neurochemical depletion of serotonin. Journal Comparative Physiology Psychology, 95, 391-405.

Waldbilling, R.J., & O'Callaghan, M. (1980). Hormones and hedonics cholecystokinin and tatste; A possible behavioural mechanism of action. Physiology Behavior, 25, 25-30.

Weiss, G.F., Papadakos, P., Knudson, K. & Leibowitz, S.F. (1986). Medial hypothalamic serotonin; Effects on deprivation and norepinephrine-induced eating. Pharmacology Biochemistry Behavior, 25, 1223-1230.

Wise, R.A. (1981). Brain dopamine and reward. In: Theory and psychopharmacology. (Eds. Cooper, S.J.) Vol. 1, pp 102-122, London, Academic Press.

Wise, R.A. & Raptis, L. (1985). Effects of pre-feeding on food-apparoach latency and food consumption speed deprived rats. Physiology Behavior, 35, 961-963.

Wise, R.A., Spindler, J., de Wit, H. & Gerber, G.J. (1978). Neuroleptic-induced 'anhedonia' in rats. Pimozide blocks the reward quality of food. Science, 201, 262-264.

Wurtman, J.J. (1985). Neurotransmitter control of carbohydrate consumption. Annals New York Academy Sciences, 443, 145-151.

Wurtman, J.J. & Wurtman, R.J. (1977). Fenfluramine and fluoxetine spare protein consumption while suppressing caloric intake by rats. Science, 198, 1178-1180.

Wurtman, J.J. & Wurtman, R.J. (1979). Drugs that enhance central serotoninergic transmission diminish elective carbohydrate consumption by rats. Life Sciences, 24, 895-904.

Wurtman, J.J. & Wurtman, R.J. (1984). d-Fenfluramine selectively decreases carbohydrate but not protein intake in obese subjects. International Journal Obesity, 8 suppl., 1, 79-84.

Wurtman, J.J., Wurtman, R.J., Growdon, J.H., Henry, P., Lipscomb, A. & Zeisel, S.H. (1981). Carbohydrate craving in obese people. Suppression by treatments affecting serotoninergic transmission. International Journal Eating Disorders, 1, 2-15.

OPPOSITE EFFECTS OF 5-HT$_{1A}$ AND 5-HT$_{1B/1C}$ AGONISTS ON FOOD INTAKE

P.H.Hutson, G.A.Kennett, T.P.Donohoe[+], C.T.Dourish[*], G.Curzon.
Department of Neurochemistry, Institute of Neurology, Queen Square, London WC1, United Kingdom

Early work largely suggested that activation of 5-HT receptors inhibited food intake (reviewed Sugrue, 1987) but revealed little about the types of receptor involved. Evidence for numerous 5-HT receptor subtypes has stimulated us to study this problem. Our results are summarised in the present paper.

Methods
Details of methods used have been previously reported in detail (see Dourish et al., 1985; Hutson et al., 1986). Briefly, singly-housed male Sprague-Dawley rats (200-260 g) on a 12 h light-dark cycle (lights on 6.00 h) were allowed free access to food (22F diet, Labsure, Poole, Dorset) and water. Before experimentation, rats were habituated overnight in individual perspex cages. A weighed amount of food and (in some experiments) 3 wood blocks were placed in the cage, drugs given and the food remaining after 2, 4 and (in some experiments) 24 h reweighed and food intake calculated. Behaviour was also recorded on videotape for subsequent behavioural analysis. In some studies on the effects of 5-HT$_{1B/1C}$ agonists, rats were deprived of food but not water for 18 h before drugs were given (Kennett and Curzon, 1988). All food intake experiments were initiated between 12.00 and 17.00 h.

Results and discussion

Effects of 5-HT$_{1A}$ agonists
Initial work showed that the 5-HT$_{1A}$ agonist 8-OH-DPAT increased food intake in freely feeding rats (Dourish et al., 1985). Intake was maximal at 500 ug/kg s.c. and declined at higher doses when the 5-HT syndrome appeared to disrupt feeding. Hyperphagia without the syndrome occurred at doses of 15-125 ug/kg. Gnawing of wooden blocks was not seen at any dose tested. Other 5-HT$_{1A}$ agonists also caused hyperphagia, i.e. ipsapirone, buspirone (Dourish et al., 1986) LY165163 (Hutson et al., 1987), gepirone (Gilbert and Dourish, 1987). These drugs were less potent than 8-OH-DPAT but LY165163 (unlike the other drugs) caused prolonged (24h) hyperphagia.

Antagonist experiments substantiated the involvement of 5-HT$_{1A}$ receptors in the hyperphagic effect of 8-OH-DPAT (Hutson et al., 1988). Thus, it was prevented by metergoline (5 mg/kg i.p.), by (-) pindolol (4 mg/kg s.c.) which binds to 5-HT$_1$ receptors and by spiperone (0.05 mg/kg i.p.) which blocks 5-HT$_{1A}$ sites. 5-HT$_2$ antagonists (methysergide, 10 mg/kg i.p.; ketanserine, 2.5 mg/kg i.p.), 5-HT$_3$ antagonists (MDL72222, 2 mg/kg i.p., ICS 205 930 1 mg/kg s.c.) and catecholamine antagonists ((+) pindolol 4 mg/kg s.c.; haloperidol 0.1 mg/kg i.p.; idazoxan, 3 mg/kg i.p.) were without effect.

Depleting 5-HT with p-chlorophenylalanine prevented induction of hyperphagia by 8-OH-DPAT but not the 5-HT syndrome (Dourish et al., 1987) which suggested that the hyperphagia but not the syndrome was presynaptically mediated. Neurochemical evidence was consistent (Hutson et al., 1986) as 8-OH-DPAT (60 ug/kg s.c.) caused hyperphagia only and decreased central 5-HIAA/5-HT especially in regions containing 5-HT cell bodies (midbrain. pons + medulla oblongata). $5-HT_{1A}$ receptors on these cell bodies (Weissmna-Nanopoulos et al., 1985) appear to be responsible as hyperphagia (without the 5-HT syndrome) occurred on infusing 8-OH-DPAT (0.25-0.5 ug) into the dorsal or medial raphe nuclei (Hutson et al., 1986). Therefore, 8-OH-DPAT (and presumably other $5-HT_{1A}$ agonists) probably cause hyperphagia by drcreasing 5-HT synthesis and release following activation of $5-HT_{1A}$ somatodendritic autoreceptors on the raphe nuclei.

Effects of agonists which act at $5-HT_{1B}$ and $5-HT_{1C}$ sites

The $5-HT_{1B}$ agonist RU24969 (1.0-10.0 mg/kg i.p.) caused up to 24 h hypophagia in freely feeding rats (Kennett et al., 1987). The involvement of $5-HT_{1B}$ receptors was indicated by the blockade of hypophagia by metergoline (5 mg/kg s.c.), (-) pindolol (3 mg/kg s.c.) and (±) cyanopindolol (8 mg/kg s.c.) but not by ketanserin (2.5 mg/kg s.c.), spiperone (0.05 mg/kg s.c.) or haloperidol (0.1 mg/kg s.c.). The lack of effect of a dose of spiperone which antagonises $5-HT_{1A}$ mediated responses (Tricklebank et al., 1984) suggest that (-) pindolol and (±) cyanopindolol which antagonise both $5-HT_{1A}$ and $5-HT_{1B}$ receptors block hypophagia at the latter sites. Mediation of the hypophagia not by terminal $5-HT_{1B}$ autoreceptors (Middlemiss et al., 1985) but by post-synaptic $5-HT_{1B}$ receptors is suggested by its enhancement by p-chlorophenylalanine pretreatment (Kennett et al., 1987). The hypophagic effects of 5-HT and of norfenfluramine when infused into the paraventricular nucleus (PVN) of the hypothalamus (Shor-Posner et al., 1987) suggest that 5-HT receptors therein are implicated. The similar hypophagic effect of infusing RU24969 (1 ug) (but not 8-OH-DPAT, 1 ug) (Hutson et al., unpublished work) into the PVN of food deprived rats indicates that the receptors are of the $5-HT_{1B}$ type. Infusion of RU24969 (unlike its systemic administration) did not cause hyperactivity.

mCPP is reported to cause hypophagia (Samanin et al., 1979). We have obtained similar results using mCPP (1.0-5.0 mg/kg) and the related TFMPP both given i.p. (1.0-10.0 mg/kg) (Kennett et al., 1987) or into the PVN (1 ug) (Hutson et al., unpublished work). These drugs were previously considered as selective $5-HT_{1B}$ agonists. However in contrast to RU 24969 they cause hypolocomotion. Since this latter response appears to be mediated by $5-HT_{1C}$ receptors (Kennett and Curzon, 1988) it seemed conceivable that they might also mediate the hypophagic response.

A subsequent study revealed that the hypophagic response to both mCPP and TFMPP in food deprived rats could be opposed by both $5-HT_{1C}$ and $5-HT_{1B}$ antagonists whilst that produced by RU 24969 was only sensitive to blockade of $5-HT_{1B}$ receptors (Kennett and Curzon, unpublished). These results suggest that both receptor subtypes may be involved in the induction of hypophagia as shown in figure 23.1.

Proposed roles of $5-HT_1$ receptor subtypes in effects of drugs on feeding

Fig. 23.1 shows the main findings integrated into a single scheme. This may be relevant to the action of appetite suppressants such as fenfluramine and to physiological and pathological aspects of appetite.

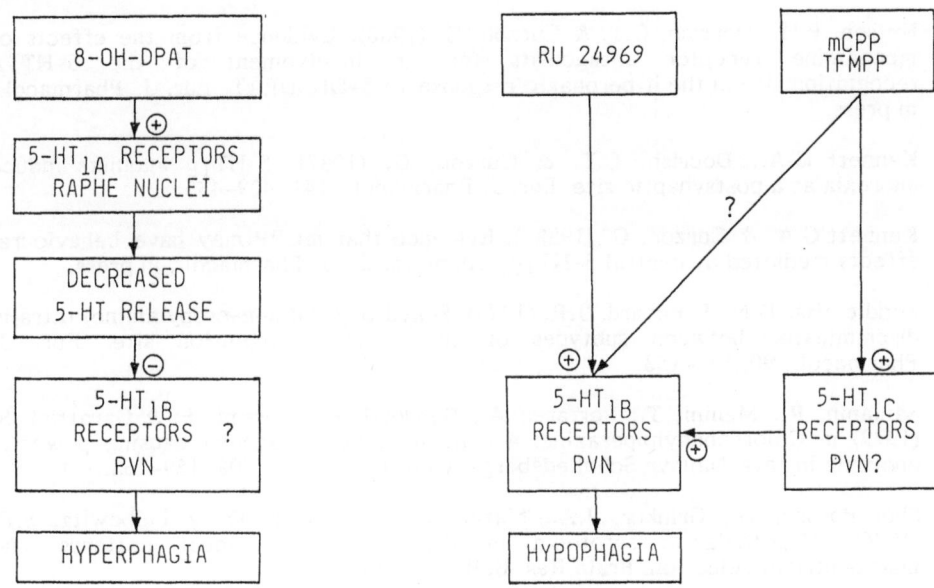

Fig. 23.1   Possible roles of 5-HT$_1$ receptor subtypes in effects of drugs on feeding

References

Dourish, C.T., Hutson, P.H. & Curzon, G. (1985). Low doses of the putative serotnin agonist 8-hydroxy-2-(di-n-propylamino) tetralin (8-OH-DPAT) elicit feeding in the rat. Psychopharmacology, **86**, 197-204.

Dourish, C.T., Hutson, P.H. & Curzon, G. (1986). Para-chlorophenylalanine prevents feeding induced by the serotonin agonist 8-hydroxy-2-(di-n-propylamino) tetralin (8-OH-DPAT). Psychopharmacology, **89**, 467-471.

Gilbert, F. & Dourish, C.T. (1987). Effects of the novel anxiolytics gepirone, buspirone and ipsapirone on free feeding and on feeding induced by 8-OH-DPAT. Psychopharmacology, **93**, 349-352.

Hutson, P.H., Dourish, C.T. & Curzon, G. (1986). Neurochemical and behavioural evidence for mediation of the hyperphagic action of 8-OH-DPAT by 5-HT cell body autoreceptors. Eur. J. Pharmacol., **129**, 347-352.

Hutson, P.H., Donohoe, T.P. & Curzon, G. (1987). Neurochemical and behavioural evidence for an agonist action of 1-[2-(4-aminophenyl)ethyl]-4-

(3-trifluoromethyl (phenyl) piperazine (LY165163) at central 5-HT receptors. Eur. J. Pharmacol., 138, 215-223.

Hutson, P.H., Dourish, C.T. & Curzon, G. (1988). Evidence from the effects of monoamine receptor antagonists for an involvement of the 5-HT$_{1A}$ recognition site in the hyperphagic response to 8-OH-DPAT. Eur. J. Pharmacol., in press.

Kennett G.A., Dourish, C.T. & Curzon, G. (1987). 5-HT$_{1B}$ agonists induce anorexia at a postsynaptic site. Eur. J. Pharmacol., 141, 429-435.

Kennett G.A. & Curzon, G. (1988). Evidence that mCPP may have behavioural effects mediated by central 5-HT$_{1C}$ receptors. Br. J. Pharmacol., in press.

Middlemiss, D.N. & Fozard, J.R. (1983). 8-hydroxy-2-(di-n-propylamino) tetralin discriminates between subtypes of the 5-HT$_1$ recognition site. Eur. J. Pharmacol., 90, 151-153.

Samanin, R., Menini, T., Ferraris, A., Bendotti, C., Borsini, F. & Garattini. S. (1979) m-Chlorophenylpiperazine: a central serotonin agonist causing powerful anorexia in rats. Naunyn Schmiedebergs Arch. Pharmacol. 308, 159-163.

Shor-Posner, G., Grinker, J.A., Marmeson, C., Brown, O. & Leibowitz, S.F. (1986). Hypothalamic serotonin in the control of meal patterns and macronutrient selection. Brain Res. Bull., 17, 663-671.

Sugrue, M.F. (1987). Neuropharmacology of drugs affecting food intake. Pharmacol. Ther., 32, 145-182.

Tricklebank, M.D., Forler, C. & Fozard, J.R. (1984). The involvement of subtypes of the 5-HT$_1$ receptor and of catecholaminergic systems in the behavioural response to 8-hydroxy-2-(di-n-propylamino) tetralin in the rat. Eur. J. Pharmacol., 106, 271-282.

Weismann-Nanopoulos D., Mach, E., Magre, J., Demassey, Y. & Pujol, J-F. (1985). Evidence for the localisation of 5-HT$_{1A}$ binding sites on serotonin containing neurons in the raphe dorsalis and raphe centralis nuclei of the rat brain. Neurochem. Int., 7, 1061.

EFFECTS OF SELECTIVE SEROTONERGIC AGONISTS ON PALATABILITY-INDUCED INGESTION

J.David Leander.
Central Nervous System Research, Lilly Research Laboratories, Eli Lilly and Company, Lilly Corporate Center, Indianapolis, IN 46285, U.S.A.

The serotonergic (5-hydroxytryptamine) neurotransmitter system has been repeatedly implicated in the control of ingestive behaviours, with most agonists producing decreases (Blundell 1977, 1986). Recently, a new potent and selective serotonin agonist, 8-OH-DPAT, has been reported to increase feeding in non-deprived rats at low doses (e.g. Dourish et al., 1985a, b).

Leander (1987) has recently demonstrated that the selective serotonin uptake inhibitor fluoxetine suppresses palatability-induced (saccharin solution) ingestion. That work with fluvoxetine led to the present question: Are the effects with fluoxetine similar to the effects produced by known selective agonists for the various serotonin receptor subtypes? The purpose of this research is to answer that question by studying the effects of 8-OH-DPAT and LY65163, purported 5-HT_{1A} specific agonists, RU-24696 and TFMPP, purported 5-HT_{1B} specific agonists, and quipazine, a purported 5-HT_2 agonist, on palatability-induced ingestion.

Methods
The methods are essentially similar to those previously used with fluoxetine (Leander, 1987), except that only one concentration of saccharin was used. Twenty-four Long-Evans hooded male rats (Charles-River Breeding Laboratories, In., Portage, MI), weighing between 400 and 500 g, were allowed access to bottles of 0.01 M sodium saccharin solution for 1 or 2 hr per weekday. The rats were individually housed in standard wire mesh cages and had free access to standard rodent lab chow and water at all times except the one hour before and the hour(s) of saccharin solution availability. The access period usually began between 10 and 11 a.m. Lights were on in the colony room between 6 a.m. and 6 p.m. each day. Injections of drugs were administered subcutaneously in a volume of 1 ml/kg of body weight either 30 min or 60 min before saccharin solution availability. Drug effects were usually studied on Tuesday and Fridays, whereas the data from Thursdays served as non-drug control data.

Results
8-OH-DPAT produced an inverted U-shaped dose response curve on saccharin ingestion -- doses of 0.04 and 0.16 mg/kg increased ingestion (159 $\pm$ 12% and 214 $\pm$ 56% of control, respectively), whereas a lower dose (0.01 mg/kg) was without effect (102 $\pm$ 9% of control), and a higher dose (0.64 mg/kg) suppresses consumption (47 $\pm$ 6% of control). LY165163 had a similar effect, with a small increase (120 $\pm$ 9% control) at 0.32 mg/kg.

Both of the 5-HT_{1B} agonists, TFMPP (0.32 - 1.25 mg/kg s.c.) and RU-24969 (0.02 - 5 mg/kg s.c.) only produced decreases in saccharin solution ingestion. These decreases were not antagonized by methysergide (5 mg/kg) or by the selective 5-HT_2 antagonist LY53857 (1 mg/kg). The 5-HT_2 agonist quipazine produced dose-related (0.64–5 mg/kg s.c.) decreases in saccharin consumption. However, in contrast to the lack of effect seen with methysergide and LY53857 against the suppressant effect of RU-24969 and TFMPP, both methysergide and LY53857 antagonized the suppressant effect of quipazine on saccharin solution ingestion. Neither methysergide (5 mg/kg) nor LY53857 blocked the decrease in saccharin consumption produced by 10 mg/kg of fluoxetine.

A dose of 8-OH-DPAT that upon acute administration produced an increase in consumption, appeared to produce tolerance over the course of five daily treatments. In contrast, there was no apparent tolerance development to the doses of RU-24969 and quipazine that decreased saccharin consumption.

Discussion
These results shown that palatability-induced (saccharin) consumption can be used to study the different effects of serotonergic agonists. $5-HT_{1A}$ selective serotonin receptor agonists produced an invested U-shaped dose-response curve with increases in consumption at low doses and then decreases at much higher doses. In contrast, $5-HT_{1B}$ selective serotonin agonists only produced dose-related decreases in consumption. Quipazine only decreased consumption, but it could be differentiated from the effects of the $5-HT_{1B}$ agonists on the basis of the studied with the serotonin antagonists. Methysergide and LY53857 antagonized the decrease in saccharin consumption produced by quipazine but not similar decreases produced by TFMPP or RU-24969.

The increases with the $5-HT_{1A}$ agonists in saccharin consumption appear similar to the increases in food consumption in sated animals that have previously been reported for 8-OH-DPAT and LY165163 (Dourish et al., 1985a,b; Hutson et al., 1987). The present results contrast with those of Fletcher (1987) who reported that 30 and 60 µg/kg of 8-OH-DPAT did not increase glucose solution consumption but did increase solid food consumption. The present results show that $5-HT_{1A}$ agonists will increase ingestion of at least one type of palatable solution. Similarity, Cooper and Desa (1987) have recently reported that $5-HT_{1A}$ agonists (8-OH-DPAT, gepirone and ipsapirone) increased hypertonic saline consumption in rehydrating rats.

The present results with the $5-HT_{1A}$, $5-HT_{1B}$ and $5-HT_2$ agonists suggest that the previously reported (Leander, 1987) effects of fluoxetine are most similar to the effects obtained with the $5-HT_{1B}$ agonists. Future work will be directed towards further delineation of the similarities between the effects of fluoxetine and $5-HT_{1B}$ agonists on ingestive behaviours.

Acknowledgements: The author thanks Mr. Don Pearson for his technical assistance.

References

Blundell, J.E. (1977). Is there a role for serotonin (5-hydroxytryptamine) in feeding? Int. J. Obes., 1, 15–42.

Blundell, J.E. (1986). Serotonin manipulations and the structure of feeding behaviour. Appetite, **7** (supplement), 39–56.

Cooper, S.J. & Desa, A. (1987). Benzodiazepines and putative 5-HT$_{1A}$ agonists increase hypertonic saline consumption in rehydrating rats. Pharmacol. Biochem. Behav., **28**, 187-191.

Dourish, C.T., Hutson, P.H. & Curzon, G. (1985a). Low doses of the putative serotonin agonist 8-hydroxy-2-(di-n-propylamino)tetralin (8-OH-DPAT) elicit feeding in the rat. Psychopharmacology, **86**, 197-204.

Dourish, C.T., Hutson, P.H. & Curzon, G. (1985b). Characteristics of feeding induced by the serotonin agonist 8-hydroxy-2-(di-n-propylamino) tetralin (8-OH-DPAT). Brain Res. Bull., **15**, 377-384.

Fletcher, P.J. (1987). 8-OH-DPAT elicits gnawing, and eating of solid but not liquid foods. Psychopharmacology, **92**, 192-195.

Hutson, P.H., Donohoe, T.P. & Curzon, G. (1987). Neurochemical and behavioural evidence for an agonist action of 1-[2-(4-aminophenyl)-ethyl]-4-(3-trifluoromethyl- phenyl) piperazine (LY165163) at central 5-HT receptors. Eur. J. Pharmacol., **138**, 215-223.

Leander J.D. (1987). Fluoxetine suppresses palatability-induced ingestion. Psychopharmacology, **91**, 285-287.

8-OH-DPAT RELIABLY INCREASES INGESTION OF SOLID BUT NOT LIQUID DIETS

A.M.J.Montgomery, P.Willner, R.Muscat
Psychology Dept., City of London Polytechnic, Old Castle St., London E1 7NT, England

The 5-HT_{1A} receptor agonist 8-OH-DPAT has been reported to elicit consumption of lab chow in non-deprived rats, over a wide range of doses between 15 and 300 µg/kg; at higher doses, feeding is displaced by the emergence of stereotyped behaviours (Dourish et al., 1985a,b; Bendotti et al., 1986). The orexic effect of low doses of 8-OH-DPAT appears to be mediated by a decrease in the release of 5-HT, brought about by a specific interaction with inhibitory presynaptic autoreceptors (Dourish et al., 1985a, b). In the present study, the effects of 8-OH-DPAT on the consumption of caloric or non-caloric liquid diets were examined, in order to assess the behavioural generality of 8-OH-DPAT induced hyperphagia.

Methods
Singly housed, male, Lister hooded rats were used in all experiments. All testing was carried out in the home cage, starting 3h into the light phase. Food and water were removed at the start of the test session; otherwise, with lab chow and water were available at all times, except in one experiment carried our after 20h. food deprivation (see Table 25.1).

The animals were first acclimatized, for 10 - 14 daily 60 mins. sessions, to one of the following test diets: wet mash (100g powdered chow + 200 ml water); sweetened wet mash (same + 10g sucrose); saline (0.9%); saccharin (0.1%); sucrose (5% and 35%); milk; high protein diet (90% milk protein made up with water to a 5.3% solution). Wet mash was presented in a spill-proof jar, and liquid diets in preweighed plastic drinking bottles.

Subsequent sessions began with a 10 min. pre-exposure to the test diet; it was then removed and the animals received one of a range of dodes of 8-OH-DPAT (see Table 25.1) by s.c. injection; 15 min. later the diet was returned for 60 min.

In three experiments, observations of consummatory and other behaviours were recorded at 15s intervals. Within-subjects repeated mesures design were used throughout, with order of doses counterbalanced between subjects and at least two drug-free days between successive tests. Results were analyzed by analysis of variance.

In a final experiment, the animals were first acclimatized to the presence of a wooden block in the cage. Episodes of chewing on the block and of grooming were then recorded for 25 min., beginning 30 min. after 8-OH-DPAT administration (0 or 60 ug/kg).

Results
8-OH-DPAT very reliably increased wet mash consumption ($p<0.001$), in both deprived and free-feeding animals (Table 25.1). 8-OH-DPAT also caused a small

Table 25.1: Effects of 8-OH-DPAT on the consumption of various diets

Diet	n	Dose (ug/kg)	Effect on intake
Solid diets:			
Wet mash	12	60	Increase (p<0.001)
Wet mash (20h food dep)	12	60	Increase (p<0.001)
Sweetened wet mash	12	60	Increase (p<0.001)
Non-caloric liquid diets:			
Saline (0.9%)	8	15/30/60	None
Saccharin (0.1%)	8	15/30/60	None
Caloric liquid diets:			
Sucrose (5%)	8	15/60	None
		30	Marginal increase
Sucrose (35%)	12	15/60	None
		30	Marginal increase
Sucrose (35%)	12	15/30/60	None
Sucrose (35%	12	60/120	None
Milk	12	30	Marginal increase
		60/120	None
High protein diet	12	15	None
		30/60/120	Decrease (p<0.001)

but significant (p<0.012) increase in the duration of chewing on wooden blocks. The only striking effect of 8-OH-DPAT on consumption of liquit diets was a reduction of high protein intake at doses of 30 ug/kg and higher (Table 25.1). In three experiments, consumption increased slightly at 30 ug/kg, but in all cases the effect was ambiguous: one was significant (p<0.05) on a paired comparison but not in the overall analysis of variance; a second (p<0.05) was confounded by differences in baseline consumption prior to 8-OH-DPAT; and the third was at the very margin of significance (p=0.05).

In the observational experiments, vehicle treated animals displayed a typical 'satiety sequence' (Antin et al., 1975) of ingestive behaviour superceded by active behaviours and grooming, followed by resting behaviours. In one experiment (wet mash consumption after food deprivation) 8-OH-DPAT (60 ug/kg) prolonged feeding and delayed resting. However, in non-deprived rats tested with sweetened wet mash the same dose delayed the onset of feeding, such that feeding was almost totally suppressed for the first 15 min. of the test. Similarly, the consumption of 35% sucrose was also suppressed (p<0.01) during the early part of the session. 8-OH-DPAT also suppressed grooming; this effect was present in all four observational experiments, being highly significant (p<0.001) in three of them though non-significant in the fourth.

Discussion
These experiments confirm that 8-OH-DPAT can reliably increase the consumption of solid foods. However, this effect does not appear to extend to

liquid diets. Increases in liquid diet consumption were at best marginal, and confined to a single low dose (30 ug/kg); this same dose abolished consumption of the high protein diet. A recent study reported that 8-OH-DPAT did increase milk consumption, while agreeing that this effect is only elicited at doses up to 30 ug/kg (Coughlan et al., 1987), a dose at which an increase in solid food intake is not always apparent. By contrast, a number of other laboratories have confirmed our finding that 8-OH-DPAT did not increase the intake of sweet solutions at higher doses that do reliably increase the intake of solid food (M.Clark, pers. comm.; Engel, 1986; Fletcher, 1987; J.Hartley, pers. comm.). These differences do not reflect an intrinsic pharmacological insensitivity of liquid diets to orexigenic agents: one study showing that 8-OH-DPAT increased the intake of lab chow but not of sucrose also found that the intake of both diets was increased by clonidine (M.Clark, pers. comm.).

It is currently unclear how best to explain the range of effects of 8-OH-DPAT on ingestive behaviour, which are clearly more complex than has previously been envisaged. The observed decreases in grooming may reflect the putative anti-anxiety effect of this compound (Engel, 1986; Engel et al., 1984), but the delayed onset of sucrose or sweetened wet mash intake is perplexing. However, the major problem arises from the behavioural specificity of the orexic effect, which would seem to preclude explanations of 8-OH-DPAT induced hyperphagia in terms of a suppression of satiety or an enhancement of appetite (Gilbert & Dourish, 1987). Indeed, the observation that 8-OH-DPAT increased the amount of chewing on a wooden block suggests instead that 8-OH-DPAT induced hyperphagia might in part reflect drug-induced chewing.

In a complementary study, we have recently observed that fenfluramine, which decreases food intake by enhancing 5-HT release, reduced the rate of consumption of wet mash, but did not affect the rate of sucrose consumption (Montgomery & Willner, 1987). Together, these two studies would not support a role for 5-HT in the motivational control of feeding, but rather, suggest that 5-HT may to some extent influence food consumption by facilitating or inhibiting specific motor components of consummatory behaviour.

References

Antin, J., Gibbs, J., Holt, J., Young, R.C. & Smith, G.P. (1975) Cholecystokinin elicits the complete behavioural sequece of satiery in rats. J. Comp. Physiol. Psychol., 89, 784-790.

Bendotti, C. & Samanin, R. (1986). 8-hydroxy-2-(di-n-propylamino) tetralin (8-OH-DPAT) elicits eating in free feeding rats by acting on central serotonin neurons. Eur. J. Pharmacol. 121, 147-150.

Coughlan, J., Dourish, C.T., Gilbert, F. & Iversen, S.D. (1987). The $5-HT_{1A}$ agonist 8-OH-DPAT increases sweetened milk consumption in rats. Brit. J. Pharmacol., in press.

Dourish, C.J., Hutson, P.H. & Curzon, G. (1985a). Low doses of the putative serotonin agonist 8-hydroxy-2(di-n-propylamino) tetralin (8-OH-DPAT) elicit feeding in the rat. Psychopharmacology, 86, 197-204.

Dourish, C.J., Hutson, P.H. & Curzon, G. (1985b). Characteristics of feeding induced by the serotonin agonist 8-hydroxy-2(di-n-propylamino) tetralin (8-OH-DPAT). Brain Res., **15**, 377-384.

Engel, J.A. (1986). Anticonflict effect of the putative serotonin receptor agonist 8-OH-DPAT. Paper presented at the European Behavioural Pharmacology Society, Antwerp, July 1986. Abstr. Psychopharmacology, **89**, S28.

Flether, P. (1987) 8-OH-DPAT elicits gnawing and eating of solid but not liquid foods. Psychopharmacology, **92**, 192-195.

Gilbert, F. & Dourish, C.T. (1987). Effects of novel anxiolytic gepirone, buspirone and ipsapirone on free feeding and on feeding induced by 8-OH-DPAT. Psychopharmacology, **93**, 349-352.

Montgomery, A.M.J. & Willner, P. (1987). Fenfluramine disrupts the behavioural satiety sequence in rats. Psychopharmacology, in press.

PHARMACOLOGICAL MANIPULATION OF 5-HT: EFFECT ON INTAKE OF DIETS SUPPLEMENTED WITH EITHER SWEET OR BLAND CARBOHYDRATES

Clare L.Lawton, J.E.Blundell
Biopsychology Group, Department of Psychology, Leeds University, Leeds, LS2 9JT, United Kingdom

It has been proposed that serotonergic neurons comprise part of a bio-behavioural feedback loop, whereby the consumption of carbohydrate diminishes the tendency to consume further carbohydrate (Wurtman, 1987). Consequently drugs which enhance central serotoninergic transmission should cause a suppression of carbohydrate intake.

Sclafani et al., (1987) have described an experimental procedure in which rats show an avoid preference for sweet (sucrose) or bland (Polycose) carbohydrates presented in solution. Therefore this paradigm appears to embrace a motivational component similar in principle to the carbohydrate craving observed in some human situations. It follows, then that these experimental circumstances should provide the best opportunity for observing a carbohydrate suppressive effect of 5-HT drugs.

The present study was designed to determine the effects of a serotoninergic agent - d-fenfluramine - on carbohydrate intake. Comparisons were also made between d-fenfluramine and other anorectic agents with no obvious action on 5-HT, namely, salbutamol and cholecystokinin (CCK).

Methods
Subjects. 24 male black hooded Lister rats were habituated for two weeks to a 12:12 hour reversed light/dark cycle (lights off 0900 hours) and a 6 hour food deprivation into three groups matched for body weight (N=8).

Laboratory chow (Labsure, ERM) in pellet form (12.6 kj/g) was used as the maintenance diet. Carboxydrate was presented as an optional supplement to standard nutritionally complete diet of chow plus wate, supplements were either a) 200 mls. water, b) 200 mls. 32% sucrose solution (5.12 kj/g), c) 200 mls. 32% Polycose solution (5.12 kj/g). Doses of 1.0, 2.0 and 4.0 mg/kg d-fenfluramine (Servier) were used for Experiment 1 and 2 mg/kg d-fenfluramine, 5 mg/kg salbutamol (Glaxo), and 5 ug/kg CCK-OP (Sigma) for Experiment 2. Physiological saline was the control in each experiment. Injections were administered i.p. in a volume of 1 ml/kg body weight. All drugs were dissolved in saline.

A within subjects design was used for each experiment. Drugs were administered double blind in a counterbalanced order via latin square. A minimum period of 48 hours separated successive treatments.
Procedure. (i) 0900 hours - Food removed from cages.
           (ii) 0430 hours (experiment 1) - 1445 hours (experiment 2) - Animals injected.
           (iii) 1500 hours - Measured amounts of chow, water and supplements presented.

(iv) Food intake measurements recorded at – 1, 2 and 18 hour intervals (experiment 1) – 30 mins, 1 and 2 hour intervals (experiment 2).

Chow and supplement intake data were analysed using 2 way ANOVA, 1 repeated measure followed by Newman Keuls post hoc 2 sample comparison test. All significance levels were set at p=0.05.

Results

In experiment 1 a clear dose dependent effect of d-fenfluramine on total energy intake (chow plus supplement) was demonstrated in the 1st. and 2nd hour after food presentation in control, sucrose and Polycose groups. In the 18 hour period a dose dependent effect was demonstrated in both sucrose and Polycose groups but in the control group 4 mg/kg d-fenfluramine was the only dose to produce a suppression of intake. In both supplemented diet groups the absolute intake of chow and the percentage of total intake consumed as chow decreased as dose of drug increased. Intake of sucrose and Polycose in the dietary supplemented groups remained relatively unaffected. Therefore the percentage of total intake consumed as supplement increased in both groups of animals. The results of experiment 2 varied slightly according to the measurement period which reflected the different half-lives of the drugs. At all time periods d-fenfluramine displayed the same effect as in experiment 1. At the 30 minute test period (see figure 26.1) CCK suppressed the percentage intake of Polycose and had a slight sparing effect on sucrose.

Salbutamol spared sucrose consumption but showed no preferential suppression of Polycose compared with chow intake. At the 1 hour test period salbutamol selectively suppressed the intake of Polycose but not sucrose. There was no marked selective suppression of sucrose (i.e. more than 50% of total energy intake) by any drug at any time period.

Discussion

The results of experiment 1 clearly demonstrated that d-fenfluramine brought about a dose related decrease in total (chow plus supplement) intake. Hence the results confirm the well known inhibitory role of serotonin containing neurons in the control of total food intake (Blundell, 1977; 1984). Administration of d-fenfluramine also brought about a dose related decrease in chow intake alone in the sucrose and Polycose groups. Contrary to expectations, however, d-fenfluramine exerted a weaker suppressive effect on intake alone in the sucrose and Polycose groups. Contrary to expectations, however, d-fenfluramine exerted a weaker suppressive effect on intake of the carbohydrate solutions (sweet and non-sweet) than on intake of chow.

These results indicate that pharmacological activation of serotoninergic neurotransmission (by d-fenfluramine) does not inevitably lead to a preferential suppression of carbohydrate intake (greater than the suppression of other foods). The results of this experiment are consistent with the findings of Orthen-Gambill (1985) which demonstrated a weaker suppressive effect of d-fenfluramine on granulated sucrose intake than on intake of chow. The results of experiment 2 confirmed the observed effects of 2 mg/kg d-fenfluramine in experiment 1 on carbohydrate and chow intake.

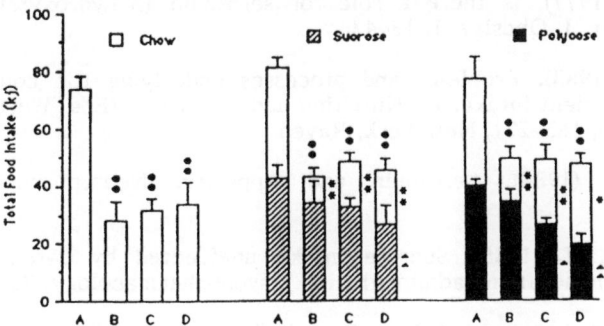

Figure 26.1 The effect of 2 mg/kg d-fenfluramine, 5 mg/kg salbutamol and 5 ug/kg CCK on Mean (+/- SE) Food intake (kj), in control, sucrose and Polycose groups, during the 30 minute period following food presentation.

However, pharmacological activation of beta adrenergic receptors by salbutamol did cause a preferential suppression of the non-sweet carbohydrate (Polycose) greater than the suppression of chow and sucrose (1 hour test period). Administration of CCK-OP (a putative satiety peptide) produced a short term selective suppression of the non-sweet carbohydrate and maintained the percentage of total intake consumed as the sweet carbohydrate (sucrose).

The results therefore indicate two main features: a) Differences exist among d-fenfluramine, salbutamol and CCK with respect to their actions on carbohydrate supplemented diets, b) It is easier for pharmacological agents to suppress Polycose than sucrose intake.

The use of the diet supplement paradigm employed in this study allows a degree of dietary choice. It does not provide a test of selection between different nutrients but does permit rats to display a strong appetite for carbohydrate. The use of this paradigm has demonstrated that carbohydrate consumption is sensitive to drug induced modifications since both preferential sparing and suppression of carbohydrate intake were demonstrated. The results obtained raise questions about the generality of the proposed role of serotonergic neurons in the control of carbohydrate consumption (Wurtman & Wurtman, 1979). This experimental diet selection strategy can therefore be used to further investigate the effects of drugs on nutrient intake. The outcome of drug nutrient interactions is likely, however, to be heavily dependent on the experimental paradigm employed (Blundell, 1983). In the behavioural analysis of dietary ingestion patterns, caution must be exercised when changes in nutrient preferences are embodied in a more general anorexic action.

References

Blundell, J.E. (1977). Is there a role for serotonin (5-hydroxytryptamine) in feeding? Internat. J. Obesity, 1, 15–42.

Blundell, J.E. (1983). Problems and processes underlying the control of food selection and nutrient intake. In: Nutrition and the brain. (Eds. Wurtman, R.J. & Wurtman, J.J.) 6, 163–221, New York, Raven.

Blundell, J.W. (1984). Serotonin and Appetite. Neuropharmacology, 23, 1537–1552.

Orthen-Gambill, N. (1985). Sucrose intake unaffected by d-fenfluramine by suppressed by amphetamine administration. Psychopharmacology, 72, 130–135.

Sclafani, A., and coworkers. (1987). Carbohydrate, Taste, Appetite and Obesity. Neuroscience and biobehavioural Reviews, 11 (2), 131–262.

Wurtman, J.J. & Wurtman, R.J. (1979) Drugs that enhance central serotoninergic transmission diminish elective carbohydrate consumption by rats. Life Sci., 24, 895–904.

Wurtman, R.J. (1987). Dietary treatments that affect brain neurotransmitters. Effects on calorie and nutrient intake. Ann. N.Y. Acad. Sci., **499**, 179–190.

5-HT, PAIN AND ANXIETY

Trevor Archer
Department of Psychology, University of Gothenburg, S 400 20 Gothenburg, Sweden

The chapter on the behavioural pharmacology of pain, by Berge et al. (Chapter 27), outlines most succinctly the procedural and methodological 'hardware'. The basic role of 5-HT in pain seems to the that stimulation of 5-HT pathways or increases in 5-HT neurotransmission lead to antinociceptive effects that may be indexed with a wide range of techniques (see table 1, Chapter 27). However, Berge et al. indicate that this simple story is much more complicated: procedurally, e.g. through test–dependency and route of administration, pharmacologically, e.g. through test drugs and lesioning techniques, and anatomically, as a result of the profound involvement of serotonergic pathways with a myriad of other neurotransmitter systems. Bearing these enormous constraints in mind, it is with humility rather than pessimism that Berge suggests that serotonergic involvement in pain and analgesia is poorly understood. What remains implicit in this 'pain' chapter is the fascinating story of how 5-HT interacts with neuropeptides, opiates, and not least of all the noradrenergic and cholinergic pathways, whether spinal or supraspinal.

The chapter by Soubrie, which attempts to relate certain aspects of depression and anxiety, provides a brief review of the 'real' evidence in favour of an involvement of 5-HT, agonists in anxiety. More important for present purposes, Soubrie (Chapter 29) seeks to enumerate the several constraints upon clear conclusions concerning these drugs and anxiety, e.g. inconsistencies over tests and testing procedures, poor replicability within a given test procedure, the narrow range of effective doses, the small or nonexistent therapeutic window, and so on. Broekkamp and Jenck (this volume) have applied several animal models of anxiety to the problem of the involvement of drugs affecting 5-HT neurotransmission in anxiety, including the social interaction test, the elevated plus maze, defensive marble burying by mice and aversive electrical stimulation of the periaqueductal gray. These authors distinguish between the effects of $5-HT_{1A}$ agonists, like 8-OH-DPAT, 5-HT antagonists and the 5-HT uptake inhibitors in the various test models. One point seems clear: None of the classes of compound tested would appear to produce as consistent or robust a pattern of anxiolytic action as the benzodiazepines.

This finding is interesting since most of the tests available have indeed been, as it were, constructed with the benzodiazepine action being the measure of predictive validity. The chapter by Tyers (this volume) provides a brief but necessary review of some actions of 5-HT receptors and the actions of the $5-HT_3$ receptor in particular. Tyers describes in detail the anxiolytic properties of some $5-HT_3$ receptor antagonists in animal models utilising mice, rats and marmosets. Although these findings are important and certainly add to the information-base, they do tend to complicate the general issue of the role of 5-HT in anxiety. Some involvement of 5-HT in anxiety, however, to be the general conclusion.

The general trend of the evidence presented to implicate the role of 5-HT in anxiety focusses upon the availability of a number of relatively selective 5-HT$_{1A}$ agonists and apparently very potent 5-HT$_3$ receptor antagonists. Fernandez-Guasti and Hong (Chapter 34) present data indicating that 8-OH-DPAT and RU 24969 were more potent than diazepam in the conditioned defensive burying procedure. The doses of 8-OH-DPAT that antagonise defensive burying compare favourably with the doses of 8-OH-DPAT and the novel 5-HT$_{1A}$ agonist, NDO 008, used in other bioassays of 5-HT receptor function, e.g. cage-leaving behaviour (see Renyi & Lewander, this volume). Lorens, Mitsushio and Van de Kar offer an important behavioural, endocrine and neurochemical analysis of the potential anxiolytic efficacy of the putative 5-HT$_{1A}$ agonist, ipsapirone. This type of painstaking analysis is quite seldom performed but ought to be of singular benefit for drug development. It is interesting to compare the results of Moser (Chapter 33) with those of Lorens et al. not least because each make some comparison of the compounds ipsapirone and buspirone, and suggest some anxiogenic effects a higher dose levels. The paper presented by Meert, Niemegeers, Gelders and Janssen (Chapter 18) describes in much detail the pharmacological profile of the 5-HT$_2$ antagonist, ritanserin, in this context.

Unfortunately, Meert et al. do not pursue the analysis of 5-HT$_2$ receptor involvement in anxiety which would have a provided invaluable dimension to the understanding of 5-HT and anxiety. Taken together, the chapters pertaining to the behavioural pharmacology of 5-HT in anxiety serve generally to indicate that 5-HT receptors are involved in anxiety (albeit in as yet poorly understood anxiolytic/anxiogenic relations) but the basic question of whether the new generation of 5-HT based anxiolytic drugs offer a suitable alternative to the benzodiazepines and other types of medication is clearly unanswered (but for a very interesting discussion on the preclinical and clinical aspects of drugs and anxiety, see Chapter 28).

THE BEHAVIOURAL PHARMACOLOGY OF SEROTONIN IN PAIN PROCESSES

Odd-Geir Berge[*], Claes Post and Trevor Archer[°].
University of Bergen[*]., Bergen, Norway, Astra Pain Control, Södertälje, Sweden
University of Gothenburg[°], Gothenburg, Sweden

Pain in humans has been defined as "an unpleasant sensory and emotional experience associated with actual or potential tissue damage, or described in terms of such damage" (Merskey et al., 1979). In animals, pain must be inferred from observed behaviour. In the overwhelming majority of studies where conscious animals are employed, stimulation is of short duration and terminates as soon as the subject displays a response taken to indicate pain. Tissue damage is usually avoided. The animals are as a rule healthy and frequently adapted to the test situation so that fear and anxiety is reduced. Thus, "pain" in the experimental situation differs significantly from pain in the clinic. Still, experiments using animal models have provide valuable insight into the physiology of pain and the mechanisms of action of analgesics and thereby contributed to the advancement of the clinical treatment of pain. A striking example is the demonstration of a spinal site of action of opiates and subsequent administration of these drugs at the level of the spinal cord for pain control (Yaksh, 1981).

That serotonin may play a clinically relevant role in pain processing is supported by observations that patients with different types of depressive disorders associated with altered serotonin neurotransmission often also have pain (von Knorring et al., 1983). Almay and collaborators demonstrated recently that patients with chronic idiopathic pain syndromes commonly have lower than normal levels of 5-HIAA in the cerebro-spinal fluid (Almay, von Knorring & Oreland, 1987). The selective serotonin reuptake inhibitor zimeldine has also been shown to relieve chronic pain (Johansson & von Knorring, 1979). However, it has not yet been convincingly shown that the analgesic effects are due to the inhibition of neuronal reuptake of the monoamines. Brodin and collaborators have recently found that in rats, long-term administration of the selective serotonin uptake inhibitor alaproclate gives rise to antinociceptive effects (Brodin & Post, unpublished), and to increases in the concentration of substance P in the periaqueductal grey area (PAG; Brodin et al., 1984).

Behavioural tests of nociception
The role of serotonergic systems in pain and nociception has mostly been studied in animal models. Several investigations have found that the consequences of manipulations of serotonergic systems differ between nociceptive tests. The differences may probably reflect specific properties of the tests, e.g. stimulus and response characteristics, as well as general factors related to the test situation, such as handling and restraint. It is therefore pertinent to start this review with a brief discussion of the most common nociceptive tests, particularly focussing on factors that may vary between implementations of the tests.

The tail-flick test employs a radient or conducted heat stimulus to the tip of the tail and measures the latency from the onset of stimulation to the occurrence of a flick. The response occurs at a critical skin temperature of approximately 42.5°C, independent of stimulus intensity (Ness & Gebhart, 1986). Although spinally integrated, the tail-flick reflex probably shares some of its neuronal substrate with supraspinally projecting nociceptive pathways and may be subjected to some of the descending control which affects pain transmission in the spinal cord.

Handling and training procedures may range from none to weeks of training and gradual adaptation to the testing environment. Restraint during testing may vary from gentle holding at the base of the tail during stimulation to immobilization in a restraining device for an entire test session lasting several hours. The test is frequently used in designs with repeated testing, sometimes with several exposures within a few minutes. The way the test is conducted is important for the outcome of the test as both the level of stress in the subjects and the local reaction of the tail skin is affected by procedural parameters.

In the hot plate test the animals are placed on a hot surface (usually about 55°C) and the latency to the occurrence of a predetermined response is recorded. Several response criteria have been employed either singly of in combination ("first response"). Common end points are licking or shaking of a paw or jumping. The responses may be altered differently by pharmacological treatment. (Fasmer et al., 1986; Fasmer et al., 1983a). Although commonly employed, fore-paw licking is unsuitable as a nociceptive end point in both mice and rats (Berge, Fasmer & Hole, 1983b; Huskaar, Berge & Hole, 1986b; Woolfe & MacDonals, 1944). The hot plate test may be employed with or without prior adaptation of the animals to the test situation and repeated testing is commonly employed. Evidence that adaptation procedures affect the outcome of the test has been reported (Hunskaar et al., 1986b; Rosland, Hunskaar & Hole, 1987).

Recently, a modified hot plate test with slowly increasing temperature has been introduced (Hunskaar et al., 1986b; Rosland et al., 1987). The temperature when a hind paw lick first occurs, is recorded as nociceptive threshold. Compared to the conventional hot plate test, by handling procedures, is sensitive to a broader range of analgesics and appears to be less sensitive to non-analgesic effects of drugs.

The tail-flick and conventional constant temperature hot plate tests are usually considered to be insensitive to mild analgesics. Another major problem with tests utilizing brief thermal stimulation is the interference of skin with the stimulus. Extensively discussed by Beecher (1957) with regard to radiant heat in human experimental research, the problem has largely been ignored in animal studies. However, it has been reported that changes in skin blood flow alter the effect of cutaneous heat stimulation in electrophysiological studies (Davies & Dray, 1980; Duggan et al., 1978) and that tail skin temperature effects the tail-flick test (Minfeng & Jisheng, 1979).

Recent investigations have shown that in both rats and mice, tail-flick latencies are inversely related to tail skin temperature and even differences in skin temperature may lead to significant differences in latency (Berge, Garcia-Cabrera & Hole, 1988: Eide et al., unpublished).

Thus, the test does not clearly distinguish between changes in pain sensitivity and changes in skin temperature. Several important implications regarding the interpretation of results obtained with this test are evident. Firstly, pharmacological tools used to study nociception may significantly affect tail skin temperature. For instance, both morphine and serotonin given intrathecally may induce a lowering of the tail skin temperature (LoPachin & Rudy, 1982; Rudy & Yaksh, 1977) which may theoretically contribute significantly to the measured antinociceptive effect of these compounds. Secondly, the degree of vasoconstriction during testing, which may be critical for the magnitude and direction of skin temperature changes, is influenced by factors like handling, immobilization, room temperature and acclimatization.

Methodological studies addressing the issue of temperature effects in the hot plate test are urgently needed. The conventional hot plate test is probably subjected to the same confounding effects as the tail-flick test, whereas the rising temperature method may be less susceptible partly because longer exposure to the hot surface will reduce the significance of the initial skin temperature and partly because the response latency is of little direct importance for the results.

In the absence of experimental data, the possibility of cardiovascular factors interfering also with the effect of chemical and electrical stimuli must be kept in mind.

The flinch-jump procedure originally described by Evans (1961) employed repeated series of shocks in ascending and descending order of intensity (amperage), administered over approximately 90 min. Flinch was defined as a "startle" or "crouch" while jump was defined as removal of two or more paws from the grid at the onset of the shock. Only the jump response was affected by analgesics. Subsequent modifications by others include changes in scoring criteria and accelerometer based registration of the responses as well as alterations in shock presentation, e.g. by discontinuing the ascending series of shocks at the first criterion response (shock titration). Thus, both the response criteria and stimulus parameters differ significantly between implementations of the flinch jump test. Unlike the tests described above, the formalin test uses a persistent stimulus of suprathreshold intensity. In man, subcutaneous injection of a small volume of 5% formalin causes a shortlasting intense pain of sharp, burning quality which after 4-5 min is replaced by a steady throbbing ache lasting 30-60 min (Dubuisson & Dennis, 1977). In mice, injection of dilute formalin into a hind paw produced a behaviour response of licking and biting of the affected paw with an initial phase of immediate onset lasting 5-10 min and a second phase starting 15-20 min after injection and lasting 30-45 min (Hunskaar, Fasmer & Hole, 1985).

Both during the initial and the second phase, the response may be suppressed by morphine and several non-opiate analgesics, while purely anti-inflammatory drugs appear to be effective during the second phase only (Hunskaar et al., 1985; Hunskaar, Berge & Hole, 1986c). Methological differences between research investigations relate to handling procedures, forepaw or hindpaw as injection site, time period studied as well as selection and weighing of criteria to be recorded. The test is sensitive to stressful or distracting manipulations

(Fanselow, 1984), and prior training and adaptation ot the test situation may critically affect the outcome of the test (Berge et al., 1987).

Anatomical Considerations

The serotonergic innervation of the central nervous system derives from the raphe area of the brainstem. In relation to results obtained by stimulation, microinjection or lesioning techniques, it should be remembered that the proportion of cells being serotonergic varies between the raphe nuclei, and is particularly low in the nucleus raphe magnus (Steinbusch & Nieuwenhuys, 1983; Wiklund, Leger & Persson, 1980).

The major ascending serotonergic projections are organized in a dorsal, a ventral and a medial pathway, originating in the raphe nuclei of the pons and mid-brain (For a comprehensive review of the anatomy of serotonergic systems, see Steinbusch & Nieuwenhuyd, 1983). The serotonergic innervation of the forebrain is extensive, and a single fiber may have terminals in anatomically separate structures. The descending serotonergic pathways originate primarily in the B1, B2 and B3 cell groups in the medulla oblongata, which roughly correspond to the raphe obscurus, raphe pallidus and raphe magnus nuclei. The bulbo-spinal serotonergic projections also contain substance P and TRH (for review see Hökfelt et al., 1986). The functional implication of the coexistence ramains to be elucidated.

The dorsal horn of the spinal cord receives projections from the nucleus raphe magnus and its paramedial extension via the dorsolateral funiculus. The terminals are located most abundantly in laminae I, II and IIa. Descending serotonergic projections also innervate the mediolateral cell column and the central horn. Thus, serotonergic terminals are found in spinal structures associated with sensory processes as well as autonomic and motor functions.

Depletion of serotonin by p-Cl-phenylalanine

A much used tool for studying serotonergic functions is the tryptophan hydroxylase inhibitor p-Cl-phenylalanine (PCPA) which produces extensive depletion of serotonin throughout the central nervous system. PCPA has largely failed to affect nociceptive thresholds in the tail-flick and hot plate tests in rats and mice (for references see Berge, Hole & Ögren, 1983c). Administration of PCPA relatively consistently enhances the response to noxious and non-noxious electrical shock, whereas both no effects and reduced thresholds have been found in the paw compression test.

The interpretation of the results obtained with PCPA must take into consideration that the compound depletes endogenous stores of serotonin throughout the central nervous system. In addition, peripheral effects may lead to reduction of receptive fields and less response to C-fiber stimulation of nociceptive neurones in the dorsal horn of the spinal cord (Dickenson et al., 1981; Le Bars et al., 1979b). In this way, PCPA presumably reduces the effect of some types of noxious stimuli.

Effects of receptor antagonists

Systemic or intrathecal administration of serotonergic receptor antagonists generally produces no change or moderate reduction in response latencies in the tail-flick and conventional hot plate test in rats and mice (Le Bars, 1988; Berge 1986). There appears to be no pattern as to which types of antagonists are effective in altering nociception. In one experiment in mice, systemic administration of the receptor antagonist metergoline produced reduced latencies in the tail-flick test but had an antinociceptive effect in the conventional hot plate test. Antinociception was also found after intracerebroventricular (i.c.v.) injection (Fasmer, Berge & Hole, 1984). Subsequent work has demonstrated antinociceptive effects also of metitepin in mice tested with the formalin and the rising temperature hot plate test (Eide, Berge & Hunskaar, 1986). The effect is enhanced after lesions of ascending serotonergic projections and seems therefore to be related to postsynaptic interactions. Administration of another serotonergic antagonist, cyproheptadine, has been reported to elevate the threshold for the jaw opening reflex in the monkey (Shyu, Lin & Wu, 1984).

Lesions of Ascending Pathways
In a series of studies, Harvey and co-workers demonstrated that lesions in the medial forebrain bundle in rats lowered the threshold for jump in the flinch-jump test and the results were taken to indicate tonic antinociceptive activity in the ascending serotonergic pathways (for reviews see Harvey & Simansky 1981, Hole & Berge 1981). However, electrolytic or neurotoxic lesions of the raphe nuclei or neurotoxic lesions of the ascending serotonergic pathways failed to alter the flinch-jump thresholds although both kinds of lesions produced extensive reductions of forebrain serotonin levels. Furthermore, none of the procedures appeared to affect responsiveness in the conventional hot plate test. Relatively extensive and selective lesions of ascending serotonergic pathways induced by administration of PCA caused a slight lowering of the jump threshold in one study (Berge et al., 1983c) but have consistently failed to affect the responsiveness of rats in the tail-flick test, the conventional and the rising temperature hot plate tests, and the formalin test (Berge et al., 1983c; Bjorkum & Berge, submitted; Örgren & Berge, 1984; Ögren & Holm, 1980, Sugrue 1979).
In mice, PCA induced lesions failed to alter the response levels in the tail-flick, formalin and conventional hot plate tests (Hunskaar et al., 1986c). Similarly, lesions of the raphe nuclei of origin of the ascending projections have largely failed to alter the responsiveness to noxious stimuli in several tests (Le Bars, 1988).

However, lesions in the raphe nuclei have been reported to elevate the thresholds for equal caused by noxious electrical stimulation of the trigeminal nerve (York & Maynert, 1978). This interesting finding has, to the best of our knowledge, not been followed up and its significance remains uncertain. Taken together, the bulk of the evidence is clearly against a tonic, antinociceptive function of the ascending serotonergic projections.

Lesions of descending pathways
Electrolytic lesioning of, or injection of a local anesthetic into the region of the medulla which contains the cell bodies giving rise to descending serotonergic pathways induced reductions in tail-flick latencies (Proudfit & Anderson, 1975; Proudfit, 1980). Selective neurotoxic lesions of these pathways have produced reductions in response latencies in the conventional hot plate test as well as in the tail flick test. The effects are temporary and tend to disappear within one

week in mice (Fasmer et al., 1983b). Recent studies have, however, shown that in both rats and mice, the reduction in tail-flick latencies after neurotoxic lesions is fully accounted for by a concomitant increase in tail temperature (Eide et al., unpublished; Tjolsen et al., 1988). On this background, experiments reevaluating the effect of lesions in the hot plate test are clearly needed. It should be pointed out that in this test, a difference between response categories with regard to the time course of the effect of lesioning has been reported, indicating that more than one mechanism may be involved (Fasmer et al., 1983b). However, at this stage it seems possible that the tonic antinoceptive functions of the descending serotonergic systems in the tail-flick and conventional hot plate test are largely secondary to changes in skin temperature.

In other tests, the effect of selective lesioning of the descending serotonergic pathways differ. In rats, lesions tend to enhance the response to noxious electrical shock in the flinch-jump test (Ögren, Berge & Johansson, 1985) but do not alter the response temperature in the rising temperature hot plate test (Björkum & Berge, unpublished). In the formalin test in mice, lesions were found to reduce the response in the early phase whereas the response in the late phase was unaffected (Fasmer, Berge & Hole, 1985). In rats, 5,6-DHT induced lesions reduced the responses during both phases of the test (Tjolsen et al., unpublished).

Three different mechanisms contributing to the results obtained with the formalin test have been suggested (Fasmer et al., 1985). 1) Exposure to stress may have antinociceptive effects, sustained or inhibited by serotonergic pathways depending on properties of the stressor (Bhattacharya, Keshary & Sanyal, 1978; Hutson, Tricklebank & Curzon, 1982; Tricklebank, Hutson & Curzon, 1982; Tricklebank, Hutson & Curzon, 1984). Thus it is possible that the stress involved in the formalin test may be more effective in reducing the response in lesioned animals. 2) Noxious stimuli of sufficient intensity may trigger the system of "diffuse noxious inhibitory controls" (DNIC), which may be important for the ability of the central nervous system to detect nociceptive messages (Le Bars, Dickenson & Besson, 1979a; Le Bars, Dickenson & Besson, 1982). DNIC is dependent on serotonergic projections and lesions may interfere with the signaling of pain in the formalin test. 3) Iontophoresis of serotonin has been found to facilitate the excitatory response of nociceptive and non-nociceptive dorsal horn neurons to non-noxious stimulation (Belcher, Ryall & Schaffner, 1978). Thus, non-nodiceptive sensory input, which may constitute part of the stimulus in the formalin test, may be reduced after lesions.

Effects of receptor agonists, precursors and releasing agents
Intraperitoneal administration of the serotonin precursor 5-hydroxytryptophan (5-HTP) leads to accumulation of serotonin both in the brain and in peripheral tissues. Injection of a high dose, which also induces generalized motor effects, increases the tail-flick latencies of intact and spinal rats (Berge, 1982; Berge & Hole, 1981; Post et al., 1986). Reduced responsiveness after 5-HTP has also been found in the conventional hot plate test and in a shock titration test (Post et al., 1986). In doses that did not affect the responsiveness of control rats, 5-HTP treatment counteracted the reduction in jump thresholds of rats with non-specific lesions of ascending serotonergic pathways (Harvey & Simansky, 1981). The interpretation was that the precursor, by replenishing serotonin in the depleted terminal areas, restored the tonic antinociceptive effect of the

ascending serotonergic pathways. Considering the evidence against such tonic activity, a more likely explanation may be that 5-HTP activates a supraspinal serotonergic antinociceptive mechanism and that the lesioned animals exhibit effects at lower doses than to controls, because of supersensitivity of denervated receptors.

Release of endogenous serotonin may be effected by systemic administration of PCA in lower doses than required for neurotoxic effects. The releasing effect is more pronounced in the brain than in the spinal cor. In rats, PCA has consistent antinociceptive effects in the conventional and rising temperature hot plate tests (Archer et al., 1986b; Berge & Ögren, 1984; Bjorkum & Berge, unpublished; Ögren & Berge, 1984, Ögren & Berge, 1985; Ögren & Holm, 1980; Ögren et al., 1985; Post et al., 1986). Antinociceptive effects have also been found in the flinch jump test (Ögren & Berge, 1984; Ögren et al., 1985) and elevated threshold to detection of electrical shock in a shock titration paradigm has been reported (Post et al., 1986). Results obtained with the tail-flick test are variable in that no effect (Ögren & Holm, 1980), biphasic effects (transient reduction in latencies followed by increased latencies; Ögren & Berge, 1984). as well as antinociceptive effects (Archer et al., 1986b; Post et al., 1986) have been reported. In mice, PCA had antinociceptive effects in the conventional hot plate and the formalin tests, but not in the tail-flick test (Hunskaar et al., 1986a). Fenfluramine, another serotonin releasing agent, had antinociceptive effects in the conventional hot plate test but reduced the latencies in the tail-flick test (Rochat et al., 1982).

Several studies have demonstrated that the effects of PCA depends on the function of serotonergic pathways. Pretreatment with several selective inhibitors of serotonin uptake prevents PCA induced antinociceptive in the conventional hot plate test whereas inhibitors of noradrenergic uptake are ineffective (Ögren & Berge, 1984; Ögren & Berge, 1985; Ögren & Holm, 1980). Depletion of serotonin throughout the central nervous system by administration of PCPA likewise attenuates the antinociceptive effect of PCA (Ögren & Berge, 1984). Pretreatment with neurotoxic doses of PCA which preferentially lesions supraspinal serotonergic structures (Ögren & Berge, 1984; Hunskaar et al., 1986a) attenuates the antinociceptive effect in rats tested with the conventional (Ögren & Holm, 1980; Ögren & Berge, 1984) and rising temperature hot plate tests (Bjorkum & Berge, unpublished). In mice, PCA induced lesions limited to the ascending pathways completely prevents the antinociceptive effect of subsequent administrations of the drug in both the conventional hot plate and the formalin tests (Hunskaar et al., 1986a).

It seems likely that the antinociceptive effect of PCA to some extent also involves the descending serotonergic pathways. In rats tested with the conventional hot plate method, selective lesions of these pathways by intrathecal injections of 5,6-DHT partly prevented the effect of PCA (Berge and Ögren, 1984; Ögren et al., 1985). The results have been confirmed and extended by experiments conducted with the rising temperature hot plate test which showed that the duration, rather than the peak effect of PCA was reduced after lesioning of the descending serotonergic pathways (Bjorkum & Berge, unpublished).

In spite of the solid evidence in favor of endogenous serotonin as the mediator of

PCA induced antinociception, most classical serotonergic antagonists have failed to block the effect (Ögren & Berge, 1984; 1985). In rats tested with the conventional hot plate test, complete and partial reversal has, however, been demonstrated after administration of metitepin and danitracene respectively (Ögren & Berge, 1985). In the same study, a series of antagonists of catecholaminergic, muscarinic, histaminergic and GABAergic receptors were without effect on the PCA induced antinociception. The available data do not allow conclusions to be drawn as to the receptor type(s) involved in the antinociceptive effect of PCA. The introduction of antagonists selective for subtypes of serotonergic receptors may offer some hope for the future elucidation of this problem.

Intrathecal injection of PCA caused a dose-related increase in tail flick latencies in mice (Berge et al., 1985b). The effect was prevented by spinalization or pretreatment with the synthesis inhibitor PCPA or the uptake inhibitor zimelidine, indicating a specific involvement of serotonergic structures. However, PCA failed to affect the response temperature in the rising temperature hot plate test. Similar results were obtained in rats after injection of either PCA or serotonin (Berge, Fasmer & Hole, 1985a and unpublished observations). Since tail skin temperatures were not recorded in these experiments, the possibility that the effects observed in the tail-flick test are due to serotonin induced changes in tail blood flow should be kept in mind.

Intracerebroventricular administration of PCA in mice caused significant elevation of the response temperature in the rising temperature hot plate test (Berge et al., 1985b). More extensive investigations in rats demonstrated dose related antinociceptive effects of intraventricular administration of PCA and relatively low doses of serotonin (0.3-10 µg) in the rising temperature hot plate test while the same treatment caused significant reductions in tail-flick latencies (Berge et al., 1985a and unpublished). Previously, it has been demonstrated that i.c.v. injection of low doses of the serotonin agonist 5-methoxy-N,N-dimethyltryptamine (5-MeODMT) may enhance the response of rats in the tail-flick test (Berge, Hole & Dahle, 1980).

Systematic administration of various serotonergic receptor agonists tend to produce antinociception (Table 27.1) but again, there is some variability. After administration of the 5-HT$_{1A}$ agonist 8-OH-DPAT, both increased latencies and no effects have been reported in the tail-flick test. Interestingly, in the hot plate test, the drug reduced jump latencies and increased hind paw lick latencies in the same animals (Fasmer et al., 1986). On basis of the available data, it seems premature to draw any conclusion as to the differential involvement of serotonergic receptor subtypes in these results. When applied intrathecally in the lumbar region, serotonin causes a dose-dependent inhibition of nociceptive responses in several tests and species (Yaksh & Wilson, 1979). The potency of serotonin intrathecally is low, since the doses required for antinociception are in the range 100 - 200 µg (Yaksh & Wilson, 1979). In comparison, equipotent doses of noradrenaline would be 1.5 - 2 µg (Archer et al., 1986b). In addition, rats administered antinociceptive doses of serotonin demonstrate the serotonin behavioural syndrome, with hind-limb abduction, tremor and fore-paw treading. It is therefore possible that the "antinociceptive effect" may be an artifact due to motor impairment, and not to an inhibition of the nociceptive signalling per se. It is well known clinically that drugs administered intrathecally may distribute rostrally, sometimes to an extent that severe supraspinal side-effects may occur. For morphine, for instance, a small intrathecal dose may cause severe respiratory depression, that sometimes may occur several hours after administration, probably because the drug diffuses to supraspinal sites. Regarding serotonin, this may also occur. A more lipophilic substance would behave differently. In recent experiments with synthetic serotonin agonists, we have found that both 5-MeODMT and 8-OH-DPAT caused no or little analgesic effects when administered intrathecally, as opposed to systemically or supraspinally (Fasmer et et., 1986; Minor et al., unpublished). Zemlan and collaborators have characterized spinal cord serotonin receptors to belong to the $5-HT_{1B}$ subtype (Harris et al., 1986). Lack of antinociception after intrathecal administration of the $5-HT_{1A}$ agonist 8-OHDPAT and the mixed agonist 5-MeODMT could then be explained by the lack of receptors for these ligands, but alternatively, the drugs may be taken up from the spinal cerebrospinal fluid too rapidly to be transported to supraspinal sites, where the analgesic action may be exerted.

Some electrophysiological evidence indicates that the serotonergic activity in the spinal cord may produce nociception. The general assumption that serotonin has depressant activity on nociceptive dorsal horn neurones has been supported by several studies where iontophoretic application of serotonin inhibits the responses of dorsal horn neurons to nociceptive stimulation (Belcher et al., 1978).

On the other hand, serotonin has also been demonstrated to exert excitatory effects on these neurones (Belcher et al., 1978). In behavioural studies, intrathecal administration of serotonin elicits responses indicative of nociceptive stimulation (Fasmer, Berge & Hole, 1983a; Hylden & Wilcox, 1983). The behavioural nociceptive reactions may not have been mediated directly by serotonin, however, since studies with substance P antagonists in mice (Fasmer & Post, 1983) indicate that serotonin exert its effect by releasing substance P.

The possibility that the various tests are affected by cardiovascular responses to stimulation of central serotonergic receptors precludes a definite interpretation of these findings. However, taken together, the available information strongly suggests a supraspinal target where serotonin may induce antinociceptive effects, consistently detectable in the hot plate tests. It is at present not possible to localize the anatomical structure(s) responsible for the serotonin related antinociception, but several sites may be possible. Microinjection of serotonin into the nucleus raphe magnus of the medulla increases the tail-flick latencies of rats (Llewelyn, Azami & Roberts, 1983). The output from the nucleus is not purely serotonergic and whether the serotonergic projections from the nucleus are important with regard to antinociception may be questioned (for references and discussion, see Berge, 1986).

Some data suggest that serotonergic pathways contribute to the analgesia elicited by manipulations of the PAG. Systemic administration of serotonergic antagonists or PCPA have been reported to reduce the antinociception induced by stimulation of some PAG sites (Akil & Mayer, 1972; Carstens, Fraunhoffer & Zimmermann, 1981). These results leave open the possibility that either raphe-spinal or supraspinal serotonergic projections play a role. Selective lesions of descending serotonergic pathways failed to affect analgesia induced by PAG stimulation or morphine microinjection in one study (Johannessen et al., 1982).

Noradrenergic involvement in serotonergic antinociception
The results from a recent series of experiments implicate a functional interaction between serotonin and noradrenaline in antinociception. Antinociceptive effects after systemic administration of the serotonin agonists 5-MeODMT and quipazine (Post et al., 1986), or intrathecally injected serotonin (Archer et al., 1986b) were studied in control animals, and in rats with selective depletions of spinal noradrenalin by 6-OHDA, or spinal as weel as supraspinal noradrenaline by DSP4. In addition, serotonin-releasing agents were used (Post et al., 1986). In general, the antinociceptive effects of all serotonergic agonists and serotonin-releasing drugs were attenuated or completely abolished by the noradrenaline depletion. The interaction was found to be due to a CNS effect, since animals where the depletions had been done in the periphery only, showed normal antinociceptive effects of 5-MeODMT (Minor et al., 1986).

The interaction appeared to take place both within the spinal cord and supraspinally. Animals administered 5-MeODMT systemically and their antinociception blocked by intrathecally injected α_2-antagonist (Archer et al., 1986a). Based on the relative effects of different types of antagonists, it was found that the receptor involved is the α_2-adrenoceptor. When injected into the locus coeruleus, 5-MeODMT also gave rise to antinociceptive effects, that again were blocked by both locus coeruleus and spinal noradrenergic lesions (Archer et al., 1986b). The most likely explanation is that serotonin and serotonin-agonists exert their analgesic effects by activating the locus coeruleus, which in turn activate descending fibers to the spinal cord, and that the analgesia ultimately therefore, is mediated by spinally released NA. This hypothesis is further supported by results obtained by Proudfit and coworkers, where phentolamine microinjected onto the nucleus raphe magnus causes hypoalgesia (Sagen, Winker & Proudfit, 1983) and release of not only serotonin but also noradrenaline into spinal cord superfusates (Sagen & Proudfit, 1987). In recent experiments, it was found that 5-MeODMT caused an activation of noradrenergic locus coeruleus cells (Minor et al., unpublished).

Serotonergic involvement in opiate analgesia
A variety of methods have been employed to study the interaction between opiate analgesia and serotonergic neurotransmission. Inhibition of serotonin synthesis by administration of PCPA has more often failed than succeeded in reducing the antinociceptive effect of morphine in the tail-flick and hot plate tests in rats while the reserve seems to the case in mice. Mixed effects have also been reported from experiments employing electrical stimuli and paw compression in rats (see Berge et al., 1983c). Several factors may contribute to the variable results obtained with PCPA. Tilson & Rech (1974) using a shock vocalization procedure found that PCPA attenuated the antinociceptive effect or morphine in one, but not in another strain of rats. Other studies have shown that in the rat hot plate test, the attenuating effect of PCPA may be altered by prior shock exposure of the subjects (Berge et al., 1983c).

Lesions affecting ascending and descending pathways either selectively or in combination have also yielded highly variable results as to the role of serotonin

Table 27.1 Effects of serotonin receptor agonists given systematically in different pain tests.

Receptor	agonist (mg/kg)	effect *	species	test (references)
$5-HT_{1,2}$	5-MeODMT (1-2)	+	rat	tail-flick (2-6, hot plate and shock titration (2-4)
	" (1-2)	+	mouse	hot plate and tail-flick (4,7)
$5-HT_{1A}$	8-OHDPAT (1)	+	rat	tail-flick, hot plate and shock titration (4)
	" (0.06-1)	+	mouse	hot plate hind paw lick (4,7,8)
	" (0.25-1)	-	mouse	hot plate jump (8)
	" (0.06)	-	"	formalin (8)
	" (1)	+	"	formalin (8)
	" (0.06-1)	0	"	tail-flick (7,8)
	" (0.25-1)	+	rat	tail-flick (4)
$5-HT_{1B}$	TFMPP (1-3)	+	rat	tail-flick, hot plate (10)
	Flupraz. (1-3)	+	rat	tail-flick, hot plate (10)
$5-HT_2$	DOB (2-4)**	(+)/-	rat	tail-flick, hot plate (1)
non-selective	quipazine (4)	+	rat	hot plate, tail-flick, shock, titration (9)

Archer et al., unpublished (1); Archer et al., 1985 (2); Archer et al., 1986b (3); Archer et al., 1987 (4); Berge, 1982 (5); Berge et al., 1983 (6); Eide et al., in press (7); Fasmer et al., 1986 (8); Post et al., 1986 (9); van der Heyden et al., unpublished (10).

* Antinociception is indicated by +, hyperanalgesia by - and no effect by 0
** Head twitches (mediated by $5-HT_2$ receptors) occur at 0.01 mg/kg. Thus, the effect on nociception may be related to a different receptor.

(Berge et al., 1983c; Berge, 1986; Le Bars, 1988) and several authors have reported test-dependent differences in this regard (Abbott & Melzack, 1982; Abbott, Melzack & Samuel, 1982; Berge et al., 1983c; Mohrland & Gebhart, 1980). An explanation for some of the many failures of lesions to alter the effect of morphine might be the rapid functional recovery that appears to occur after such lesions (Berge et al., 1983c; Romandini et al., 1986).

On the other hand, it should be stressed that procedures that appear to attenuate morphine induced antinociception frequently alters the responsiveness of the animals also in the absence of the analgesic (see for instance Berge et al., 1983b,

Berge et al., 1983c, Romandini et al., 1986). Underestimation of the functional importance of such shifts in responsiveness may lead to erroneous conclusions as to the role of serotonergic systems.

Another methodological problem deserving future attention is the sensitivity of nociceptive tetst to alterations in skin blood flow, which may be caused by opiates as well as by drugs and lesions acting on the serotonergic systems. In spite of the uncertainty that arises from methodological problems, it seems likely that serotonergic systems contribute to the analgesic effects of opiates. However, the highly variable relationship between serotonergic manipulations and changes in morphine induced antinociception indicates that the serotonergic involvement may be indirect. In this context it should be noted that serotonergic pathways may be important for the interaction between stress and morphine induced antinociception rather than for morphine antinociception per se (Kelly & Franklin, 1984).

Conclusions

Although intensively investigated for more than two decades, the physiology and pharmacology of the serotonergic systems with regard to pain processes and analgesia are still poorly understood. Manipulations that lead to activation of serotonergic structures in the brain or in the spinal cord tend to cause analgesia as measured by simple nociceptive tests, and conversely, procedures that block the activity may induce apparent hyperalgesia or antagonize the effect of narcotic analgesics. There are also some clinical observations compatible with a positive correlation between serotonin and antinociception. However, other results indicate that serotonergic systems are supporting signalling of pain. Furthermore, due to the anatomical, biochemical and functional diversity of the serotonergic systems, their role in pain and nociception can hardly be isolated from different functions which may interfere with behavioural testing, and recent work has demonstrated that the robustness of some of the standard tests must be questioned. Thus, it may be necessary to reevaluate the generally accepted concept of serotonergic neurotransmission as exclusively antinociceptive.

References

Abbott, F.V. & Melzack, R. (1982). Brainstem lesions dissociate neural mechanisms of morphine analgesia in different kinds of pain. Brain Res., 251, 149–155.

Abbott, F.V., Melzack, R. & Samuel, C. (1982). Morphine analgesia in the tail-flick and formalin pain tests is mediated by different neural systems. Exp. Neurol., 751, 644–651.

Akil, H. & Mayer, D.J. (1972). Antagonism of stimulation-produced analgesia by p-CPA, a serotonin synthesis inhibitor, Brain Res., 44, 692–697.

Almay, B.G.L., Von Knorring, L. & Oreland, L. (1987). Platelet MAO in patients with idiopathic pain disorders. J. Neurl. Transm., 69, 243–253.

Archer, T., Arweström, E., Minor, B.G., Persson, M.-L., Post, C., Sundström, E.

& Jonsson, G. (1987). (+)-8-OH-DPAT and 5-MeODMT induced analgesia is antagonised by noradrenaline depletion. Physiol. Behav., 39, 95-102.

Archer, T., Danysz, W., Jonsson, G., Minor, B.G. & Post, C. (1986a). 5-Methoxy-N-N-dimethyltryptamine-induced analgesia is blocked by α-adrenoceptor antagonists in rats. Br. J. Pharmacol., 89, 293-298.

Archer, T., Jonsson, G., Minor, B.G. & Post, C. (1986b). Noradrenergic-serotonergic interactions and nociception in the rat. Eur. J. Pharmacol., 120, 295-308.

Archer, T., Minor, B.G. & Post, C. (1985). Blockade and reversal of 5-methoxy-N.N-dimethyltryptamine-induced analgesia following noradrenaline depletion. Brain Res., 333, 55-61.

Beecher, H. (1957). The measurement of pain. Pharmacol. Rev., 9, 59-209.

Belcher, G., Ryall, R.W. & Schaffner, R. (1978). The differential effects of 5-hydroxytryptamine, noradrenaline and raphe stimulation on nociceptive and non-nociceptive dorsal horn interneurones in the cat. Brain Res., 151, 307-321.

Berge, O.-G. (1982). Effects of 5-HT receptor agonists and antagonists on a reflex response to radiant heat in normal and spinally transected rats. Pain, 13, 253-266.

Berge, O.-G. (1986). Regulation of pain sensitivity, influence of prostaglandins. Cephalagia Suppl., 46, 21-31.

Berge, O.-G., Fasmer, O.B., Flatmark, T. & Hole, K. (1983a). Time-course of changes in nociception after 5,6-dihydroxytryptamine lesions of descending 5-HT pathways. Pharmacol. Biochem. Behav., 18, 637-643.

Berge, O.-G., Fasmer, O.B. & Hole, K. (1983a). Serotonin receptor antagonists induce hyperalgesia without preventing morphine antinociception. Pharmacol. Biochem. Behav., 19, 873-878.

Berge, O.-G. & Hole, K. (1985a). Antinociceptive effect of systemic, intrathecal and intracerebro-ventricular administration of PCA, a serotonin releasing compound. Neurosci. Lett. suppl., 22, s 470.

Berge, O.-G., Fasmer, O.B., Jorgensen, H.A. & Hole, K. (1985b). Test-dependent antinociceptive effect of spinal serotonin release induced by intrathecal p-chloroamphetamine in mice. Acta Physiol. Scand., 123, 35-41.

Berge, O.-G., Furset, K. & Garcia-Cabrera, I. (1987). The formalin test in rats: an experimental pain model aplicable to hyperbaric conditions. EUBS Proceedings, 13, 29-31.

Berge, O.-G., Garcia-Cabrera, I. & Hole, K. (1988). Response latencies in the tail-flick test depend on tail skin temperature. Neurosci. Lett. In press.

Berge, O.-G. & Hole, K. (1981). Tolerance to the antinociceptive effect of morphine in the spinal rat. Neuropharmacology, 20, 653–657.

Berge, O.-G., Hole, K. & Dahle, H. (1980). Nociception is enhanced after low doses and reduced after high doses of the serotonin receptor agonist 5-methoxy-N,N-dimethyl-tryptamine. Neurosci. Lett., 19, 219–223.

Berge O.-G., Hole, K. & Ögren, S.O. (1983c). Attenuation of morphine-induced analgesia by p-chlorophenylalanine and p-chloroamphetamine: test-dependent effects and evidence for brainstem 5-hydroxytryptamine involvement. Brain Res., 271, 51–64.

Berge O.-G. & Ögren, S.O. (1984). Selective lesions of the bulbospinal serotonergic pathways reduce the analgesia induced by p-chloroamphetamine in the hot-plate test. Neurosci. Lett., 44, 25–29.

Bhattacharya, S.K., Keshary, P.R. & Sanyal, A.K. (1978). Immobilisation stress-induced antinociception in rats: possible role of serotonin and prostaglandins. Eur. J. Pharmacol., 50, 83–85.

Brodin, E., Peterson, L.L., Ögren, S.O. & Bartfai, T. (1984). Chronic treatment with the serotonin uptake inhibitor zimelidine elevates substance P levels in rat spinal cord. Acta Physiol Scand., 122, 209–211.

Carstens, E., Fraunhoffer, M. & Zimmermann, M. (1981). Serotonergic mediation of descending inhibition from midbrain periaqueductal grey, but not reticular formation, of spinal nociceptive transmission in the cat. Pain, 10, 149–167.

Davies, J. & Dray, A. (1980). Vascular effects of substance P change synaptic responsiveness of cat dorsal horn neurons. Life Sci., 26, 1851–1856.

Dickenson, A.H., Rivot, J.P., Chaouch, A., Besson, J.M. & Le Bars, D. (1981). Diffuse noxious inhibitory controls (DNIC) in the rat with or without pCPA pretreatment. Brain Res., 216, 313–321.

Dubuisson, D. & Dennis, S.G. (1977). The formalin test: a quantitative study of the analgesic effects of morphine, meperidine, and brain stem stimulation in rats and cats. Pain, 4, 161–174.

Duggan, A.W., Griersmith, B.T., Headley, P.M. & Maher, J.B. (1978). The need to control skin temperature when using radiant heat in tests of analgesia. Exp. Neurol., 61. 471–478.

Eide, P.K., Berge, O.G. & Hunskaar, S. (1987). Test-dependent changes in nociception after administration of the putative serotonin antagonist metitepin in mice. Neurpharmacol., 26, 1121–1126.

Eide, P.K., Hole, K., Berge, O.G. & Broch O.J. (1988). 5-HT depletion with 5,7-DHT, PCA and PCPA in mice: differential effects on the sensitivity to 5-MeODMT, 8-OH-DPAT and 5-HTP as measured by two nociceptive tests. Brain Res. (In press).

Evans, W.O. (1961). A new technique for the investigation of some analgesic drugs on a reflexive behaviour in the rat. Psychopharmacologia, 2, 318–325.

Fanselow, M.S. (1984). Shock-induced analgesia on the formalin test: effects of shock severity, naloxone, hypophysectomy, and associative variables. Behav. Neurosci., 98, 79–95.

Fasmer, O.B., Berge, O.G. & Hole, K. (1983a). Similar behavioural effects of 5-hydroxytryptamine and substance P injected intrathecally in mice. Neuropharmacology, 22, 485–487.

Fasmer, O.B., Berge, O.G. & Hole, K. (1984). Metrgoline elevates or reduces nociceptive thresholds in mice depending on test method and route of administratrion. Psychopharmacology, 82, 306–309.

Fasmer, O.B., Berge, O.G. & Hole, K. (1985). Changes in nociception after lesions of descending serotonergic pathways induced with 5,6-dihydroxytryptamine. Different effects in the formalin and tail-flick tests. Neuropharmacology, 24, 729–734.

Fasmer, O.B., Berge, O.G., Post, C. & Hole, K. (1986). Effects of the putative 5-HT$_{1A}$ receptor agonist 8-OH-2-(di-n-propylamino) tetralin on nociceptive sensitivity in mice. Pharmacol. Biochem. Behav., 25, 883–888.

Fasmer, O.B., Berge, O.G., Walther, B. & Hole, K. (1983b). Changes in nociception after intrathecal 5,6-dihydroxytryptamine in mice. Neuropharmacology, 22, 1197–1201.

Fasmer, O.B. & Post, C. (1983). Behavioural responses induced by intrathecal injection of 5-hydroxytryptamine in mice are inhibited by a substance P antagonist, D-Pro2, D-Tr7,9-substance P. Neuropharmacology, 22, 1397–1400.

Harris, G.D., Zemlan, F.P., Murphy, R.M. & Behbehani, M.M. (1986). Spinal cord 5-HT 1A and 1B receptor subtypes: relation to pain transmission. Soc. Neurosci. Abst., 12, 1015.

Harvey, J.A. & Simansky, K.J. (1981). The role of serotonin in modulation of nociceptive reflexes. Adv. Exp. Med. Biol., 133, 125–151.

Hökfelt, T., Holets, V.R., Staines, W., Meister, B., Melander, T., Schalling, M., Schutzberg, M., Freedman, J., Björklund, H., Olson, L., Lindh, B., Elfvin, L.G., Lundberg J., Lindgren, J.A., Samuelsson, B., Pernow, B., Terenius, L., Post, C., Everitt, B. & Goldstein, M. (1986). Coexistence of neuronal messengers – an overview. Progr. Brain Res., 68, 33–70.

Hole, K. & Berge, O.G. (1981). Regulation of pain sensitivity in the central nervous system. Cephalagia, 1, 51–59.

Hunskaar, S., Berge, O.G., Broch, O.J. & Hole, K. (1986a). Lesions of ascending serotonergic pathways and antinociceptice effects after systemic administration of p-chloroamphetamine in mice. Pharmacol. Biochem. Behav., 24, 709–714.

Hunskaar, S., Berge, O.G. & Hole, K. (1986b). A modified hot-plate test sensitive to mild analgesics. Behav. Brain., 21, 101-108.

Hunskaar, S., Berge, O.G. & Hole, K. (1986c). Dissociation between antinociceptive and anti-inflammatory effects of acetylsalicylic acid and indometacin in the formalin test. Pain, 25, 125-132.

Hunskaar, S., Fasmer, O.B. & Hole, K. (1985). Formalin test in mice, a useful technique for evaluating mild analgesics. J.Neurosci. Meth., 14, 69-76.

Hutson, P.H., Tricklebank, M.D. & Curzon, G. (1982). Enhancement of footshock-induced analgesia by spinal 5,7-dihydroxytryptamine lesions. Brain Res., 237, 367-372.

Hylden, J.L.K. & Wilcox, G.L. (1983). Intrathecal serotonin in mice: analgesia and inhibition of a spinal action of substance P. Life Sci., 33, 789-795.

Johannessen, J.N., Watkins, L.R., Carlton, S.M. & Mayer, D.J. (1982). Failure of spinal cord serotonin depletion to alter analgesia elicited from the periaqueductal gray. Brain Res., 237, 373-386.

Johansson, F. & Von Knorring, L. (1979). A double-blind controlled study of a serotonin uptake inhibitor (zimelidine) versus placebo in chronic pain patients. Pain, 7, 69-78.

Kelly, S.J. & Franklin, K.B.J. (1984). Electrolytic raphe magnus lesions block analgesia induced by a stress-morphine interaction but not analgesia induced by morphine alone. Neurosci. Lett., 52, 147-152.

Le Bars, D. (1988). Serotonin and pain: In: Neuronal serotonin. (Eds. (Osborne, N.N. & Hamon, M.), New York: Wiley.

Le Bars, D., Dickenson, A.H. & Besson, J.M. (1979). Diffuse noxious inhibitory controls (DNIC). 1. Effects on dorsal horn convergent neurones in the rat. Pain, 6, 283-304.

Le Bars, D., Dickenson, A.H. & Besson, J.M. (1982). The triggering of bulbo-spinal serotonergic inhibitory controls by noxious inputs: In: Brain Stem Control Spinal Mechanisms. (Eds. Sjölund, B. & Björklund, A.), pp. 381-410. Elsevier: Amsterdam.

Le Bars, D., Rivot, J.P., Guilbaud, G., Menetrey, D. & Besson J.M. (1979). The depressive effect of morphine on the C fibre response of dorsal horn neurones in the spinal rat pretreated or not by pCPA. Brain Res., 176, 337-353.

Llewelyn, M.B., Azami, K. & Roberts, M.H.T. (1983). Effects of 5-hydroxytryptamine applied into nucleus raphe magnus on nociceptive thresholds and neuronal firing rate. Brain Res., 258, 59-68.

Lopachin, R.M. & Rudy, T.A. (1982). The thermoregulatory effects of noradrenaline, serotonin and carbachol injected into the rat spinal subarachnoid space. J. Physiol., **333**, 511–529.

Merskey, H., Albe–Fessard, D.G., Bonica., J.J., Carmon, A., Dubner, R., Kerr, F.W.L., Lindblom, U., Mumford, J.M., Nathan, P.W., Noordenbos, W., Pagni, C.A., Renaer, M.J., Sternbach, R.A. & Sunderland, S. (1979). Pain terms: a list with definitions and notes on usage. Pain, **6**, 249–252.

Minfeng, R. & Jisheng, H. (1979). Rat tail flick acupunture analgesia model. Chin. Med., **92**, 576–582.

Minor, B.G., Archer, T., Post, C., Jonsson, G., Mohammed, A.K. (1986). 5–HT agonist induced analgesia modulated by central but not peripheral noradrenaline depletion in rats. J. Neural. Transm., **66**, 243–260.

Mohrland, J.S. & Gebhart, G.F. (1980). Effects of selective destruction of serotonergic neurons in nucleus raphe magnus on morphine–induced antinociception. Life Sci. **27**, 2627–2632.

Ness, T.J. & Gebhart, G.F. (1986). Centrifugal modulation of the rat tail flick reflex evoked by graded noxious heating of the tail. Brain Res., **386**, 41–52.

Ögren, S.O. & Berge, O.G. (1984). Test–dependent variations in the antinociceptive effect of p-chloroamphetamine–induced release of 5–hydroxytryptamine. Neuropharmacology, **23**, 915–924.

Ögren, S.O. & Berge, O.G. (1985). Evidence for selective serotonergic receptor involvement in p-chloroamphetamine–induced antinociception. Naunyn-Schmiedeb. Arch. Pharmacol., **329**, 135–140.

Ögren, S.O., Berge, O.G. & Johansson, C. (1985). Involvement of spinal serotonergic pathways in nociception but not in avoidance learning. Psychopharmacology, **87**, 260–265.

Ögren, S.O. & Holm, A.C. (1980). Test–specific effects of the 5–HT reuptake inhibitors alaproclate and zimelidine on pain sensitivity and morphine analgesia, J. Neural. Transm., **47**, 253–271.

Post, C., Minor, B.G., Davies, M. & Archer, T. (1986). Analgesia induced by 5–hydroxytryptamine receptor agonists is blocked or reversed by noradrenaline–depletion in rats. Brain Res., **363**, 18–27.

Proudfit, H.K. (1980). Reversible inactivation of raphe magnus neurons: effects on nociceptive threshold and morphine–induced analgesia. Brain Res., **201**, 459–464.

Proudfit, H.K. & Anderson, E.G. (1975). Morphine analgesia: blockade by raphe magnus lesions. Brain Res., **98**, 612–618.

Rochat, C., Cervo, L., Romandini, S. & Samanin, R. (1982). Differences in the effect of d–fenfluramine and morphine on various responses of rats to painful stimuli. Psychopharmacology, **76**, 188–192.

Romandini, S., Pich, E.M., Esposito, E., Kruszewska, A.Z. & Samanin, R. (1986). The effect of intracerebroventricular 5,7-dihydroxytryptamine on morphine analgesia is time dependent. Life Sci., 38, 869-875.

Rosland, J.H., Hunskaar, S. & Hole, K. (1987). The effect of diazepam on nociception in mice. Pharmacol. Toxicol., 61, 111-115.

Rudy, T.A. & Yaksh, T.L. (1977). Hyperthermic effects of morphine: set point manipulation by a direct spinal action. Brit. J. Pharmacol., 61, 91-96.

Sagen, J. & Proudfit, H.K. (1987). Release of endogenous monoamines into spinal cord superfusates following the microinjection of phentolamine into the nucleus raphe magnus. Brain Res., 406, 246-254.

Sagen, J., Winker, M.A. & Proudfit, H.K. (1983). Hypoalgesia induced by the local injection of phentolamine in the nucleus raphe magnus: blockade by depletion of spinal cord monoamines. Pain, 16, 253-263.

Shyu, K.W., Lin, M.T. & Wu, T.C. (1984). Possible role of central serotonergic neurons in the development of dental pain and aspirin-idncued analgesia in the monkey. Exp. Neurol., 84, 179-187.

Steinbusch, H.W.M. & Nieuwenhuys, R. (1983). The raphe nuclei of the rat brainstem: A cytoarchitectonic and immunohistochemical study. Chem. Neuroanat., 131-207.

Sugrue, M.F. (1979). Effect of depletion of rat brain 5-hydroxyrtyptamine on morphine-induced antinociception. J. Pharm. Pharmacol., 31, 253-255.

Tilson, H.A. & Rech, R.H. (1974). The effects of p-chlorophenylalanine in two strains of rats.. Psychopharmacologia, 35, 45-60.

Tjolsen, A., Berge, O.G., Eide, P.K., Broch, O.J. & Hole, K. (1988). Apparent hyperalgesia after lesions of the descending serotonergic pathways is due to increased tail skin temperature. Pain. (In press).

Tricklebank, M.D., Hutson, P.H. & Curzon, G. (1982). Analgesia induced by brief foot shock is inhibited by 5-hydroxytryptamine but unaffected by antagonists of 5-hydroxytryptamine or by naloxone. Neuropharmacology, 21, 51-56.

Tricklebank, M.D., Hutson, P.H. & Curzon, G. (1984). Analgesia induced by brief or more prolonged stress differs in its dependency on naloxone, 5-hydroxytryptamine and previous testing of anaglesia. Neuropharmacology, 23, 417-421.

Von Knorring, L., Perris, C., Eisemann, M., Eriksson, U. & Perris, H. (1983). Pain as a symptom in depressive disorders. I. Relationship to diagnostic subgroup and depressive symptomatology. Pain, 15, 19-26.

Wiklund, L., Leger, L. & Persson, M. (1980). Monoamine cell distribution in the cat brain stem. A fluorescence histochemical study with quantification of indolaminergic and locus coeruleus cell groups. J. Comp. Neurol., 203, 613-647.

Woolfe, G. & MacDonals, A.D. (1944). The evaluation of the analgesic action of pethidine hydrochloride (demerol). J. Pharmacol. Exp. Ther., 80, 300-307.

Yaksh, T.L. (1981). Spinal opiate analgesia: characteristics and principles of action. Pain, 11, 293-346.

Yaksh, T.L. & Wilson, P.R. (1979). Spinal serotonin terminal system mediates antinociception. J. Pharmacol. Exp. Ther., 208, 446-453.

York, J.L. & Maynert, E.W. (1978). Alterations in morphine analgesia produced by chronic deficits of brain catecholamines or serotonin: role of analgesimetric procedure. Psychopharmacol., 56 119-125.

THE RELATIONSHIP BETWEEN VARIOUS ANIMAL MODELS OF ANXIETY, FEAR-RELATED PSYCHIATRIC SYMPTOMS AND RESPONSE TO SEROTONERGIC DRUGS

C.L.Broekkamp, F.Jenck.
Department of CNS Pharmacology, Organon, Oss, P.O.Box 20, 5340 BH, The Netherlands.

There are several ways to discuss the interrelation between animal models, anxiety disorders and serotonergic drugs. We choose here to discuss first the effects of the serotonin-related drugs in putative animal models for anxiety, then to discuss the effect of such drugs in anxiety disorders and then to see whether this information provides evidence to link a particular model to a particular anxiety disorder. The subject of serotonin related drugs and animal models for anxiety has been reviewed recently by different authors (Gardner, 1986, 1987; Johnston and File, 1986; Chopin and Briley, 1987). We will discuss the effect of serotonin-related drugs in five animal models for anxiety: the tests based on conflict behaviour, the social interaction test, the elevated plus-maze, aversive brain stimulation and defensive burying.

Conflict tests

Under the heading of conflict tests we grouped the data obtained with tests based on appraoch-avoidance conflicts, wherein the aversive stimulus is electric shock and the approach is induced by water or food reward for deprived animals or the spontaneous tendency to explore the environment.

The data allow two opposing views on the effect of serotonin related compounds in such procedures. One group of investigators concludes that serotonin antagonists behave as anxiolytics in these models and one group concludes that serotonin antagonists are inactive in conflict tests. Anti-conflict effects are reported for cyproheptadine (Sepinwall and Cook, 1980; Graeff, 1974; Brady and Barrett, 1985; Meert and Colpaert, 1986a, Deacon and Gardner, 1986), metergoline (Leone et al., 1983; Brady and Barrett, 1985; Gardner, 1985), methysergide (Stein et al., 1973; Brady and Barrett, 1985; Graeff, 1974; Winter, 1972; Meert and Colpaert, 1986a), mianserin (Van Riezen et al., 1981; Mason et al., 1987; Brady and Barrett, 1985), pizotifen (Mert and Colpaert, 1986a), cinanserin (Cook and Sepinwall, 1975; Geller et al., 1974; Kilts et al., 1982, Brady and Barrett, 1982; Winter et al., 1972), pirenpirone (Colpaert et al., 1985) and ritanserin (Colpaert et al., 1985; Meert and Colpaert, 1986a).

On the other hand an equally impressive amount of information points to negative effects of anti-serotonin compounds in conflict tests. Non-significant effects were reported for cyproheptadine (Gardner, 1985; Kilts et al., 1982; Petersen and Buus Lassen, 1981), metergoline (Kilts et al., 1982; Meert and Colpaert, 1986a; Gardner, 1985; Deacon and Gardner, 1986), methysergide (Kilts et al., 1982; Petersen and Buus Lassen, 1981) mianserin (Meert and Colpaert, 1986a), pizotifen (Gardner, 1985), cinanserin (Petersen and Buus Lassen, 1981), pirenperone (Brady and Barrett, 1985) and ritanserin (Meert and Colpaert, 1986b; Deacon and Gardner, 1986). However, the data are not as conflicting as would appear from the above summary. It is observed that, when positive findings are found, these effects are smaller and less robust than the effects of benzodiazepines (Meert and Colpaert, 1986a).

The best interpretation of the data appears to be that there is a disinhibitory effect of serotonin antagonists in conflict tests but that the effect is small and requires hitherto undescribed special conditions in order to become clearly visible. The factor of the rat strain which is used, should not be underestimated since differences in the responsiveness of the serotonergic system among rat strains is well demonstrated (Gudelsky et al., 1985). Additional complications for drawing conclusions from data with antagonists arise from the different profile of activity on various receptors for these drugs. Not only they often affect different serotonin receptors but also sometimes dopaminergic, adrenergic, histaminergic or cholinergic receptors. Whatever mechanisms are involved, we conclude that for those disorders for which conflict tests are suitable models, it is predicted that serotonin antagonists will be drugs with lesser efficacy than benzodizapines.

A similar situation emerges from the data obtained with the $5-HT_{1A}$-related compounds 8-OH-DPAT, buspirone and ipsapirone. 8-OH-DPAT was found active by Engel et al. (1984) and Meert and Colpaert (1986b), but inactive by Deacon and Gardner (1986). Buspirone was found active by Mason et al. (1987), Geller and Hartmann (1982), Barrett et al. (1986), McColskey et al. (1986) and Merlo Pich & Samanin (1986), but inactive by Meert and Colpaert (1986b), Gardner (1986), Goldberg (1983) and Budhram et al. (1986). Ipsapirone was found active by Schuurman et al. (1986) and Young et al. (1987), but inactive by Deacon and Gardner (1986) and Meert and Colpaert (1986b). Several investigators, reporting on the efficacy of buspirone in a conflict test, remark on the weak activity in comparison to the effect of benzodiazepines (Merlo Pich and Samanin, 1986; McCloskey et al., 1987; Young et al., 1987). Merlo Pich and Samanin (1986) found that haloperidol and sulpiride had anti-conflict activity in the same paradigm as buspirone. This is also our experience with a water-lick conflict test wherein buspirone is active and ipsapirone, a compound devoid of anti-dopaminergic activity, is inactive (unpublished). It is therefore likely that in several instances positive effects are found with buspirone which should be ascribed to the antidopaminergic property of this compound. From these data we conclude that also for $5-HT_{1A}$-related compounds, tests based on conflict do not predict strong therapeutic effects in those anxiety disorders for which the conflict tests are relevant.

Concerning the serotonin uptake inhibitors the results obtained with the conflict test predict, if anything anxiogenic effects. Petersen and Buus Lassen (1981) found paroxetine to be inactive in their conflict paradigm. In their procedure an enhancement of shock-induced inhibition was not possible. In our procedure we clearly observed an enhancement of the shock-induced suppression of licking by citalopram (Fig. 28.1).

The $5-HT_{1B}$ agonist mCPP also enhanced shock-induced suppression of drinking but unpunished drinking was also suppressed (Kilts et al., 1982). The conflict test does therefore not reveal any utility for serotonin-mimetic compound in anxiety disorders.

Social interaction

The data on the effect of serotonin–related compounds on social interaction are less extensive. Methysergide (File, 1981) and metergoline (File, 1981; Gardner, 1985) are inactive. Buspirone is active according to results of Gardner (1985), but inactive according to results of File (1985). Schuurmans and Spencer (1987) report activity for ipsapirone. Clomipramine is inactive (File, 1985). Although

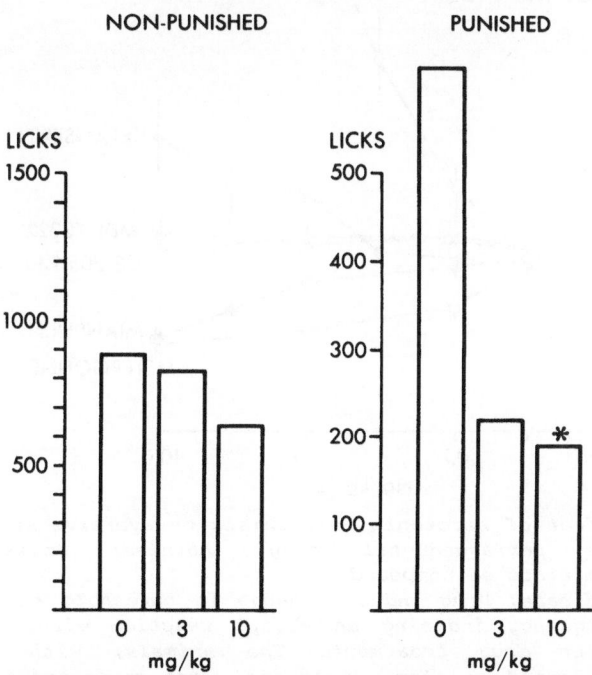

Figure 28.1 *Effect of citalopram s.c. on conflict behaviour. Rats were 24 hr water deprived and trained to lick for water from a metal tube for three days. On the fourth day of the expriment rats were drug treated and every twentieth lick a shock puls of 0.6 mA was delivered via the drinking spout and the grid floor of the cage. The compound was given 45 minutes before the start of a 20 min drinking session. The number of licks during the second ten minutes of the drinking period was counted and compared with the number of licks on the third day for the comparable period. Note the different scales in the figure for the number of licks for punished and unpunished drinking.*
*: $p<0.05$.

complete reports are still to be published on some of these results there seems to emerge from this that serotonin does not play a very important role in this behaviour.

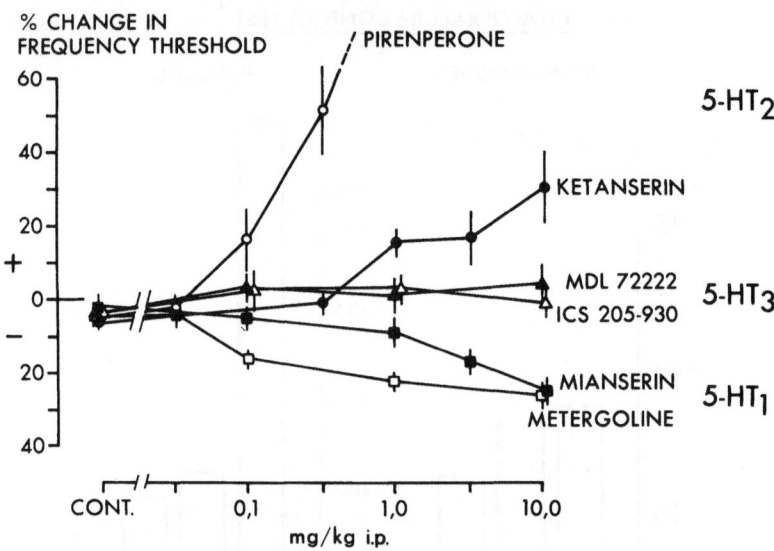

Figure 28.2 *Effect of serotonin antagonists on aversive stimulation in the periaqueductal gray. Absciss: Doses of the administered compounds.*
Ordinate: Drug induced change in threshold of stimulation frequency inducing an escape reaction within 20 seconds after drug treatment. The animals, with permanently implanted bipolar electrodes, were connected to leads for electrical stimulation and moved freely in a rectangular cage. A passage over a hurdle to the other side of the cage was the learned escape response. The frequency of the monophasic square wave pulses was adjusted after each trial (one trial per minute) in order to titrate to the effective frequency for escape. The labels on the right of the figure refer to our interpretation of the data: $5-HT_2$-receptor blockade diminishes the sensitivity of the animals for the aversive stimulation, $5-HT_1$-receptor blockade enhances sensitivity and $5-HT_3$-receptor blockade has no effect. Details are reported in Jenck et al. (In Press).

Elevated plus maze
In the elevated plus maze the results with compounds activating the 5-HT$_{1A}$ receptor, such as 8-OH-DPAT or the partial agonists buspirone and ipsapirone, and the non selective activators, such as 5-MeODMT, RU 24969 and quipazine are consistently pointing against anxiolytic-like effects (Critchley and Handley, 1987; Pellow et al., 1987).

Rather, in this model anxiogenic-like effects are found with these compound. The situation is less clear for the antagonists. Ketanserin and ritanserin were found to have clear anxiolytic-like effects by Critchley and Handley (1987), but Pellow et al. (1987) found an anxiolytic-like effect with metergoline and an anxiogenic-like effect with ritanserin. Strain differences or other non-conspicuous experimental differences are not yet identified enabling to explain the different results with ritanserin. In the related emergence test used by Colpaert et al. (1985) disinhibitory effects are found for pirenperone, ritanserin (Colpaert et al., 1985) and cyproheptadine (Meert, 1986).

The information on drugs in the elevated plus maze indicates that disorders for which this type of model is predictive will aggravate in response to 5-HT$_{1A}$ agonists and might respond well to 5-HT antagonists such as metergoline, ritanserin and the successor of pirenperone, risperidone.

Aversive brain stimulation
Early studies reported that serotonin-mimetic drugs such as clomipramine and 5-HTP (Kiser et al., 1978) were reducing the aversive effect of electrical brain stimulation and the serotonin antagonists cyproheptadine and methysergide were potentiating aversion in only a few individual rats (Schenberg and Graeff, 1978).
We have studied the effect of various serotonin antagonists on aversive periaqueductal gray brain stimulation. In figure 28.2 data are presented showing that the 5-HT$_2$ blockers pirenperone and ketanserin, diminish the efficacy of aversive periaqueductal gray stimulation. The 5-HT$_3$-blockers, MDL 72222 and ICS 205-930F, do not affect aversive brain stimulation. Metergoline and mianserin, two compounds which also block 5-HT$_1$-receptors, facilitate escape from PAG-stimulation. Thus, blockade of the various receptor types differentially contributes to the control of aversive brain stimulation. Apparently, blockade of a particular type of 5-HT$_1$, maybe the 5-HT$_{1C}$ receptor, overcomes the effect of blockade of 5-HT$_2$ receptors on aversive brain stimulation. Although several of the antagonists mentioned here also block other than 5-HT receptors, such as α_2, dopamine or histamine receptors such antagonisms could not explain these effects as could be concluded from testing specific blockers of these receptors (Jenck et al., unpublished).

We found no effect with ipsapirone at doses up to 10 mg/kg. From the antagonist data we conclude that serotonin plays an important role in the control of escape from aversive brain stimulation and the data indicate particular drugs of choice for those disorders for which aversive PAG stimulation will ultimately prove to be a suitable model.

Defensive burying
Our experience is with a particular variant of defensive burying. Instead of conditioning rats and measuring the height of the mound of bedding material

against the shock prod, we count the number of marbles buried by mice which display burying spontaneously when placed in a novel cage with marbles on the bedding material (Broekkamp et al., 1986). As control for the specificity of the inhibition of burying with any drug we measured also the effect of grooming behaviour. The uptake-inhibitors of serotonin were found to block this behaviour selectively (Table 28.1a).

Table 28.1a Antidepressants on burying and grooming in mice.

Compound	ID_{50} burying in mg/kg s.c.	ID_{50} grooming in mg/kg s.c.	Ratio gro/bur
Paroxetine	2	60	30
Sertraline	5	>100	>200
Citalopram	5	>46	10
Org 6997	2	18	9
Fluvoxamine	15	>100	>6.5
Trazodone	4	20	5
mCPP	1.5	10	6.7
Fluoxetine	10	40	4
Indalpine	10	35	3.5
Clomipramine	10	25	2.5
Zimelidine	25	40	1.6
Nortriptyline	8	15	2
Amitriptiline	8	7	1
Imipramine	12	17	1.4
Maprotyline	15	15	1
Mianserin	6	5	1
Ninaprine	20	30	1.5

We did not find selective inhibition with buspirone, 8-OH-DPAT, RU 24969, 5-MeODMT or 5-HT$_3$ antagonists (table 28.1b). Thus, the disease for which burying stands as a model will be characteristically and especially responsive to serotonin-uptake inhibitors.

Anxiety disorders
Now that we have summarized a number of proposed and used models for anxiety and discussed the effect of serotonin-related drugs therein, it is of interest to consider various fear-related psychiatric diseases and then see whether some links can be made between a particular pathological state and a particular animal model.

As pharmacotherapy differs for the various anxiety disorders, attempts to find relations between animal models and anxiety disorders will benefit from comparisons of the sensitivity and responsiveness to drugs. We will summarize here information concerning the pharmacotherapy of the separate anxiety

Table 28.1b Other serotonin-related agents on burying and grooming.

Compound	ID_{50} burying in mg/kg s.c.	ID_{50} grooming in mg/kg s.c.	Ratio gro/bur
GR38032F	>1	>1	
ICS205930	>1	>1	
Buspirone	>2.2	>2.2	
8-OHDPAT	0.8	1	0.8
Ru 24969	1.3	1	1.3
5-MeODMT	7	7	1

disorders as differentially classified according to the DSM III, even though the information on drug responsivity is limited in comparison to the animal data.

This is in part because many drugs used in animal research are not yet available for human application, and in part because clinical trials require a lor more time than animal experiments.

Generalized anxiety disorder
The generalized anxiety disorder (GAD) is characterized by feelings of anxiety throughout the day and difficulties falling asleep. Benzodiazepines provide acute relief (Hollister, 1986). This effect is persistent but often insufficient or concomitant with sedative side effects. Of more recent data is evidence that antidepressants are effective in this disorder (Kahn et et., 1987). In a well controlled large trial it was even found than on long term treatment imipramine is superior over chlordiazepoxide in the treatment of anxiety (Kahn et al., 1986). The serotonin-releated antidepressant drugs mianserin and trazodone have also been shown to be efficacious in anxiety (Conti and Pinder, 1979; Wheatley, 1976) and recently it was found that the selective $5-HT_2$ antagonist, ritanserin, appears to have efficacy for treatment of anxiety. Buspirone has been introduced for the treatment of GAD (Ortiz et al., 1987). This compound has, contrary to the benzodiazepines, an effect which becomes apparent after several weeks of treatment and sedative side effects are insignificant. In view of these characteristics the action of buspirone in anxiety is comparable to the effect of antidepressants in anxiety. In agreement with this view is that buspirone and the related compound gepirone seem to be active as antidepressants (Schweizer et al., 1986; Amsterdam et al., 1987).

Agoraphobia and panic disorders
In agoraphobia the feelings of anxiety are not always present but arise in particular situations, usually outside the house of the patient. In agoraphobia with panic attacks the agoraphobia may sometimes have developed as

anticipatory fear for a panic attack. In panic disorder, which can occur without agoraphobia, panic attacks occur with a particular frequency and only very loosely related to particular situations. Drug treatment is an important aid in successful therapy, whereby behavioural techniques are often used as the main therapy (Pecknold, 1987). There is no evidence that drug treatment should be different for panic disorder and agoraphobia. The antidepressants are the drugs of choice for both disorders (Lydiard and Ballenger, 1987; Rizley et al., 1986). Positive clinical experience is described for phenelzine, imipramine and tradozone, and the serotonin uptake inhibitors clomipramine and fluoxetine (Gorman et al., 1987; Gloger et al., 1981; Lydiard, 1987; Mavissakalian et al., 1987). The beneficial effect if not acute but develops after a few weeks of treatment. In the initial treatment period some worsening of symptoms might even occur in some patients. This phenomenon was reported for clomipramine (Liebowitz et al., 1986), imipramine (Lydiard, 1987) and buspirone (Frazer and Lapierre, 1987).

Worsening of panic attacks of anxiety in man was also reported for yohimbine and mCPP (Charney et al., 1984, 1987). There are reasons to be careful to interpret the findings with yohimbine in terms of a putative role for noradrenaline in anxiety or panic attacks. Although yohimbine is often seen as antagonist of α_2-adrenoceptors it has also strong interactions with serotonin receptors (Feuerstein et al., 1985; Clineschmidt et al., 1985). The observations with mCPP are made with high intravenous doses. Lower and gradually rising doses of this compound as occurs when the precursor of mCPP, tradazone, is given to man, do not seem to have similar adverse effects. There is more information needed to ascribe an exclusive role for serotonin or noradrenaline in the treatment of these disorders.

In considering drug treatment of agoraphobia and panic disorder it is striking to see that antidepressants rather than benzodiazepines are the drugs of choice whereas in animal models for anxiety the antidepressants are mostly inactive, have anxiogenic-like activity or are insufficiently investigated.

Obsessive-compulsive disorders
Obsessive-compulsive disorder (OCD) has been classified as a distinct subtype of anxiety disorders. It has a prevalence of more than 2% of the population (Robins et al., 1984). Clomipramine has for a long time been the most effective medication for OCD patients and its anti-obsessional effects were found independent of its antidepressant activity. Several controlled studies have demonstrated clomipramine's superiority over placebo treatment as well as over treatment with other antidepressants, such as imipramine, desipramine, nortryptiline, amitriptilyne and clorgyline (Zohar and Insel, 1987). The efficacy of serotonin uptake inhibitors in compulsive disorder is further confirmed with fluvoxamine (Price et al., 1987), fluoxetine (Turner et al., 1985) and zimelidine (Kahn et al., 1984). The serotonin precursor L-tryptophan has been reported to reduce obsessional symptoms and to potentiate the effects of clomipramine (Rasmussen, 1984). This information clearly points to an important role for potentiation of serotonergic transmission in the pharmacotherapy of obsessive-compulsive disorder.

Posttraumatic stress disorder
Post-traumatic stress is an anxiety disorder arising from strong psychological trauma. This syndrome is characterized by recurring reliving experiences of stressful life event(s). Again, the antidepressants seem to be the preferred drugs for supplementary pharmacotherapy (Van Der Kolk, 1987).

Social phobia
Social phobia is an anxiety disorder that has recently attracted more interest because it is often complicated by alcohol and drug abuse. β-blocking drugs and MAO inhibitors appear to be useful for social phobics (Gorman and Gorman, 1987). Benzodiazepines are not very effective and are not recommended for this disorder in view of risks for abuse by these patients. Some indication for beneficial effects were obtained for clomipramine (Gungras, 1977). In healthy individuals metergoline potentiated the anxiety induced by public speaking (Graeff et al., 1985). It is too early to conclude about any role for serotonin in the treatment of social phobia.

Conclusion
As expressed earlier it would be useful if we could relate a particular animal model with a particular human anxiety disorder. Similarities in the sensitivities for drugs between an animal model and an anxiety disorder could be helpful for this and supplement other analogies. Although there is both from the human as well as from the animal side insufficient knowledge to make firm statements on such relations, there are nevertheless pieces of information which allow some speculations on this issue.

At the level of face validity it would be attractive to associate panic attacks with aversive brain stimulation, the social interaction test with social phobia and the elevated plus maze with agophobia. However, in the social interaction test the beta-blockers are not as effective as in social phobia, The antidepressants, which are drugs of choice for panic syndrome and agoraphobia, are not yet tested on aversive brain stimulation and in the elevated plus maze.

The behavioural tests which are most sensitive to benzodiazepines, an aspect which was not discussed before in this text, are the conflict test and the social interaction test. The anxiety syndrome wherein the benzodiazepines are clearly benefial is the generalized anxiety syndrome. This suggests that the conflict test and the social interaction test are relevant for the generalized anxiety syndrome. As the serotonin receptor blockers are only weakly active in these models it will be crucial to see whether or not treatments based on serotonin receptor blockade will gain acceptance as pharmacotherapeutic treatment for this disorder. Before reasoning further along these lines it should be mentioned that even if we find comparable drug efficacy in an animal model and an anxiety syndrome it is to be realized that in the clinical situation most treatments are effective after at least a week of treatment, whereas in the animal experiments most drugs are given acutely. Also in the treatment of psychosis and depression a slow recovery is observed with effective drug treatment. Possibly there is a conditioned component in human pathological states which can only gradually be overcome during recovery when a causal factor for the disease is removed. In most animal models the animals are acutely confronted with the test environment so that drugs can more easily reveal their immediate effects than in the human situation.

Bearing this in mind, it is reasonable to search for a relation between the model of defensive burying and obsessive compulsive disorder. Both are selectively sensitive for compounds which block the reuptake of serotonin and thereby give the clearest indication for a strong involvement of serotonin.

References

Amsterdam, J.D., Berwish, N., Potter, L. & Rickels, K. (1987). Open trial of gepirone in the treatment of major depressive dororder. Curr. Ther. Res. Clin. Exptl., **41**, 185-193.

Barrett, J.E., Witkin, J.M., Mansbach, R.S., Skolnick, P. & Weissman, B.A. (1986). Behavioral studies with anxiolytic drugs. III. Antipunishment actions of buspirone in the pigeon do not involve benzodiazepine receptor mechanisms. J. Pharmacol. Exptl. Ther., **238**, 1009-1013.

Brady, L.S. & Barrett, J.E. (1985). Effects of serotonin receptor antagonists on punished responding maintained by stimulus-shock termination or food presentation in squirrel monkeys. J. Pharmacol. Exp. Ther., **234**, 106-112.

Broekkamp, C.L., Rijk, H.W., Joly-Gelouin, D. & Lloyd, K.L. (1986). Major tranquillizers can be distinghuished from minor tranquillizers on the basis of effects on marble butying and swin-induced grooming in mice. Eurpean J. Pharmacol., **126**, 223-229.

Budhram, P., Deacon, R. & Gardner, C.R. (1986). Some putative non-sedating anxiolytics in a conditioned licking conflict. Br. J. Pharmacol., **88**, 331P.

Ceulemans, D.L.S., Hoppenbrouwers, H.J.A., Gelders, Y.G. & Reyntjens, A.J.M. (1985). The influence of ritanserin, a serotonin antagonist, in anxiety disorders: a double-blind placebo-controlled study versus lozarepam. Pharmacopsychiatry, **18**, 303-305.

Charney, D.S., Heninger, G.R. & Breier, A. (1984). Noradrenergic function in panic anxiety: effects of yohimbine in healthy subjects and patients with agoraphobia and panic disorder. Arch. Gen. Psychiat., **41**, 751-763.

Charney, D.S., Woods, S.W., Goodman, W.K. & Heninger, G.R. (1987). Serotonin in anxiety. II Effects of the serotonin agonist MCPP in panic disorder patients. Psychopharmacol., **92**, 14-24.

Chopin, Ph. & Briley, M. (1987). Animal models of anxiety: the effect of compounds that modify 5-HT neurotransmission. Trends in Pharmacol. Sci., **8**, 383-388.

Clineschmidt, B.V., Reiss, D.R., Pettibone, D.J. & Robinson, J.L. (1985). Characterization of 5-hydroxytryptamine receptors in rat stomach fundus. J. Pharmacol. Exptl. Ther., **235**, 696-708.

Colpaert, F.C., Meert, T.F., Niegemeers, C.J.E. & Janssen, P.A.J. (1985). Behavioural and 5-HT antagonist effects of ritanserin: a pure and selective antagonist of LSD discrimination in rat. Psychopharmacol., 86, 45-54.

Conti, L. & Pinder, R.M. (1979). A controlled comparative trial of mianserin and diazepam in the treatment of anxiety states in psychiatric out-patients. J. Int. Med. Res., 7, 285-289.

Cook, L. & Sepinwall, J. (1975), Behavioral analysis of the effects and mechanisms of action of the benzodiazepines. In: Mechanisms of actions of benzodizepines. (Eds. Costa, E. & Greengard, P.), pp. 1-28. New York: Raven Press.

Critchley, M.A.E. & Handley, S.L. (1987). Effects in the X-maze of agents acting at $5-HT_1$ and $5-HT_2$ receptors. Psychopharmacol., 93, 502-506.

Deacon, R. & Gardner, C.R. (1986). Benzodiazepine and 5-HT ligands in a rat conflict test. Br. J. Pharmacol., 88, 330 P.

Engel, J.A., Hjorth, S., Svenson, K., Carlsson, A. & Liljequist, S. (1984). Anticonflict effect of the putative serotonin receptor agonist 8-hydroxy-2-(di-n-propylamino) tetralin (8-OH-DPAT). Eur. J. Pharmacol., 105, 365-368.

Feuerstein, T.J., Hertting, G. & Jackisch, R. (1985). Endogenous noradernaline as modulator of hippocampal serotonin (5-HT)-release; dual effects of yohimbine, rauwolscine and corynanthine as α-adrenoceptor antagonists and serotonin receptor agonists. Naunyn Schmiedeb. Arch. Pharmacol., 329, 216-221.

File, S.E. (1981). Behavioural effects of serotonin depletion. In: Metabolic disorders of the nervous system. (Ed. Clifford Rose, E.), pp. 429-445. London: Pitmans.

File, S.E. (1985). Animal models for predicting clinical efficacy of anxiolytic drugs: Social behaviour. Neuropsychobiology, 13, 55-62.

Frazer, G.A. & Lapierre, Y.D. (1987). The effect of buspirone on panic disorder: a case report. J. Clin. Psychopharmacol., 7, 118-119.

Gardner, C.R. (1985). Pharmacological studies on the role of serotonin in animal models of anxiety. In: Neuropharmacol., of serotonin.(Ed. Green, A.R.),pp. 325-326. Oxford: Oxford University Press.

Gardner, C.R. (1986). Recent developments in 5-HT-related pharmacology of animal models of anxiety. Pharmacol. Biochem. Behav., 24, 1479-1485.

Gardner, C.R. & Guy, A.P. (1985). Pharmacological characterisation of a modified social interaction model of anxiety in the rat. Neuropsychobiol., 13, 194-200.

Geller, I. & Hartmann, R.J. (1982). Effects of buspirone on operant behavior of laboratory rats and cynomolgus monkeys. J. Clin. Psychiatry, **43**, 25-33.

Geller, I., Hartmann, R.J. & Croy, D.J. (1974). Attenuation of conflict behavior with cinanserin, a serotonin antagonist: reversal of the effect with 5-hydroxytryptophan and α-methyltryptamine. Res. Commun. Chem. Pathol., **7**, 165-174.

Gloger, S., Grunhaus, L., Birmacher, B. & Troudart, T. (1981). Treatment of spontaneous panic attacks with clomipramine. Am. J. Psychiatry, **138**, 1215-1217.

Goldberg, M.E., Salama, A.I., Patel, J.B. & Maleck, J.B. (1983). Novel non-benzodiazepine anxiolytics. Neuropharmacol., **22**, 1499-1504.

Gorman, J.M. & Gorman, L.K. (1987). Drug treatment of social phobia. J. Affective Disorders, **13**, 183-192.

Gorman, J.M., Liebowitz, M.R., Fyer, A.J., Goetz, D., Campeas, R.B., Fyer, M.R., Davies, S.O. & Klein, D.F. (1987). An open trial of fluoxetine in the treatment of panic attacks. J. Clin. Psychopharmacol., **7**, 329-332.

Graeff, F.G. (1974). Tryptamine antagonists and punished behavior. J. Pharmacol. Exp. Ther., **189**, 344-350.

Graeff, F.G., Zuardi, A.W., Giglio, J.S., Lima Filho, E.C. & Karniol, I.G. (1985). Effect of metergoline on human anxiety. Psychopharmacology, **86**, 334-338.

Gudelsky, G.A., Koenig, J.I. & Meltzer, H.Y. (1985). Altered responses to serotonergic agents in Fawn-Hooded rats. Pharmacol. Biochem. Behav., **22**, 489-492.

Gungras, M. (1977). An uncontrolled trial of clomipramine (Anafranil) in the treatment of phobic and obsessional states in general practice. J. Int. Med. Res., **5** (Suppl. 5), 111-115.

Hollister, L.E. (1986). Pharmacotherapeutic considerations in anxiety disorders. J. Clin. Psychiat., **47** (Suppl), 33-36.

Johnston, A.L. & File, S.E. (1986). 5-HT and anxiety: Promises and pitfalls. Pharmacol. Biochem. Behav., **24**, 1467-1470.

Kahn, R.J., McNair, D.M. & Frankenthaler, L.M. (1987). Tricyclic treatment of generalized anxiety disorder. J. Affect. Disorders, **13**, 145-151.

Kahn, R.J., McNair, D.M., Lipman, R.S., Rickels, K., Downing, K., Fisher, S. & Frankenthaler, L.M. (1986). Imipramine and chlordiazepoxide in depressive and anxiety disorders. 2. Efficacy in anxious outpatients. Arch. Gen. Psychiat., **43**, 79-85.

Kahn, R.J., Westenberg, H.G. & Jolles, J. (1984). Zimelidine treatment of obsessive-compulsive disorder; biological neuropsychological aspects. Acta

Psychiatr. Scand., 69, 259-261.

Kilts, C.D., Commissaris, R.L., Cordon, J.J. & Rech, R.H. (1982). Lack of central 5-Hydroxytryptamine influence on the anticonflict activity of diazepam. Psychopharmacology, 78, 156-164.

Kiser, J.R.R.S., German, D.C. & Lebovitz, R.M. (1978). Serotonergic reduction of dorsal central gray area stimulation-produced aversion. Pharmacol. Biochem. Behav., 9, 27-31.

Leone, C.M.L., De Aquiar, J.C. & Graeff, F.G. (1983). Role of 5-hydroxytryptamine in amphetamine effects on punished and unpunished behaviour. Psychopharmacology, 80, 78-82.

Lydiard, R.B. (1987). Successful utilization of maprotiline in a panic disorder. Patient intolerant of tricyclics. J. Clin., Psychopharmacol., 7, 113-114.

Lydiard, R.B. & Ballenger, J.C. (1987). Antidepressants in panic disorder and agoraphobia. J. Affective Disorders, 13, 153-168.

Mason, P., Skinner, J. & Luttinger, D. (1987). Two tests in rats for antianxiety effect of clinically anxiety attenuating antidepressants. Psychopharmacology, 92, 30-34.

Mavissakalian, M., Perel, J., Bowler, K. & Dealy, R. (1987). Trazodone in the treatment of panic disorder and agoraphobia with panic attacks. Am. J. Psychiat., 144, 785-787.

McCloskey, T.C., Paul, B.K. & Commissaris, R.L. (1987). Buspirone effects in an animal conflict procedure: comparison with diazepam and phenobarbital. Pharmacol. Biochem. Behav., 27, 171-175.

Meert, T.F. (1986). A comparative study of the effects of ritanserin (R 55 667) and chlordiazepoxide on rat open field behavior. Drug Developm. Research, 8, 197-204.

Meert, T.F. & Colpaert, F.C. (1986a). The shock probe conflict procedure. A new assay responsive to benzodiazepines, barbiturates and related compounds. Psychopharmacol., 88, 445-450.

Meert, T.F. & Colpaert, F.C. (1986b). The effects of agonists and antagonists at putative 5-HT receptor subtypes in the shock-probe conflict procedure. Psychopharmacol., 89, S23.

Merlo Pich, E. & Samanin, R. (1986). Disinhibitory effects of buspirone and low dosis of sulpiride and haloperidol in two experimental anxiety models in rats: possible role of dopamine. Psychopharmacol., 89, 125-130.

Ortiz, A., Pohl, R. & Gershon, S. (1987). Azaspirodecanediones in generalized anxiety disorder: buspirone. J. Affect. Disorders, 13, 131-143.

Pellow, S., Johnston, A.L. & File, S.E. (1987). Selective agonists and antagonists for 5-hydroxytryptamine receptor subtypes, and interactions with yohimbine and FG 7142 using the elevated plus-maze test in the rat. J. Pharm. Pharmacol., 39, 917-928.

Pecknold, J.C. (1987). Behavioural. Biol. Psychiat., 11, 97-104.

Petersen, E.N. & Buus Lassen, J. (1981). A water lick conflict paradigm using drug experienced rats. Psychopharmacol., 75, 236-239.

Rasmussen, S.A. (1984). Lithium and tryptophan augmentation in clomipramine-resistent obsessive-compulsive disorder. Am. J. Psychiatr., 141, 1283-1285.

Redmond, D.E. Jr, Huang, Y.K., Snyder, D.R., Maas, J.W. & Baulu, J. (1976). Behavioral effects of stimulation of the locus coeruleus in the stumptail monkey (Macaca arcoides). Brain Research, 116, 502-510.

Rizley, R. Kahn, R.J., McNaur, D.M. & Frankenthaler, L.M. (1986). A comparison of alprazolam and imipramine in the treatment of agoraphobia and panic disorder. Psychopharmacol. Bull., 22, 167-172.

Robins, L.N., Helzer, Y.E., Weissman, M.H., Orvaschel, H., Gruenberg, E., Burke, J.D. & Regier, D.A. (1984). Lifetime prevalence of psychiatric disorders in three communities. Arch. Gen. Psychiatry, 41, 949-967.

Schenberg, L.C. & Graeff, F.G. (1978). Role of periaqueductal gray substance in the antianxiety action of benzodiazepines. Pharmacol. Biochem. Behav., 2, 287-295.

Schweizer, E.E., Amsterdam, J., Rickels, K., Kaplan, M. & Droba, M. (1986). Open trials of buspirone in the treatment of major depressive disorder. Psychopharmacol. Bull., 22, 183-185.

Schuurman, T. & Spencer 1_A-receptor ligand ipsapirone TVX-Q-7821, A comparison with 8-hydroxy-2-di-N-propylaminotetralin and diazepam. Psychopharmacol., 89, S54.

Sepinwall, J.L. & Cook, L. (1980). Mechanism of action of the benzodiazepines: behavioral aspects. Fed. Proceed., 39, 3024-3031.

Stein, L., Wise, C.D. & Berger, B.D. (1973). Anti-anxiety action of benzodiazepine: decrease in activity of serotonin neurones in the punishment system. In: The benzodiazepines. (Eds. Garattini, S., Mussini, E. & Randall, L.O.), pp. 299-326. New York: Raven Press.

Turner, S.N., Jacobs, R.G., Beidel, D.C. & Himmelhock, J. (1985). Fluoxetine treatment of obsessive-compulsive disorder. J. Clin. Psychopharmacol., 5, 207-212.

Van der Kolk, (1987). The drug treatment of post-traumatic stress disorder. J.

Affective Disorders, 13, 203–213.

Van Riezen, H., Pinder, R.M., Nickolson, V.J., Hobbelen, P., Zayed, I. & Van der Veen, F. (1981). Mianserin. In: Pharmacological and biochemical properties of drug substances. Vol. 3, (Ed. Goldberg, M.E.), pp.56–93. Washington: American Pharmaceutical Association.

Wheatley, D. (1976). Evaluation of trazodone in the treatment of anxiety. Curr. Ther. Res., 20, 74–83.

Winter, J.C. (1972). Comparison of chlordiazepoxide, methysergide, and cinanserin as modifiers of punished behavior and as antagonists of N,N–dimethyltryptamine. Arch. Int. Pharmacodyn., 197, 147–159.

Young, R., Urbancic, A., Emrey, T.A., Hall, P.C. & Metcalf, G. (1987). Behavioral effects of several new anxiolytics and putative anxiolytics. European J. Pharmacol., 143, 361–371.

Zohar, J. & Insel, T.R. (1987). Drug treatment of obsessive–compulsive disorder. J. Affect. Disorders., 13, 193–202.

5-HT$_{1A}$ RECEPTORS: A BRIDGE BETWEEN ANXIETY AND DEPRESSION?

P. Soubrie
Neurobiology, Sanofi Recherche, Montpellier, France

The recent development of compounds selective for the different serotonin (5-HT) receptor subtypes has led to a renewed interest in the field of 5-HT in the control of anxiety.

When examining the effects in different animal models of anxiety of compounds exerting global actions on 5-HT systems, it appears that a reduction in 5-HT neurotransmission tends to reduce anxiety (Iversen, 1984; Thiebot, 1980).

Radioligand binding studies initially revealed two 5-HT$_1$ (^{3}H-5-HT labeled) recognition sites in the rat brain, 5-HT$_{1A}$ and 5-HT$_{1B}$. It has been suggested that the stimulation of 5-HT$_{1A}$ receptors located on the cell bodies and dentrites of neurons in the raphe nuclei result in a reduced 5-HT function i.e. decreases in firing and release of the transmitter (Middlemiss & Fozard, 1983; Pazos & Palacios, 1985; Trulson & Arasteh, 1986; Verge et al., 1985).

Hence 5-HT$_{1A}$ agonists are expected to behave as anxiolytics (benzodiazepines) in animal models of anxiety. In fact the effects of 5-HT$_{1A}$ agonists on anxiety have not been particularly consistent.

On this account, the first aim of this study is to briefly review the most significant results obtained with these drugs in animal models of anxiety. Our second aim is to examine some of the factors which may account for the main features of the effects of 5-HT$_{1A}$ agonists in these models: inconsistency across tests, poor reproducibility in a particular test, narrow range of effective doses. Special attention will be paid to the evidence that 5-HT$_{1A}$ binding sites may exist as pre- and post-synaptic receptors (Pazos & Palacios, 1985), stimulation of both receptors by 5-HT$_{1A}$ agonists being likely to result in opposing functional and behavioral consequences. Hence, variables which might disrupt the balance between pre/post receptor stimulation towards an enhanced post-synaptic 5-HT$_{1A}$ function will be discussed.

On the other hand, when considering both the weak efficacy of 5-HT$_{1A}$ agonists in animal models optimized for the benzodiazepines and the reported anti-anxiety activity of buspirone and related drugs in humans (see File, 1987) it may be suggested that 5-HT$_{1A}$ agonists produce an anxiety reduction differing in nature from that produced by benzodiazepines. Thus, the effects of 5-HT$_{1A}$ agonists on uncontrollable stress-induced behavioral deficits in animals will be presented.

5-HT autoreceptor stimulation-induced negative feedback as a mechanism for producing anti-anxiety effect
The hypothesis linking decreased serotonin transmission to reduced anxiety stemmed from the observations that the 5-HT depletor, parachlorophenylalanine (PCPA), or 5-HT receptor blockers as well as lesion studies with more or less selective neurotoxins such as 5,7-dihydroxytryptamine (5,7-DHT) produced a

significant attenuation of punishment-induced behavioral suppression (Iversen, 1984; Thiebot, 1986).

Drugs such as lysergic acid diethylamide known to inhibit 5-HT neuronal activity by activating 5-HT autoreceptors were also found to produce anti-punishment effects. More recently it was reported that microinjections of exogenous 5-HT at the level of the dorsal raphe cells released behavioral suppression elicited by the presentation of a stimulus previously paired with punishment (see Iversen, 1984 and Thiebot, 1986). Together with the possibility that benzodiazepines might enhance dendritic release of 5-HT, thus reducing 5-HT function (Soubrie et al., 1983), these findings clearly suggest that stimulation of autoreceptor-mediated feedback control could be a particularly efficient mechanism in producing anti-anxiety effects in animals.

Radioligand binding studies have led to the characterization of the autoreceptors which may play a significant role in the control of 5-HT function in revealing that they belong to the $5-HT_{1A}$ subtype (Verge et al., 1985). Further observations demonstrated that drugs such as buspirone, 8-hydroxy-2(di-n-propylamino) tetralin (8-OH-DPAT), and gepirone behave as agonists at those sites.

Electrophysiological studies clearly indicated that these drugs when either applied with very small iontophoretic currents to raphe cells or injected systemically depressed firing of neurons assumed to be serotonergic in nature (Blier & De Montigny, 1987; Sprouse & Aghajanian, 1987; Trulson & Arasteh, 1986, Vandermaelen et al., 1986). Furthermore, biochemical studies are consonant with these drugs reducing 5-HT function (Hjorth & Carlsson, 1982; Huston et al., 1986). Finally, $5-HT_{1A}$ agonists have been reported to elicit hyperphagia (i.e. increase feeding), this effect being abolished by PCPA which prevents pre-synaptic control of 5-HT function (Dourish et al., 1986). That the neurons of the raphe nuclei play a significant role in $5-HT_{1A}$ agonist-induced hyperphagia was substantiated by the observation of increased feeding in rats after injection of nanogram amounts of 8-OH-DPAT into the raphe nuclei (Bendotti & Samanin, 1986; Hutson et al., 1986). However, although there is clear electrophysiological, biochemical and behavioral evidence that buspirone, 8-OH-DPAT and related compounds may behave as agonists at $5-HT_{1A}$ autoreceptors, data obtained on animal models of anxiety are conflicting.

Effects of $5-HT_{1A}$ agonists on experimentally induced anxiety
First, it should be recalled that not all of the actions of 5-HT on raphe neurons are necessarily mediated by $5-HT_{1A}$ receptors (Sprouse & Aghajanian, 1987). Second, it should be mentioned that any attempt at clarifying the effects of $5-HT_{1A}$ agonists is constricted by the fact that in several cases, few models of anxiety have been investigated. For instance, these drugs have practically not been studied in animal models of frustrative non reward, frustration being assumed to show considerable equivalence with fear or anxiety. The sole available report mentions no effect of 8-OH-DPAT on withdrawal of reward (time out) at a dose (0.25 mg/kg) producing a clear though limited release of punished behaviour (Hodges et al., 1987).

In classical conflict procedures (Vogel lick-test of Geller & Seifter model) the

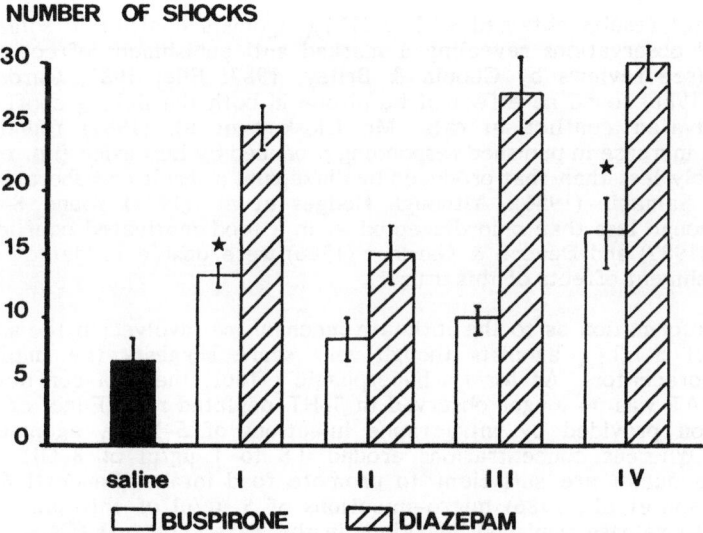

Fig. 29.1 Effects of buspirone and diazepam on the water-lick conflict test in the course of 4 independent experiments (I-IV). During preceding the test-session free access to water in the home cage was restricted to 2 hours a day (between 3 and 5 p.m.) and the rats were trained (one 15 min session a day, between 10 a.m. and 1 p.m.) to drink in the test-apparatus (30x30x32 cm Plexiglas box with one transparent wall and electified grid-floor). Each box was equipped with one bottle located in a corner, the end of the drinking tube being set 5 cm above the floor. During the test-session, the animals were allowed to drink water during an initial 15 sec period and were subsequently given one electric foot-shock (0.40 mA, 45 msec) every time a 3 sec period of licking was completed. The number of shocks received during 3 min following the first shock was recorded, and presented here as mean ± SEM (columns + Vertical bars). Buspirone and diazepam were administered i.p. 30 min before testing at 2 mg/kg.
The effects of buspirone were significant in only two experiments (* $p<0.05$) whereas the effects of diazepam were significant ($p<0.02$) in each cases (ANOVA).

most recent results obtained with 5-HT$_{1A}$ agonists conflict in some way with the initial observations revealing a marked anti-punishment effect of 5-HT$_{1A}$ agonists (see reviews by Chopin & Briley, 1987; File, 1987; Gardner, 1986). Gardner (1986) found no effect of buspirone in both the licking conflict and the food-motivated conflict in rats. Mc Closkey et al. (1987) found that the maximum increase in punished responding produced by buspirone (i.p. or s.c.) was considerably less than that produced by diazepam, a result corroborated by Merlo Pich and Samanin (1986). Although Hodges et al. (1987) found 8-OH-DPAT active, though less than chlordiazepoxide, in a food-motivated conflict. Carli & Samanin (1987) and Deacon & Gardner (1986) were unable to detect significant anti-punishment effects of this drug.

Current information as to the intimate mechanisms involved in the anti-anxiety activity of 5-HT$_{1A}$ agonists though very scare suggests the implication of 5-HT autoreceptors. As for its hyperphagic effect, the anti-conflict effect of 8-OH-DPAT was no longer observed in 5-HT-depleted rats (Engel et al., 1984). Information provided by intra raphe injections of 5-HT$_{1A}$ agonists are less clearcut. Whereas concentrations around 0.5 to 1 µg/µl of 8-OH-DPAT into the raphe nuclei are sufficient to promote food intake (Bendotti & Samanin, 1986; Hutson et al., 1986) micro-injections of 5 µg/µl of this same drug were required to release punished behaviour in the water lick test (Carli & Samanin, 1987).

Other animal procedures often used to detect anxiolytic (Benzodiazepine-like) actions such as the social interaction test, the elevated plus-maze or the two-compartment exploratory tests, generally reveal only weak effects of buspirone, 8-OH-DPAT, or ipsapirone. It should be mentioned, however, that the evidence in support of the involvement of 5-HT neurons in most of these tests is far from being compelling (see review by Soubrie, 1986).

A series of experiments undertaken in our laboratory in rats subjected to the water lick test revealed that whereas the effects of a reference benzodiazepine were easily reproducible, this was not the case for buspirone.
Indeed buspirone produced a significant release of punished behaviour only in 50% of the trials (Fig. 29.1). However, when the results obtained in the course of this study were pooled, a significant anti-punishment effect was obtained (Fig. 29.2), although the effective dose-range was extremely narrow (1 dose) and the magnitude of the effect was inferior to that produced by the reference benzodiazepine.

To summarize, it appears that the effects of 5-HT$_{1A}$ agonists on animal models of anxiety:
* show less consistency across testing procedures than benzodiazepines,
* are, when they are observed (mainly in punishment procedures) detected in a narrow dose-range,
* are less marked and reproducible than those generally elicited by benzodiazepines.

Several factors may account for such a conflicting picture with regard to the 5-HT hypothesis of anxiety.

First of all, it should be recalled that even though a substantial amount of information militates in favor of the notion that a reduced 5-HT transmission may underlie attenuation of anxiety, it has not been established that anti-anxiety effects of benzodiazepines relate to their anti-5-HT effects (Soubrie, 1986; Thiebot, 1986).

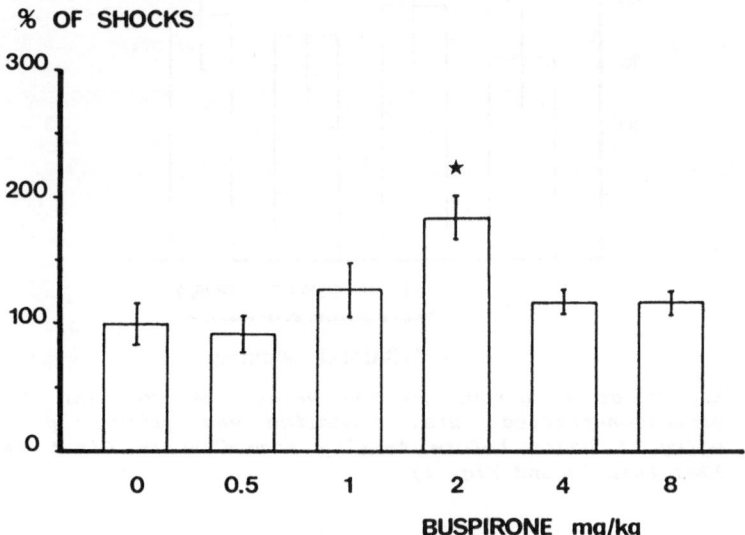

Fig. 29.2 Dose-response curve of buspirone in the water-lick conflict test (data obtained in the course of the independent experiments presented in Fig. 1 were pooled).
* significantly different from control at p<0.02 (ANOVA).

Second the anti-anxiety efficacy of $5-HT_{1A}$ agonists may vary across species, perhaps in relation with possible inter species variations in pharmacokinetics. This may particularly be relevant for buspirone, this drug being found as active or slightly less active than benzodiazepines in pigeons and monkeys. Consistent with the significant "first-pass" effect of this compound, relative to i.p. administration, buspirone was somewhat more potent when administered s.c. in rodents (McCloskey et al., 1987). One of the main metabolites of buspirone and gepirone is 1-(2-pyridinyl)-piperazine (1-PmP), which has been reported to exert moderate or no anti-conflict effect (Gardner, 1986, personal unpublished results). Variations in the rate of formation of 1-PmP may contribute to the reported inconsistencies in the effects of parent $5-HT_{1A}$ agonists. Since we found that proadifen may prevent some 1-PmP-related effects of buspirone (Giral et al., 1987), we decided to study the effects of buspirone in the presence of proadifen on the water lick test in rats. If rapid hepatic metabolisation of buspirone is responsible for its weak or perhaps inconsistent effects in tests of anxiety, its efficacy should be improved in proadifen-treated rats. In fact, no such a change in buspirone activity was observed (unpublished results, fig. 29.3).

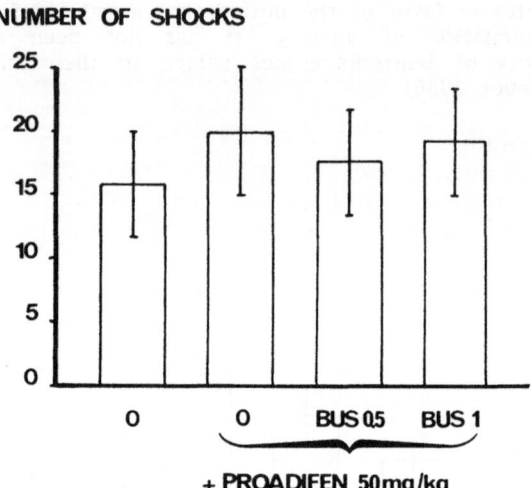

Fig. 29.3 Effect of buspirone in the water-lick conflict test in proadifen-treated rats. Proadifen was given i.p. at 50 mg/kg, 5 hours before testing according to Girat et al., 1987 (see legend Fig. 1).

Third, evidence suggests that the activity of 5-HT$_{1A}$ agonists is dependent on several procedural variables such as: the degree of previous experience with the situation, the drive state of the animals and the functional status of 5-HT systems (see Carli & Samanin, 1987). I would like to consider some of these factors in relation to the variations they may induce in the balance between the sensitivity of two different types of 5-HT$_{1A}$ receptors which may have opposing functional effects.

Factors able to modulate the effects of 5-HT$_{1A}$ agonists in animal models of anxiety

We have already seen that considerable information has accumulated to suggest that 5-HT autoreceptors located on dendrite and cell bodies of raphe cells belong to the 1A subtype. Biochemical and electrophysiological evidence suggests that there are also 5-HT$_{1A}$ receptors located post-synaptically, especially in the hippocampus and septum (Pazos & Palacios, 1985; Peroutka et al., 1987). When 5-HT$_{1A}$ agonists are injected systemically, stimulation of such post-synaptic receptors may mitigate the reduction in serotonergic transmission related to the activation of 5-HT$_{1A}$ autoreceptors.

Previous experience with 5-HT$_{1A}$ agonists could be an important variable. Indeed, two recent studies clearly indicate that pretreatment with 5-HT$_{1A}$ agonists rapidly desensitises 5-HT$_{1A}$ pre-synaptic receptor-mediated response.

Blier & De Montigny (1987) showed that after seven days gepirone treatment at the mammoth dose of 15 mg/kg/day, inhibition of firing of 5-HT cells of the dorsal raphe was almost abolished, whereas the responsiveness of hippocampal pyramidal neurons to gepirone or 8-OH-DPAT was not altered by a simular one week treatment. Kennett et al. (1987) reported that a single priming injection of 8-OH-DPAT, buspirone or ipsapirone, attenuated the hyperphagic response to 8-OH-DPAT administered on the next day. The ability of 8-OH-DPAT to reduce raphe 5-HIAA levels was also impaired by previous 8-OH-DPAT treatment. In addition, the authors showed that the pre-synaptic $5-HT_{1A}$-desensitizing effect of pretreatment with 8-OH-DPAT required between 5 to 10 days to vanish. Conversely, $5-HT_{1A}$ post-synaptic receptor-mediated behavioral responses were unaffected by the above pretreatments.

This may have considerable importance when animals (as frequently occurs in the Geller & Seifter procedure) are challenged a few days apart with either various dosages of the same drug (let's say a $5-HT_{1A}$ agonist) or with different drugs. Indeed, the above mentioned results would suggest that a rapid tolerance to the anxiolytic effects of $5-HT_A$ agonists may occur as a result of prior experience with these compounds.

Another crucial factor in altering the balance between post- vs pre-synaptic $5-HT_{1A}$ receptor stimulation could be the animal level of food-deprivation. Bendotti & Samanin (1987) showed that the hyperphagic response produced by 8-OP-DPAT in free-feeding rats was converted into a marked reduction in food intake in food-deprived animals. Conversely, indexes of post-synaptic $5-HT_{1A}$ receptor activation (5-HT motor syndrome) were enhanced in food-deprived rats relative to free-feeding rats. These data may suggest that depending on the food deprivation level of the animals $5-HT_{1A}$ agonists may exert differential effects on food-motivated conflict behaviour.

Taken together, these findings tend to indicate that stimulation of $5-HT_{1A}$ post-synaptic receptors in brain structures enriched in those receptors may cancel or attenuate the behavioral effects of $5-HT_{1A}$ agonists that depend on a reduced 5-HT function linked to a stimualtion of 5-HT autoreceptors. Two series of experiments may substantiate such a hypothesis.

Engel et al. (1984) reported that pretreatment with PCPA was able to antagonize the anti-conflict effect of 8-OH-DPAT but also that the anti-conflict effect of PCPA was reversed by 8-OH-DPAT administration. The same reciprocal antagonism between $5-HT_{1A}$ agonists and serotonin depletion was reported by Dourish et al. (1986) in food-intake related studies. On the other hand, Hodges et al. (1987) observed that intra amygdaloid injections of 8-OH-DPAT not only did not release punished behaviour but produced a pro-conflict effect.

These observations suggest that the anxiolytic effects ("pre-synaptic") of $5-HT_{1A}$ agonists may be favoured when the activity of 5-HT neurons is increased, and that stimulation of certain post-synaptic $5-HT_{1A}$ receptors may counteract the "anxiolytic" effects of $5-HT_{1A}$ agonists. What could be the consequences of this on the psychopharmacological profile of $5-HT_{1A}$ agonists? The fact that $5-HT_{1A}$ agonists are actually not able to reduce 5-HT function at $5-HT_{1A}$ post-synaptic receptor synapses may contribute to their inability to produce anxiolytic effects as marked and as global as manipulations known or thought to produce a more uniform reduction in 5-HT function (lesions, PCPA, benzodiazepine). Thus, it appears important to attempt a delineation of the consequence that inhibition or facilitation of 5-HT function at 5-HT_{1A} post-synaptic receptors may have upon behaviour. We would like to speculate that among other factors, blockade of post-synaptic 5-HT_{1A} receptors – perhaps of the hippocampus or septal nuclei – may reduce the waiting capacity in animals.

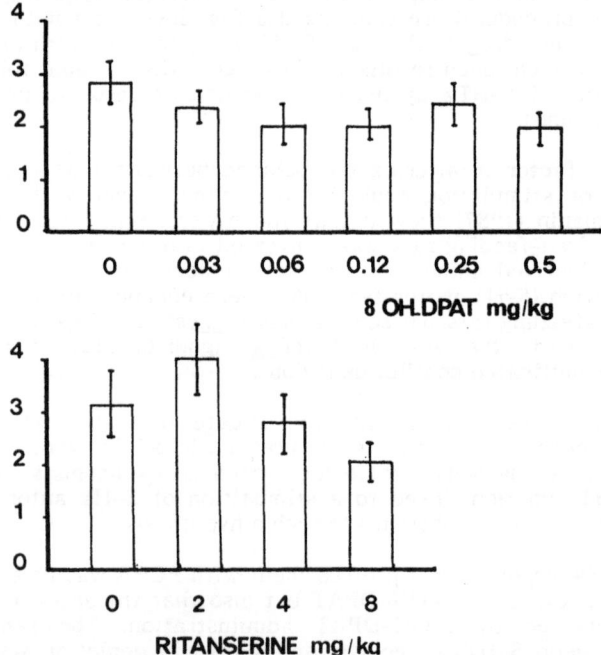

Fig. 29.4 Effects of 8-OH-DPAT and ritanserine on waiting capacity for food reward. Rats were trained to run a T-maze (two daily sessions of five trails) for reward with left and right goal boxes provided with ten and two pellets, respectively. Within ten sessions rats chose the arm associated with the large reward in 80-100% of the trials. Then, the rats were detained (15 sec) in the arm associated with the large reward before having access to the ten pellets. Columns represent the number of choises (mean ± SEM) of the immediate reward over ten trials within two sessions of 5 trials each. Drugs were given i.p. 30 min before test. No significant drug effect was detected (ANOVA).

Serotonin and tolerance to reward delay
A behavioral model without any nociceptive components was designed to study the possible role of 5-HT processes on the waiting capacity inferred from the behavioral modifications induced by delayed reward (Thiebot et al., 1985a).

Rats placed in a T-maze were alowed to choose between 2 magnitudes of reward: small (2 pellets) vs large (10 pellets). After training, a delay (15 of 25 sec depending on the experiments) was introduced before access to the large reward. Under this alternative, saline-rats selected the large but delayed reward in 80% (15 sec waiting) or less than 40% (25 sec waiting) of the trials. Serotonin uptake blockers: clomipramine, indalpine, zimeldine increased the number of choices of the large-but- (25 sec) delayed reward. Conversely, 5,7-DHT lesions or benzodiazepine administration reduced the number of choices of the large-but- (15 sec) delayed reward. On this model, neither 8-OH-DPAT (Fig. 29.4), ipsapirone, nor gepirone were found to exert significant effect, although buspirone was observed to reduce waiting capacity (Thiebot, 1986).

In addition, ritanserine (Fig. 29.4) a relatively selective $5-HT_2$ receptor antagonist was devoid of effect on tolerance to reward delay. Although further studies are required to draw definite conclusions, it is conceivable that inhibition of 5-HT function at post-synaptic $5-HT_{1A}$ receptor synapse is a prerequisite for reduction in waiting capacity. This may have some relevance with regard to the limited anti-anxiety effects of $5-HT_{1A}$ agonists and $5-HT_2$ antagonists relative to benzodiazepines.

Indeed, it has been suggested that in conflict procedures delay of reward mediated by shock-avoidance behaviour could be as critical a target as shock-induced fear for anti-punishment effects of benzodiazepines in animals (Thiebot et al., 1985b).

Implications for future research
Buspirone has been reported to be as effective as diazepam in alleviating anxiety, and, preliminary clinical studies conducted with gepirone concluded as to its efficacy in generalized anxiety disorders (Jacobson et al., 1985; Csanalosi et al., 1987). If these clinical results are confirmed, it would appear that drugs having minimal effects on several benzodiazepine-highly sensitive animal models (perhaps as a result of their lack of ability to reduce 5-HT function at post-synaptic $5-HT_{1A}$ receptor) may notwithstanding exert anti-anxiety effect in humans. This possibility would lead to a reconsideration of the animal models of anxiety which (or most of which), as pointed out by Chopin and Briley (1987), have been optimized for the benzodiazepines. It should be added that, even for benzodiazepines, the exact significance of the so-called anti-anxiety effects (release of punished, exploratory or feeding behaviours) in predicting an effect on anxiety states in humans remains to be fully established. More especially, dramatic increases in punished responding such as those produced by some benzodiazepines or barbiturates seem to be not associated with a proportional heigthening of the therapeutic action of these molecules. On the other hand, failure to reduce 5-HT function at post-synaptic $5-HT_{1A}$ receptors might be an advantage for $5-HT_{1A}$ agonists or $5-HT_2$ antagonists as far as these drugs might promote less impulsive conduct or paradoxical behaviours than benzodiazepines.

In any case, it is clear that drugs acting on particular 5-HT receptor subtypes will prompt psychopharmacologists to a remodelling of their animal paradigms of anxiety. A first illustration of this are the recently developed tests that appear to be able to reveal an anxiolytic profile of drugs acting on the $5-HT_3$ receptors, such as GR 380032 F (Jones et al., 1987), although nothing is yet known about the CNS localization of $5-HT_3$ binding sites.

However, when one considers the classical tests of anxiety there is no evidence that the results obtained with $5-HT_{1A}$ agonists or $5-HT_2$ antagonists, are in any way determined by the particular test or group of tests used. Every thing happens as though most existing tests are almost exclusively sensitive to the type of anxiety reduction produced by benzodiazepines and marginally to that elicited by 5-HT drugs. The observed borderline activity of these latter drugs could be the reflection of a limited overlap between these two putative types of anxiety reduction. Hence the current strategy consisting in investigating the effects of 5-HT related compounds on the benzodiazepine-sensitive models, could be a wrong one. In this context it becomes crucial to analyse what could be the psychobiological of psychopathological dimensions (anxiety or others) specifically affected by drugs acting at particular 5-HT binding sites.

$5-HT_{1A}$ agonists and stress-induced depression in animals
Classical animal tests of anxiety are not sensitive to the action of antidepressants whereas several clinical reports have pointed to the efficacy of antidepressants in reducing clinical anxiety. For instance, antidepressants have consistently been found to exert a favourable effect in disorders, classified in the DSM-II as anxiety disorders (obsessive compulsive disorders, panic attacks). Interestingly, a recent review suggested buspirone could be efficacious in the treatment of anxiety in patients suffering from depression (Goa & Ward, 1986) and a preliminary open trial reported its utility in the treatment of non-melancholic depressed subjects (Schweizer et al., 1986).

This prompted us to study the potential action of $5-HT_{1a}$ agonists on animal models of stress-induced deficits such as those observed in the learned helplessness paradigm.

This study was designed to investigate whether $5-HT_{1A}$ agonists would eliminate, as classical antidepressants do (but not benzodiazepines), the escape deficits produced by prior inescapable stress. Rats were first exposed to 60 inescapable shocks (15 sec duration, 0.8 mA, every 1 min $\pm$ 15 sec), and 48 hr later, they were subjected to daily shuttlebox sessions. Daily intraperitoneal injection of buspirone (total daily dose 0.5 and 1 mg/kg), gepirone (0.06 and 0.125 mg/kg), 8-OH-DPAT (0.03, 0.06, 0.125 and 0.25 mg/kg), and ipsapirone (0.03 and 0.06 mg/kg) eliminated escape failures (Giral et al., 1987).

These findings are consonant with reports indicating that $5-HT_{1A}$ agonists exert behavioral effects common to antidepressants, in that they alter behaviours that follow incontrollable or inescapable stresses. For instance, Kennett et al. (1987) reported that reduced locomotor activity following a 2 h restraint period was prevented by a single post-restraint administration of $5-HT_{1A}$ agonists. Desipramine attenuated the behavioral effect of restraint but only after subacute administration whereas benzodiazepines were inactive.

Likewise, although 8-OH-DPAT was found inactive in the conflict water-lick test and in the two compartment exploratory test in otherwise unstressed rats, this substance was able to stimulate exploratory activity in rats previously subjected to immobilization stress (Carli & Samanin, 1987). At variance, however with the report of Kennett et al. (1987), Carli & Samanin (1987) refer to unpublished results to mention that acute diazepam but not desipramine was found to stimulate locomotor activity in stressed rats.

It is therefore conceivable that these findings suggest a potential antidepressant profile of 5-HT$_{1A}$ agonists. Alternatively, as there probably is a relationship between anxiety and the development of learned helplessness, the reversal by 5-HT$_{1A}$ agonists of escape failures may relate to their anti-anxiety activity. When administered before the uncontrollable stress, classical anxiolytics (such as benzodiazepines), but also buspirone have been shown to prevent escape deficits (Drugan et al., 1984, 1987). However, post-stress administration of benzodiazepines was consistently ineffective in eliminating escape failures.

Hence, it cannot be excluded that uncontrollable stress is associated with a particular type of anxiety that is resistant to benzodiazepines but sensitive to 5-HT$_{1A}$ agonists (see also Van der Kar et al 1985). The mirror immage of this would be that anxiety sensitive to benzodiazepines mainly linked to uncertainty and anticipation of aversive events (anticipatory anxiety) is practically insensitive to 5-HT$_{1A}$ agonists. It is clear that the nature of the anxiety putatively associated with uncontrollability needs to be characterized. As indicated above one likely source that generates (anticipatory) anxiety is the presence or deliviery of signals, events or stimuli which signify danger. An additional, far less studied, source of anxiety could relate to the disappearance of signals, objects or individuals which ensure or relate to safety. It is not clear whether these two putative types of anxiety differ as to their neurobiological substrates, and whether uncontrollability might approximate some of the features associated with the latter rather than the former type of anxiety. Even in humans, the exact profile of the antianxiety effects of 5-HT$_{1A}$ agonists is far from being fully characterized. What is the significance of reports suggesting that the resistance to the anti-anxiety effect of buspirone in male patients or in subjects with a history of benzodiazepine use (Schweizer et al., 1986)?

Interestingly, we found (Fig. 29.5) that the reversal of helpless behaviour by 8-OH-DPAT was preserved in animals whose 5-HT neurons were destroyed by prior injection of 5,7-DHT into the raphe nuclei, such a lesion being sufficient to suppress indalpine-induced reversal of escape failures. This clearly suggests that, as distinct from their effects in classical tests of anxiety, the antidepressant-like effects of 8-OH-DPAT and analogues involved a stimulation of post-synaptic rather than pre-synaptic 5-HT$_{1A}$ receptors.

This may be compatable with the important role of the hippocampus, one of the most enriched brain structures in 5-HT$_{1A}$ post-synaptic receptors, in recovery from stress-induced deficits (Soubrie et al., 1987).

In conclusion, from the material reviewed in the present study it would be tempting to speculate that the two types of anxiety: the benzodiazepine-sensitive and the 5-HT$_{1A}$-sensitive may somewhat overlap,

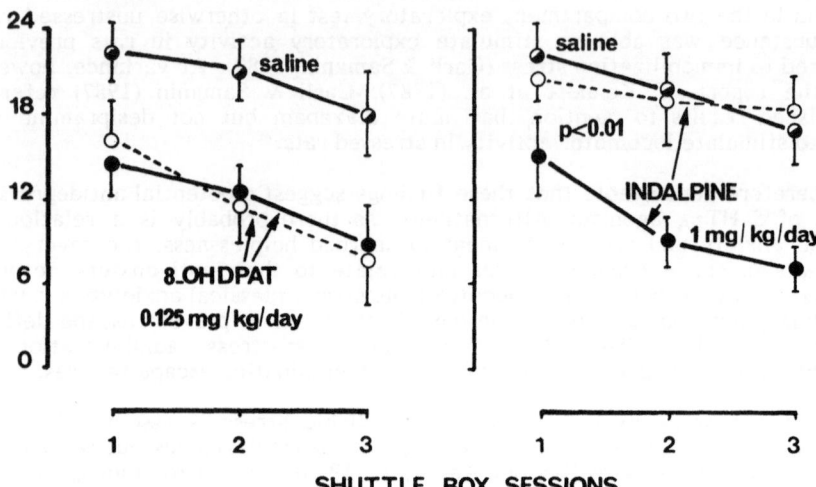

Fig. 29.5 Reversal by 8-OH-DPAT and indalpine of escape deficits produced by uncontrollable stress in sham or 5,7-DHT lesioned rats.

Rats were first subjected to 60 inescapable, randomized electric foot shocks (15 sec duration, 0.8 mA, every min ± 15 sec). Escape deficits were estimated on three consecutive 15 min daily shuttle-box sessions (30 two-way avoidance trials/session), the first session being performed 48 h after inescapable shocks. Drugs or saline were injected i.p. 6 h after inescapable shocks and then, twice a day (morning and afternoon).

Three weeks before being subjected to the learned helplessness paradigm, rats were either sham-operated or given 5,7-DHT into the midbrain raphe nuclei (3 µg in 0.4 µl) after desipramine pretreatment (25 mg/kg i.p.). After behavioral testing, animals were sacrificed and [^{3}H]serotonin uptake (15 nM; 26.2 Ci/mmol) was assayed in synaptosomal preparations from the hippocampus, striatum and cerebral cortex.

Data are the mean ± SEM of escape failures, referring to failure of the rat to change compartments during the electric foot shock (0.8 mA; 3 sec duration).

ANOVA revealed that there was a statistically significant difference (p<0.01) in thge ability of indalpine, but not of 8-OH-DPAT, to reverse escape failures in 5,7-DHT lesioned rats as compared with sham operated rats.

such overlapping accounting for the efficacy albeit weak of 5-HT$_{1A}$ agonists in the benzodiazepine-sensitive animal models of anxiety. Alternatively, but with a similar reasoning in mind, the efficacy of 5-HT$_{1A}$ agonists in tests of anxiety and of depression, could be taken as a further support to the notion that there is considerable overlapping in the neurobiological substrates of anxiety and depression.

Neither of these alternatives are supported by the animal data reviewed here. Indeed, although the 5-HT$_{1A}$ agonists promote behavioral changes in either type of models (depression or anxiety linked to uncontrollability vs anticipatory anxiety), the underlying neurobiological substrates are different and even opposite. This is clearly suggested by the aforementioned experiments dealing with 5-HT-depleted or -lesioned rats.

In a recent review entitled: "Denosologization of biological psychiatry or the specificity of 5-HT disturbances in psychiatric disorders", Van Praag et al. (1987) argue that "the greater the biochemical specificity of a drug, the greater the chance it will be nosologically non-specific, but effective against certain pathopsychological dimensions, across diagnoses".

Many scientists or clinicians will probably endorse the opinion that drugs' could more readily be explained in terms of (pathological) dimentions than in terms of nosography. However, concluding that disorders A and B share a common dimension on the basis that both are cured by the same drug might be the source of major misinterpretations if the greaters care if not taken to establish the identity of the intimate neurobiological mechanisms involved in each of these curative actions.

Finally some of the data and speculations presented in this review might prompt us to address the question of inescapable stress-induced deficits as a model of depression. In spire of its remarkable sensitivity to a large variety of antidepressants, the valitiy of the learned helplessness paradimg as a model of depression has been questioned (Willner, 1984). In light of the results presented here with 5-HT$_{1A}$ agonists, one may wonder whether the ability of classical antidepressants to reverse escape failures could be a reflection of their efficacy in the treatment of certain anxiety disorders (panic attacks, obsessive compulsive disorders) rather than of their classical antidepressant property.

References

Bendotti, C. & Samanin, R. (1986). 8-Hydroxy-2-(DI-n-Propylamino) tetralin (8-OH-DPAT) elicits eating in free-feeding rats by acting on central serotonin neurons. Eur. J. Pharmacol., 121, 147-150.

Bendotti, C. & Samanin, R. (1987). The role of putative 5-HT$_{1A}$ and 5-HT$_{1B}$ receptors in the control of feeding in rats. Life Sci., 41, 635-642.

Blier, P. & De Montigny, C. (1987). Modification of 5-HT neuron properties by sustained administration of the 5-HT$_{1A}$ agonist gepirone: electrophysiology studies in the rat brain. Synapse, 1, 470-480.

Carli, M. & Samanin, R. (1987). Potential anxiolytic properties of 8-Hydroxy-2-(D-N-Propylamino), a selective serotonin$_{1A}$ receptor agonist. Psychopharmacology, (In Press).

Chopin, P. & Briley, M. (1987). Animal models of anxiety: the effect of compounds that modify 5-HT neurotransmission. TIPS Reviews, 8.

Csanalosi, I, Schweizer, E., Case, W. & Rickels, K. (1987). Gepirone in anxiety: a pilot study. J. Clin. Psychopharmacol., 7, 31-193.

Deacon R. & Gardner, C.R. (1986). Benzodiazepine and 5-HT ligands in a rat conflict test. Brit. J. Pharmacol., 88, suppl., 330 P.

Dourish, C.T., Hutson, P.H. & Curzon, G. (1986). Para-chlorophenylalanine prevents feeding induced by the serotonin agonist 8-Hydroxy-2-(di-n-propylamino) tetralin (8-OH-DPAT). Psychopharmacol., 89, 367-471.

Drugan, R.C., Ryan, S.M., Minor, T.H. & Maier, S.F. (1984). Librium prevents the analgesia and shuttlebox escape deficit typically observed following inescapable shock. Pharmacol. Biochem. Behav., 21, 749-754.

Drugan, R.C., Crawley, J.N., Paul, S.M. & Skolnick, P. (1987). Buspirone attenuates learned helphellness behavior in rats. Drug Dev. Res., 10, n° 2, 63-67.

Engel, J.A, Hjorth, S., Svensson, K., Carlsson, A. & Liljequist (1984). Anticonflict effect of the putative serotonin receptor agonist 8-Hydroxy-2-(Di-n-Propylamino) tetralin (8-OH-DPAT). Eur. J. Pharmacol., 105, 365-368.

File, S.E. (1987). The search for novel anxiolytics. TINS, 10, n° 11, 461.

Gardner, C.R. (1986). Recent developments in 5-HT-related pharmacology of animal models of anxiety. Pharmacol. Biochem. & Behav., 24, 1479-1485.

Giral, P., Martin, P., Soubrie, P. & Simon, P. (1987). Reversal of helpless behavior in rats by putative 5-HT$_{1A}$ agonists. Biol. Psychiatry, (in press).

Giral, P., Soubrie, P. & Puech, A.J. (1987). Pharmacological evidence for the involvement of 1-(2-pyridinyl)-piperazine (1-PmP) in the interaction of buspirone of gepirone with noradrenergic systems. Eur. J. Pharmacol., 134, 113-116.

Goa, K.L. & Ward, A. (1986). Buspirone: a preliminary review of its pharmacological properties and therapeutic efficacy as an anxiolytic. Drugs, 32, 114-129.

Hjorth, S. & Carlsson, A. (1982). Buspirone: effects on central monoaminergic transmission - possible relevance to animal experimental and clinical findings. Eur. J. Pharmacol., 83, 299-303.

Hodges, H., Green, S. & Glenn, B. (1987). Evidence that the amygdala is involved in benzodiazepine and serotonergic effects on punished responding but not on discrimination. Psychopharmacology, 92, 491-504.

Hutson, P.H., Dourish, C.T. & Curzon, G. (1986). Neurochemical and behavioural evidence for mediation of the hyperphagic action of 8-OH-DPAT by 5-HT cell body autoreceptors. Eur. J. Pharmacol., 129, 347-352.

Iversen, S.D. (1984). 5-HT and anxiety. Neuropharmacol., 23 N° 12B, 1553-1560.

Jacobson, A.F., Dominguez, R.A., Goldstein, B.J. & Steinbook, R.M. (1985). Comparison of buspirone and diazepam in generalized anxiety disorder. Pharmacotherapy, 5, 290-296.

Jones, B.J., Oakley, N.R. & Tyers, M.B. (1987). The anxiolytic activity of GR38032F, a $5-HT_3$ receptor antagonist in the rat and cynomolgus monkey. Br. J. Pharmacol., 90, Mar, Suppl. 88 P.

Kennett, G.A., Dourish, C.T. & Curzon, G. (1987). Antidepressant-like action of $5-HT_{1A}$ agonists and conventional antidepressants in an animal model of depression. Eur. J. Pharmacol., 134, 265-274.

Kennett, G.A., Marcou, M., Dourish, C.T. & Curzon, G. (1987). Single administration of $5-HT_{1A}$ agonists decreases $5-HT_{1A}$ presynaptic, but not postsynaptic receptor-mediated responses: relationship to antidepressant-like action. Eur. J. Pharmacol., 138, 53-60.

McCloskey, T.C., Paul, B.K. & Commissaris, R.L. (1987). Buspirone effects in an animal conflict procedure: comparison with diazepam and phenobarbital. Pharmacol. Biochem & Behav., 27, 171-175.

Merlo, M., Pich, E.M. & Samanin, R. (1986). Disinhibitory effects of buspirone and low doses of sulpiride and haloperidol in two experimental anxiety models in rats: possible role of dopamine. Psychopharmacology, 89, 125-130.

Middlemiss, D.N., Fozard, J.R. (1983). 8-Hydroxy-2-(di-n-propylamino) tetralin discriminates between subtypes of the 5-HT recognition site. Eur. J. Pharmacol,. 90, 151-153.

Pazos, A. & Palacios, J.M. (1985). Quantitative autoradiographic mapping of serotonin receptors in the rat brain. I: serotonin-1 receptors. Brain Res., 346, 205-230.

Peroutka, S.J., Mauk, M.D. & Kocsis, J.D. (1987). Modulation of neuronal activity in the hippocampus by 5-hydroxytryptamine and 5-hydroxytryptamine$_{1A}$ selective drugs. Neuropharmacol., 26, N° 2/3, 139-146.

Schweizer, E., Amsterdam, K., Rickels, K., Myron, K. & Droba, M. (1986a). Open trial of buspirone in the treatment of major depressive disorder.

Psychopharmacol. Bull., 22, 183–185.

Schweizer, E., Rickels, K. & Lucki, I. (1986b). Resistance to the anti-anxiety effect of buspirone in patients with a benzodiazepine use. New Engl. J. Med., 314, 719–720.

Soubrie, P., Blas, C., Ferron, A. & Glowinski, J. (1983). Chlordiazepoxide reduces in vivo serotonin release in the basal ganglia of encephale isole but not anesthetized cats: evidence for a dorsal raphe site of action. J. Pharmacol. Exp. Ther., 226, N° 2, 526.

Soubrie, P. (1986). Reconciling the role of central serotonin neurons in human and animal behavior. The Behav. and Brain Sci., 9, 319–364.

Soubrie, P., Martin, P., El Mestikawy, S. & Hamon, M. (1987). Delayed behavioral response to antidepressant drugs following selective damage to the hippocampal noradrenergic innervation in rats. Brain Res., (In Press).

Sprouse, J.S. & Aghajanian, G.K. (1987). Electrophysiological responses of serotoninergic dorasal raphe neurons to $5-HT_{1A}$ and $5-HT_{1B}$ agonists. Synapse, 1, 3–9.

Thiebot, M.H., Le Biahn, C., Soubrie, P. & Simon, P. (1985a). Benzodiazepines reduce the tolerance to reward delay in rats. Psychopharmacol., 86, 147–152.

Thiebot, M.H., Soubrie, P. & Simon, P. (1985b). Is delay of reward mediated by shock–avoidance behavior a critical target for anti–punishment effects of diazepam in rats? Psychopharmacol., 87, 473–479.

Thiebot, M.H. (1986). Are serotonergic neurons involved in the control of anxiety and in the anxiolytic activity of benzodiazepines? Pharmacol. Biochem. Behav., 24, 1471–1477.

Trulson, M.E. & Arasteh, K. (1986). Buspirone decreases the activity of 5–hydroxytryptamine–containing dorsal raphe neurons in vitro. J. Pharm. Pharmacol., 38, 380–382.

Van de Kar, L.D., Urban, J.H., Lorens, S.A. & Richardson, K.D. (1985). The non–benzodiazepine anxiolytic buspirone inhibits stress–induced renin secretion and lowers heart rate. Life Sci., 36, 1149–1155.

Van Praag, H.M., Kahn, R.S., Asnis, G.M., Wetzler, S., Brown, S.L., Bleich, A. & Korn, M.L. (1987). Denosologization of biological psychiatry or the specificity of 5–HT distrubances in psychiatric disorders. J. Affective Disord., 13, 1–8.

Verge, D., Daval, G., Patey, A., Gozlan, H., El Mestikawy, S. & Hamon, M. (1985). Presynaptic 5–HT autoreceptors on serotonergic cell bodies and/or dendrites but not terminals are of the $5-HT_{1A}$ subtype. Eur. J. Pharmacol., 113, 463–464.

Willner, P. (1984). The validity of animal models of depression. Psychopharmacol., 38, 1–16.

A REVIEW OF THE EVIDENCE SUPPORTING THE ANXIOLYTIC POTENTIAL OF 5-HT$_3$ RECEPTOR ANTAGONISTS

M.B.Tyers.
Department of Neuropharmacology, Glaxo Group Research Ltd., Ware, Herts., SG12 ODJ, United Kingdom

5-Hydroxytryptamine receptors exist in at least three different types (Bradley et al., 1986), known at '5-HT$_1$-like', '5-HT$_2$' and '5-HT$_3$'. The absence of selective antagonists for the various effects of 5-HT attributable to actions on the 5-HT$_1$-like receptors precludes further characterisation within this class. However, it is quite clear that the 5-HT$_1$-like receptors are not homogeneous and can mediate distinct pharmacological effects of 5-HT including both contraction (Apperley et al. 1980) and relaxation (Feniuk et al., 1983) of vascular smooth muscle and inhibition of neurotransmitter release from peripheral nerves (Cox & Ennis, 1982; McGrath, 1977; North et al., 1980). The relationships between 5-HT$_1$-like functional receptors and 5-HT$_1$ binding sub-sites (Peroutka & Snyder, 1979; Pedigo et al., 1981) are not clear. The 5-HT$_{1A}$ site may be linked to adenylate cyclase while the 5-HT$_{1B}$ site appears to have an autoreceptor role on 5-HT release (Engel et al., 1986). The 5-HT$_2$ receptor is probably the same as the 5-HT$_2$ binding site since ketanserin, methysergide, pizotifen and cyproheptadine have similar affinities for both (Humphrey et al., 1982). 5-Hydroxytryptamine produces many actions via the 5-HT$_2$ receptor including smooth muscle contraction, platelet aggregation (Fozard, 1984) and various 5-HT related behaviours such as the 'head twitch' and 'wet dog shake' in rats (Green et al., 1983). Antagonists activity for the 5-HT$_3$ receptor has been described for norcocaine (Fozard et al., 1979), MDL 72222 (Fozard, 1984b), ICS 205-930 (Richardson et al., 1985), BRL 24924 (Cooper et al., 1986), BRL 43694 (Fake et al., 1987) and GR 38032F (Brittain et al., 1987). Activation of 5-HT$_3$ receptors induces some well characterised effects such as the Bezold-Jarisch reflex (Collins & Fortune, 1983; Fozard, 1984), excitation of primary afferent nociceptors and depolarisation of the membrane potential of the isolated superior cervical ganglion or vagus nerve preparations (Collins & Fortune, 1983; Ireland et al., 1982; Ireland et al., 1983, Ireland & Tyers, 1987). Antagonists of the 5-HT$_3$-receptors may be useful in treating migraine (Fozard, 1980). gastroparesis (Richardson, 1985) and certain forms of emesis (Costall et al., 1987d; Miner & Sanger, 1986, Stables et al., 1987).

The potential anxiolytic properties of 5-HT$_3$ antagonists have been described from experiments in several models of anxiety (Jones et al., 1987; Costall et al., 1987e; Tyers et al., 1987). In the mouse, the effects on suppressed spontaneous behaviour in the light/dark discrimination test (Costall et al., 1987e) have been determined. In this test, mice, taken from a dark environment, have a choice between a light and dark compartment and can move freely within and between them. The behaviour of mice which are naive to these conditions is suppressed in the white, brightly lit environment, and they prefer to stay mostly in the dark compartment.

In this test the suppressed behaviour is disinhibited by GR38032F, 0.01-10 μg/kg i.p., ICS 205-930, 0.01-10 μg/kg i.p. BRL43694, 0.01-0.1 μg/kg i.p.

and diazepam, 0.125–5 mg/kg i.p. All drugs increase exploratory activity in the light area with corresponding decreases in the dark area. The highest doses of ICS 205-930, 0.1 mg/kg. and BRL43694, 0.1 and 1.0 mg/kg, are less active than lower doses (Costall et al., 1987e) while diazepam, 10 mg/kg, causes overt sedation shown as reduced activity in both the light and dark areas. In the rat, social interaction (i.e. sniffing, following, tumbling, etc.) between pairs of rats (File & Hyde, 1979) is also disinhibited by 5–HT_3 antagonists when examined under conditions where social interaction is markedly suppressed, i.e. high light/unfamiliar. Pre-treatment orally with single doses of GR38032F, 0.5–100 µg/kg, ICS 205-930, 0.01–1.0 µg/kg, BRL43694, 0.1–10 µg/kg or MDL 72222, 1–100 µg/kg, increases social interactions without affecting locomotor activity (Figure 30.1). Diazepam, 2 mg/kg orally, also increases social interaction; but only slightly higher doses of diazepam reduce locomotor activity and social interaction indicative of its known sedative action. The variability of both the amount of social interaction and the effects of diazepam between tests makes it difficult to compare the maximum achieved by each 5–HT_3 antagonists. Comparisons with the reference drug, diazepam, provide more reliable indications of relative effect. In contrast to their effect in the social interaction and light/dark discrimination tests, in a water–lick conflict test in the rat (Vogel, 1971), neither GR38032F, 0.01–1 mg/kg i.p., BRL43694,1 mg/kg, nor ICS 205-930, 0.001–100 µg/kg i.p., are effective.

In the marmoset (Callithrix jacchus) characteristic behaviours of apprehension and anxiety are provoked when a human observer stands in front of the housing cage. Of these behaviours the most prominent are frequent aggressive postures, changes in vocalisation and a tendency to stay in the back of the cage rather than on the cage front; other responses include anal and vibrissae marking and piloerection. Pre-treatment with GR38032F, 0.1–10 µg/kg i.p. BRL43694, 0.1–10 µg/kg or ICS 205-930, 0.1–10 µg/kg., markedly reduces the number of aggressive postures and vocalisations; they spend more time on the front of the cage and those marmosets which were previously apprehensive of human contact can be handled or even fed by hand. Other territorial behaviours and piloerection are also inhibited. In these tests the anxiolytic activity of the 5–HT_3 antagonists appears to be similar to that of the benzodiazepines. However, they differ from the benzodiazepines in being ineffective in the water–lick conflict test as well as lacking anticonvulsant, muscle relaxant and hypnotic activities even at very high dose levels (Jones et al., 1987). The lack of action in the water–lick conflict test is similar to that found for 5–HT neurotoxins which have little or no effect in conflict models of anxiety (Peterson & Lassen, 1981; Commissaris et al., 1981). The clinical significance of the pharmacological profile of 5–HT_3 antagonists remains to be established.

The localisation of 5–HT in the brain is consistent with a role in the control of emotional behaviour. 5–Hydroxytryptamine cell bodies in the raphe nucleus project to limbic structures, cortical regions and the neostriatum. A role for 5–HT in anxiety and in the mechanism of action of benzodiazepines has been suggested previously (Wise et al., 1972). Benzodiazepines reduce both 5–HT turnover (Stein et al., 1973; Jenner et al., 1975) and the release of ^{3}H–5–HT from cortical slices in vitro (Collinge & Pycock, 1983) or from striatum and substantia nigra in vivo (Soubrie et al., 1983).

30: 5-HT$_3$ & Anxiety

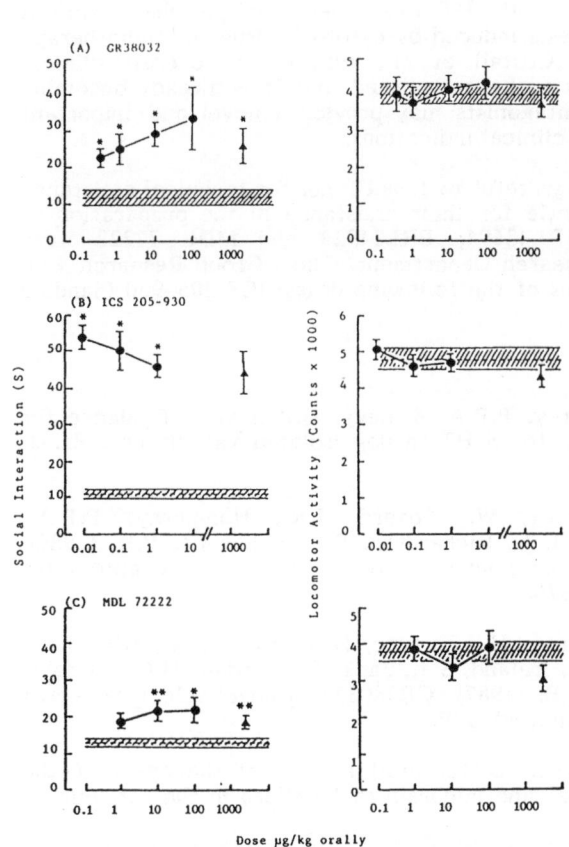

Fig. 30.1
Anxiolytic effects of 5-HT$_3$ antagonists in the social interaction test in the rat. The social interaction between pairs of rats was recorded under the experimental conditions of high light and unfamiliarity with the test arena. Prior to the test, animals were housed in groups of 5 rats under low light and quiet environment for at least 3 days. The time spent in active social interaction during a period of 10 min was scored from a video recording. Evcaluation was carried out blind. Drugs were given orally 45 min before testing. Control values ($\pm$ SEM) for social interaction and locomotor activity are shown as hatches areas. Significant increases/decreases in responding were indicated as *p=<0.01, **p=<0.05. (A) GR38032F, (B) ICS 205-930, (C) MDL 72222. Effects of diazepam, 2 mg/kg p.o., are indicated as Δ in each test.

Until very recently the presence of 5-HT$_3$ receptors in the CNS could only be inferred from behavioural studies such as those described above. Using a novel, highly selective 5-HT$_3$ antagonist, ^{3}H-GR65639. Kilpatrick et al. (1987) have now identified a binding site in rat brain which appears indistinguishable from functional 5-HT$_3$ receptors in the rat isolated vagus nerve. Furthermore, the location of 5-HT$_3$ receptors in limbic terminal and cortical areas, particularly the entorhinal cortex, compliments the behavioural effects of 5-HT$_3$ antagonists. The identification of 5-HT$_3$ receptors in human brain remains to be demonstrated.

The results obtained with 5-HT$_3$ antagonists suggest that these drugs may have important roles to play in the therapeutic management of anxiety. There is also extensive evidence from animal studies that 5-HT$_3$ antagonists may also have antipsychotic properties (Costall et al., 1987c; Hagan et al., 1987). Other studies (Richardson, 1985; Miner & Sanger, 1986; Costall et al., 1987; Cunningham et al., 1987a; Stables et al., 1987) have described peripheral actions showing that they inhibit the emesis induced by cytotoxic drugs or radiotherapy and promote gastric emptying (Costall et al., 1987f). The overall clinical potential of these drugs remains to be established but it is already becoming apparent that 5-HT_3 receptor antagonists may provide a novel and important advance in the therapy of several clinical indications.

Acknowledgements: We are very grateful to Lois Ullmer for technical assistance and to Tessa Parker and Ken Mantle for their assistance in the preparation of this manuscript. GR38032F, BRL43694, BRL24924 and MDL 72222 were synthesised in the Chemistry Research Department, Glaxo Group Research Ltd. Ware. We are grateful for supplies of the following drugs: ICS 205-930 (Sandoz) and diazepam (Roche).

References

Apperley, E., Feniuk, W., Humphrey, P.P.A. & Levy, G.P. (1980). Evidence for two types of excitatory receptor for 5-HT in dog isolated vasculature. Br. J. Pharmac., **68**, 215-224.

Bradley, P.B., Engel, G., Feniuk, W., Fozard, J.R., Humphrey, P.P.A., Middlemiss, D.N., Mylecharane, E.J., Richardson, B.P. & Saxena, P.R. (1986). Proposals for the classification and nomenclature of functional receptors for 5-HT. Neuropharmacol., **25**, 563-576.

Brittain, R.T., Butler, A., Coates, I.H., Fortune, D.H., Hagan, R., Hill, J.M., Humber, D.C., Humphrey, P.P.A., Ireland, S.J., Jack, D., Jordan, G.C., Oxford, A., Straughan, D.W. & Tyers, M.B. (1987). GR38032F, a novel selective 5-HT_3 receptor antagonist. Br. J. Pharmac., **90**, 87P.

Collinge, J. & Pycock, C.J. (1983). Differential actions of diazepam of the release of ^3H-5-HT from cortical and midbrain raphe slices in the rat. Eur. J. Pharmac., **85**, 9.

Collins, D. & Fortune, D.H. (1983). Phenylbiguanide mimics the Bezold-Jarisch effect of 5-HT in the rat. Br. J. Pharmac., **80**, 570P.

Commissaris, R.L., Lynes, W.H. & Rech, R.H. (1981). The effects of LSD, DOM, pentobarbital and methaqualone on punished responding in control and 5,7,DHT-treated rats. Pharmac. Biochem. Behav., **14**, 617.

Cooper, S.M., McClellan, C.M., McRitchie, B. & Turner, D.H. (1986). BRL24924: a new and potent gastric motility stimulant. Br. J. Pharmac., **88**, 383P.

Costall, B., Domeney, A.M., Gunning, S.J., Naylor, R.J., Tattersall, F.D. & Tyers, M.B. (1987a). GR38032F: a potent and novel inhibitor of cisplatin-induced emesis in the ferret. Br. J. Pharmac., **90**, 90P.

Costall, B., Domeney, A.M., Naylor, R.J. & Tyers, M.B. (1987b). Effects of the 5-HT_3 antagonists GR38032F, BRL43694 and ICS 205-930 in tests for anxiolytic activity. This volume.

Costall, B., Domeney, A.M., Naylor, R.J. & Tyers, M.B. (1987c). Effects of the 5-HT$_3$ antagonist, GR38032F, on raised dopaminergic activity in the mesolimbic system of the rat and marmoset brain. Br. J. Pharmac., (In press.)

Costall, B., Domeney, A.M., Gunning, S.J., Naylor, R.J., Tattersall, F.D. & Tyers, M.B. (1987d). GR38032F: a potent and novel inhibitor of cisplatin-induced emesis in the ferret. Br. J. Pharmac., 90, 90P.

Costall, B., Domeney, A.M., Gerrard, P.A. Jones, B.J., Kelly, M.E., Oakley, N.R. & Tyers, M.B. (1987e). The anxiolytic activities of 5-HT$_3$ antagonists BRL43694, ICS 205-930 and GR38032F. This volume.

Costall, B., Gunning, S.J., Naylor, R.J. & Tyers, M.B. (1987f). The effect of GR38032F, a novel 5-HT$_3$ antagonists on gastric emptying in the guinea-pig. Br. J. Pharmac., 91, 263-5.

Cox, B. & Ennis, C. (1982). Characterisation of 5-HT autoreceptors in the rat hypothalamus. J. Pharm. Pharmac., 34, 439-441.

Cunningham, D., Hawthorn, J., Pople, A., Gazet, J.C., Ford, H.T., Challoner, T. & Coombes, R.C. (1987). Prevention of emesis in patients receiving cytoxic drugs by GR38032F, a selective 5-HT$_3$ receptor antagonist. Lancet, i, 1461-1463.

Engel, G., Göthert, M., Hoyer, D., Schliker, E., Sitonen, L. & Stadler, P.A. (1986). Identity of inhibitory presynaptic 5-HT autoreceptors in the rat brain cortes with 5-HT$_{1B}$ binding sites. Naunyn-Schmiedebergs Arch. Pharmac., 324, 116-124.

Fake, C.S., King, F.D. & Sanger, G.J. (1987). BRL43694: a potent and novel 5-HT$_3$ receptor antagonist. Br. J. Pharmac., 92, 335P.

Feniuk, W., Humphrey, P.P.A. & Watts, A.D. (1983). 5-HT-induced relaxation of isolated mammalian smooth muscle. Eur. J.Pharmac., 96, 71-78.

File, S.E. & Hyde, J.R.G. (1979) Can social interaction be used by measure anxiety? Pharmac. Biochem. Behav., 11, 65-69.

Fozard, J.R. (1984b). MDL 72222: a potent and highly selective antagonist at neuronal 5-HT receptors. Naunyn-Schied. Arch. Pharmac., 326, 36-44.

Fozard, J.R., Mobarok Ali, A.T.M. & Newgrosh, G. (1979). Blockade of serotonin receptors on autonomic neurones by (-)-cocaine and some related compounds. Eur. J. Pharmac., 59, 195-210.

Fozard, J.R. (1984a). Neuronal 5-HT receptors in the periphery. Neuropharmacol., 23, 1473-86.

Fozard, J.R. (1980). In Proc. Int. Symp. Headache, Florence 1980, Raven Press.

Gardner, C.R. (1985). Pharmacological studies of the role of serotonin in animal models of anxiety. In: Neuropharmacology of Serotonin. pp. 296-297 Oxford: Univ. Press.

Green, A.R., O'Shaughnessy, K., Hammon, M., Schachter, M. & Grahame-Smith, D.G. (1983). Inhibition of 5-HT mediated behaviour by the putative 5-HT$_2$ antagonist pirenpirone. Neuropharmacol., 22, 573-578.

Gunning, S.J., Hagan, R.M. & Tyers, M.B. (1987). Cisplatin induces biochemical and histological changes in the small intestine of the ferret. Br. J. Pharmac., 90, 135P.

Hagan, R.M., Butler, A., Hill, J.M., Jordan, C.C., Ireland, S.J. & Tyers, M.B. (1987). Effect of the 5-HT$_3$ receptor antagonist, GR38032F, on responses to injection of a neurokinin agonist into the ventral tegmental area of the rat brain. Eur. J. Pharmac., 138, 303-305.

Humphrey. P.P.A., Feniuk, W. & Watts, A.D. (1982). Ketanserin: a novel hypertensive drug? J. Pharm. Pharmac., 34, 541.

Ireland, S.J., Straughan, D.W. & Tyers, M.B. (1982). Antagonism by metoclopramide and quipazine of 5-HT-induced depolarisations of the rat isolated vagus nerve. Br. J. Pharmac., 75, 16P.

Ireland, S.J., Fortune, D.H. & Tyers, M.B. (1983). Influence of 5-HT uptake onto the 5-HT antagonist activity of metoclopramide on the rat superior cervical ganglion. Br. J. Pharmac., 78, 68P.

Ireland, S.J. & Tyers, M.B. (1987). Pharmacological characterisation of 5-HT-induced depolarisation of the rat isolated vagus nerve. Br. J. Pharmac., 90, 229-238.

Jenner, P., Chadwick, D., Reynolds, E.J. & Marsden, C.D. (1975). Altered 5-HT metabolism with clonazepam, diazepam and diphenylhydrantoin. J. Pharm. Pharmac., 27, 707-710.

Jones, B.J., Costall, B., Domeney, A.M., Kelly, M.E., Naylor, R.J., Oakley, N.R. & Tyers, M.B. (1987). The potential anxiolytic activity of GR38032F a 5-HT$_3$ receptor antagonist. Br. J. Pharmac., (In Press.)

Kilpatrick, G.K., Jones, B.J. & Tyers, M.B. (1987). The identification and distribution of 5-HT$_3$ receptors in the brain. Nature, (In Press.)

McGrath, M.A. (1977). 5-HT and neurotransmitter release in canine blood vessels. Circulation Res., 41, 428-435.

Miner, W.D. & Sanger, G.J. (1986). Inhibition of cisplatin-induced vomiting by selective 5-HT M-receptor antagonism. Br. J. Pharmac., 88, 497-499.

North, R.A., Henderson, G., Katayama, Y. & Johnson, S.M. (1980). Electrophysiological evidence for presynaptic inhibition of acetylcholine release by 5-HT in the enteric nervous system. Neurosci., 5, 581-586.

Pedigo, N.W., Yamamura, H.I. & Nelson, D.L. (1981). Discrimination of multiple ^{3}H-5-HT binding sites by the neuroleptic spiperone in rat brain. J. Neurochem., 36, 220-226.

Petersen, E.N. & Lassen, J.B. (1981). A water lick conflict paradigm using drug experienced rats. Psychopharmacol., 75, 236.

Peroutka, S.J. & Snyder, S.H. (1979). Multiple serotonin receptors: differential binding of ^{3}H-LSD and ^{3}H-spiroperidol. Molec. Pharmac., 16, 687-699.

Richardson, B.P., Engel, G., Donatsch, P. & Stadler, P.A. (1985). Identification of 5-HT M-receptor sub-types and their specific blockade by a new class of drugs. Nature, 316, 126-131.

Richardson, B.D. (1985). 5-HT receptor antagonism by metoclopramide and ICS 205-930 in the guinea-pig leads to enhancement of stomach strip contractions induced by electrical field stimulation and facilitation of gastric emptying in vivo. J. Pharm. Pharmac., 37, 664-667.

Soubrie, P., Blas, C., Ferron, A. & Glowinski, J. (1983). Chlordiazepoxide reduced in vivo serotonin release in the basal ganglia of encephale isole but not anaesthetised cats. Evidence for a dorsal raphe site of action. J. Pharmac. Exp. Ther., 226, 526.

Stables, R., Naylor, P.L.R., Bailey, H.E., Costall,. B., Gunning, S.J., Hawthorn, J., Naylor, R.J. & Tyers, M.B. (1987). Antiemetic properties of the 5-HT$_3$ antagonist, GR38032F. Cancer Treatment Review. (In press.)

Stein, L., Wise, C.D. & Berger, B.D. (1973). In: The Benzodiazepines. pp. 299, Raven Press.

Tyers, M.B., Costall, B., Domeney, A.M., Jones, B.J., Kelly, M.E., Naylor, R.J. & Oakley, N.R. (1987). The anxiolytic activities of 5-HT$_3$ antagonists in laboratory animals. Neurosci. letters, Suppl., 29, S68.

Vogel, J.R., Beer, B. & Clody, D.E. (1971). A simple and reliable conflict procedure for testing anti-anxiety agents. Psychopharmacologia (Berl.), 21, 1-7.

Wise, C.D., Berger, B.D. & Stein, L. (1972). Benzodiazepines: anxiety-reducing activity by reduction of serotonin turnover in the brain. Sci., 177, 180-183.

ULTRASONIC VOCALIZATIONS BY RAT PUPS AS AN ANIMAL MODEL FOR ANXIOLYTIC ACTIVITY: EFFECTS OF SEROTONERGIC DRUGS

J.Mos & B.Olivier, Department of Pharmacology, Duphar B.V., P.O.Box 2, 1380 AA Weesp, The Netherlands

An interesting development in the psychopharmacology of anxiolytics is the study of 'distress calls' of young animals in response to separation of the mother and/or siblings (Hofer and Shair, 1978). Rat pups react with ultrasonic calls of 35-55 kHz frequency to such separation. The number of these calls is reduced after treatment with benzodiazepines and this effect was not caused by sedation (Gardner and Brudham, 1987). In the present study we tested the effects of various serotonergic agents as there is ample evidence that 5-HT is involved in anxiety (Broekkamp, 1988; Soubrie, 1988). We therefore have adopted an experimental set-up in which the ultrasound production by rat pups of 9-11 days old (Hard et al., 1982) was measured under two conditions with varying stress levels. Additionally, we used a simple psychomotor test to verify whether drugs affected sensory and/or motor capabilities.

Methods
Wistar rats were used in all experiments. After birth, pups were left undisturbed with their mother except for replacement of cage bedding. Within 24 hours of birth, litters were culled to eight, without differentiation between male and female pups. All pups were tested once at an age of 9-11 days. On the experimental day the mother was separated from her pups, which were transported in their home cage to the experimental room. Pups were randomly allocated to vehicle or one of the drug doses according to a balanced block design. Pups were weighed, marked, and dosed intraperitoneally with either vehicle or drug and replaced in the home cage. Thirty minutes after dosing the animals were tested for 5 minutes to record their ultrasonic calls. Animals of a litter were placed on one of two circular aluminium plates which were kept at a temperature of either 18°C (cold plate) or 37°C (warm plate) respectively by circulation of water through a reservoir below the plate, similar to the set up described by Gardner (1985a). A circular plexiglass wall of 25 cm height was placed on the aluminum plate and covered with a plexiglass top on which the microphone was mounted. Ultrasounds were recorded with a Bruel & Kjaer 2619 preamplifier, connected to a B & K 2608 type amplifier using a B & K 4135 condensor microphone. Filters (Interelectronic AF 173) were adjusted to selectively count ultrasonic cries between 30 and 60 kHz. The ultrasonic calls were fed through an electronic device with variable trigger level, prior to processing on a computer. The number of calls was counted per minute and the total number of calls was summed over the whole test period. When the signal intensity dropped below the trigger level, a new call was only registered when it was separated by at least 20 ms or longer from the previous one.

A natural response of rat pups when they are placed on an inclination with their head down is to turn around against this negative geotaxic direction. This behaviour needs a certain stage of development of sensory-motor maturity. At an age of 9 days all rat pups are capable of performing appropriately in this negative geotaxis test within a couple of seconds. Pups of the litter which were not used for the ultrasound recording were injected with drug or vehicle and 30 minutes later five consecutive negative geotaxis tests were performed. If the animal failed to turn around within 60 seconds a new trial was started and also immediately after a successful response. The median value in seconds of the five tests was used for statistical comparison. Apart from simple slowing down of the negative geotaxis response, two other disturbed reactions could occur. After drug treatment some pups failed to move when placed on the platform. This led to a cut-off value of 60 seconds. Other pups started to turn but were unsuccessful, either because of muscle relaxation (sliding down the platform) or due to incoordinated movements.

Statistical analysis
The sum score over the first five test minutes proved to be the most sensitive measure to show drug effects. The indicated statistical analysis for the drug effects is an analysis of variance (ANOVA) which assumes normality of the data. For the warm plate, this assumption is approximately met for the log transformation of the sum score, whereas for the cold plate the raw sum score suffices. Analysis of variance was followed by Tukey's multiple comparisons at the 5% level of significance to compare cohorts of treatment groups and to study monotonicity to dose level. The negative geotaxis response was analysed by Kruskal-Wallis analysis of variance followed by Mann Whitney U tests to compare differences from vehicle treatment.

Results
Table 31.1 briefly summarizes the results. As found earlier benzodiazepines (BDZ) suppressed ultrasounds under both testing conditions. Chlordiazepoxide dose-dependently reduced ultrasonic calling, but the level under the cold condition remained higher than under the warm condition. The doses selected for diazepam were clearly too high because the maximal response was already present at 2.5 mg/kg. At the same dose the negative geotaxis response was distorted due to muscle relaxation; in contrast, the negative geotaxis response under chlordiazepoxide was not affected. Oxazepam reduced ultrasounds similar to chlordiazepoxide and diazepam and was intermediate at the negative geotaxis test. It can thus be concluded that reduced ultrasound production is not necessarily linked to sedation or muscle relaxation as revealed in the geotaxis test.

$5-HT_{1A}$ agonists were also quite potent and effective in reducing ultrasonic calls. 8-OH-DPAT, buspirone, ipsapirone and flesinoxan reduced the number of ultrasounds under both test conditions. Apart from the increased potency, the main difference from BDZ is the fact that at the highest dose tested the cold and warm plate did not differ in ultrasounds evoked. While with supramaximal doses of diazepam a certain level of calling remained under cold conditions, this disappeared using $5-HT_{1A}$ agonists. The latter virtually abolished ultrasounds under both conditions at the highest dose. Except flesinoxan, $5-HT_{1A}$ agonists did not affect geotaxis responding. The $5-HT_2$ agonist DOI also reduced ultrasounds but the animals were severely sedated in the negative geotaxis test at all doses.

Another class of response was caused by treatment with TFMPP, a $5-HT_{1B}$ agonist, by eltoprazine (DU 28853) and RU 24969, mixed $5-HT_{1A/B}$ agonists

Table 31.1 Effects of benzodiazepines and serotonergic drugs on the number of ultrasonic cries and on negative geotaxis. Under the warm and cold conditions the doses used (mg/kg, ip.) are listed. When a dose is underlined, that drug dose decreased the ultrasonic calls significantly.
For the negative geotaxis the lowest effective dose disrupting the behaviour is given. *An arrow depicts that the behaviour is either enhanced or decreased at the dose given.

Drug:	warm condition	cold condition	geotaxis
Benzodiazepines			
chlordiazepoxide	2.5 _5.0_ _10.0_	2.5 _5.0_ _10.0_	>10.0
diazepam	_2.5_ _5.0_ _10.0_	_2.5_ _5.0_ _10.0_	2.5
oxazepam	_1.0_ _3.0_ _10.0_	_1.0_ _3.0_ _10.0_	10.0
5-HT1a agonists			
8-OH-DPAT	0.05 _0.1_ _0.2_	0.05 _0.1_ _0.2_	>0.2
buspirone	0.3 _1.0_ _3.0_	0.3 _1.0_ _3.0_	3.0
ipsapirone	0.3 _1.0_ 3.0	0.3 _1.0_ _3.0_	3.0
flesinoxan	_0.3_ _1.0_ _3.0_	_0.3_ _1.0_ _3.0_	0.3
5-HT2 agonist			
DOI	_0.3_ _1.0_ _3.0_	_0.3_ _1.0_ _3.0_	0.3
5-HT1b agonist			
TFMPP	0.3 1.0 3.0	0.3 1.0 _3.0_	>3.0* ↑
5-HT1a/b agonists			
RU 24969	0.3 1.0 3.0	0.3 _1.0_ _3.0_	0.3↑ 1.0↓
eltoprazine	0.3 1.0 3.0	0.3 _1.0_ _3.0_	>3.0* ↑
5-HT reuptake blockers			
fluvoxamine	5.0 10.0 20.0	_5.0_ _10.0_ _20.0_	>20.0
zimeldine	1.0 3.0 10.0	1.0 _3.0_ _10.0_	3.0
fluoxetine	5.0 10.0 20.0	5.0 10.0 _20.0_	>20.0
5-HT1/2 antagonists			
methysergide	0.3 1.0 3.0	0.3 1.0 3.0	n.t.
ritanserine	0.3 1.0 3.0	0.3 1.0 3.0	n.t.
5-HT3 antagonist			
GR 38032F	0.001 0.01 0.1	0.001 0.01 0.1	>0.1
GR 38032F	0.1 0.3 1.0	0.1 0.3 1.0	>1.0

and by serotonin reuptake blockers. No significant changes were observed under warm conditions, whereas dose-dependent reductions in pup vocalizations were found at the cold plate.

The decrease in ultrasonic calls reached the level of the calls produced at the warm plate, but was never significantly lower than the latter condition. Interestingly, the drugs differed remarkably at the geotaxis test. TFMPP and eltoprazine increased the speed of turning at all doses. RU 24969 also improved the response at 0.3 mg/kg but its strong stimulatory activity precluded appropriate responses at higher doses. The serotonin reuptake blockers fluvoxamine, zimeldine and fluoxetine did not deteriorate the negative geotaxic response or only affected it at the highest dose tested. The ratio between inhibition of pup calling and disturbed motor responses was greatest for fluvoxamine. The serotonin antagonists methysergide (a mixed $5-HT_{1/2}$ antagonist), ritanserine, a specific $5-HT_2$ antagonist, and GR 38032F, a $5-HT_3$ antagonist were ineffective in this animal model.

Discussion
The benzodiazepines, which are known for their clinical efficacy in alleviating anxiety symptoms, reduce separation-induced ultrasounds in rat pups (Gardner, 1985b; Insel and Hill, 1987). This effect is consistent for all benzodiazepines tested so far and seems reasonably specific, i.e. the ratio to a specific disturbance (geotaxis, motor coordination) is sufficiently large. This has also been reported by Gardner and Brudham (1987) for several (potential) anxiolytics interacting with the GABA/benzodiazepine receptor complex. The benzodiazepines reduce ultrasounds under both testing conditions, abolishing it almost completely at the warm plate at the higher doses but reducing it to a certain baseline level at the cold plate. The new generation non-benzodiazepine anxiolytics such as buspirone and ipsapirone also reduce separation-induced ultrasonic calling albeit in a slightly different way. Both under the warm and the cold condition ultrasound production goes down after drug treatment. However, the ultrasounds under the cold condition do not remain stable at a certain baseline level, but they are almost completely abolished as are the ultrasounds under the warm condition. An as yet unexplored possibility is that hypothermia, caused by drug treatment interferes. Many $5-HT_{1A}$ agonists cause hypothermia (Hjorth 1985; Hutson et al., 1987), which may interfere with ultrasound production.

It is premature to speculate upon the exact influence of hypo- or hyperthermia on ultrasonic responses. Other $5-HT_{1A}$ agonists such as 8-OH-DPAT and flesinoxan also reduce pup ultrasounds. Although it thus seems tempting to ascribe anxiolytic properties to $5-HT_{1A}$ receptor stimulation, it needs to be said that most drugs are not devoid of significant activity on other receptors.

Some caution is needed but it looks promising to explore other $5-HT_{1A}(-ERROR-)$ and try to antagonize their action. The pup ultrasonic calls are in any case sensitive to both classical benzodiazepine anxolytics and newer drugs, which certainly does not apply to all conventional models for anxiety in which benzodiazepines are found.

Other serotonin agonists such as the rather specific $5-HT_{1B}$ agonist TFMPP and the mixed $5-HT_{1A/1B}$ agonists eltoprazine and RU 24969 effectively block ultrasounds only at the cold plate, TFMPP being the weakest. Serotonin reuptake blockers also affect pup ultrasounds only at the cold plate. The significance of this finding is not yet clear. Reuptake blockers which are clinically effective antidepressants may reduce symptoms of anxiety (Lader, 1985; Wakelin, 1988). It could be hypothesized that these drugs exert a positive influence under more stressful conditions, but again interactions with temperature cannot be excluded. Not all antidepressants however, exert similar actions on pup vocalizations (Mos and Olivier, 1988). DOI, a putative purported $5-HT_2$ agonist was not very specific in ultrasound reduction as geotaxis responses were affected in the same dose range. Since DOI also has potent $5-HT_{1C}$ affinity it remains to be established which receptor interaction is responsible for which effect. The serotonin $5-HT_2$ and $5-HT_3$ antagonists were devoid of significant activity in the pup vocalization test under both conditions. This is somewhat surprising in view of the reported anxiolytic activity of ritanserin in humans (Reyntjen et al., 1986) and the results of the $5HT_3$ antagonists by Costall et al. (1988). We have no explanation for these observations. Future research aimed at measuring duration, frequency and temporal distribution of the calls may unravel subtle modification of the behaviour undetectable by counting the number of calls alone.

A final remark pertains to the question of specificity of drug effects. Although a distorted response in the negative geotaxis test is indicative of (non-specific) side effects, it is not necessarily the cause of a decrease in pup vocalizations. Some reference drugs (haloperidol, clonidine, prazosin, methiothepin) inhibit a normal response in the negative geotaxic test but do not decrease pup vocalizations significantly. Although several experiments have to be performed to elaborate the evidence that this model is indeed an improvement to present models for anxiolytic drugs, the initial results look promising and the validity of separation induced vocalizations as an animal model seems to extend to primates (Suomi et al., 1978) and perhaps to humans (Hinde, 1974).

References

Broekkamp, C.L. & Jenck, F. (1988). The relationship between various animal models of anxiety, fear-related psychiatric symptoms and response to serotonergic drugs. This volume.

Costall, B., Domeney, A.M., Gerrard, P.A., Jones, B.J., Kelly, M.E., Oakley, N.R. & Tyers, M.B. (1988). The anxiolytic activities of the $5-HT_3$ receptor antagonists GR38032F, ICS 205-930 and BRL 43694. This volume.

Gardner, C.R. (1985a). Inhibition of ultrasonic distress vocalizations in rat pups by chlordiazepoxide and diazepam. Drug Development Research, 5, 185-193.

Gardner, C.R. (1985b). Distress vocalization in rat pups. A simple screening method for anxiolytic drugs. Journal of Pharmacological Methods, 14, 181-187.

Gardner, C.R. & Budham, P. (1987). Effects of agents which interact with central benzodiazepine binding sites on stress-induced ultrasounds in rat pups. European Journal of Pharmacol., 134, 275-283.

Hard, E., Engel, J. & Musi, B. (1982). The ontogeny of defensive reactions in the rat: influence of the monoamine transmitter systems. Scandinavian Journal of Psychology, suppl 1. 90-96.

Hinde, R.A. (1974). Biological bases of human social behaviour. New York: McGraw-Hill.

Hjorth, S. (1985). Hypothermia in the rat induced by the potent serotonergic agent 8-OH-DPAT. Journal of Neural Transmission, 61, 131-135.

Hofer, M.A. & Shair, H. (1978). Ultrasonic vocalization during social interaction and isolation in 2-week-old rats. Developmental Psychobiol., 11, 495-504.

Hutson, P.H., Donohoe, T.P. & Curzon, G. (1987). Hypothermia induced by the putative 5-HT_{1A} agonists LY165163 and 8-OH-DPAT is not prevented by 5-HT depletion. European Journal of Pharmacol., 143, 221-228.

Insel, T.R., Hill, J.L. & Mayor, R.B. (1986). Rat pup ultrasonic isolation calls: possible mediation by the benzodiazepine receptor complex. Pharmacol., Biochem. & Behav., 24: 1263-1267

Lader, M.H. (1985). The borderline between anxiety and depression: drug treatment. In: the borderline between anxiety and depression, biological and diagnostic aspects. (Eds. Verhoeven, W.M.A., Westenberg, H.G.M. & Knottnerus, J.G.), pp. 78-86. Leusden: Medidact.

Mos, J. &, Olivier, B. (1988). Ultrasonic vocalizations by rat pups as an animal model for anxiolytic activity: effects of antidepressants drugs. In: the second depression, anxiety and aggression conference. (Eds. Blijleven, W. & Snethlage, C. Amsterdam: Medidact. (In Press.)

Reyntjens, A., Gelders, Y.G., Hoppenbrouwers, M.L.J.A. & Van den Bussche, G. (1986). Thymostenic effects of ritanserin (R 55 667), a centrally acting serotonin-S_2 receptor. Drug Dev. Res., 8, 205-211.

Soubrie, P. (1988). Serotonin 1A receptors: a bridge between anxiety and depression. This volume.

Suomi, S.J., Seaman, S.F., Lewis, J.K., Delizio, R.D. & McKinney, W.T. (1978). Effects of imipramine treatment of separation-induced social disorders in rhesus monkeys. Archives of General Psychiatry,35, 321-325.

Wakelin, J.S. (1988). The role of serotonin in depression and suicide; do serotonin reuptake inhibitors provide the key? In: selective 5-HT reuptake inhibitors: novel or commonplace agents? (Eds. Mastpar, G. & Wakelin, J.S.), Basle: Karger. (In Press.)

EFFECTS OF THE 5-HT$_{1A}$ AGONIST IPSAPIRONE ON THE BEHAVIOURAL, ENDOCRINE AND NEUROCHEMICAL RESPONSES TO CONDITIONED FEAR

S.A.Lorens, H.Mitsushio, L.D. Van de Kar
Department of Pharmacology, Stritch School of Medicine, Loyola University of Chicago, Maywood, Illinois 60153, U.S.A.

The anxiolytic benzodiazepines, chlordiazepoxide and midazolam, block the corticosterone and prolactin but not the renin response to stress (Van de Kar et al., 1985a). In contrast, the non-benzodiazepine anxiolytic 5HT-1A agonist, buspirone, attenuates the corticosterone, prolactin and renin responses to conditioned fear (Van de Kar et al., 1985b; Urban et al., 1986). High doses of buspirone and ipsapirone, however, elevate plasma corticosterone and renin concentrations in non-stressed male rats (Urban et al., 1986; Lorens & Van de Kar, 1987). These observations suggest that distinct types of anxiolytic drugs differentially affect the endocrine responses to stress. In the present study we examined the dose-dependent effects of the putative anxiolytic 5HT-1A agonist, ipsapirone, on the behavioural, endocrine and neurochemical responses to conditioned fear.

Method
Male Sprague-Dawley rats, weighing 300-325 g at the time of testing, were used. The animals were housed in a temperature and illumination (12 h light-dark cycle; lights on at 7.00 h) controlled room.

Beginning two weeks after their arrival in the laboratory, the rats were subjected to a 15 min conditioned emotional response (CER) procedure (Van de Kar et al., 1985a). Fifteen min after being transported to the testing room and ten min after being placed in the experimental chamber, the stressed rats received an inescapable foot shock (0.8 mA for 10 s), then were returned to their home cage. This procedure was repeated, once per day between 13.00-15.00 h, for 3 consecutive days. The behaviour of the rats was videotaped for subsequent analysis. On the fourth day the rats received injections of vehicle (2.0 ml/kg, i.p. of isotonic saline) or different doses of ipsapirone (0.5, 1.0, 2.5 and 5.0 mg/kg, i.p.) 15 min prior to being transported to the testing room. No shock was delivered and the animals were sacrificed by decapitation. Trunk blood was collected for hormonal analysis by radioimmunoassay, and regional CNS samples were obtained for analysis by HPLC-EC. Control animals were treated in the same manner as the stressed rats, except that they never received foot shock. The data were analyzed by a 2x5 way analysis of variance (ANOVA) followed by Duncan's multiple range test where appropriate.

Results
Compared to the vehicle treated control rats, the stressed animals showed 100% increases in plasma corticosterone, renin and prolactin concentrations. Low doses (0.5-2.5 mg/kg) of ipsapirone attenuated but did not block the corticosterone and renin responses to stress, whereas only the high doses (2.5-5.0 mg/kg) of ipsapirone reduced the prolactin response to stress. The highest dose of ipsapirone (5.0 mg/kg), furthermore, significantly increased (100%) the corticosterone levels of the non-stressed rats.

The vehicle treated stressed animals showed significant elevations in medial frontal cortex (MFC; 56%) and amygdaloid (35%) DOPAC concentrations, as well as in DOPAC/DA ratios. By itself, the CER procedure did not significantly affect DOPAC levels in either the nucleus accumbens or hypothalamus. DA concentrations were slightly reduced in the MFC but not in other areas.

Ipsapirone attenuated the stress induced increases in amygdaloid but not MFC DOPAC content and DOPAC/DA ratio. In contrast, the highest dose of ipsapirone (5.0 mg/kg) significantly elevated the DOPAC contents and DOPAC/DA ratios in all four regions of the vehicle treated control rats. The effects of this dose of ipsapirone and stress, however, were not additive.

Conditioned fear did not affect 5-HT and 5-HIAA levels, or 5-HIAA/5-HT ratios, in any of the areas assayed. However, the highest dose (5.0 mg/kg) of ipsapirone significantly decreased the 5-HIAA concentrations (25-35%) and 5-HIAA/5-HT ratios (25-35%) in all four regions studied in both stressed and non-stressed animals.

Although ipsapirone dose-dependently reduced the number of fecal boli (defecation score) produced during stress, it did not otherwise detectably alter the behaviour of the fear conditioned rats. Thus, the duration of immobilization and the number of rears were not significantly affected.

Discussion
The present observations suggest that ipsapirone can attenuate the magnitude of several responses to fear evoking stimuli and possesses an anxiolytic property.

Ipsapirone is a $5-HT_1$ agonist with binding characteristics similar to the non-benzodiazepine anxiolytic drug buspirone (Peroutka, 1985). Buspirone effectively inhibits the effect of conditioned fear on renin, prolactin and corticosterone secretion (Urban et al., 1986; Van de Kar et al., 1985b). Our present data suggest that ipsapirone and buspirone have similar neuroendocrine effects, and support the hypothesis that ipsapirone has anxiolytic characteristics. One significant difference between ipsapirone and buspirone is that ipsapirone does not increase plasma prolactin levels in unstressed rats. This probably indicates a lack of dopamine antagonistic properties, and suggests that ipsapirone is a more selective $5-HT_{1A}$ agonist than buspirone.

The highest dose (5.0 mg/kg) of ipsapirone significantly increased DOPAC concentrations and DOPAC/DA ratios, and significantly reduced 5-HIAA levels and 5-HIAA/5-HT ratios in the four brain areas studied in both stresses and non-stressed rats. This dose of ipsapirone also produced a 100% increase in plasma corticosterone levels and significantly reduced rearing scores in non-stressed rats. These observations suggest that high doses of ipsapirone may be anxiogenic. Interestingly, in this regard, the anxiogenic beta-carboline drug, FG 7142, and high doses of chlordiazepoxide also has been reported to produce significant increases in MFC DA turnover and in plasma corticosterone levels (Roth et al., 1987; and our unpublished observations).

A few differences between ipsapirone and benzodiazepines should be noted. Ipsapirone prevented the stress-induced increase in amygdaloid DA turnover but did not affect MFC DA turnover. The benziodiazepines have been reported to prevent stress induced increases in MFC DA turnover. To our knowledge, the effects of benzodiazepines on stress induced amygdaloid DA turnover have not been examined. Low doses (0.5-2.5 m/gkg) of ipsapirone attenuated the renin and corticosterone response to conditioned fear. In contrast, benzodiazepines inhibit the effect of stress on corticosterone but not renin secretion (Van de Kar, et al., 1985).

Overall, the results suggest that ipsapirone affects the behavioural, endocrine, and neurochemical responses to conditioned fear in a manner which is distinct from the benzodiazepine anxiolytic drugs.

References

Lorens, S.A. & Van de Kar, L.D. (1987). Differential effects of serotonin (5-HT1A and 5-HT2) agonists and antagonists on renin and corticosterone secretion. Neuroendocrinology, 45, 305-310.

Peroutka, S.J. (1985). Selective interaction of novel anxiolytics with 5-hydroxytryptamine1A receptors. Biol. Psychiat., 20, 971-979.

Roth, R.H., Wolf, M.E. & Deutch, A.Y. (1987). Neurochemistry of midbrain dopamine systems. In: Psychopharmacology: The Third Generation of Progress. (ed. Meltzer, H.Y.), pp. 81-94. New York: Raven Press.

Urban, J.H., Van de Kar, L.D., Lorens, S.A. & Bethea, C.L. (1986). Effect of the anxiolytic drug buspirone and prolactin and corticosterone secretion in stressed and unstressed rats. Pharmacol. Biochem. Behav., 25, 457-462.

Van de Kar, L.D., Lorens, S.A., Urban, J.H., Richardson, K.D., Paris, J.M. & Bethea, C.L. (1985a). Pharmacological studies on stress-induced renin and prolactin secretion: Effects of benzodiazepines, naloxone, propranolol and disopropylfluorophosphate (DFP). Brain Res., 345, 257-263.

Van de Kar, L.D., Urban, J.H., Lorens, S.A. & Richardson, K.D. (1985b). The nonbenzodiazepine anxiolytic buspirone inhibits stress-induced renin secretion and lowers heart rate. Life Sci., 36, 1149-1155.

BUSPIRONE, IPSAPIRONE AND 5-HT$_{1A}$-RECEPTOR AGONISTS SHOW AN ANXIOGENIC-LIKE PROFILE IN THE ELEVATED PLUS-MAZE

Paul Moser
Merrell Dow Research Institute, Strasbourg Centre, 16 rue d'Ankara, 67084 Strasbourg Cedex, France

Much interest has focussed on the role that 5-hydroxytryptamine (5-HT) plays in anxiety and the actions of anxiolytic compounds, particularly following the appearance of compounds such as buspirone and ipsapirone (e.g. Gardner, 1985; Chopin & Briley, 1987). These putative anxiolytic, non-benzodiazepine, compounds have high affinity for the 5-HT$_{1A}$ subtype of CNS 5-HT receptors, which has led to the suggestion that 5-HT$_{1A}$ receptors are the primary target site for their anxiolytic effects (Peroutka, 1985).

Clinically, buspirone appears to be equipotent with diazepam in the relief of anxiety (Schuckit, 1984), but in animal tests the results have been inconsistent (Chopin & Briley, 1987). In 1955, Montgomery showed that rats have a preference for enclosed arms in an elevated maze and this observation has recently been used as the basis for an animal test of anxiolytic compounds using an elevated maze in the shape of a cross with two open and two enclosed arms (Handley & Mithani, 1984; Pellow et al., 1985). However, the reported effects of buspirone, ipsapirone and 8-hydroxy-2-(di-n-propylamino)tetralin (8-OH-DPAT) in this test have been conflicting (File et al., 1987; Critchley & Handley, 1986, 1987). The experiments described here have further examined the effects of buspirone and ipsapirone in the elevated plus-maze test and the role that their activity at 5-HT$_{1A}$ receptors may play in the behavioural changes they induce in this test. This was done by comparing their actions with other 5-HT$_{1A}$ receptor agonists such as 8-OH-DPAT and LY 165163 (1-(m-trifluoromethyl-phenyl)-4-p-aminophenylethyl piperazine). Also used were the optical isomers of MDL 72832 (8-[4-(1,4-benzo-dioxan-2-yl methylamino)butyl] 8-azaspiro [4,5]decane-7,9-dione), which have been shown to be useful tools in ascribing actions at the 5-HT$_{1A}$ receptor (Fozard et al., 1987).

Methods
Male Sprague-Dawley rats (200-300 g, Charles River, France) were used throughout. Prior to experimental use they were housed in groups of 6 in temperature (22 ± 1°C) and humidity (55% R.H.) controlled animal quarters under a 12 h light/dark cycle (lights on 06.00) with free access to food and water. All experimental procedures were carried out between 13.00 and 18.00 h. The apparatus consisted of a plus maze constructed in grey plastic to the design of Pellow et al. (1985), with arms 50 cm long and 10 cm wide. The two enclosed arms had side walls 50 cm high and the maze was raised 50 cm off the floor.

Drugs were dissolved or suspended in saline (following sonication and addition of surfactant if necessary) and given to rats subcutaneously 30 min (15 min for 8-OH-DPAT) before testing in the maze. For testing, the rats were placed individually in the centre of the maze in a darkened room and observed for 5 min by an observer in another room via a closed circuit T.V. camera.

The number of open and closed arm entries (defined as entry of all four limbs into the arm) during the observation period were recorded and after each rat the maze was cleaned.

Results are expressed as the mean of the open/total ratios for individual rats and as the mean number of open arm entries and mean total entries made in the plus-maze by the rats during the 5 min observation period. In all experiments diazepam (2 mg/kg s.c.) was used as a positive control.

Results were analysed using the Mann-Whitney U-test following a significant (p <0.05) result using the Kruskall Wallis one way analysis of variance for non-parametric data.

Results

Buspirone and ipsapirone did not significantly increase any of the measures of activity in the plus-maze. Instead, they dose-dependently decreased both the percentage of open arm entries and the total arm entries made by the rats (table 33.1).

Table 33.1 The effect of buspirone, ipsapirone and 8-OH-DPAT on the behaviour of rats in the elevated plus-maze. All values are the mean ± S.E. of values obtained for individual animals. The significance of the differences from the corresponding control value were calculated using the Mann-Whitney U-test* $p < 0.05$, ** $p < 0.01$.

Treatment	Dose (mg/kg sc)	n	% Open arm entries	Open arm entries	Total arm entries
Saline		16	8.3 ± 2.1	0.8 ± 0.2	9.0 ± 0.6
Diazepam	2	16	33.6 ± 2.7**	4.3 ± 6.4**	13.1 ± 1.0**
Buspirone	0.25	8	11.6 ± 5.3	1.0 ± 0.4	7.6 ± 1.6
	0.5	8	10.7 ± 4.8	0.8 ± 0.3	5.3 ± 1.0*
	1	8	0*	0*	2.3 ± 0.7**
	2	8	0*	0*	0.8 ± 0.3**
Saline		10	13.6 ± 3.2	1.9 ± 0.5	11.9 ± 1.1
Diazepam	2	10	39.1 ± 2.3**	6.2 ± 0.5**	16.0 ± 0.9*
Ipsapirone	1.25	8	7.1 ± 2.5	0.6 ± 0.2	7.8 ± 1.4*
	2.5	8	15.5 ± 5.5	1.5 ± 0.6	8.8 ± 1.7
	5	6	5.5 ± 5.5	0.2 ± 0.2*	4.7 ± 1.6*
	10	6	0*	0*	2.3 ± 0.8**
Saline		12	16.3 ± 4.7	1.8 ± 0.6	9.9 ± 0.7
Diazepam	2	10	34.4 ± 3.4*	4.4 ± 0.9*	11.9 ± 1.2
Ipsapirone	0.025	10	9.6 ± 3.2	1.3 ± 0.5	10.7 ± 1.6
	0.05	14	7.2 ± 2.7	0.9 ± 0.4	8.9 ± 1.6
	0.1	17	10.6 ± 5.4	0.9 ± 0.5	7.5 ± 1.4
	0.2	8	2.1 ± 2.1*	0.1 ± 0.1	5.5 ± 0.7**

This was in marked contrast to diazepam which consistently increased these two measures, as shown by the results of the positive control. The effects seen with buspirone and ipsapirone did not appear to be due to a general depression of locomotor activity as the rats still moved around in one of the closed arms. The rats would frequently come to the centre of the maze and show a stretched attend posture directed at one of the other three arms but without crossing into them. Rats treated with diazepam appeared far more sedated but an increase in all of the maze measures was still seen.

Rats pretreated with 8-OH-DPAT behaved in a similar manner in the plus-maze to those pretreated with buspirone and ipsapirone, although 8-OH-DPAT had a small effect on total arm entries. This latter result may be due to the hyperactivity that 8-OH-DPAT induces (Tricklebank et al., 1985), an effect not seen with buspirone or ipsapirone. Higher doses of 8-OH-DPAT could not be tested due to the appearance of the 5-HT syndrome which made it impossible for the rats to walk on the open arms.

Similar patterns of activity in the plus-maze were also seen for the $5-HT_{1A}$ agonists LY 165163 (0.5–4 mg/kg), (–)MDL 72832 (0.05–0.8 mg/kg) and (+) MDL 72832 (0.4–3.2 mg/kg) (data not shown). The two isomers of MDL 72832 were found to have a difference in potency of 10–30 fold (depending on measure used) which is similar to their difference in affinity for the $5-HT_{1A}$ binding site (Fozard et al., 1987).

Discussion

Benzodiazepine anxiolytics have been shown to reliably increase the percentage of open arm entries made by rats in an elevated plus maze, an effect interpreted as showing that they have an anxiolytic effect (Pellow et al., 1985). At none of the doses tested did either buspirone or ipsapirone show such an anxiolytic-like profile in the elevated maze, but instead were found to reduce the percentage of open arm entries and to reduce the total number of entries.

This profile is more in keeping with that of anxiogenic agents such as yohimbine and caffeine (Pellow, 1986). The interpretation of an anxiogenic effect has previously relied on showing a decrease in the proportion of open arms visited but with no corresponding decrease in total arm entries. Decreases in both measures have been interpreted as a non-specific depression of activity. From the present results, however, it seems that this is not necessarily the case. Firstly, diazepam treated animals appear markedly sedated in their home cages but in the plus-maze they are more active than vehicle treated rats as shown by the significant increases in the total number of arm entries. Secondly, rats treated with the $5-HT_{1A}$ agonists were seen to move about one closed arm but avoided entering the other arms (including the other enclosed arm) while still showing an interest in them, as shown by the stretched attend posture frequently taken up at the entrance of the closed arm they had first visited. It would seem that maze behaviour is affected by too many factors to make total arm entries useful as a measure of activity, and under the conditions of the present experiments this may even be useful as an additional measure of anxiety, provided appropriate controls are performed to study non-specific effects on locomotor activity using tests other than the elevated plus-maze.

The similarity of the effects of the 5-HT$_{1A}$ to those of buspirone and ipsapirone suggests that it is this action that is responsible for the effects of buspirone and ipsapirone in the elevated plus-maze test. The absence of anxiolytic activity of buspirone in this test suggests that the elevated plus-maze test is insensitive to some types of anxiolytic compounds. This could be a result of 5-HT$_{1A}$ agonist effects interfering with buspirone's anxiolytic action in the rat, or perhaps the elevated plus-maze test is measuring something other than anxiety.

Acknowledgements: I would like to acknowledge the excellent technical assistance of Thierry Fischer. Buspirone was kindly supplied by Bristol Myers (Indiana, USA) and ipsapirone by Troponwerke (Köln, FRG).

References

Chopin, P. & Briley, M. (1987). Animal models of anxiety: the effect of compounds that modify 5-HT neurotransmission. Trends Pharmacol. Sci., **8**, 383-388.

Critchley, M.A.E. & Handley, S.L. (1986). Anxiogenic-like effects of three 5-HT agonists. Psychopharmacol., **89**, 556.

Critchley, M.A.E. & Handley, S.L. (1987). 5-HT$_{1A}$ ligand effects in the x-maze anxiety test. Br.J. Pharmacol., In press.

File, S.E., Johnston, A.L. & Pellow, S. (1987). Effects of compounds acting at CNS 5-hydroxytryptamine systems on anxiety in the rat. Br.J. Pharmacol., **90**, 265P.

Fozard, J.R., Hibert, M., Kidd, E.J., Middlemiss, D.N., Mir, A.K. & Tricklebank, M.D. (1987). MDL 72832: a potent selective and stereospecific ligand for 5-HT$_{1A}$ receptors. Br.J. Pharmacol., **90**, 273P.

Gardner, C.R. (1985). Pharmacological studies of the role of serotonin in animal models of anxiety. In: Neuropharmacology of Serotonin (Ed. Green, A.R.), pp. 281-325. Oxford: Oxford University Press.

Handley, S.L. & Mithani, S. (1984). Effects of alpha-adrenoceptor agonist and antagonists in a maze exploration model of "fear"-motivated behaviour. Naunyn-Schmiedeberg's Arch. Pharmacol., **327**, 1-5.

Montgomery, K.C. (1955). The relation between fear induced by novel stimulation and exploratory behaviour. J. Comp. Physiol. Psychol., **48**, 254-260.

Pellow, S., Chopin, P., File, S.E. & Briley, M. (1985). Validation of open/closed arm entries in an elevated plus-maze as a measure of anxiety in the rat. J. Neurosci. Methods, **14**, 149-167.

Pellow, S. (1986). Anxiolytic and anxiogenic drug effects in a novel test of anxiety: are exploratory models of anxiety in rodents valid? Meth. Find. Exptl. Clin. Pharmacol., **8**, 557-565.

Schuckit, M.A. (1984). Clinical studies of buspirone. Psychopathol., 17, 61–68.

Tricklebank, M.D., Forler, C. & Fozard, J.R. (1985). The involvement of subtypes of the 5-HT$_1$ receptor and of catecholaminergic systems in the behavioural response to 8-hydroxy-2-(di-n-propylamino)tetralin in the rat. Eur.J. Pharmacol., 106, 271–282.

ANTIANXIETY EFFECT OF VARIOUS PUTATIVE 5-HT$_1$ RECEPTOR AGONISTS ON THE CONDITIONED DEFENSIVE BURYING PARADIGM

A.Fernandez-Guasti, E.Hong.
Seccion de Terapeutica Experimental, Departamento de Farmacologia y Toxicologia, CINVESTAV, Mexico, D.F. and Division de Investigationes en Neurosiencias, Insituto Mexicano de Psiquiatria, Mexio, D.F., Mexico

Introduction

The role of the serotonergic system in the regulation of anxiety is not known. In the pharmacological analysis of this neurotransmitter system several inconsistencies and even contradictions have appeared (cf. Graeff, 1986; Johnston & File, 1986). Thus, some authors have found anxiolytic actions after serotonin agonists administration (Engel et al., 1984; Graeff & Schoenfeld, 1970; Traber et al., 1984), while others have found no effect or even anxiogenic actions (Johnston & File, 1986; Shephard & Broadhurst, 1982; Shephard et al., 1982). The use of different animal models to test anxiety and the lack of selectivety of the drugs for the different serotonergic receptor subtypes, may at least partly, underly the controversies in this area.

Recently it has been reported that the very selective 5-HT$_{1A}$ receptor agonist, 8-OH-DPAT [8-hydroxy-2(di-n-propylamino) tetralin] induces anxiolytic effects in a licking conflict test (Engel et al., 1984). Furthermore, it has been proposed that the non-benzodiazepine anxiolytic used in clinics, buspirone (Goldberg & Finnerty, 1979), reduces anxiety through its agonistic action on the 5-HT$_1$ receptor subtype (Barrett et al., 1986; Glaser & Traber, 1983).

Pharmacological studies in several systems have indicated that ipsapirone (Dompert et al., 1985; Glaser et al., 1987), indorenate (Dompert et al., 1985; Fernandez-Guasti et al., 1987), RU 24969 (Green et al., 1985; Tricklebank et al., 1986) and 8-OH-DPAT (Arvidsson et al., 1981; Hjorth et al., 1982) are serotonergic 5-HT$_1$ receptor agonists. Therefore the purpose of the present study was no analyze whether these 5-HT$_1$ receptor agonists reduce anxiety in the conditioned defensive burying paradigm. This animal model was selected on the basis of its advantages over other paradigms designed to test anxiety (cf. Treit 1985).

Material and Methods

Male adult Wistar rats (280-350) were used. All animals were individually housed and kept in a room under controlled light-dark cycle (12h light: 12h dark. Lights on at 0300 h). All animals had free access to commercial rat chow and tap water all over the experiment.

The 5-HT$_1$ receptor agonists and the doses used were: 8-OH-DPAT (purchased from Research Biochemical Inc., 0.0625-0.5 mg/kg, -15 min); ipsapirone (Tropon Chemistry Department*, 0.625-10.0 mg/kg, -30 min); indorenate (Department of Pharmacology, CINVESTAV, 3.1-10.0 mg/kg, -90 min), RU24969 [5-methoxy-3-(1,2,3,6 tetrahydro-4-pyridinyl)-1H- indole, Roussel UCLAF*, 0.125-0.5 mg/kg, -15 min].

Diazepam (Roche*, 0.55 & 1.0 mg/kg, -30 min) and saline (2.0 ml/kg) were used as positive and negative controls repectively. All drugs and saline were injected i.p.

The procedure followed to test anxiety in the burying behaviour paradigm was the same as previously described by Treit et al. (1981). The parameters registered were the latency to the appearance of the burying behaviour display and the cumulative duration of the burying behaviour during a 15 min test.

Results

Results are shown in Figure 34.1. Control animals (injected with saline) showed a cumulative burying behaviour of 5.96 ± 0.59 min (X + S.E.) (shaded area). The administration of diazepam resulted in a dose dependent inhibition of the burying behaviour as previously reported by Treit et al. (1981). The systemic administration of all $5-HT_1$ receptor agonists resulted in a similar effect to that produced by diazepam, i.e. reduction in the cumulative time of burying behaviour accompanied by an increase in the burying behaviour latency. The $ED_{50}s$ for the different drugs were: Diazepam, 0.69 mg/kg; 8-OH-DPAT, 0.12 mg/kg; RU24969, 0.20 mg/kg; ipsapirone, 3.16 mg/kg and indorenate, 6.91 mg/kg. It is clear from these results that lower doses of 8-OH-DPAT and RU24969 were required to produce the same effects than diazepam. In all cases the latency to the appearance of the burying behaviour correlated inversely with the cumulative duration of the burying behaviour during the test (data not shown).

Discussion

Present data show that 8-OH-DPAT, RU24969, ipsapirone and indorenate induce antianxiety effects in the conditioned defensive burying paradigm. The conditioned defensive burying paradigm involves motor coordination; therefore, from present data it could be argued that the effect of these various $5-HT_1$ receptor agonists is mediated through an unspecific action on motor coordination. However, it has been described that at the doses of the drugs used in this study, no motor impairment occurs (Fernandez-Guasti et al., 1987; Gardner & Guy, 1983; Traber et al., 1984; Tricklebank et al., 1984).

Recently it has been proposed that ipsapirone, a selective $5-HT_{1A}$ receptor agonist (Dompert et al., 1985), is equipotent to diazepam in reducing anxiety in various animal models (Traber et al., 1984). Present results further support the idea of ipsapirone possessing anxiolytic actions, however, in the animal model used in this study, higher doses of ipsapirone were required to produce similar effects to those induced by diazepam.

It has been proposed that indorenate effects various behavioural components (Fernandez-Guasti et al., 1987). Thus, this drug induces some aspects of the serotonin syndrome and stimulates the copulatory behaviour and the spontaneous motor activity. Moreover, recently we found that indorenate induced anxiolytic actions in a licking conflict test and in a taste aversion paradigm, further support the idea of indorenate possessing anxiolytic properties. Additionally it has been reported that indorenate induces a reduction in 5-hydroxy-indolacetic acid (5-HIAA) levels in the brian stem (Hong et al., 1987). This reduction in 5-HIAA levels has been interpreted in the light of an agonistic action on 5-HT receptors leading to an antihypertensive effects of indorenate (Hong et al.,

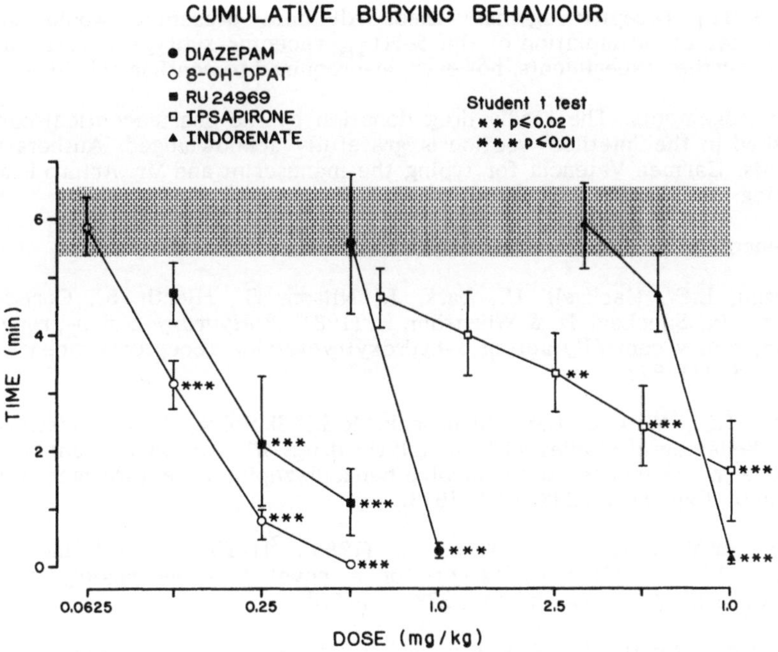

CUMULATIVE BURYING BEHAVIOUR

- ● DIAZEPAM
- ○ 8-OH-DPAT
- ■ RU 24969
- □ IPSAPIRONE
- ▲ INDORENATE

Student t test
** p ≤ 0.02
*** p ≤ 0.01

Fig. 34.1 Effect of various $5-HT_1$ receptor agonists on the cumulative burying behaviour during a 15 min test. Dashed area represents saline control values. Student t test, **p <0.02; ***p <0.01. Figure shows mean ± S.E. values.

1983). However, it has been reported that brain stem structures may be involved in the regulation of anxiety processes (Schenberg & Graeff, 1978). Thus, it is possible to suggest that the actions of indorenate on 5-HIAA brain levels might be associated with its antianxiety actions. Further experiments, however, are required to test this proposition.

It has been shown that of the various $5-HT_1$ receptor agonists used in this study, 8-OH-DPAT (Dompert et al., 1985; Hoyer et al., 1985) ipsapirone (Dompert et al., 1985) and indorenate (Dompert et al., 1985; Hoyer et al., 1985) selectively bind to the $5-HT_{1A}$ receptor subtype. By contrast, RU24969 possesses equally high affinity for the $5-HT_{1A}$ and $5-HT_{1B}$ receptor subtypes (Hoyer et al., 1985). In the present study all $5-HT_1$ receptor agonists used exerted similar antianxiety actions. Interestingly, a relation between the affinity for the $5-HT_{1A}$ receptor subtype of the drugs used and the doses required to induce anxiolytic effects has been found. Thus, it has been reported that 8-OH-DPAT is the most potent and selective agonist for the $5-HT_{1A}$ receptor subtype (Hoyer et al., 1985) and hence very low doses of this drug were required to induce anxiolytic effects. A similar relation was obtained for the other 5-HT$_1$ receptor agonists used. All data presented would therefore suggest that the stimulation of the 5-HT$_{1A}$ receptor subtype causes anxiolytic effects. Further experiments, however, are required to confirm this idea.

Acknowledgements: The kindly drug donation by the pharmaceutical companies asterisked in the "method" section is gratefully acknowledged. Authors wish to thank Ms. Carmen Valencia for typing the manuscript and Mr. Arturo Franco for preparing the figure.

References

Arvidsson, L.E., Hacksell, U., Lars, J., Nilson, G., Hjorth, S., Carlsson, A., Lindberg, P., Sanchez, D. & Wikström, H. (1981). 8-Hydroxy-2(di-n-propylamino) tetralin, a new centrally acting 5-hydroxytryptamine receptor agonist. J. Med. Chem., **24**, 921-923.

Barrett, J.E., Witkin, J.M., Mausback, R.S., Skolnick, P. & Weissman, B.A. (1986). Behavioural studies with anxiolytic drugs III. Antipunishment actions of buspirone in the pigeon do not involve benzodiazepine receptore mechanisms. J. Pharmacol. Exp. Ther., **238**, 1009-1013.

Dompert, W.V., Glaser, T. & Traber, J. (1985). ^{3}H-TVX Q 7821: identification of 5-HT binding sites as target for a novel putative anxiolytic. Naunyn Schmiedeberg's Arch. Pharmacol., **328**, 467-470.

Engel, J.A., Hjorth, S., Svensson, K., Carlsson, A. & Liljeqvist, S. (1984). Anticonflict effect of the putative setotonin receptor agonist 8-hydroxy-2(di-n-propylamino) tetralin (8-OH-DPAT). Eur. J. Pharmacol., **105**, 365-368.

Fernandez-Guasti, A., Oscos, A., Meneses, A., Escalante, A. & Agmo, A. (1987). Behavioural actions of indorenate, a new putative 5-HT agonist, Eur. J. Pharmacol. submitted.

Gardner, C.R. & Guy, A.P. (1983). Behavioural effects of RU24969, a 5-HT receptor agonist, in the mouse. Br. J. Pharmacol., **78**, 96P.

Glaser, T. & Traber, J. (1983). Buspirone: action on serotonin receptors in calf hippocampus. Eur. J. Pharmacol., **88**, 137-138.

Glaser, T., Dompert, W.U., Schuurman, T., Spencer Jr., D.G. & Traber, J. (1987). Differential pharmacology of the novel 5-HT$_{1A}$ receptor ligands 8-OH-DPAT, BAY R 1531 and ipsapirone. In: Brain serotonergic mechanisms: The Pharmacology, Biochemical and Potential Therapeutic Actions of 8-OH-DPAT and other 5-HT$_{1A}$ agonists. (Eds. Dourish, C.T., Ahlenius, S. & Hutson, P.).London: Ellis-Horwood Ltd.

Goldberg, H.L. & Finnerty, R.J. (1979). Tryptaminergic mechanisms in punished and non-punished behaviour. J. Pharmacol. Exp. Ther., **173**, 277-283.

Graeff, F.G. & Schoenfeld, R.I. (1970). Tryptaminergic mechanisms in punished and non-punished behaviour. J. Pharmacol. Exp. Ther., 173, 277-283.

Graeff, F.G. (1987). Ansioliticos e serotonina. Rev. Asoc. Bras. Psiquiatr. (In Press.)

Green, A.R., Guy, A.P. & Gardner, C.R. (1984). The behavioural effects of RU 24969, a suggested 5-HT$_1$ receptor agonist in rodents and the effects on the behaviour of treatment with antidepressants. Neuropharmacol, 23, 655-661.

Hjorth, S., Carlsson, A., Lindberg, P., Sanchez, D., Wikström, H., Arvidsson, L.E., Hacksell, U. & Nilson, J.L.G. (1982) 8-Hydroxy-2(di-n-propylamino) tetralin, 8-OH-DPAT, a potent and selective simplified ergot congener with central 5-HT receptor stimulating activity. J. Neural Trans., 55, 169-188.

Hong, E., Rion, R. & Vidrio, H. (1983). Stimulation of central serotonin receptors as a novel mechanism of antihypertensive activity. In: Vascular Neuroeffector Mechanisms. 4th. International Symposium. (Eds. Bevan, J.A., Fujimara, M., Maxwell, R.A., Mohri, K., Shibata, S. & Treda, N.), pp. 2273-2277. New York: Raven Press.

Hong, E., Rion, R., Aceves, J., Benitez-King., G. & Anton Tay, F. (1987). Further evidence for a central antihypertensive effect of indorenate. Proc. West. Pharmacol. Soc., 30, 1-3.

Hoyer, D., Engel, G. & Kalman, H.O. (1985). Molecular pharmacology of 5-HT$_1$ and 5-HT recognition sites in rat and pig membranes: radioligand binding studies with [^{3}H]5-HT, [^{3}H]8-OH-DPAT, (-) (^{125}I) iodocyanopindolol, [^{3}H] medulergine and [^{3}H] ketanserin. Eur. J. Pharmacol., 118, 13-23.

Johnston, A.L. & File, S.E. (1986). 5-HT and anxiety: promises and pitfalls. Pharmacol. Biochem. Behav., 24, 1467-1470.

Schenberg, L.C. & Graeff, F.G. (1978). Role of the periaqueductal gray substance in the antianxiety action of benzodizaepines. Pharmacol. Biochem. Behav., 9, 287-295.

Shephard, R.A., Buxton, D.A. & Broadhurst, P.L. (1982). Drug interactions do not support the reduction in serotonin turnover as the mechanism of action of benzodiazepines. Neuropharmacol., 21, 1027-1032.

Shephard, R.A. & Broadhurst, P.L. (1982). Effects of diazepam and of serotonin agonists on hyponeophagia in rats. Neuropharmacol., 21, 337-340.

Traber, J., Davies, M.A., Dompert, W.U., Glaser, T., Schuman, T. & Seidel, P.R. (1984). Brain serotonin receptors as a target for the putative anxiolytic TVX Q 7821. Brain Res. Bull., 12, 741-744.

Treit, D., Pinel, P.J.P. & Finiger, H.C. (1981). Conditioned defensive burying: a new paradigm for the study of anxiolytic agents. Pharmacol. Biochem. Behav., 15, 619-626.

Treit, D. (1985). Animal models for the study of antianxiety agents: a review. Neurosci. Biobehav. Rev., 9, 203–222.

Tricklebank, M.D., Middlemiss, D.N. & Neill, J. (1986). Pharmacological analysis of the behavioural and thermoregulatory effects of the putative 5-HT_1 receptor agonist, RU 24969, in the rat. Neuropharmacology, 25, 877–886.

THE ANXIOLYTIC ACTIVITIES OF THE 5-HT$_3$ RECEPTOR ANTAGONISTS GR38032F, ICS 205-930 AND BRL 43694

B.Costall, A.M.Domeney, P.A.Gerrard, B.J.Jones, M.E.Kelly, N.R.Oakley[o], M.B.Tyers.
Postgraduate School of Studies in Pharmacology, University of Bradford, Bradford, BD7 1DP and [o] Neuropharmacology Department, Glaxo Group Research Ltd., Ware, SG12 ODJ, United Kingdom

The potential anxiolytic activities of the selective 5-HT$_3$ receptor anatgonists, GR38032F (Costall et al. 1987; Jones et al. 1987), ICS 205-930 and MDL 72222 (Tyers et al. 1987) have been shown in both rodent and primate models. We have now extended these studies to include evaluation of BRL 43694 (Fake et al. 1987).

Methods
Three test procedures were used, (1) a black:white two-compartment box in which mice taken from the dark show aversion to the light compartment, (2) rat social interaction in which two unfamiliar rats based on the retreat from, or posturing towards a human observer (reduced by anxiolytic agents).

The first test used male BKW albino mice (Bradford bred (25-30g)) housed under dark illumination between 07.00h and 19.00h. They were taken from their dark holding room (housed in groups of 10) in a dark container to a dark test room (red illumination) in which they were allowed to acclimatise for 1h before test. The test box was open-topped, 45cm long, 27cm wide and 27cm high, divided into a small (2/5) and a large (3/5) area by a partition that extended 20 cm above the walls. The large compartment was painted white and was brightly illuminated (60W tungsten bulb) whilst the small compartment was painted black and subject to dim illumination (60W red bulb). Movement between the two compartments was enabled by a 7.5 x 7.5 cm opening at floor level in the partition.

Tests were carried out between 30.00 h and 18.00 h when each mouse was tested by placing it in the centre of white area and allowing it to explore the novel environment for 5 min. Its behaviour was recorded on videoptape and the behavioural analysis was performed subsequently from the recording. Four parameters were measured: the latency for initial movement from the white to the black compartment, the % of time spent in the black, the numbers of exploratory rears in the white and black, and the numbers of crossings of lines (9 cm square) marked on the floor of the white and black compartments.

In the black:white test box the action of diazepam and the 5-HT$_3$ receptor antagonists were tested after acute administration (45 min pretreatment), after subchronic administration (b.d. dosing for 7 days) and then after withdrawal from 7 days of b.d. dosing. Mice were used on one occasion only. For social interaction studies, male Lister Hooded rats (Glaxo bred, 2000-250 g) were housed 5 to a cage and kept in the laboratory environment for at least a week before testing. Rats paired in the test were taken from separate cages.

The method was based on that described by File (1980). The test arena consisted of an open-topped box, 62 x 62 x 33cm with a 7 x 7 matrix of infra-red photocell beams in the walls, 2.5 cm from the floor. The light intensity at the floor of the arena was 380 lux (high light conditions).

Drug were tested by treating both members of a pair of rats with the same treatment 45 min before testing. During the 45 min pretreatment time, the rats were placed singly in small cages immediately after dosing until they were tested. Testing involved placing each member of a pair of rats in opposite corners of the arena and then leaving them undisturbed for 10 min while recording their behaviour remotely on videotape. The behavioural assessments were made subsequently from the recordings. The time spent in social interaction was measured and expressed as a cumulative total for the 10 min session. The behaviours that comprised social interaction were: following with contact, sniffing (but not sniffing of the hindquarters), crawling over and under, tumbling, boxing and grooming.

The marmoset 'human threat' test used common marmosets (Callithrix jacchus), body weights 315 ± 20g, of both sexes, housed in single sex pairs. They were tested in their home cages. Only marmosets which gave consistent and reliable responses to either no treatment or vehicle injection were selected for the studies. It was essential to allow 1-2 day breaks between test days, and no marmoset was tested more than 3 times in one week. Animals were subject to one test only on each test day.

Drugs were injected subcutaneously, each member of the pair receiving the same treatment. On any one occasion the behaviour of at least 2 pairs of animals was assessed and it was important that the 2 animals in one cage were not tested consecutively to avoid influence of the behavioural responding of the first test animal on the second.

45 min after drug treatment, the marmosets were confronted by an observer standing in close proximity to the cage, with the cage door closed. Over a 2 min period the number of "aggressive" postures was recorded (tail erect with exposure of the genitals, slit stare facial expression, scenting, arch piloerect locomotion where marmoset moves to and for along perch with back arched and full piloerection). The amount of time spent on the wire cage front and the number of jumps between the front and back of the cage were also recorded. The marmosets were also observed carefully for any overt behavioural changes. Drugs were injected by the intraperitoneal route in the mouse studies (vehicle saline, or minimum PEG for diazepam), orally in the rat (vehicle acacia) and subsutaneously in the marmoset (dissolved or suspended in saline). Drugs used were diazepam, GR38032F, ICS 205-930 and BRL 43694.

Results
Diazepam, GR38032F, ICS 205-930 and BRL 43694 were all shown to reduce mouse aversion to a brightly lit environment. Control animals exhibited 69.6 ± 6.3 rears/5 min in the black, but only 30.1 ± 3.7 rears/5 min in the white. Similarly, line crossings of control animals were 81.1 ± 9.2/5 min in the black, but only 37.1 ± 3.9/5 min in the white. All agents tested changed this behaviour to a preference for the white, with similar maxima achieved by all agents. Thus, under the influence of diazepam, GR38032F, ICS 205-930 and BRL 43694 rearing

in the which increased to 71.2 ± 7.6 to 83.6 ± 8.9/5 min, with reduction in the black to 13.7 ± 1.6 to 22.4 ± 2.4/5 min. The magnitude of change in line crossings was similar, and latency to move from the white to the black compartment was delayed by all compounds from control values of 10–12 sec to 22–44 sec. % time spent in the black also decreased and this is shown on Fig. 35.1. It should be noted that there was a loss of effect at the higher doses of ICS 205-930 and BRL 43694: this was also seen in all other measures of anxiolytic potential in the mouse black:white box.

The anxiolytic actions of GR38032F (10 µg/kg i.p.) and ICS 205-930 (10 µg/kg i.p.) were maintained on subchronic treatment (7 days, b.d. dosing) and such action slowly waned on drug withdrawal after 7 days. In contrast, on subchronic treatment tolerance developed to the anxiolytic effects of low doses of BRL 43694 (1 µg/kg). Further, anxiogenesis followed the abrupt withdrawal from treatments with BRL 43694. Similar anxiogenesis can be seen to result from withdrawal of diazepam treatment. For example, 8h after withdrawal 7 days treatment with 0.001 mg/kg b.d. BRL 43694, or 10 mg/kg b.d. diazepam, behaviour in the black was significantly enhanced above control values (e.g. rears elevated by 80–95% in the black, $p<0.001$, with corresponding decreases in the white). Diazepam, GR38032F, ICS 205-930 and BRL 43694 were all shown to increase social interaction in the rat. As in the mouse test the 5-HT$_3$ receptor antagonists were shown to be considerably more potent than diazepam although the maximum response attained was equal for all compounds (Fig. 35.1). There was some loss of activity at the highest dose of BRL 43694 used. Also, this compound caused some locomotor stimulation which was significant at 0.01 mg/kg (see legend Fig. 35.1). Locomotor activity was not significantly altered by any other compound tested, with the exception of sedation at the highest dose of diazepam (4 mg/kg p.o.).

In the marmoset human threat test all compounds significantly reduced the numbers of postures in response to the human presence (Fig. 35.1). As in the other tests, the 5-HT$_3$ receptor antagonists were more potent than diazepam, and there was a loss of activity at the highest dose of BRL 43694 tested (Fig. 35.1). All compounds causes significant increases in the % of time spent on the cage front (control values 15–30%, increased to 38–80% by maximally effective doses of diazepam, GR38032F, ICS 205-930 and BRL 43694). Of the compounds tested, only BRL 43694 consistently caused a stimulation of locomotor activity, exhibited as increased numbers of jumps (control 6.7 ± 0.8, increased to 19.6 ± 2.1 and 29.4 ± 3.2 at 1 ng/kg and 0.1 mg/kg respectively).

Discussion
In mouse, rat and mamoset, tests which are sensitive to detecting behavioural change indicative of anxiolytic activity, the selective 5-HT$_3$ receptor antagonists GR38032F, ICS 205-930 and BRL 43694 would appear to be effective and potent anxiolytic agents. The actions of diazepam, GR38032F and ICS 205-930 are maintained on subchronic treatment, but tolerance develops to the actions of BRL 43694. Also, on withdrawal from subchronic treatment with BRL 43694 the behavioural change which occurs in the mouse is similar to with that seen on treatment with an anxiogenic β-carboline (e.g. FG 7142, Costall et al. 1987) or withdrawal from subchronic treatment with diazepam (Barry et al. 1987). withdrawal of continued treatment with GR38032F or ICS 205-930. Thus,

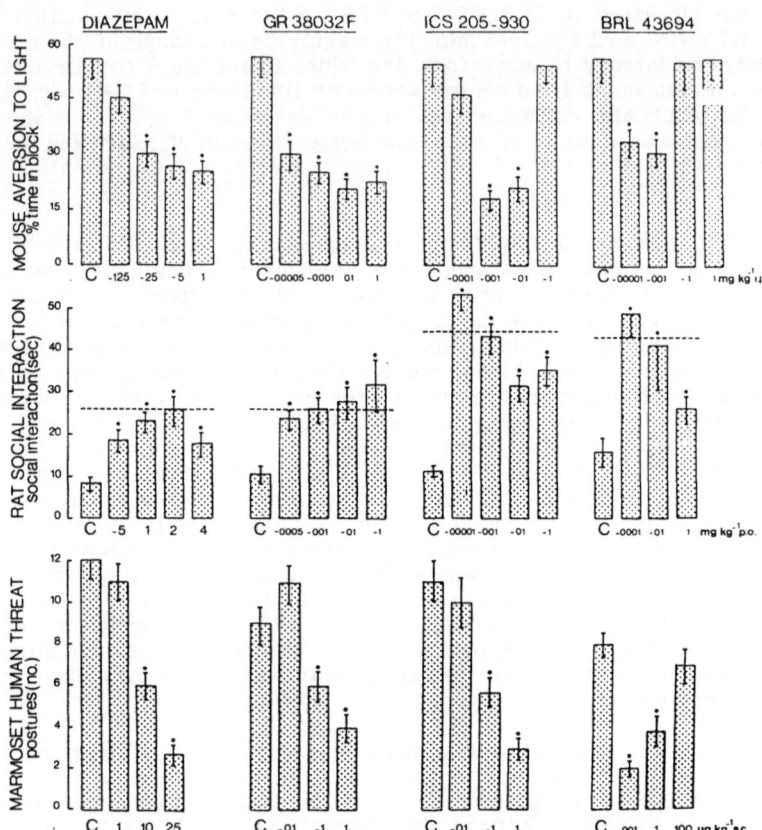

Figure 35.1 Activity of diazepam, GR38032F, ICS 205-930 and BRL 43694 in three animal models sensitive to the detection of behavioural change indicative of anxiolytic potential. 1. reduction in mouse aversion ot light (indicated as reduced % time spent in the black compartment of a two-compartment black:white box), 2. increased social interaction of unfamiliar paired rats under high illumination and 3. reduced "aggressive" posturing of marmosets in a situation of threat caused by a human presence. In the mouse test n=5, S.E.M.s are calculated from original data, drugs were given intraperitoneally 45 min before test. In the rat social interaction test n=8 pairs, S.E.M.s given, drugs were administered orally 45 min before test, ----- indicates maximum response of the rats on the day of test to 2 mg/kg p.o. diazepam, ° indicates change in locomotor activity (sedation caused by 4 mg/kg diazepam, stimulation by 0.01 mg/kg BRL 43694, control value 257.8 ± 11.8, increased to 313.0 ± 9.0 by BRL 43694, p<0.05). In the marmoset test n = 4-6, S.E.M.s given, drugs were administered subcutabneously 45 min before test. For all tests significant change from vehicle control values (C) is indicated as *p<0.05 - p<0.001 (one-way ANOVA followed by Dunnett's t test).

whilst the selective 5-HT$_3$ receptor antagonists consistently cause changes in animal behaviour indicative of anxiolytic potential, differences do exist, not only on long-term treatment and withdrawal, but also apparently in terms of loss of activity at higher doses. The bases for such differences are presently being investigated.

References

Barry, J.M., Costall, B., Kelly, M.E. & Naylor, R.J. (1987). Withdrawal syndrome following subchronic treatment with anxiolytic agents. Pharmac. Biochem. Behav., **27**, 239-245.

Costall, B., Domeney, A.M., Hendrie, C.A., Kelly, M.E., Naylor, R.J. & Tyers, M.B. (1987). The anxiolytic activity of GR38032F in the mouse and marmoset. Br. J. Pharmac., **90**, 257P.

Costall, B., Jones, B.J., Kelly, M.E. & Naylor, R.J. (1987). Investigations into the use of a two compartment black and white test box to measure changes in mouse exploratory behaviour relevant to assessment of anxiety. In preparation.

Fake, C.S., Kind, F.D. & Sanger, G.J. (1987). BRL34694: a potent and novel 5-HT$_3$ receptor antagonist. Br. J. Pharmac., **91**, 335P.

File, S.E. (1980). The use of social interaction as a method for detecting anxiolytic activity of chlordiazepoxide-like drugs. J. Neurosci. Meth., **2**, 219-238.

Jones, B.J., Oakley, N.R. & Tyers, M.B. (1987). The anxiolytic activity of GR38032F, a 5-HT$_3$ receptor antagonist, in the rat and cynomolgus monkey. Br. J.Pharmac., **90**, 88P.

Tyers, M.B. Costall, B., Domeney, A.M., Jones, B.J., Kelly, M.E., Naylor, R.J. & Oakley, N.R. (1987). The anxiolytic activities of 5-HT$_3$ antagonists in laboratory animals. Neurosci. Lett., Suppl., **29**, 568.

BEHAVIOURAL EFFECTS OF 5-HT$_3$ ANTAGONISTS IN ANIMAL MODELS FOR AGGRESSION, ANXIETY AND PSYCHOSIS

Jan Mos, Jan v.d.Heyden, Berend Olivier
Duphar B.V., Department of Pharmacology, P.O. Box 2, 1380 AA Weesp, Holland

Introduction
The development of specific 5-HT$_3$ antagonists with high receptor affinity has recently led to the discovery that such drugs are involved in the modulation of anxiety, psychosis and aggression. Costall et al (1987) and Tyers (1988) have described significant efficacy of 5-HT$_3$ antagonists on the behaviour of animals in models thought to be predictive for anxiolytic and antipsychotic activity. Part of their evidence is based on a model of dopamine infusion in the nucleus accumbens. Since this is neither a simple model nor generally used, we tried to replicate the effects of two 5-HT$_3$-antagonists, GR38032F and MDL 72222, in two more routinely used animal models of antipsychotic activity, conditioned avoidance response (CAR) and (functional) antagonism of dopaminergic behaviour, apomorphine-induced climbing. As anxiolytic tests we used the four-plate test, the light/dark exploration model and a recently developed model for separation-induced anxiety-ultrasonic calling of rat pups. Because of our interest in the role of serotonin in aggression we also investigated 5-HT$_3$ antagonists in aggression; as models we used isolation-induced aggression in mice and maternal aggression in lactating female rats. For comparative reasons, we used the reference compounds haloperidol (neuroleptic), diazepam (anxiolytic) and eltoprazine (serenic).

Methods
Antipsychotic test: The CAR procedure and the apomorphine climbing test have been described previously (Van der Heyden, 1988). For the CAR test, male Wistar rats were housed three at a cage and trained extensively before use. For each dose, 9 animals were used; drugs were given orally (60 minutes before the test) or intraperitoneally (30 minutes). For the apomorphine climbing test male DAP mice received 1 mg/kg apomorphine (sc) 10 minutes before the test. Vehicle or drug were administered either orally 60 min or intraperitoneally 30 min before apomorphine.

Anxiolytic tests
Anxiolytic activity was determined in the 4-plate test and the light-dark model. The 4-plate cage has a floor which consists of four metal plates of 12 x 8 cm which are electrically insulated and through which shocks can be delivered (Aron et al., 1971; Boissier et al., 1968). Thirty minutes before testing, female DAP-mice, nine per treatment group, were injected s.c. During the first non-punished minute, crossings from one plate to another were counted. In the second minute, crossings were inhibited by punishment with a mild footshock (0.4 mA). Crossings were again counted for two minutes and the latency to the fifth shock was registered.

The light-dark cage is divided in two compartments of 20x26x28 cm connected by a gate of 6x8 cm. One compartment is completely dark, the other is brightly illuminated (Crawley & Goodwin, 1980). Thirty minutes after receiving a s.c.

injection of drug or vehicle, a naive male DAP-mouse is placed in the dark compartment. Nine mice per treatment group were used. The time spent in the light compartment during five minutes was measured and afterwards expressed as a percentage of the total testing-time. Moreover the total activity (Animex) was measured as well as the latency to enter the light compartment. Pup vocalizations were measured as described by Mos and Olivier (1988). Rat pups, 9 - 12 days old were treated with drugs 30 minutes before the test. Separation from the littermates and mother lasted 5 minutes, during which the number of ultrasonic calls was recorded under two temperature conditions, warm and cold. The aluminium plate on which the animal was tested was kept either 18°C or 37°C, the former resulting in considerably more ultrasounds with a different pattern over test time than under the warm test condition. Drug effects on the total number of ultrasounds were evaluated with Friedman analysis of variance followed by Wilcoxon matched pair tests.

Antiaggressive activity: Isolation-induced aggression and maternal aggression have been described by Olivier et al. (1987). Briefly, male DAP-mice made aggressive by prolonged isolation are confronted with a group-housed opponent. The ED_{50} to inhibit attacks is calculated. Drugs are given orally 60 minutes before testing to the isolated males. Female lactating rats are very aggressive against male intruder rats. Female TMD-S_3 rats were tested on postpartum days 3, 5, 7 and 9. Wistar rats (200-220 g) served as opponents. Treatment was given according to a Latin square design. Friedman analysis was performed on the total number of bite attacks during the 5 minute test.

Results

Antipsychotic activity

Conditioned avoidance response: An animal model in which most clinically effective antipsychotic drugs are active is the Conditioned Avoidance Response (CAR). The efficacy of the drugs is expressed in the ED_{50} to suppress avoidance responses upon the conditioning signal. Non specific effects are noted when the drug suppresses escape responses during (unconditioned) electric shocks, at doses near the ED_{50} avoidance (ratio escape/avoidance <2). Table 36.1 summarizes the data. From 0.001 mg/kg up to 10 mg/kg (ip or po) GR38032F did not affect avoidance. MDL 72222 inhibited avoidance significantly at 16 mg/kg ip, an extremely high dose compared to the reported dose ranges for peripheral 5-HT_3 antagonistic effects in anti-emetic activity (Miner & Sanger, 1986). MDL 72222 was not very specific as the ED_{50} for escape was close to the ED_{50} avoidance. In contrast, both the efficacy and the specificity of haloperidol were much higher. Diazepam was ineffective, whereas eltoprazine was only moderately effective after ip treatment, and even less after oral administration. We therefore conclude that 5-HT_3 antagonists are either not active or not very specific in this model for antipsychotic activity.

Apomorphine-induced climbing in mice: Because of their blocking action on dopamine receptors, neuroleptics antagonize apomorphine (DA-receptor agonist) induced behaviours (e.g. gnawing and climbing) in mice. The ED_{50} value of 5-HT_3 antagonists to inhibit such climbing was compared to the reference compounds. As expected haloperidol inhibited apomorphine induced climbing at low doses. In the dose-range 0.1 to 10 mg/kg the 5-HT_3 antagonists did not functionally counteract apomorphine induced climbing, again indicating the absence of antipsychotic activity.

Anxiolytic activity
4-plate test: Benzodiazepines modify the behaviour of mice who are subjected to punished exploration (4-plate test). Simultaneously, punished exploration was also measured. Selective anxiolytic activity was determined by the differential activation of punished and unpunished responses. After oral administration GR38032F was not active in this test. At 10 mg/kg activity was decreased, as unpunished exploration was lower than control. After subcutaneous administration a very different profile emerged: the lowest doses selectively increased the number of shocks taken and reduced the latency to the first five shocks. However, 0.1 mg/kg GR38032F was no longer effective, confirming the reported bell-shaped dose effect curves (Costall et al., 1988). Diazepam exhibited anxiolytic activity, both after oral and subcutaneous administration, although the unpunished activity was also somewhat enhanced. Haloperidol decreased overall activity while eltoprazine undifferentially increased both punished and unpunished activity.

Light-dark test: The natural aversion of rodents to avoid brightly lit places is reduced by anxiolytic drugs (Crawley and Goodwin, 1980). In a two compartment box, one dark and one brightly lit, the LED changing the total activity (Animex), the time spent in the light and the latency to enter the light compartment were scored. GR38032F reduced total activity at 10 mg/kg (po), which is close to toxic doses in mice. GR38032F had no effect on the latency to enter the light nor on the time spent in the light compartment (both indicative of putative anxiolytic action). After subcutaneous administration again a different profile emerged. At all doses the latency to enter the light compartment increased. Total activity was decreased at 0.01 and 0.1 mg/kg, and the time spent in the light compartment was reduced at the highest dose. Haloperidol sedated the animals already at 0.1 mg/kg. Eltoprazine and diazepam increased the general activity at low doses, but only diazepam reliably increased the time spent in the light and reduced the latency to enter the light. Eltoprazine marginally influenced the time spent in the light box at the highest dose tested.

Ultrasonic vocalizations: Rat pups emit ultrasounds when separated from their mother and littermates (Hofer & Shair, 1978). Benzodiazepines reduce these vocalizations. Table 36.1 shows the LED to reduce pup ultrasounds tested at 18°C and 37°C. As expected, diazepam reduced ultrasounds under both conditions. GR38032F had no effect at any of the doses tested. Haloperidol similarly failed to affect pup ultrasounds. Eltoprazine reduced ultrasounds only under the more stressful cold condition.

Antiaggressive activity
Isolation induced aggression in mice: Mice made aggressive by prolonged isolation will rapidly attack a strange male conspecific. In a simple screening model the ED_{50} to inhibit these attacks was determined. Table 36.1 indicates the efficacy of eltoprazine and haloperidol. 5-HT_3 antagonists (po or sc administered) did either not inhibit attacks (GR38032F) or did so only at high doses (MDL 72222). The serenic eltoprazine and the neuroleptic haloperidol were very effective anti-aggressive drugs.

Table 36.1.a Summary of test results for the $5-HT_3$ antagonists and reference compounds. For details of procedure see text.

Conditioned Avoidance Response (n= 9/group)

	ED_{50} avoidance	ED_{50} escape	ratio E/A
GR38032F	>10.0 po	-	-
	>10.0 ip	-	-
MDL 72222	16.0 ip	21.0 ip	1.3
haloperidol	0.9 po	8.2 po	9.1
	0.45 ip	>1.2 ip	>2.7
eltoprazine	100.0 po	>100.0 po	-
	9.4 ip	>10.0 ip	>1.1
diazepam	>270.0 po	-	-

Antagonism of Apomorphine-induced climbing (n= 9/group)

	ED_{50} mg/kg
GR38032F	>10.0 po
MDL 72222	>10.0 ip
haloperidol	0.3 po
eltoprazine	31.0 po
diazepam	not tested

Four-plate test in mice (n=9-18/group)

	LED* unpunished mg/kg	LED punished mg/kg	(dose range-route)
GR38032F	no effect	no effect	(0.001-10.0 po)
GR38032F	no effect	0.001 ↑	(0.001-0.1 sc)
haloperidol	0.2 ↓	0.2 ↓	(0.025-3 sc)
eltoprazine	0.3 ↑	0.3 ↑	(0.1-3.2 sc)
diazepam	0.3 ↑	0.3 ↑↑↑	(0.3-10.0 sc)
diazepam	0.3 ↑	0.3 ↑↑↑	(0.3-10.0 po)

36: 5-HT$_3$ Antagonists, Aggression, Anxiety & Psychosis

Table 36.1.b Summary of test results for the 5-HT$_3$ antagonists and reference compounds. For details of procedure see text.

Light/dark cage exploration (n=9-18/group)

	LED total activity	LED time spent in light	latency to enter light	dose range tested
GR38032F	10.0 ↓	>10.0	>10.0	(0.001-10 po)
GR38032F	0.01 ↓	0.1 ↓	0.1 ↑	(0.001-0.1 sc)
haloperidol	<0.1 ↓	>1.0	1.0 ↑	(0.1-1.0 sc)
eltoprazine	0.4 ↑	>0.8	0.8 ↓	(0.2-0.8 sc)
diazepam	0.3 ↑	0.3 ↑↑↑	0.3 ↓	(0.1-3.0 sc)

Rat pup ultrasonic vocalizations (n= 8/group)

	LED 37°C	LED 18°C	dose range tested
GR38032F	>10.0	>10.0	(0.001-1.0 ip)
haloperidol	>0.3	>0.3	(0.03-0.3 ip)
eltoprazine	no effect	1.0	(0.3-3.0 ip)
diazepam	2.5	2.5	(2.5-10.0 ip)

ED$_{50}$ to inhibit isolation-induced aggression in mice
ED$_{50}$ (mg/kg)

GR38032F	>4.6 po and sc
MDL 72222	7.5 ip
haloperidol	0.75 po
eltoprazine	0.4 po
diazepam	not tested

- irrelevant because of lack of activity in avoidance: LED Lowest Effective Dose.
↑ activity or latency increased, ↓ activity or latency decreased

Maternal aggression in rats: Lactating female rats fiercely attack intruders during the postpartum period. Drug effects have been described previously (Olivier and Mos, 1986). A concise summary of the results of the 5-HT$_3$ antagonists reveals that GR38032F (0.1–10.0 mg/kg ip) did not reduce aggression. MDL 72222 (1.25–5.0 mg/kg ip) did not reduce maternal attacks either. The neuroleptic haloperidol (0.05–0.2 mg/kg ip) reduced aggression, at doses which also increased sedation and decreased social interest and exploratory activities.

Eltoprazine (2.0–8.0 mg/kg po) specifically reduced aggression, leaving social interest and exploration intact. Diazepam (1.25–5.0 mg/kg po) increased aggression at low doses, but at higher doses aggression returned to control levels. This finding is well in line with the previously reported pro-aggressive actions of low doses of benzodiazepines (Miczek, 1987; Mos et al., 1987).

Discussion

In two generally accepted animal models of antipsychotic activity, 5-HT$_3$ antagonists were not active, neither in very low, nor in relatively high doses. This is in contrast to the putative antipsychotic effects found by Costall et al., 1987 (functional activation caused by chronic dopamine infusion in the nucleus accumbens). In one aggression model some indication of central activity was found. However, a consistent role for 5-HT$_3$ antagonists in aggression cannot be inferred from these experiments.

The results from the experiments on anxiety are difficult to interpret. There seems to be an important variable in the route of administration. Oral administration was ineffective in the four plate test, whereas after subcutaneous administration a bell-shaped dose response curve indicative of anxiolytic activity was found. In the light/dark model and separation induced ultrasonic calls by rat pups anxiolytic activity could not be confirmed, despite the wide dose range tested and the differences in the route of administration.

It remains to be established what causes these differences in response in anxiolytic and other models, compared to the reported effects by Costall et al. (1988) and Tyers et al. (1988). In analogy to the benzodiazepines and the new non-benzodiazepine anxiolytics, it might be hypothesized that conventional models do not predict activity of a new class of drugs. Another possibility that needs to be explored is the effect of the route of administration on drug effects, whereas also the time of administration might be a critical factor.

References

Aron C., Simon P., Larousse C. & Boissier J.R. (1971) Evaluation of a rapid technique for detecting minor tranquillizers. Neuropharm., **10**, 459–569.

Boissier J.R., Simon P. & Aron C. (1968) A new method for rapid screening of minor tranquillizers in mice. Eur. J. Pharmacol., **4**, 145–151.

Costall, B., Domeney, A.M., Gerrard, P.A., Jones, B.J., Kelly, M.E., Oakley, N.R. & Tyers, M.B. (1988). The anxiolytic activities of the 5-HT$_3$ receptor antagonists GR38032F, ICS 205-930 and BRL 43694. This volume.

Costall, B., Domeney, A.M., Naylor, R.J. & Tyers, M.B. (1987). Effects of the 5-HT$_3$ receptor antagonist, GR38032F, on raised dopaminergic activity in the mesolimbic system of the rat and marmoset brain. Br. J. Pharmac., **92**, 881-895.

Crawley J. & Goodwin F.K.(1980) Preliminary report of a simple animal behavior model for the anxiolytic effects of benzodiazepines. Pharmacol Biochem. Behav., 13, 167-170.

Hofer, M.A. & Shair, H. (1978). Ultrasonic vocalization during social interaction and isolation in 2-week-old rats. Developmental Psychobiology, 11, 495-504.

Miczek, K.A. (1987). The psychopharmacology of aggression. In: Iversen, L.L., Iversen, S.D. & Snyder, S.H. (eds). Handbook of Psychopharmacology. Vol. 19: Behavioural Pharmacology. New York: Plenum Press, pp. 183-327.

Miner, W.D. & Sanger, G.J. (1986). Inhibition of cisplatin-induced vomiting by selective 5-hydroxytryptamine M-receptor antagonism. Br. J. Pharmac., **88**, 497-499.

Mos, J., Olivier, B. & Van der Poel, A.M. (1987). Modulatory actions of benzodiazepine receptor ligand on agonistic behavior. Physiol. & Behav., 41, 265-278.

Mos, J. & Olivier, B. (1988). Ultrasonic vocalizations by rat pups as an animal model for anxiolytic activity: effects of serotonergic drugs. This volume.

Olivier, B.,, Mos, J., Van der Heyden, J.A.M., Schipper, J., Tulp, M., Berkelmans, B. & Bevan, P. (1987). Serotonergic modulation of agonistic behaviour. In: Olivier, B., Mos, J. & Brain P.F. (eds). Ethopharmacology of agonistic behaviour in animals and humans. Martinus Nijhoff, Publishers, pp. 162-186.

Tyers, M.B. (1988). A review of the evidence supporting the anxiolytic potential of 5-HT$_3$ receptor antagonists. This volume.

Van der Heyden, J.A.M. (1988). Modulation of the 5-HT system results in antipsychotic activity. This volume.

5-HT: DRUG DISCRIMINATION AND AVERSION

T.U.C.Järbe and T.Archer
Uppsala and Södertäjle, Sweden

Serotonin and aversion

The use of discrimination methodologies to discern drug action(s) has progressed during the last decade, the reason probably being that drug classes can be compared in a uniform manner using a single, pharmacologically specific response, the discriminative choice. We shall consider aspects of the behavioural pharmacology of 5-hydroxytryptamine (5-HT) in the light of novel experimental data involving compounds affecting 5-HT neurotransmission and their involvement in drug stimulus discrimination and aversion. Two central concepts in drug discrimination learning (DDL) are discrimination and generalization. Discrimination refers to the ability to differentiate between one stimulus value and other stimulus values along the same dimension or other dimensions. Generalization reflects the inverse of discrimination, i.e. whether or not other stimulus values are perceived as being similar to the training (reference) stimulus (Kalish, 1969). In DDL animals commonly are trained to respond in one way in the presence of the reference (training) drug (D), and in another way in nondrug (N) sessions, e.g. by pressing one of two different levels (rats and monkeys) or pecking keys (pigeons) in an operant chamber (Skinner-box). Which manipulandum is correct during a given training session in dependent upon whether the drug stimulus was administered prior to the session or not. Hence the correct manipulandum correlates in a one-to-one fashion with the imposed training state (drug or nondrug), and commonly a double alternation sequence of stimulus presentation is followed (D,D,N,N, etc.). Even though the drug versus nondrug discrimination is the one situation most often used, animals can also be trained to discriminate between the effects of two different drugs or two different doses of the same drug in two choice experiments. Discrimination between two drug states and a nondrug condition in three-choice settings has also been reported (for references, see Järbe & Swedberg, 1982). These latter approaches may be useful when teasing out the stimulus characteristics of subclasses of drugs (for examples, see e.g. Holtzman, 1985; Järbe, 1986).

During the discrimination training, responding in the presence or absence of the drug stimulus is stabilized during several weeks. Commonly the incentive is food or water; hence the animals are deprived of food or water to maintain their weights around 80% of their free-feeding weights. To obtain a reward the animals have to press the injection appropriate lever/key a specified number of times to get access to the reinforcement. Fixed ratio procedures are commonly used (e.g. Koek & Colpaert, this volume) but several other schedules of reinforcement are also used such as the variable interval schedule of reinforcement (e.g. Glennon, Young & Pierson, this volume). The use of a novel procedure, i.e. discriminated conditioned taste aversion to study the stimulus properties of 5-HT agonists is described by Lucki and South (this volume). Methodological aspects of DDL has recently been discussed by Järbe (1987).

After having mastered the DDL, the animals may be tested with different drugs and doses. This is usually referred to as generalization testing. The choice behaviour is then taken as the index of the degree of correspondence or similarity in the stimulus effects between the test and drug training condition. The index may be based on the degree of correspondence expressed as the proportion drug associated responding out of the total number of responses emitted during the session. Tests may occur in extinction (i.e. unreinforced) or they may be reinforced. In the next figure typical dose generalization curves are depicted. The animals had been trained to discriminate between delta-9-tetrahydrocannabinol (THC) and the vehicle condition, the THC training doses being 1 mg/kg (pigeons) and 3 mg/kg (rats). From these graphs it can be seen that the animals emit progressively fewer drug related responses when tested with amounts lower than the training drug. The data pertaining to the ascending portion of the curve may be used to determine an ED_{50} value. In the present cases these values, according to logarithmic regression analyses, were: 0.43 mg/kg (pigeons) and 1.17 mg/kg (rats), respectively. Thus the dose expected to induce 50% drug appropriate responding is a fraction of the training dose and consequently varies with the drug training dose employed.

In this regard ED_{50} values derived from DDL experiments differ from those derived from more regular pharmacological preparations where there is a more obvious end-point for the reaction such as maximum contractibility of, e.g. the guinea pig ileum preparation, in response to the application of a certain pharmacologically active substance. Note also that in tests with doses higher than the training doses, there is no change in the curvature for either of the two generalization gradients. This is reminiscent of the findings from experimental psychology that sensory signals like brightness and white noise (intensity continuea) produce similar generalization gradients (Heinemann & Chase, 1975; Mackintosh, 1974) which contrasts with the pattern obtained using e.g. frequency or wave length as the discriminative stimuli. In the latter instances the target value is surrounded by fairly equal generalization decrement curves (see De Witte, 1978, for an example encompassing both in these features using intracranial stimulation).

DDL techniques allow determinations of commonalities and differences in the perceptible qualities of drug induced stimuli. Thus compounds can be categorized into various major drug classes according to their stimulus attributes (Barry, 1974). Specific pharmacological tools (antagonists, agonists, uptake and synthesis inhibitors. and neurotoxins) can be employed to characterize the mediation of neuropharmacological substrates [type of receptor and mechanism(s) of action]. For example, the commonality of the stimulus effects of CNS motor stimulants has been related to dopaminergic transmission as evinced by cross-generalization between such stimulants and antagonism by neuroleptics such as haloperidol, even though the modes of action may differ; pretreatment with the synthesis inhibitor alpha-methyl-p-tyrosine attenuates the amphetamine cue, but not that of, e.g. cocaïne, the latter agent rather being attenuated by reserpine (Silverman & Ho, 1977) [for further examples, see Colpaert & Rosecrans (1977); Colpaert & Slangen (1982); Järbe (1986; 1987); Overton (1984)].

The early DDL studies concerning drugs purportedly affecting serotonergic transmission often employed psychedelic drugs such as lysergic acid diethylamine (LSD) as the training agent (Winter, 1974).

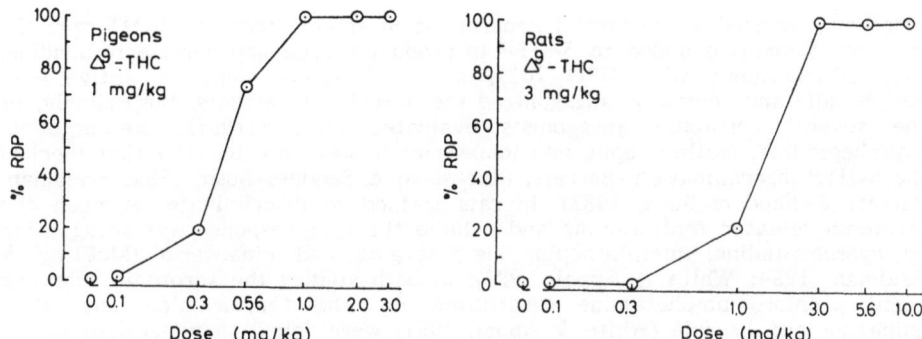

Fig. VII.1 Dose-generalization gradients in pigeons (left) and rats (right) trained to discriminate between 1 and 3 mg/kg, respectively, of delta-9-tetrahydrocannabinol (THC) and vehicle. Injections were given 90 (i.m.) and 30 (i.p.) min, respectively, prior to training and testing. Y-axis, percent responding to drug-associated position (% RDP); x-axis, dose of delta-9-THC (adapted from Järbe & McMillan, 1980).

Generally it was found i) that drugs capable of producing a psychedelic reaction in man also substituted for each other's discriminable effects (e.g. LSD, mescaline, and psilocybin), ii) that serotonin antagonists such as methiothepin, cyproheptadine and pizotifen (BC-105) attenuated the drug induced reaction, and iii) that several, but certainly not all non-psychedelic compounds failed to disclose a substantial transfer effect; e.g. the non-psychedelic agent quipazine consistently produced effects quite similar to the above mentioned psychedelics (Appel, White & Kuhn, 1978; Rosecrans et al., 1978; Silverman & Ho, 1978). The correlation between "halucinogenic" potency in man and the median effective dose (ED_{50}) estimates derived from DD1 studies was emphasized by Glennon, Rosecrans and Young (1982). Attempts to identify a specific anatomical locus as responsible for mediating the effects generally were met with limited success (Järbe, in press, see, however, also Nielson & Scheel-Kruger, 1986). Of concern is also the finding that rats could be trained to reliably discriminate the serotonin precursor l-5-hydroxytryptophan (5-HTP) from the nondrug condition (Barrett, Blackhear & Sanders-Bush, 1982). Following the observations of Carter, Dykstra, Leander and Appel (1978) that disruption of bar-pressing behaviour appeared to be due to peripherally activated serotonin rather than a centrally mediated serotonergic effect, the strategy of Barrett and colleagues (1982) was to block the peripheral "side" effects of 5-HTP by means of a peripherally acting decarbocylase inhibitor. The authors reported that fenfluramine, a releaser of serotonin, evinced a substantial generalization effect; that the 5-HT reuptake inhibitor fluoxetine added to 5-HTP in producing drug-appropriate responding; that pretreatment with NSD 1015, a decarboxylase inhibitor active both peripherally and centrally, antagonized the 5-HTP cue effects. Surprisingly, of the several serotonin antagonists evaluated (methysergide, methergoline, cyproheptadine, methiothepin, and mianserin) it was only BC-105 that blocked the 5-HTP discrimination (Barrett, Blackshear & Sanders-Bush, 1982; Friedman, Barrett & Sanders-Bush, 1983). In rats trained to discriminate between the serotonin releaser fenfluramine and vehicle the drug response was antagonized by cyproheptadine, methiothepin, methysergide and cinanserin (McElroy & Feldman, 1984; White & Appel, 1981); in both studies the serotonin releasing agent p-chloro amphetamine substituted for the fenfluramine cue. Both quipazine and lisuride (White & Appel, 1981) were found to generalize to the fenfluramine stimulus. 5-HTP only partially substituted for the fenfluramine cue (McElroy & Feldman, 1984). Taken together, these data might indicate the modeling of different serotonergic receptor types depending on the training agent used. More recent work concerning the behavioural pharmacology of 5-HT has been founded on the proposed classification of serotonin receptors into $5-HT_1$, $5-HT_2$ and $5-HT_3$ subclasses; the $5-HT_1$ subclass has been further divided into the subgroups of A, B, and C (see Peroutka, 1987; and also this volume). More or less specific agents are now available to analyze further the in vivo characteristics of the various 5-HT binding sites.

Koek and Colpaert (ibid) provide an overview of recent DDL data pertaining to this nomenclature (see also Appel & Cunningham, 1986). Although Koek and Colpaert (ibid) acknowledge a differentiation of 5-HT agonist drug stimuli they, however, stress the need for more thorough examination using uniform methodology. From their tabulations it is clear that the reported transfer test results are not totally concordant with the binding data presented as derived from different sources. Hence, the transfer test results so far published at first glance do not seem to be that easily accomodated within the conceptual framework presented above of the selectivity of various agents for different subtypes of serotonin binding sites/subclasses. It is suggested that the idea of a single receptor model might be more profitable for the study of 5-HT agonists at the behavioural level; this is reminiscent of the idea proposed earlier by Colpaert (1986) as regards opioid compounds. Essentially what is argued is that partial generalization does not necessarily refute selectivity of agonists of different binding sites but rather that a given training compound is activating one site in particular, and that subsequent transfer testing with various agents reflects their different degrees of intrinsic activity for that receptor site. Indeed a quite plausible postulate. Such a model should take into account also the effects of 5-HTP and 5-HT releasing drugs (see above) as well as that of 5-HT uptake blockers, i.e. increased presynaptic serotonergic activity.

Additionally, the importance of the drug training/test dosage has not been thoroughly investigated. As an example, Young, Rosecrans and Glennon (1986) trained rats to discriminate between 5-methoxy-N,N-dimethyltryptamine (5-MeODMT) and saline. They then examined various doses of different 5-HT antagonists in combination with 5-MeODMT to test the efficacy of these compounds in attenuating the 5-MeODMT stimulus. Differential antagonism, dependent on 5-MeODMT dose and 5-HT antagonist, were obtained leading these authors to advice caution in drawing conclusions regarding the direct acting 5-HT agonists and antagonists.

5-HT has been implicated in anxiety (see relevant chapters, this volume; Chopin & Briley, 1987). Thus, several putative 5-HT$_{1A}$ agonists have been found to exert antianxiety effects in behavioural screening and also to be able to relieve anxiety in man as assessed in clinical trials (Goa & Ward, 1986). Glennon et al. (ibid, this volume) describe experiments showing that the discriminative stimulus effects of diazepam and several 5-HT agonists, with purported selectivity for the 5-HT$_{1A}$ binding site, are dissimilar. Hence these second generation anxiolytics do not seem to act at the benzodiazepine receptor to produce the antianxiety effects. In addition experiments, Glennon and Pierson (this volume) more extensively studied the purported 5-HT$_{1B}$ agonist 1-[3-(Trifluoromethyl)phenyl]piperazine (TFMPP) and concluded that the discriminative stimulus effects of TFMPP may involve both 5-HT$_{1B}$ and 5-HT$_{1C}$ components. Furthermore Glennon and Pierson (ibid, this volume) identified mesulergine as a 5-HT$_{1C}$ agonist and suggested, based on its binding profile, that this agent may be the first known 5-HT agonist with selectivity for 5-HT$_{1C}$ sites. Barrett, Olmstead, Lei, Harrod, Hoffamn, Glover, Nader and Weissman (this volume) have examined the effects of spiroxatrine, a putative 5-HT$_{1A}$ antagonist, in pigeons. These authors used three procedures in different groups of birds.

One procedure was a mixed appetitive fixed-interval (FI), fixed-ratio (FR) schedule of reinforcement, the second procedure essentially constituted a conflict situation (punishment), and the third set-up was a discrimination between spiroxatrine and saline using DDL methodology. In general all of these experiments suggested that spiroxatrine exerts its activity through a 5-HT$_{1A}$ receptor mechanism, the agonist activity profile, however, being predominant with behavioural functions more similar to other 5-HT$_{1A}$ agonists such as 8-OH-DPAT and buspirone. The importance of the order of transfer testing was emphasized because generalization to 8-OH-DPAT did not occur if this drug was studied after buspirone. Such profound effects of the order of treatment have not been described in the DDL literature previously even though it has been opbserved that the generalization gradients of caffeine varied depending upon whether the training session prior to caffeine dose-generalization testing had been drug (caffeine) or no drug (saline); sensitivity was greater after saline sessions as reflected by a lower ED$_{50}$ value (Modrow, Holloway & Carney, 1981; see also Modrow & Holloway, 1985).

Attempts to use the conditioned taste aversion (CTA) procedure to assess discriminative stimulus effects of drugs have been reported earlier (D'Mello & Stolerman, 1978) but the data by Lucki and South (ibid, this volume) appear to be the first convincing demonstration of drug induced CTA-DDL as a bio-behavioural assay in psychopharmacological research. One advantage of CTA-DDL appears to be efficiency in terms of speed of acquisition. One potential drawback is the use of lithium chloride (LiCl) in the conditioning process as LiCl might alter the stimulus properties of drugs. Drugs affecting thirst mechanisms might also pose problems. Nevertheless the data concerning 5-HT agonists generated by Lucki and South (ibid, this volume) are reassuring and only future work can delineate the pros and cons of this procedure in the study of the discriminative stimulus properties of drugs.

Representative agents from the major drug classes have been shown to successfully serve a discriminative function in laboratory animals. However, agents from some drug classes such as the "classical" antidepressants desimipramine (Shearman, Miksic & Lal, 1978) and imipramine (Schechter, 1983; Overton & Batta, 1977) appear to be more difficult to establish as discriminative cues than the majority of psychotropic compounds investigated. Recent experiments by Järbe, Hiltunen, Kamkar and Archer (in preparation) indicated that the clinically effective antidepressant zimeldine, primarily a 5-HT reuptake blocker, could be established as a discriminative cue in rats.

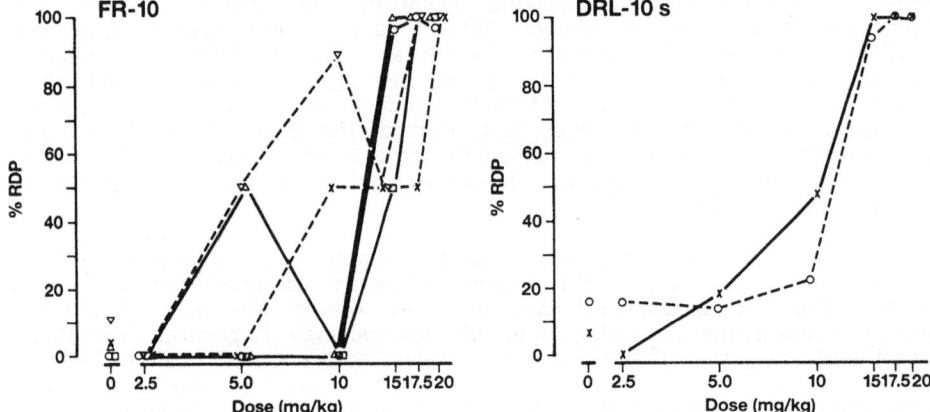

Fig. VII.2 Dose-generalization gradients of individual rats trained to discriminate between 20 mg/kg zimeldine and saline. The right section refers to differential reinforcement of low rates of responding (DRL, n=2) and to the left the fixed ration 10 (FR=10, n=5) schedule of reinforcement is indicated.

Two schedules of reinforcements (FR-10 and DRL-10 sec) were used to control behavior in different groups of rats (n=5 and n=2, respectively). The discriminative conditions were 20 mg/kg zimeldine versus 1 ml/kg saline, injected i.p. 30 min prior to sessions. Figure VII.2 shows the percent responding to drug associated position (%RDP) for individual animals when tested with different doses of the training drug (range: 2.5 - 20 mg/kg) under both schedules of reinforcement employed. The animals were tested according to a descending/ascending order of dose presentation and data reflect the mean of these two observations in each individual animal.

DDL procedures can be used to discern similarities and dissimiliarities in the stimulus effects of drugs and other sensory events. However DDL does not convey an assessment of the "quality" of effect. It is known that electrical stimulation of the dorsal midbrain central grey (DCG) as well as other parts of the "Papez circuit" elicits defensive behavior and generates aversive effects in experimental animals as well as induces unpleasant feelings in humans (Graeff et al., 1986). Graeff (this volume) summarizes experiments pointing towards 5-HT being involved in the inhibition of neurons responsible for the elaboration and/or expression of aversive states. Graeff suggests that $5-HT_2$ receptors are mainly involved because $5-HT_2$ receptor antagonists do not elicit flight behavior when microinjected into the DCG. Furthermore, "5-HT modulation is likely to be phasic, rather than tonic". Graeff concludes his overview by noting that it seems likely "to implicate serotonergic mechanisms modulating the neuronal network responsible for the elaboration and/or expression of aversive states in the pathophysiology of several types of anxiety disorders as well as in the mode of action of drugs used for their clinical treatment". Jenck, Broekkamp and Van Delft (personal communication) examined various 5-HT antagonists in rats trained to interrupt a periaqueductal gray matter aversive stimulation by escaping from one compartment to the opposite compartment. The conclusions reached were that "a) blockade of 5-HT2 receptor subtypes suppresses centrally induced aversion; b) blockade of 5-HT1 subtypes facilitates aversion with dominance over the 5-HT2 mediated suppression; c) blockade of 5-HT3 subtypes seems to have no effect". Examining buspirone, Rowan, Cullen, Moulten and Anwyl (in preparation) concluded thet "Buspirone, at a 'non-sedative' dose, appeared to impair learning only on the passive avoidance task" (the other task was the Morris Water Maze). "This may be due to the ability of buspirone to reduce the suppressive effect on activity of a mild, uncontrolled stressor".

Acknowledgements: Preparation of this chapter was supported by grants from the Swedish Medical Rerearch Council, The Bank of Sweden Tercentenary Foundation, The Swedish Council for Research in the Humanities and the Social Sciences, The Magnus Bergvall's Foundation, The Gustav Lundahl's Donation for Research and Albert Nilsson's Research Fund (the two latter funds being administered by the Swedish Social Welfare Board).

References

Appel, J.B., & Cunningham, K.A. (1986). The use of drug discrimination procedures to characterize hallucinogenice drug actions. Psychopharmacology bulletin, 22, 959-967.

Appel, J.B., White, F.J., & Kuhn, D.M. (1978). The use of drugs as discriminative stimuli in behavioral pharmacology. In: Stimulus properties of drugs: Ten years of progress. (Eds. Colpaert F.C. & Rosecrans J.A.) Amsterdam, Elsevier/North-Holland Biomedical Press.

Barrett, R.J., Blackshear, M.A., & Sanders-Bush, E. (1982). Discriminative stimulus properties of L-5-hydroxytryptophan: behavioral evidence for multiple serotonin receptors. Psychopharmacology, 76, 29-35.

Barry, H., III. (1974). Classification of drugs according to their discriminable effects in rats. Federation Proceedings, 33, 1814-1824.

Carter, R.B., Dykstra, L.A., Leander, J.D., & Appel, JH.B. (1978). Role of peripheral mechanisms in the behavioral effects of 5-hydroxytryptophan. Pharmacology, Biochemistry & Behavior, 9, 240-253.

Chopin, P., & Briley, M. (1987). Animal models of anxiety: The effect of compounds that modify 5-HT neurotransmission. Trends in Pharmacological Sciences, 8, 383-388.

Colpaert, F.C. (1986). Drug discrimination: behavioral, pharmacological and molecular mechanisms in discriminative drug effects. In: Behavioral analysis of drug dependence. (Eds. Goldberg S.R. & Stolerman I.P.) New York, Academic Press.

Colpaert, F.C., & Rosecrans, J.A. (Eds.), (1978). Stimulus properties of drugs: Ten years of progress. Amsterdam: Elsevier/North-Holland Biomedical Press.

Colpaert, F.C., & Slangen, J.L. (eds.), (1982). Drug discrimination: applications in CNS pharmacology. Amsterdam: Elsevier/North-Holland Biomedical Press.

De Witte, PH. (1978). Hierarchy of the perceptuel dimensions of hypothalamic stimulation: The psychophysics of rewarding cues, and effects of sensory stimulation on self-stimulation. In: Stimulus properties of drugs: Ten years of progress. (Eds. Colpaert F.C. & Rosecrans J.A.) Amsterdam, Elsevier/North-Holland Biomedical Press.

D'Mello, G.D., & Stolerman, I.P. Methodological issues in drug discrimination research. (1978). In: Stimulus properties of drugs: Ten years of progress. (Eds. Colpaert F.C. & Rosecrans J.A.) Amsterdam, Elsevier/North-Holland Biomedical Press.

Friedman, R., Barrett, R.J., & Sanders-Bush, E. (1983). Additional evidence that the L-5-hydroxytryptophan discriminaiton models a unique serotonin receptor. Psychopharmacology, 80, 209-213.

Glennon, R.A., Rosecrans, J.A., & Young, R. (1982). The use of the drug discrimination paradigm for studying hallucinogenic agents. A review. In: Drug discrimination: Applications in CNS pharmacology. (Eds. Colpaert F.C. & Slangen J.L.) Elsevier/North-Holland, Biomedical Press.

Goa, K.L., & Ward, A. Buspirone: A preliminary review of the pharmacological properties and therapeutic efficacy as an anxiolytic. (1986). Drugs, 32, 114-129.

Graeff, F.G., Brandao, M.L., Audi, E.A., & Schuetz, M.T.B. Modulation of the brain aversive system by GABAergic and serotonergic mechanisms (1986). Behavioral Brain Research, 21, 65-72.

Heinemann, E.G., & Chase, S. (1975). Stimulus generalization. In: Conditioning and behaviour theory. (Eds. Estes W.A.) Hillsdale, Lawrence Erlbaum ass. Publ.

Holtzman, S.G. Discriminative stimulus properties of opioids that interact with mu, kappa and PCP/sigma receptors (1985). In: Behavioral pharmacology: The current status. (Eds. Seiden L.S. & Balster R.L.) New York, Alan R. Liss Inc.

Järbe, T.U.C. State-dependent learning and drug discriminative control of behaviour: An overview (1986). Acta Neurologica Scandinavica, 74 (Suppl. 109), 37-59.

Järbe, T.U.C. Drug discrimination learning: Cue properties of drugs (1987). In: Experimental psychopharmacology. (Eds. Greenshaw A.J. & Dourish C.T.) Clifton, Humana Press.

Järbe, T.U.C. Discrimination learning with drug stimuli. Methods and applications (in press). In: Neuromethods. (Eds. Boulton A.A., Baker G.B. & Greenshaw G.J.) Volume 13. Psychopharmacology. Clifton, Humana Press.

Järbe, T.U.C., & McMillan, D.E. Delta-9-THC as a discriminative stimulus in rats and pigeons: Generalization to THC-metabolites and SP-111 (1980). Psychopharmacology, 71, 281-289.

Järbe, T.U.C., & Swedberg, M.D.B. A conceptualization of drug discrimination learning (1982). In: Drug discrimination: Applications in CNS pharmacology. (Eds. Colpaert F.C. & Slangen J.L.) Amsterdam, Elsevier/North-Holland. Biomedical Press.

Kalish, H.T. Stimulus generalization: Method and theory (1969). In M.H. Marx (Ed.), Learning: Processes. London: MacMillan Co.

Mackintosh, N.J. The psychology of animal learning (1974). New York: Academic Press.

McElroy, J.F., & Feldman, R.S. (1984). Discriminative stimulus properties of fenfluramine: Evidence for serotonergic involvement. Psychopharmacology, 83, 172-178.

Modrow, H.E., & Holloway, F.A. (1985). Drug discrimination and cross generalization between two methylxanthines. Pharmacology, Biochemistry & Behaviour, 23, 425-429.

Modrow, H.E., Holloway, F.A., & Carney, J.M. (1981). Caffeine discrimination in the rat. Pharmacology, Biochemistry & Behaviour, 14, 683-688.

Nielsen, E.B., & Scheel-Kruger, J. (1986). Cueing effects of amphetamine and LSD: Elicitation by direct microinjection of the drugs into the nucleus accumbens. European Journal of Pharmacology, 125, 85-92.

Overton, D.A. State dependent learning and drug discrimination (1984). In: Handbook of psychopharmacology. (Eds. Iverson L.L., Iverson S.D. & Snyder S.H.) vol. 18, New York, Plenum Press.

Overton, D.A. & Batta, S.K. Relationship between abuse liability of drugs and their degree of discriminability in the rat (1977). In: Predicting dependence liability of stimulant and depression drugs. (Eds. Thompson T. & Unna K.R.)

Baltimore: University Park Press.

Peroutka, S.J. Serotonin receptors (1987). In: Psychopharmacology: The third generation of progress. (Ed. Meltzer H.Y.) New York, Raven Press.

Rosecrans, J.A., Krynock, G.M., Newlon, P.G., Chance, W.T., & Kallman, M.J. (1978). Control mechanisms of drugs as discriminative stimuli: Involvement of serotonin pathways. In: Stimulus properties of drugs: Ten years of progress. (Eds. Colpaert F.C. & Rosecrans J.A.) Amsterdam, Elsevier/North-Holland Biomedical Press.

Schechter, N.D. Discriminative stimulus control with imipramine: Transfer to other anti-depressants (1983). Pharmacology, Biochemistry & Behavior, 19, 751-754.

Shearman, G., Miksic, S., & Lal, H. Discriminative stimulus properties of desipramine (1978). Neuropharmacology, 17, 1045-1948(.

Silverman, P.B., & Ho, B.T. Characterization of discriminative response control by psychomotor stimulants (1977). In: Discriminative stimulus properties of drugs. (Ed. Lal, H.) New York, Plenum Press.

Silvermann, P.B., & Ho, B.T. Stimulus properties of DOM: Commonality with other hallucinogens (1978). In: Stimulus properties of drugs: Ten years of progress. (Eds. Colpaert F.C. & Rosecrans J.A.) Amsterdam, Elsevier/North-Holland Biomedical Press.

White, F.J., & Appel, J.B. A neuropharmacological analysis of the discriminative stimulus properties of fenfluramine (1981). Psychopharmacology, 73, 110-115.

Winter, J.C. Hallucinogens as discriminative stimuli (1974). Federation Proceedings, 33, 1825-1832.

Young, R., Rosecrans, J.A., & Glennon, R.A. (1986). Further studies on the dose-dependent stimulus properties of 5-methoxy-N,N-dimethyltryptamine. Pharmacology, Biochemistry & Behavior, 25, 1207-1210.

RECEPTOR MECHANISMS OF THE DISCRIMINATIVE STIMULUS PROPERTIES OF PUTATIVE SEROTONIN AGONISTS

Wouter Koek, Francis C. Colpaert
Neurobiology Division, FONDAX - Groupe de Recherches servier, 7 rue Ampere, France

Drug discrimination is a reliable, sensitive and specific behavioural procedure that provides a powerful tool in analyzing neuropharmacological mechanisms underlying the behavioural effects of diverse classes of compounds (e.g., Colpaert & Slangen, 1982; Colpaert & Balster, 1987). The specificity of drug discrimination is apparent from the fact that, in general, compounds from different pharmacological classes produce different discriminative stimulus effects, whereas compounds from the same class produce similar discriminative stimulus effects. Drug discrimination procedures may even differentiate between compounds within a single class. The opiates, for example, comprise a heterogenous group of compounds that can be partitioned into subgroups on the basis of their discriminative stimulus effects (e.g., Holtzman, 1985; Woods et al., 1987). Because of this high specificity, drug discrimination procedures are suitable not only to evaluate hypotheses about receptors at the behavioural level, but also to evaluate hypotheses about receptor subtypes.

Various serotonin (5-HT) agonists can be discriminated by laboratory animals (see review by Glennon, 1987). In recent years, radioligand binding experiments have distinguished distinct binding sites that can conceivably operate as 5-HT receptors: these incluse the $5-HT_{1A}$, $5-HT_{1B}$, $5-HT_{1C}$ and $5-HT_2$ binding sites (see review by Peroutka, 1987). As a result, studies are being aimed at establishing which of these binding sites may mediate the discriminative stimulus effects of 5-HT agonists. For example, the discriminative stimulus effects of d-lysergic acid diethylamine (LSD) and of 1-(2,5-dimethoxy-4-methylphenyl)-2-amino-propane (DOM) have been found to be mediated primarily by a $5-HT_2$ mechanism (e.g., Appel and Cunningham, 1986; Glennon et al., 1986). The discovery of multiple 5-HT binding sites prompted also the development of novel drugs that possess a high degree of selectivity for these sites. Some of these novel agents have been used as training drug in drug discrimination procedures, i.e., the $5-HT_{1A}$-selective agonists 8-hydroxy-2 (di-n-propylamino) tetralin (8-OH-DPAT) and TVX Q 7821 (ipsapirone), and the $5-HT_{1B}$ selective drug 1-(3-trifluoromethylphenyl) piperazine (TFMPP). Results of generalization and antagonism tests suggest that the discriminative stimulus effects of 8-OH-DPAT and TFMPP are mediated primarily by $5-HT_{1A}$ and $5-HT_{1B}$ mechanisms, respectively (for a review, see Glennon, 1987). Thus, it has been proposed that the discriminative stimulus effects of 8-OH-DPAT, TFMPP and LSD can be used for the evaluation of effects elicited at $5-HT_{1A}$, $5-HT_{1B}$ and $5-HT_2$ receptors, respectively (e.g. Arnt, 1987).

In this review, results obtained in recent drug discrimination studies with the novel binding site-selective drugs will be discussed. Although the discriminative stimulus effects of the "prototypical" 5-HT agonists appear to be different to a large extent, partial generalization occurs with some agents.

In LSD-discriminating rats, partial generalization produced by various serotonergic drugs has been characterized in detail (Colpaert et al., 1982); here, we will indicate possible ways in which partial generalization in 8-OH-DPAT and TFMPP-discriminating animals can perhaps be analyzed further. Studies of the discriminative stimulus effects of opiates may serve to indicate fruitful approaches to such an analysis (e.g., Colpaert & Janssen, 1984; Woods et al., 1987).

Serotonergic mechanisms have been implicated in a number of human disorders, such as anxiety and depression. Because serotonergic agents that produce partial generalization in LSD-discriminating animals may produce only limited putative anxiolytic effects (Colpaert et al., 1985), a thorough characterization and analysis of partial generalization produced by serotonergic agents is not only important from a mechanistic point of view, but perhaps also from a clinical perspective.

Characteristics of 5-HT binding sites

Radioligand binding studies of 5-HT sites have been reviewed recently by Peroutka (1987). Briefly, Peroutka and Snyder (1979) concluded that at least two distinct 5-HT membrane recognition sites exist in the central nervous system: 5-HT$_1$ binding sites labeled by [^{3}H]5-HT and 5-HT$_2$ binding sites labeled by [^{3}H]spiperone. Nonsigmoidal displacement of [^{3}H]5-HT suggested the presence of 5-HT$_1$ binding site subtypes, i.e., 5-HT$_{1A}$ and 5-HT$_{1B}$ (Pedigo et al., 1981). More recently, evidence has been obtained suggesting a third subtype of 5-HT$_1$ binding sites (Pazos et al., 1984); as yet, this subtype has been less well characterized than the 5-HT$_{1A}$ and 5-HT$_2$ sites. The 5-HT$_{1A}$ site can be labeled with [^{3}H]8-OH-DPAT, the 5-HT$_{1B}$ site with [^{125}I]cyanopindolol, and the 5-HT$_2$ site with [^{3}H]ketanserin. Table 37.1 presents a summary of drug affinities at each of the aforementioned 5-HT binding subtypes.

Table 37.1 Drug affinities K_i (in nM) for different 5-HT binding sites

Drugs	Binding Site		
	5-HT$_{1A}$	5-HT$_{1B}$	5-HT$_2$
8-OH-DPAT	1	24000	15000
Ipsapirone	2.9	52000	10000
Buspirone	15	68000	2100
5-Me-ODMT*	5	70	1530
RU 24969	2.5	0.38	1000
TFMPP*	160	10	445
Quipazine*	2120	130	810
d-LSD	3.9	170	13

Data are from Peroutka (1987); data obtained with drugs that are marked with an astrick are from Spencer et al. (1987).

At $5-HT_{1A}$ sites, 8-OH-DPAT, ipsapirone and buspirone are potent, selective agents, whereas 5-methoxydimethyltryptamine (5-MeODMT) is potent but less selective. TFMPP, RU 24969 and quipazine have higher affinity for the $5-HT_{1B}$ site than for the other sites. LSD has about equally high affinity for $5-HT_{1A}$ and $5-HT_2$ sites, but only moderate affinity for the $5-HT_{1B}$ site.

Recent drug discrimination studies using selective serotonergic agonists
Table 37.2 lists drug discrimination studies in which selective 5-HT agonist were used as training drug and/or as test drug. The following discussion is largely based on the results obtained in these investigations.

Receptor mediation
Evidence that the discriminative stimulus effects of drugs are receptor-mediated can be provided by showing that 1) relatively small changes of a drug's activity, 2) drugs that differ only in the arrangement of groups around an optically active carbon show differences in activity, and 3) effects can be antagonized by drugs that interact also with the receptor but have low intrinsic activity. Support for receptor-mediation of the discriminative stimulus effects of 8-OH-DPAT comes from studies on several 8-OH-DPAT analogs (Glennon, 1987): in 8-OH-DPAT-trained animals, 8-OH-DEAT produces full generalization, 8-OH-DMAT produces partial generalization, and 8-OH-DBAT produces exclusively saline-appropriate responding. Further, the discriminative stimulus effects of 8-OH-DPAT can be antagonized (Arnt, 1987) by spiroxatrine (1-oxo-4-phenyl-8-(1,4-benzodioxan-2-ylmethyl)-2,4, 8-triazaspiro[4,5]decane); spiroxatrine has been proposed as a pure antagonist at $5-HT_{1A}$ receptors (Nelson and Taylor, 1986).

As yet, structure-activity studies characterizing the TFMPP-like discriminative stimulus effects of TFMPP analogs have not been reported. The discriminative stimulus effects of TFMPP may be antagonized by (-)alprenolol (Arnt, 1987).

LSD has been studied extensively as a training drug in drug discrimination (see review by Appel and Cunningham, 1986). The discriminative stimulus effects of LSD can be antagonized completely by the $5-HT_2$ antagonists pirenperone (Colpaert et al., 1982) and ketanserin (e.g. Cunningham & Appel, 1987).

Structure-activity studies have characterized the DOM-like discriminative stimulus effects of DOM derivatives (Glennon et al., 1986). In addition to demonstrating structure-activity relations, these studies have shown that the discriminative stimulus effects of DOM are stereoselective: (-)-DOM was found to be about 8 times more potent that (+)-DOM in producing DOM-like discriminative stimulus effects (Glennon et al., 1986). Further, pirenperone and ketanserin potently antagonized the discriminative stimulus effects of DOM (Glennon et al., 1983).

In summary, although there is substantial evidence that the discriminative stimulus effects of 8-OH-DPAT, TFMPP, LSD and DOM are receptor-mediated, this evidence appears to be less than complete for all of these drugs, except DOM. Further, although the discriminative stimulus effects of each of the agonists can be antagonized, the nature of the antagonism has as yet not been fully characterized. Obviously, such a characterization is important because the antagonism may be functional (i.e., based on counter-regulatory effects mediated by other mechanisms) rather than strictly competitive. Thus, one should be able to demonstrate that the antagonism is surmountable, i.e., the antagonist should shift the generalization gradient of the training drug to the right in a parallel manner, without a reduction of the maximum response produced by the training drug.

Table 37.2 Survey of data obtained in drug discrimination studies utilizing novel 5-HT agents

Training Drug	Training Dose (mg/kg)	Reference
8-OH-DPAT	0.2	Glennon (1986)
	0.05	Tricklebank et al. (1987)
	0.4	Cunningham et al. (1987)
	0.4	Arnt (1987)
Ipsapirone	10	Spencer and Traber (1987)
TFMPP	1	Glennon et al. (1984)
	1	McKenney and Glennon (1986)
	0.8	Cunningham and Appel (1986)
	1	Arnt (1987)
d-LSD	0.08	Cunningham and Appel (1987)
	0.16	Arnt (1987)
DOM	1	Glennon et al. (1983)
	1	Glennon and Hauck (1985)
5-Me-ODMT	1.5	Young et al. (1983)
	3.0	Young et al. (1983)
	1.25	Spencer et al. (1987)

Only those studies are included in which novel agents have been used, that are relatively selective for 5-HT binding subtypes, either as training and/or as test drug.

In this respect, note that ritanserin, a pure antagonist of LSD discrimination in rats (Colpaert et al., 1985), may not act as a competitive 5-HT$_2$ receptor antagonist in rat tail artery, but may act through an allosteric site which is distinct from the 5-HT$_2$ receptor (Frenken & Kaumann, 1987). Indeed, in LSD-discriminating rats both ritanserin and pirenperone caused the LSD dose-effect curve to shift to the right in a manner that failed to meet the requirements for a reversible competitive interaction (Colpaert et al., 1985; Colpaert & Janssen, 1983).

Mediation through one or more receptors
Actions of different drugs are likely mediated by different receptors if 1) they produce entirely different, non-overlapping effects, and if 2) the effects of one agonist are blocked and the effects the other agonists are not modified by a specific antagonist. Data relevant to the first point are summarized in Table 37.3. 8-OH-DPAT and ipsapirone show symmetrical generalization; TFMPP, LSD and DOM do not generalize completely to 8-OH-DPAT and ipsapirone. Thus, 8-OH-DPAT and ipsapirone may produce similar discriminative stimulus effects that are not identical to the discriminative stimulus effects of TFMPP, LSD and DOM. LSD and DOM show symmetrical generalization and do not substitute completely for 8-OH-DPAT, ipsapirone and TFMPP. TFMPP appears to produce discriminative stimulus effects that to some extent differ from the effects produced by the other drugs. Therefore, the data in Table 37.3 suggest three groups of distinct, but to some extent overlapping, discriminative stimulus effects that may be mediated by different receptors. Because generalization not only occurs within subgroups of drugs, but also between subgroups, one could attempt to explain the results in Table 37.3 by assuming that all these drugs interact with a single, common receptor, but differ in their intrinsic activity at this receptor (Colpaert, 1986, 1987). The data that are currently available on the discriminative stimulus effects of serotonergic agents are not sufficient to enable an adequate test of this assumption.

Table 37.3 Maximal percentage of training drug-appropriate responding induced by various test drugs

Test Drug	8-OH-DPAT	Training Drug Ipsapirone	TFMPP	d-LSD	DOM
8-OH-DPAT	_100_	_85_	5-46	17-32	46
Ipsapirone	_85-88_	_100_	14	71	NT
TFMPP	31-34	33	_100_	40-51	28
d-LSD	20-50	NT	21-65	_100_	_90_
DOM	10-32	NT	28-70	_88_	_100_

Data are from the studies referred to in table 37.2. Percentages greater than 80% are underlined; NT: not tested.

Table 37.4 describes the results that are obtained in antagonism tests. Ketanserin antagonized selectively the stimulus effects of LSD and of DOM, suggesting that the stimulus effects of LSD and DOM are 0ediated by receptors that are different from the receptors underlying the discriminative stimulus effects of 8-OH-DPAT and TFMPP. Spiroxatrine antagonized the effects of 8-OH-DPAT, but not the effects of TFMPP, suggesting that different receptors mediate the discriminative stimulus effects of 8-OH-DPAT and TFMPP. As yet, it is not clear whether spiroxatrine is selective for the receptor underlying the stimulus effects of 8-OH-DPAT, because the effects of spiroxatrine on the discriminative stimulus effects of LSD and DOM have not been studied. (-)-Alprenolol may have a moderate selectivity for the receptor underlying the stimulus effects of TFMPP, as this drug failed to antagonize LSD and did not

Table 37.4 Maximal percentage of drug-appropriate responding, produced by the training dose of various training drugs, after pretreatment with different antagonists

Test Drug	Training Drug			
	8-OH-DPAT	TFMPP	d-LSD	DOM
Ketanserin	67-98	90-98	*25*	*0*
(-)Alprenolol	*14-50*	*<50*	>50	NT
Spiroxatrine	*<50*	>50	NT	NT

Data are from the studies referred to in table 37.2, except*; Glennon et al. (1983). Percentages smaller than 50 are underlined; NT: not tested.

completely antagonize the stimulus effects of 8-OH-DPAT. Taken together, the data in Table 37.4 provide further evidence for the involvement of multiple receptors in the discriminative stimulus effects of 8-OH-DPAT, TFMPP, and LSD.

Clearly, the availability of a selective TFMPP antagonist would greatly facilitate studies aimed at studying multiple receptor involvement further.

Relationship to functional 5-HT receptor subtypes

As a result of controversy in the literature regarding the classification and nomenclature of functional receptors for 5-HT, Bradley et al. (1986) have proposed a classification of 5-HT receptors into three main groups, based largely on data from studies in isolated peripheral tissues. Bradley et al. (1986) suggest that this classification may be relevant to functional 5-HT-receptor-mediated responses in the central nervous system.

Briefly, the following classification criteria are proposed: 1) a response is mediated by "5-HT$_1$-like" receptors when a) the response can be antagonized by methiothepin, but not by antagonists that are selective for 5-HT$_2$ or for 5-HT$_3$ receptors (e.g., ketanserin and ICS 205-930, respectively), and when b) the response can be mimicked by 5-carboxamidotryptamine.

Bradley et al. (1986) adopted the term "5-HT-like" primarily because of the absence of selective antagonists that would enable one to discriminate between sub-populations of receptors. A response is mediated by a 5-HT$_2$ receptor when the response can be antagonized by 5-HT$_2$ selective antagonists, such as ketanserin, but not by 5-HT$_3$ selective antagonists, such as ICS 205-930. A response is mediated by a 5-HT$_3$ receptor when a) the response can be antagonized by 5-HT$_3$ selective antagonists, but not by 5-HT$_2$ selective antagonists, and b) when the response is mimicked by 2-methyl-5-HT, a selective 5-HT$_3$ agonist.

The receptor that mediates the discriminative stimulus effects of LSD and of DOM appears to correspond to the 5-HT$_2$ receptor proposed by Bradley et al., because ketanserin is an antagonist of their discriminative stimulus effects. However, evidence that selective 5-HT$_3$ antagonists do not antagonize the discriminative stimulus effects of LSD and DOM has yet to be demonstrated. The receptors underlying the discriminative stimulus effects of 8-OH-DPAT and TFMPP fulfill one of the criteria for being considered "5-HT1-like", i.e., their stimulus effects are not fully antagonized by $5-HT_2$ selective antagonists. Further, the stimulus effects of 8-OH-DPAT are not antagonized by the $5-HT_3$-selective antagonist ICS 205-930. However, methiothepin does not fully antagonize their discriminative stimulus effects either, and the effects of the $5-HT_1$ selective agonist 5-carboxytryptamine have not been evaluated. Thus, at present there is some evidence for the identification of the receptors that mediate the discriminative stimulus effects of 8-OH-DPAT and TFMPP on the one hand, and LSD and DOM on the other, with the "$5-HT_1$-like" and the $5-HT_2$ receptor, respectively. However, the evidence is incomplete. The role of the $5-HT_3$ receptor has yet to be studied with drug discrimination procedures.

If the receptors that underlie the discriminative stimulus effects of 8-OH-DPAT, TFMPP and LSD are indeed serotonin receptors, one would expect that fenfluramine, a 5-HT-releasing agent, and 5-HTP, a precursor of 5-HT, will produce drug-appropriate responding in 8-OH-DPAT-trained animals, in TFMPP-trained animals, and in LSD-trained animals.

Fenfluramine substitutes completely for TFMPP (Cunningham & Appel, 1986), produced 80% LSD-appropriate responding (Winter, 1980), but produces only 24% 8-OH-DPAT-appropriate responding (Glennon, 1986). 5-HTP does not elicit more than 50% LSD-like responding; 5-HTP has not been studied in 8-OH-DPAT and in TFMPP-discriminating animals.

Relationship to 5-HT binding sites
Table 37.5 indicates the relative $5-HT_{1A}$ binding-site affinities of drugs that have been tested in 8-OH-DPAT-trained animals. All drugs that produce full 8-OH-DPAT-appropriate responding (i.e., 8-OH-DPAT, ipsapirone, and buspirone) have high affinity for the $5-HT_{1A}$ binding site; 8-OH-DPAT is the most potent of these agents, both in producing 8-OH-DPAT-like discriminative stimulus effects and in displacing [^{3}H]-8-OH-DPAT. Note, however, that not all drugs that have a high affinity for the $5-HT_{1A}$ site produce complete 8-OH-DPAT-appropriate responding (i.e., RU 24969, LSD, 5-MeODMT, TFMPP).

Table 37.6 shows that RU 24969, a compound with high affinity for the $5-HT_{1B}$ binding site, produces complete generalization in animals trained to discriminate TFMPP from saline. RU 24969 has higher affinity for the $5-HT_{1B}$ site than TFMPP and was found to be somewhat more potent than TFMPP in producing TFMPP-like stimulus effects in two out of the three studies cited in Table 37.6. Note that 5-MeODMT, quipazine and LSD have an appreciable affinity for $5-HT_{1B}$ sites, yet fail to produce full TFMPP-appropriate responding.

The rank order of potency of LSD, quipazine and 5-MeODMT in producing LSD-like and DOM-like discriminative stimulus effects (i.e., LSD > quipzine > 5-MeODMT) correlates positively with their affinity for $5-HT_2$ sites (Table 37.7); however, TFMPP and RU 24969 have affinities for the $5-HT_2$ site that are intermediate between the affinities of LSD and 5-MeODMT, yet did not substitute completely for LSD and DOM. To date, only in the study of the discriminative stimulus effects of DOM has a strong positive correlation, based on a large number of drugs, between binding affinity and ED_{50} to produce DOM-like discriminative stimulus effects been reported (Glennon et al., 1986).

To summarize the foregoing discussion, agents that produce complete generalization in animals trained to discriminate 8-OH-DPAT, TFMPP or LSD/DOM from saline have high affinity for $5-HT_{1A}$-, $5-HT_{1B}$- and $5-HT_2$-binding sites, respectively. The converse, however, does not appear to be the case: even for agents that appear to be efficacious in producing discriminative stimulus effects, high affinity for the aforementioned binding sites does not necessarily imply the compound to induce complete generalization.

Partial generalization produced by binding-site selective serotonin agonists
Many of the aforementioned drugs do not exclusively produce either full 8-OH-DPAT-appropriate responding, TFMPP-appropriate responding, of LSD/DOM-appropriate responding (Table 37.8).

Table 37.5 $5-HT_{1A}$ binding site affinities of various drugs and the maximal percentage of 8-OH-DPAT - appropriate responding they have been reported to induce. Numbers between parentheses: doses, in mg/kg, at which the maximal effect was attained.

	$5-HT$ $1A$ binding site affinity (nM)	8-OH-DPAT - saline discrimination			
		Arnt (1987)	Cunningham et al. (1987)	Tricklebank et al. (1987)	Glennon (1986)
8-OH-DPAT	1	100 (0.1)	100 (0.4)	100 (0.5)	95 (0.2)
RU 24969	2.5	NT	26 (1)	0 (1)	NT	
Ipsapirone	2.9	<u>86 (1.8)</u>	<u>85 (4)</u>	<u>88 (3)</u>	NT
d-LSD	3.9	50 (0.5)	30 (0.24)	20 (0.1)	NT
5-MeODMT	5	67 (2.1)	38 (3)	50 (0.5)	38 (0.7)
Buspirone	15	<u>97 (0.62)</u>	90 (4)	<u>96 (1)</u>	NT
TFMPP	160	31 (4.1)	21 (0.5)	NT	35 (0.5)
Quipazine	2120	32 (15)	2 (1)	0 (1)	NT

Percentages greater than 80 are underlined; NT: not tested.

For example, ipsapirone produces, in addition to 8-OH-DPAT-appropriate responding, substantial LSD-appropriate responding; 5-MeODMT produces complete LSD-appropriate responding, yet produces also substantial 8-OH-DPAT-appropriate responding; LSD substitutes completely for DOM, but substitutes also, albeit partially, for 8-OH-DPAT and for TFMPP. In general, full substitution appears to be produced only when a test drug is relatively selective for the binding site that is thought to underlie the discriminative stimulus effects of a particular training drug; i.e., the only drugs that substitute completely for 8-OH-DPAT are ipsapirone and buspirone, drugs that have a higher affinity for $5-HT_{1A}$ binding sites than for $5-HT_{1B}$- or $5-HT_2$ binding sites. RU 24969, the only drug that substitutes completely for TFMPP, has a higher affinity for $5-HT_{1B}$ sites than for $5-HT_{1A}$ of $5-HT_2$ sites. (Note, however, that 5-MeODMT, quipazine and LSD produce complete LSD/DOM-like responding, yet appear to have a higher affinity for $5-HT_1$ sites than for $5-HT_2$ sites). Thus, one has attempted to explain the occurrence of partial generalization in terms of lack of selectivity, by assuming that actions at another receptor than the one underlying the discriminative stimulus effects of the training drug interfere with the production of complete generalization.

For example: 1) 5-MeODMT substitutes partially for 8-OH-DPAT, because 5-MeODMT has, in addition to its affinity for $5-HT_{1A}$ sites, a strong affinity for $5-HT_2$ sites and this latter activity somehow prevents 5-MeODMT from substituting completely; 2) LSD and quipazine substitute only partially for TFMPP, because both drugs display a high affinity for $5-HT_2$ receptors. Full instead of partial substitution should be produced by these drugs after pretreatment with a selective $5-HT_2$ antagonist, if the hypothesis is correct. However, whereas ketanserin increased the extent to which LSD substituted for TFMPP, the substitution was incomplete (Glennon, 1987).

Analysis of partial generalization: applications of receptor theory
Partial generalization is not necessarily a result of lack of selectivity for the receptors that underlie the discrimination, but may also result from a low or different level of intrinsic activity at these receptors (Colpaert, 1986, 1987). Thus, instead of evoking actions at other receptors in an attempt to explain partial generalization, it maybe helpful to assume initially that drugs that produce partial generalization are true partial agonists and to determine their receptor mechanism. Such an approach has proven fruitful in the characterization of partial generalization effects that are observed in the study of discriminative stimulus effects of opiates (Colpaert & Janssen, 1984). For example, cyclazocine does not substitute partially for morphine because this drug is a full agonist at sigma receptors, but because it it an only partial agonist at mu receptors. Cyclazocine appears to produce its morphine-like stimulus effects through its actions at mu receptors; however, because of its low intrinsic activity at these receptors, cyclazocine is not able to produce full morphine-appropriate responding.

Evidence for this explanation of the partial generalization of morphine to cyclazocine is 1) the observation that cyclazocine partically antagonizes the discriminative stimulus effects of morphine (Holtzman, 1983; Colpaert & Janssen, 1984), in agreement with the theory of competitive dualism (Ariens, 1964), and 2) the observation that cyclazocine produces more morphine-appropriate responding as the training dose of morphine is lowered and consequently less intrinsic activity is required to produce a full effect (Holtzman, 1983; Colpaert & Janssen, 1984).

In accordance with receptor theory (Ariens, 1964), the following predictions (Colpaert, 1982, 1987; Koek & Woods, 1987) can be derived from the hypothesis that partial generalization is based on partial agonist properties at the receptors that underlie the discriminative stimulus effects of the training drug:

1) the generalization gradient of a partial agonist is less steep than the generalization gradient of a full agonist.
2) the partial generalization is blocked by a full antagonist of the discriminative stimulus effects of the training drug.
3) a partial agonist antagonizes to some extent the discriminative stimulus effects of the training drug.
4) a partial agonist produces less generalization and more antagonism of the effects of the training drug as the training dose is increased, and produces more generalization and less antagonism of the training drug as the training dose is decreased.
5) the discriminative stimulus effects of a partial agonist can be blocked by another partial agonist that has lower intrinsic activity.

Although the confirmation of any of these predictions alone can not be held as conclusive evidence that partial generalization is indeed based on partial agonist properties at the receptor that underlies the discriminative stimulus effects of the training drug, confirmation of all of these predictions offers converging evidence for this hypothesis.

Table 37.6 $5-HT_{1B}$ binding site affinities of various drugs and the maximal percentage of TFMPP-appropriate responding they have been reported to induce. Number between parentheses: doses, in mg/kg, at which the maximal effect was attained.

	5-HT 1B binding site affinity (nM)	TFMPP - saline discrimination			
		Arnt (1987)	McKenney and Glennon (1986)	Cunnningham and Appel (1986)	Glennon (1984)
8-OH-DPAT	0.38	100 (0.21)	NT	100 (1.6)	91 (0.5)
TFMPP	10	100 (0.4)	98 (1)	100 (0.8)	97 (1)
5-Me-ODMT	70	34 (2.9)	19 (1)	NT	NT
Quipazine	130	70 (5)	71 (2.2)	30 (3.2)	NT
d-LSD	170	21 (0.24)	65 (0.1)	52 (0.8)	NT

Percentages greater than 80 are underlined; NT: not tested.

In pigeons that were trained to discriminate morphine from saline, partial generalization was produced by cyclazocine, by 1-NANM and by ketamine; the results obtained with cyclazocine and with 1-NANM were in agreement with all of the aforementioned predictions (Koek & Woods, 1987).

However, the partial generalization effect produced by ketamine failed to meet any one of the aforementioned criteria. Thus, the results suggest that cyclazocine and 1-NANM, but not ketamine, produce partial generalization because of their partial agonist actions at mu-opiate receptors.

In the study of serotonergic agents, the first detailed characterization of partial agonist and partial antagonist effects of putative serotonin antagonists has been conducted by examining each of the putative antagonists in terms of their ability both to mimic and to antagonize the discriminative stimulus effects of LSD (Colpaert et al., 1982).

From the studies of the discriminative stimulus effects of serotonin agonists reviewed here, it is apparent that 5-MeODMT, TFMPP and LSD may produce a substantial amount of partial generalization in animals trained to discriminate 8-OH-DPAT from saline. The slope of the generalization gradient obtained with each of these drugs in 8-OH-DPAT-trained animals appears to be less steep than the slope of the generalization gradient of the training drug, thus confirming the first of the aforementioned predictions. The availability of a selective antagonist at this receptor will of course greatly facilitate the further characterization of partial generalization effects in 8-OH-DPAT-trained animals. Spiroxatrine, and perhaps the buspirone analog BMY 7378 that appears to have high affinity and selectivity but low intrinsic activity at $5-HT_{1A}$ receptors (Yocca et al., 1987), may prove to be useful tools for such an analysis.

At present, there is some additional evidence that drugs that produce 8-OH-DPAT-appropriate responding may differ in intrinsic activity at the $5-HT_{1A}$ receptor. 8-OH-DPAT induces discriminative stimulus effects and, at higher doses, directly observable behavioural effects (i.e., reciprocal forepaw treading and flat body posture) that are probably mediated by $5-HT_{1A}$ receptors (e.g. Tricklebank, 1985). Ipsapirone mimics the discriminative stimulus effects of 8-OH-DPAT, but antagonizes 8-OH-DPAT-induced reciprocal forepaw treading and flat body posture (Maj et al., 1987); these observations suggest that ipsapirone may have a lower intrinsic activity at $5-HT_{1A}$ receptors that 8-OH-DPAT.

If drugs act at the same receptor, but differ in their intrinsic activity at this receptor, one would expect that drugs with a high intrinsic activity are able to produce the effects that can also be produced by drugs with lower intrinsic activity, but not the reverse. The maximum percentage of 8-OH-DPAT-appropriate responding produced by ipsapirone is 88, and the maximum percentage produced by 5-MeODMT and by LSD is 67 and 50, respectively (Table 37.9). If the maximum percentage of 8-OH-DPAT-appropriate responding that can be produced by these drugs offers some indication of their intrinsic activity, this would suggest the following rank order of intrinsic activity: LSD < 5-MeODMT < ipsapirone < 8-OH-DPAT. The generalization results, obtained in 8-OH-DPAT-, ipsapirone-, and

5-MeODMT-discriminating animals appear to be in agreement with this hypothesized rank order: 5-MeODMT produces 38-67% drug-lever responding in 8-OH-DPAT-trained animals, but produces 100% drug-appropriate responding in animals trained to discriminate ipsapirone, presumably a weaker agonist than 8-OH-DPAT. Further, LSD produces only 20-50% generalization in 8-OH-DPAT-trained animals, but substitutes completely for 5-MeODMT, presumably a weaker agonist at $5-HT_{1A}$ receptors than 8-OH-DPAT. All three discriminations that are described in Table 37.9 are likely to be mediated by $5-HT_{1A}$ receptors, given that 8-OH-DPAT and ipsapirone produce full generalization under all three conditions.

Partial generalization in animals trained to discriminate TFMPP from saline has been reported with quipazine (30-71% TFMPP-appropriate responding), DOM (28-70%), LSD (21-65%), buspirone (46%), and with 8-OH-DPAT (5-46%). In all cases, the generalization gradient obtained with the afrementioned drugs in TFMPP-discriminating animals appears to be less steep that the generalization gradient of TFMPP itself, in agreement with the assumption that these drugs may be partial agonists at TFMPP receptors.

Table 37.7 $5-HT_2$ binding site affinities of various drugs and the maximal percentage of LSD-appropriate responding and of DOM-appropriate responding they have been reported to induce. Numbers between parentheses: doses, in mg/kg, at which the maximal effect was attained.

	$5-HT\ 2$ binding site affinity (nM)	LSD-saline discrimination		DOM-saline discrimination	
		Arnt (1987)	Cunningham and Appel (1987)	Glennon et al. (1983)	Glennon & Hauck (1985)
d-LSD	13	<u>84 (0.05)</u>	<u>92 (0.08)</u>	<u>90 (0.1)</u>	NT
DOM	100*	NT	NT	<u>96 (1.0)</u>	<u>93 (1.0)</u>
TFMPP	445	51 (4)	40 (1.6)	28 (0.5)	NT
Quipazine	810	<u>96 (1.3)</u>	<u>90 (3.2)</u>	<u>95 (3.2)</u>	NT
RU 24969	1000	50 (2.2)	45 (3.2)	22 (0.6)	NT
5-Me-ODMT	1530	<u>97 (1.7)</u>	NT	<u>90 (3)</u>	NT
Ipsapirone	10000	71 (4.9)	NT	NT	NT
8-OH-DPAT	15000	17 (0.24)	32 (0.32)	NT	46 (0.33)

Percentages greater than 80 are underlined; NT: not tested; Data from Glennon (1986).

Table 37.8 Ranges of the maximal percentage of training drug-appropriate responding induced by various test drugs in 8-OH-DPAT, TFMPP, d-LSD, and DOM-trained rats

Test drug	Training drug			
	8-OH-DPAT	TFMPP	d-LSD	DOM
8-OH-DPAT	95-100	5-46	17-32	46
Ipsapirone	85-88	14	71	NT
Buspirone	90-97	46	NT	NT
5-MeODMT	38-67	19-34	97	90
RU 24969	0-26	91-100	45-50	22
TFMPP	21-35	97-100	40-51	28
Quipazine	0-32	30-71	90-96	95
d-LSD	20-50	21-65	84-92	95
DOM	10-32	28-70	88	96

Data are from the studies referred to in table 37.2; NT: not tested

A partial agonist will, to some extent, antagonize the effects of a full agonist.

Cunningham and Appel (1986) reported, however, that while 8-OH-DPAT produced 30% TFMPP-appropriate responding, no evidence was obtained that 8-OH-DPAT antagonized the stimulus effects of TFMPP. Currently, no other data that bear on this or any of the other predictions mentioned in the forgoing are available. Obviously, the development of a specific $5-HT_{1B}$ antagonist would facilitate greatly the further characterization of the receptors underlying the discriminative stimulus effects of TFMPP, and of partial generalization effects in TFMPP-discriminating animals. In rats that were trained to discriminate LSD from saline, partial generalization has been observed in tests of ipsapirone (71% LSD-appropriate responding), with TFMPP (40-51% LSD-appropriate responding), and with RU 24969 (45-50%).

Again, generalization gradients for these drugs were shallower than the gradients of the drugs that produced full generalization; and again, no more information is currently available that is relevant to the characterization of these partial generalization effects. Finally, TFMPP and RU 24969 have been reported to produce a degree of partial generalization in DOM-trained animals (i.e., 22-28%) that is less than the degree of partial generalization produced by these compounds in LSD-discriminating animals. However, the converse holds true for partial generalization produced by 8-OH-DPAT, i.e., 8-OH-DPAT produced

Table 37.9 Ranges of the maximal percentage of drug-appropriate responding in 8-OH-DPAT-, Ipsapirone- and 5-Me-ODMT-discriminating rats induced by drugs that produce partial generalization in 8-OH-DPAT - trained animals

	Training drug		
Test drug	8-OH-DPAT	Ispapirone	5-MeODMT
8-OH-DPAT	95-100	85	100
Ipsapirone	85-88	100	100
5-MeODMT	38-67	100	100
d-LSD	20-50	NT	100

Data from the studies referred to in table 37.2

17-32% LSD-appropriate responding, but 46% DOM-appropriate responding. Thus, from the data reviewed here, it appears that the discriminative stimulus effects of DOM may not be more selective than the discriminative stimulus effects of LSD, in terms of the overall degree of partial generalization produced by various test drugs. Recently, it has been suggested that certain novel derivatives of DOM (i.e., DOB and DOI) may not only be more selective that DOM in binding studies, but perhaps also in drug discrimination procedures (Glennon, 1987).

Analysis of partial generalization: clinical relevance
It is important to analyze partial generalization data thoroughly, perhaps along the lines indicated above, not only to advance our understanding of the molecular mechanisms underlying behavioural effects of serotonergic drugs, but also in an effort to contribute to the development of clinically useful compounds. In this latter respect, it seems especially relevant to characterize more fully the partial generalization produced by certain compounds in LSD-discriminating animals, given the suggestion that agents that appear to act as partial LSD agonists (e.g., R 56413) may, because of their partial LSD-like properties, produce only limited putative anxiolytic effects, as compared to full LSD antagonists (Colpaert et al., 1985). Thus, it may very well be that drugs that act as partial LSD agonists, which may be true for the purported anxiolytic buspirone and ipsapirone, will therefore have limited anxiolytic potential.

Summary and conclusions
The discovery of multiple 5-HT binding sites prompted the development of novel serotonergic drugs that have a high degree of selectivity for these sites. Some of these novel agents have been used as training drug in drug discrimination studies. Based on these studies, it has been suggested that the discriminative stimulus effects of 8-OH-DPAT, TFMPP and LSD can be used for evaluation of effects elicited at $5-HT_{1A}$, $5-HT_{1B}$, and $5-HT_2$ receptors, respectively.

The generalization results reviewed herein are consistent with the hypothesis that 8-OH-DPAT, TFMPP and LSD produce discriminative stimulus effects that are mediated by different receptors and suggest that these receptors may correspond to $5-HT_{1A}$, $5-HT_{1B}$, and $5-HT_2$ binding sites. To elevate the validity of these suggestions further, much remains to be explored: in particular, the observation that many drugs have been found to produce partial generalization in animals that discriminate the "prototypic" 5-HT agonists. It is proposed herein to proceed with the analysis of partial generalization along the lines that have proven to be useful in the analysis of discriminative stimulus effects of opiates.

Such an approach is likely to advance our understanding of the discriminative stimulus effects of serotonergic agents, and may perhaps contribute to the development of more efficacious, clinically useful compounds.

References

Appel, J.B. & Cunningham, K.A. (1986). The use of drug discrimination procedures to characterize hallucinogenic drug actions. Psychopharmacology Bulletin, 22, 959-967.

Ariens, E.J. (1964). The mode of action of biologically active compounds: In Molecular pharmacology, Vol. 1, New York: Academic Press.

Arnt, J. (1987). Characterization of the discriminative stimulus properties induced by $5-HT_1$ and $5-HT_2$ agonists in rats. Serotonin Symposium, Heron Island.

Bradley, P.B., Engel, G., Feniuk, W., Fozard, J.R., Humphrey, P.P.A., Middlemiss, D.N., Mylecharane, E.J., Richardson, B.P. & Saxena, P.R. (1986). Proposals for the classification and nomenclature of functional receptors for 5-hydroxytryptamine. Neuropharmacology, 25, 563-567.

Colpaert, F.C. (1987). Intrinsic activity and discriminative effects of drugs. In: Transduction mechanisms of drug stimuli. (Eds. Colpaert F.C. & Balster R.L.) Berlin, Springer Verlag (in press).

Colpaert, F.C. (1986). Drug discrimination: behavioural, pharmacological and molecular mechanisms of discriminative drug effects. In: Behavioural analysis of drug dependence. (Eds. Goldberg S.R. & Stolerman I.P.) pp. 161-193, New York, Academic Press.

Colpaert, F.C. (1982). The pharmacological specificity of opiate drug discrimination: In: Drug discrimination: applications in CNS pharmacology. (Eds. Colpaert F.C. & Slangen J.L.) pp. 3-16, Amsterdam, Elsevier Biomedical Press.

Colpaert, F.C. & Balster, R.L. (1987). Transduction mechanisms of drug stimuli. Berlin: Springer Verlag (in press).

Colpaert, F.C. & Janssen, P.A.J. (1984). Agonist and antagonist effects of prototype opiate drugs in rats discriminating fentanyl from saline:

characteristics of partial generalization. Journal of Pharmacology and Experimental Therapeutics, 230, 193–199.

Colpaert, F.C. & Janssen, P.A.J. (1983). A characterization of LSD-antagonist effects of pirenperone in the rat. Neuropharmacology, 22, 1001–1005.

Colpaert, F.C., Meert, T.F., Niemegeers, C.J.E. & Janssen, P.A.J. (1985). Behavioural and 5-HT antagonist effects of ritanserin: a pure and selective antagonist of LSD discrimination in rat. Psychopharmacology, .., 45–54.

Colpaert, F.C., Niemegeers, C.J.E. & Janssen, P.A.J. (1982). A drug discrimination analysis of lysergide acid diethylamine (LSD): in vivo agonist and antagonist effects of purported 5-hydroxytryptamine antagonists and of pirenpirone, a LSD-antagonist. Journal of Pharmaclogy and Experimental Therapeutics, 221, 206–214.

Colpaert, F.C. & Slangen, J.L. (1982). Drug discrimination: applications in CNS pharmacology. Amsterdam: Elsevier Biomedical Press.

Cunningham, K.A. & Appel, J.B. (1986). Possible 5-hydroxytryptamine (5-HT$_1$) receptor involvement in the stimulus properties of 1-(m-trifluoromethylphenyl) piperazine (TFMPP). Journal of Pharmacology and Experimental Therapeutics, 237, 369–377.

Cunningham, K.A. & Appel, J.B. (1987). Neuropharmacological reassessment of the discriminative stimulus properties of d-lysergic acid diethylamide (LSD). Psychopharmacology, 91, 67–73.

Cunningham, K.A., Callahan, P.M. & Appel, J.B. (1987). Discriminative stimulus properties of 8-hydroxy-(di-n-propylamino) tetralin (8-OH-DPAT): implications for understanding the actions of novel anxiolytics. European Journal of Pharmacology, 138, 29–36.

Frenken, M. & Kaumann, A.J. (1987). Allosteric properties of the 5-HT$_2$ receptor system of the rat tail artery. Naunyn-Schmiedeberg's Archives of Pharmacology, 335, 359–366.

Glennon, R.A. (1986). Discriminative stimulus properties of the 5-HT$_{1A}$ agonist 8-hydroxy-(di-n-propylamino) tetralin (8-OH-DPAT). Pharmaoclogy, Biochemistry and Behaviour, 25, 135–139.

Glennon, R.A. (1987). Site-selective serotonin agonists as discriminative stimuli. In: Transduction mechanisms of drug stimuli. (Eds. Colpaert F.C. & Balster R.L.), Berlin, Springer Verlag (in press).

Glennon, R.A. & Hauck, A.E. (1985). Mechanistic studies on DOM as a discriminative stimulus. Pharmacology, Biochemistry and Behaviour, 23, 937–941.

Glennon, R.A., McKenney, J.D. & Young, R. (1984). Discriminative stimulus properties of the serotonin agonist 1-(3-trifluoromethylphenyl) piperazine

(TFMPP). Life Sciences, 35, 1475-1480.

Glennon, R.A., Titeler, M. & Young, R. (1986). Structure-activity relationships and mechanism of action of hallucinogenic agents based on drug discrimination and radioligand binding studies. Psychopharmacology Bulletin, 22, 953-958.

Glennon, R.A., Young, R. & Rosecrans, J.A. (1983). Antagonism of the effects of the hallucinogenic DOM and the purported 5-HT agonist quipazine 5-HT$_2$ antagonists. European Journal of Pharmacology, 91, 189-196.

Holtzman, S.G. (1983). Discriminative stimulus properties of opioid agonists and antagonists. In: Theory in psychopharmacology. (Eds. Cooper S.J.) Vol. 2, pp. 1-45. London, Academic Press.

Holtzman, S.G. (1985). Drug discriminative studies. Drug and Alcohol Dependence, 14, 263-282.

Koek, W. & Woods, J.H. (1987). Partial generalization in pigeons trained to discriminate morphine from saline: applications of receptor theory. Journal of Pharmacology and Experimental Therapeutics (submitted).

Maj, J., Chojnmacka-Wojcik, E., Tatarczynska, E. & Klodzinska, A. (1987). Central action of ipsapirone, a new anxiolytic drug, on serotonergic, noradrenergic and dopaminergic functions. Journal of Neural Transmission, 70, 1-17.

McKenney, J.D. & Glennon, R.A. (1986). TFMPP may produce its stimulus effects via a 5-HT$_{1B}$ mechanism. Pharmacology, Biochemistry and Behaviour, 24, 43-47.

Nelson, D.L. & Taylor, E.W. (1986). Spiroxatrine: a selective serotonin 1A receptor antagonist. European Journal of Pharmacology, 124, 207-208.

Pazos, A., Hoyer, D. & Palacios, J.M. (1984). European Journal of Pharmacology, 106, 531-538.

Pedigo, N.W., Yamamura, H.I. & Nelson, D.L. (1981). Discrimination of multiple [^{3}H]-5-hydroxytryptamine binding sites by the neuroleptic spiperone in rat brain. Journal of Neurochemistry, 36, 220-226.

Peroutka, S.J. (1987). Serotonin receptors. In: Psychopharmacology: the third generation of progress. (Ed. Meltzer H.Y.) pp. 303-311, New York, Raven Press.

Peroutka, S.J. & Snyder, S.H. (1979). Multiple serotonin receptors: differential binding of [^{3}H]-5-hydroxytryptamine, [^{3}H]-lysergic acid diethylamide and [^{3}H]-spiroperidol. Molecular Pharmacology, 16, 687-699.

Spencer, D.G., Glaser, T. & Traber, J. (1987). Serotonin receptor subtype mediation of the interoceptice discriminative stimuli induced by 5-methoxy-N,N-dimethyltryptamine. Psychopharmacology, 93, 158-166.

Spencer, D.G. & Traber, J. (1987). The interoceptive discriminative stimuli induced by the novel putative anxiolytic TVX Q 7821: behavioural evidence for the specific involvement of serotonin 5-HT_{1A} receptors. Psychopharmacology, 91, 25-29.

Tricklebank, M.D. (1985). The behavioural response to 5-HT receptor agonists and subtypes of the central 5-HT receptor. Trends in Pharmacological Sciences, October, 403-407.

Tricklebank, M.D., Neill, J., Kidd, E.J. & Fozard, J.R. (1987). Mediation of the discriminative stimulus properties of 8-hydroxy-(di-n-propylamino) tetralin (8-OH-DPAT) by the putative 5-HT_{1A} receptor. European Journal of Pharmaoclogy, 133, 47-56.

Winter, J.C. (1980). Effects of the phenethylamine derivatives, BL-3912, fenfluramine, and SCH-12679, in rats trained with LSD as a discriminative stimulus. Psychopharmacology, 68, 159-162.

Woods, J.H., Bertalmio, A.J., Young, A.M., Essman, W.D. & Winger, G. (1987). Receptor mechanisms of opioid drug discrimination. In: Transduction mechanisms of drug stimuli. (Eds. Colpaert F.C. & Balster R.L.) Berlin, Springer Verlag (in press).

Yocca, F.D., Hyslop, D.K., Smith, D.W. & Maayani, S. (1987). BMY 7378, a buspirone analog with high affinity, selectivity and low intrinsic activity at the 5-HT_{1A} receptor in rat and guinea-pig hippocampal membranes. European Journal of Pharmacology, 137, 293-294.

Young, R., Rosecrans, J.A. & Glennon, R.A. (1983). Behavioural effects of 5-methoxy-N,N-dimethyltryptamine and dose-dependent antagonism by BC-105. Psychopharmacology, 80, 156-160.

SEROTONIN AND AVERSION

F.G.Graeff
Laboratory of Psychobiology, FFCLRP, Campus of University of Sao Paulo, 14049 Ribeirao Preto, SP, Brazil

Introduction
Serotonin (5-HT) has been implicated in pathological pain, anxiety and clinical depression, as well as in the mechanism of action of analgesic, anxiolytic and antidepressant drugs (Iversen, 1984; Johnston & File, 1986; Soubrie, 1986; Willner, 1985). Dysphoria or psychological suffering is a prominent complaint in all these conditions, and its alleviation a primary therapeutic goal. Thus, the question may be asked whether brain 5-HT affects the functioning of neuronal networks responsible for the elaboration of aversive states.

The nervous structures related to aversion have been identified mainly by electrically stimulating limbic brain areas. In this way, behavioural reactions such as affective-defense and flight have been reported following electrical stimulation of central and corticomedial amygdala, medial hypothalamus (MH), and dorsal midbrain central grey (DCG) of experimental animals (Fernandez de Molina & Hunsperger, 1959; Hess & Bruegger, 1943; Panksepp, 1971; Wolfle et al., 1971). In monkeys, the behaviours elicited by brain electrical stimulation were similar to natural patterns of defense and escape (Lipp & Hunsperger, 1978) and aggression was directed against socially meaningful targets (Alexander & Perachio, 1973; Robinson, Alexander & Browne, 1969), indicating that a true emotional state was generated by the electrical stimulation. Autonomic changes characteristic of the neurovegetative defense reaction have also been evoked from the same subcortical structures (Abrahams, Hilton & Zbrozyna, 1960; Mancia & Zanchetti, 1981). In addition, animals can learn to switch off electrical stimulation applied to the MH or dorsal CG, indicating that an aversive motivational state is produced (Delgado, Roberts & Miller, 1954; Nakao, 1958; Olds & Olds, 1962, Schmitt, Abou-Hamed & Karli, 1977).

Further testifying to the unpleasant character of such brain stimulation, verbal reports of fear, rage of nonlocalized pain were obtained by electrically stimulating the amygdala, hypothalamus and DCG of neurosurgical patients (Amano et al., 1978; Chapman et al., 1954; Kim & Umbach, 1973; Nashold, Wilson & Slaughter. 1974; Schwarcz, 1977). As a complement, lesions of the DCG (Blanchard, Williams, Lee & Blanchard, 1981) or medial amygdala (Kemble et al., 1984) were shown to markedly decrease defensive of fear-motivated aggression in wild rats, and in human patients with chronic pain, therapeutic lesion in or neat the lateral edge of the midbrain CG resulted in calming and relief of pain suffering, without affecting pain perception (Nashold et al., 1974). On the basis of such experimental evidence, the concept of a longitudinally organized brain aversive system (BAS), responsible for the elaboration and expression of aversive states was formulated (for reviews see Graeff, 1981, 1987, 1988).

Although the neurochemistry of the BAS has been comparatively less explored than the neurochemical basis of reward (Wise & Bozarth, 1984); there is pharmacological evidence implicating gamma-aminobutryc acid (GABA), endogenous opioids, acetylcholine, excitatory aminoacids and 5-HT as neuromediators operating in the neuronal network integrating aversive states (Carobrez, 1987; Graeff, 1984, 1987, 1988; Graeff et al., 1986; Graeff et al., 1986; Schmitt et al., 1984; Schmitt et al., 1986).

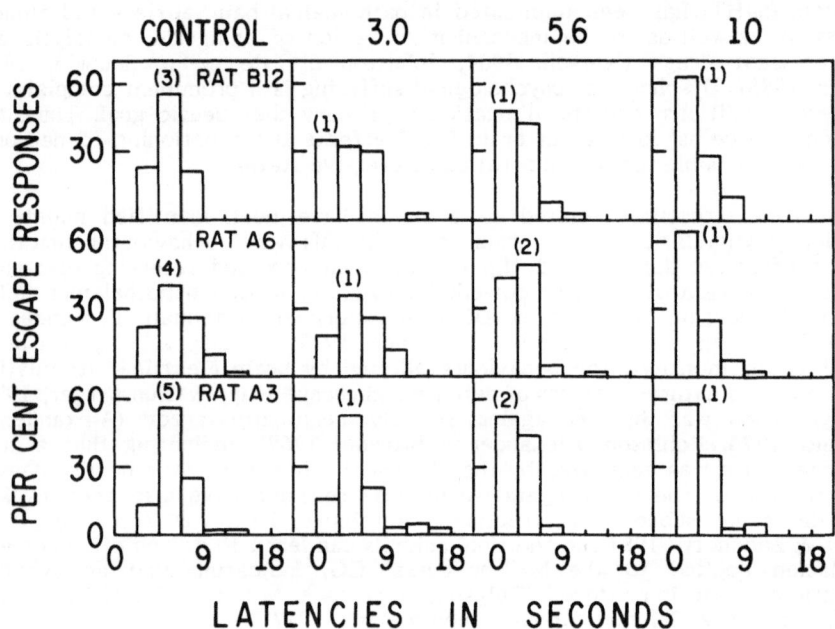

Fig. 38.1 Decreases of escape latency caused by cyproheptadine in three rats trained to press a lever to switch off electrical stimulation of the DCG. Each panel shows the frequency histograms of escape latencies in experimental sessions consisting of 40 discrete escape trials. Responses were classified within successive 3-second time bins from stimulus onset. Each column is the percentage of responses within a time bin. Figures in parenthesis are the number of sessions. Notice the progressive displacement to the left of the frequency histogram caused by increasing doses of cyproheptadine, resulting in shortening of the average escape latency. Cyproheptadine was injected i.p., 25 min before the experimental session (Schenberg & Graeff, 1978, with permission).

The results concerning 5-HT will be reviewed in the following sections. The first section deals with the effect of systemically administered drugs affecting 5-HT neurotransmission in a more or less selective way on behaviour either maintained or suppressed by electrical stimulation of the DCC of the rat. In the second section, more direct evidence obtained with electrode-cannulae or chemitrodes chronically implanted into the MH or DCG will be presented. In the final section, the implications of three experimental findings for the understanding of the pathophysiology of anxiety disorders as well as of the mode of action of anxiolytic drugs will be discussed.

Systemic drug injection
Using a decremental escape procedure in which rats were trained to press a lever in order to reduce by 5% the intensity of electric current stimulation the DCG, Kiser and Lebowitz (1975) reported that intraperitoneal injection of the 5-HT synthesis inhibitor para-chorophenylalanine (PCPA, 316 mg/kg) facilitated escape responding. In the same way, Schenberg and Graeff (1978) observed that two 5-HT receptor antagonists, methysergide (10 and 30 mg/kg, i.p.) and cyproheptadine (5.6 and 10 mg/kg, i.p.), decreased the latency of lever-pressing behaviour switching off electrical stimulation applied to the DCG of the rat. The effect of cyproheptadine on the distribution of escape latencies is illustrated in Fig. 38.1. Similar results have been reported by Clarke and File (1982). Therefore, decrease in 5-HT activity seems to enhance the aversive effect of DCG electrical stimulation.

Nevertheless, the above method does not discriminate between proaversive and behavioural stimulant drug actions, both tending to facilitate escape. In order to separate the two influences, Morate de Carvalho, de Aguiar, and Graeff (1981) used DCG electrical stimulation as punisher. Thus, during the punished component of a multiple schedule of reinforcement each lever press was immediately followed by stimultaneous presentation of water reward and aversive brain stimulation. If central punishment were equivalent to peripheral punishment, 5-HT antagonists should release lever pressing suppressed by CG electrical stimulation, since previous studies had shown that cyproheptadine as well as methysergide increased responding punished by electric foot shock, in rats (Graeff, 1974) or by electric shock applied around the punis bones, in pigeons (Graeff & Schoenfeld, 1970). However, the same range of doses of methysergide (3-30 mg/kg) and cytroheptadine (3-17 mg/kg) failed to release lever-pressing behaviour punished by brain stimulation, in contrast to the anxiolytics chlordiazepoxide and pentobarbital which facilitated centrally punished behaviour as they characteristically to with responding punished by peripheral nociceptive stimuli (Sepinwall & Cook, 1978).

It may be concluded that systemically administered 5-HT antagonists exert a true proaversive action, since they enhance responding maintained by decrease or termination of electrical stimulation of the DCG of the rat as well as fail to release behaviour punished by the same brain stimulation, in contrast to peripherally punished responding.

Complementary evidence also exists, showing that drugs which are believed to increase the availability of 5-HT at postsynaptic receptors, namely the synthesis precursor 5-hydroxytryptophan (5-HTP, 75 and 150 mg/kg, i.p.) and the reuptake inhibitor chlorimipramine (15 mg/kg, i.p.), depress decremental escape from electrical stimulation of the DCG of the rat (Kiser, German & Lebowitz, 1978). Nevertheless, drug-induced sedation may have contributed to their effect, since neither 5-HTP nor chlorimipramine have been tested on behaviour suppressed by central punishment.

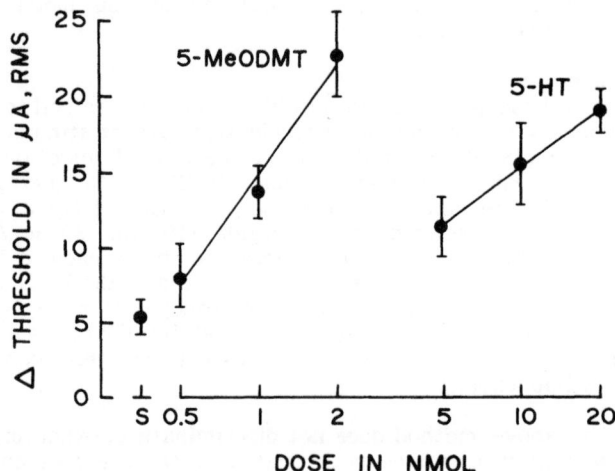

Fig. 38.2 *Dose-effect functions of the increases in aversive threshold caused by microinjections (0.5 µl in 30 s) of serotonin (5-HT) and 5-methoxy-N,N-dimethyltryptamine (5-MeODMT) in the DCG. The aversive threshold was the lowest current intensity inducing midline crossing responses in three successive trials of electrical stimulation, applied to the same brain area of rats placed inside a shuttle-box. The change in threshold (Δ) was the difference between the threshold determined 10 min after intracerebral injection and the basal threshold in the same animal. The initial threshold value was 20.2 ± 0.8 µA (RMS) for the 167 rats used in the whole study. Each point in the figure represents the mean and the vertical bars the SEM of 11 rats. S stands for saline (0.9% NaCl) injection. The straight lines are calculated linear regressions (Schuetz et al., 1985, with permission).*

Intracerebral drug injection

In order to investigate the role of 5-HT in the BAS more directly, a series of experiments was performed in rats with chronically implanted chemitrodes, allowing microinjection of drugs at the same brain sites where electrical stimulation produced aversive effects. In the first study, Leroux and Meyers (1975) showed that 5-HT itself caused a fourfold increase in tolerance to versive electrical stimulation of the rat hypothalamus following its local microinjection (5 µg in 0.4 to 0.7 µl). In this experiment, rats were electrically stimulated with increasing current intensities until they pressed a lever switching off brain stimulation. Following escape, stimulation was reinstated at the lowest intensity.

Subsequent experiments were conducted by our research group, then at the Department of Pharmacology, School of Medicine of Ribreirao Preto, University of Sao Paulo, and concentrated in a more purely aversive area, the DCG of the rat, as shown by Schmitt, Eclancher, and Karli (1974). In the first study, Schuetz, de Aguiar & Graeff (1985) demonstrated that microinjection of 5-HT into the DCG raised the minimum current intensity necessary to maintain switch off responding in rats placed inside a shuttle-box. The direct 5-HT receptor agonist 5-methoxy-N,N-dimethyltryptamine (5-MeODMT), which is not as rapidly inactivated by neuronal uptake as 5-HT, was about ten times more potent than the latter, generating a steeper dose-effect line placed at the left side of the 5-HT dose-effect (Fig. 38.2).

The above results indicate that stimulation of 5-HT receptors in aversive areas of the hypothalamus and midbrain CG attenuates the negative motivational state generated by electrical stimulation. The role of 5-HT receptors in aversion was strongly supported by addition results obtained by Schuetz et al. (1985), showing that the antiaversive effect of 10 nmol of 5-HT was antagonized by pretreatment with two 5-HT receptor blockers, metergoline and ketanserin, also microinjected into the same brain area 30 min before 5-HT.

Because the former drug is a nonselective 5-HT antagonist while ketanserin has selective affinity for type-2 5-HT receptors and the same molar dose (100 nmol) was used for both agents, we suggested that the antiaversive effect of 5-HT is exerted through activation of 5-HT$_2$ receptors existing in the DCG. The midbrain CG also contains serotonergic terminals of nerve fibers originating from cell bodies placed mainly in the dorsal raphe nucleus, but also in the median raphe nucleus and in the nucleus raphe magnus (Nieuwenhuys, 1985). Given this anatomical arrangement, it was predicted that blocking presynaptic amine uptake would enhance the antiaversive effect of 5-HT injected into the DCG. This result was indeed obtained by Schuetz et al. (1985) with the selective 5-HT uptake inhibitor zimeldine (100 nmol), given 30 min before 5-HT. Furthermore, it was also observed that zimelidine caused an antiaversive effect of its own, indicating that 5-HT may be released from nerve endings by aversive electrical stimulation. Consistent with this conclusion are previously reported results showing that electrical stimulation of the dorsal raphe nucleus depressed decremental escape from DCG of MH stimulation, an effect presumably mediated by 5-HT release (Kiser et al., 1980; Schmitt et al., 1983). On the basis of the aforementioned results it was suggested that endogenous 5-HT plays an inhibitory role in the midbrain CG integrating aversion (Schuetz et al., 1985).

In order to verify whether the above antiaversive effect of zimelidine was actually mediated by 5-HT, Audi, de Aguiar, and myself (unpublished results) preceded the administration of zimelidine with a microinjection of another 5-HT$_2$ receptor blocker, ritanserin. The results of this experiment showed that ritanserin blocked the antiaversive effect of zimeldine (Fig. 38.3).

In contrast, a similar effect induced by the benzodiazepine receptor agonist midazolam (40 nmol) was unaffected by ritanserin (10 nmol). Previously reported work has already shown that the antiaversive effect of midazolam as well as of its congener chlordiazepoxide was blocked by the competitive benzodiazepine receptor antagonist Ro 15-1788 (Audi & Graeff, 1984). Thus, pharmacological antagonism with drugs microinjected into the CG seems to be specific, validating their use for investigating the neurochemical basis of aversion.

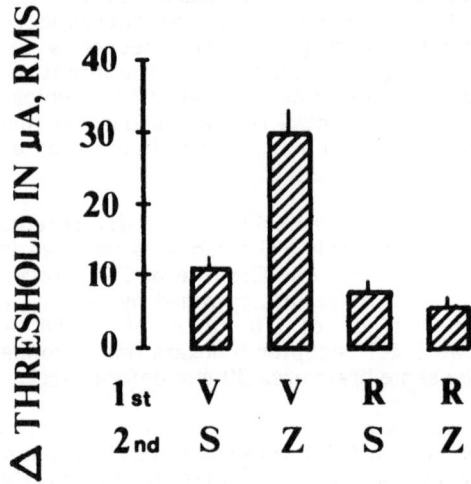

Fig. 38.3 Antagonism by ritanserin (R,10 nmol) of the antiaversive effect of zimeldine (Z, 100 nmol). The 5-HT antagonist or vehicle (V, 2% Tween 80) was injected 10 min before administration of zimelidine or saline (S) into the DCG). The aversive threshold was the lowest intensity causing the rat to cross the midline of the shuttle-box (switch-off response) at least 9 times in 10 successive trials with electrical stimulation of the DCG. The second threshold determination was made 20 min after the second injection. N=9. Other specifications in the legend of Fig. 2. A two-factor ANOVA showed a significant interaction between zimelidine and ritanserin ($F_{1,32}=17.6$, $p<0.001$). The following group comparisons were significant: VS X VZ ($t_{16}=4.6$, $p<0.001$) and VZ X RZ ($t_{16}=6.3$, $p<0.001$).

A strong interaction of beta-adrenoceptor blockers with type-1 5-HT receptors has been recently demonstrated (Pazos, Engel & Palacios, 1985). In particular, Middlemiss (1984) reported that propranolol prevented the inhibitory effect of 5-HT on K^+-evoked release of $[^3H]$-5-HT from perfused cortical slices of the rat brain, which is likely to be enacted through presynaptic receptors of the $5-HT_{1B}$ subtype. Therefore, propranolol may enhance 5-HT release from nerve terminals by blocking inhibitory presynaptic autoreceptors. If this action were to occur in the DCG, one would expect propranolol to cause antiaversive effects mediated by endogenous 5-HT following its intracerebral injection.

This prediction was fulfilled by additional results obtained by E.A.Audi, J.C.de Aguiar, and myself, showing that microinjection of 2.2, 4.4 and 8.8 nmol of propranolol into the DCG caused dose-dependent increases of the aversive threshold determined 10 min after the injection, using the procedure described in the legend of Fig. 38.3. As with 5-HT and 5-MeODMT (Fig. 38.2), a significant ($F_{1,25}$ = 4.7, p<0.05) linear regression was obtained when log dose of propranolol was plotted against drug-induced increases in aversive threshold. The antiaversive effect of the intermediated dose of propranolol was antagonized by ritanserin, as shown in the upper panel of Fig. 38.4.

The last result suggest mediation of the effect of propranolol by 5-HT, through activation of $5-HT_2$ receptors. Furthermore, the effect of propranolol added to that of zimelidine when both drugs were injected in close succession into the DCG (lower panel of Fig. 38.4), supporting the hypothesis that zimelidine and propranolol enhance the concentration of 5-HT in the synaptic cleft through complementary mechanisms, namely inhibition of neuronal uptake and blockade of presynaptic autoreceptors which decrease 5-HT release from nerve endings.

In contrast to the above drugs, the selective $5-HT_{1A}$ receptor agonist (or partial agonist) ipsapirone (10, 20 and 40 nmol) was little effective in raising the aversive threshold of electrical stimulation of the dorsal DCG of the rat, causing a maximum threshold increase of only 10 μA RMS) at the dose of 20 nmol. Therefore, this subtype of 5-HT receptor does not seem to participate to any important degree in the mediation of aversive effects generated in this region by 5-HT-like drugs.

The modulatory role likely to be exerted by 5-HT in the DCG integrating version seems to be phasic, rather than tonic, since local administration of 5-HT receptor blockers did not induce flight behaviour. Under the same experimental conditions, GABA-A receptor antagonists as well as inhibitors of GABA synthesis elicit vigorous running and jumping, indicating that GABA tonically inhibits the neurnal substrate ot aversion in the DCG (Graeff et al., 1986; Schmitt et al., 1986).

In the amygdaloid area, the third major component of the BAS, the effect of 5-HT-like drugs on the aversive threshold of electrical stimulation has not been measured so far. Besides, the effect of intraamygdalar injection of drugs affecting 5-HT neurotransmission on aversively motivated behaviour is conflicting. Thus, it has been reported that injection of the 5-HT receptor blockers methysergide and ketanserin into the lateral and basolateral amygdala released lever-pressing suppressed by foot-shock, in the rat (Hodges, Green & Glenn, 1987; Petersen & Schel -Krueger, 1984).

Another 5-HT receptor antagonist, cyproheptadine has also been shown to facilitate punished behaviour following its microinjection into the rat's central amygdala, while its injection into the mamillary body was ineffective (Kataoka et al., 1987). Therefore, impairing 5-HT neurotransmission in the amygdala

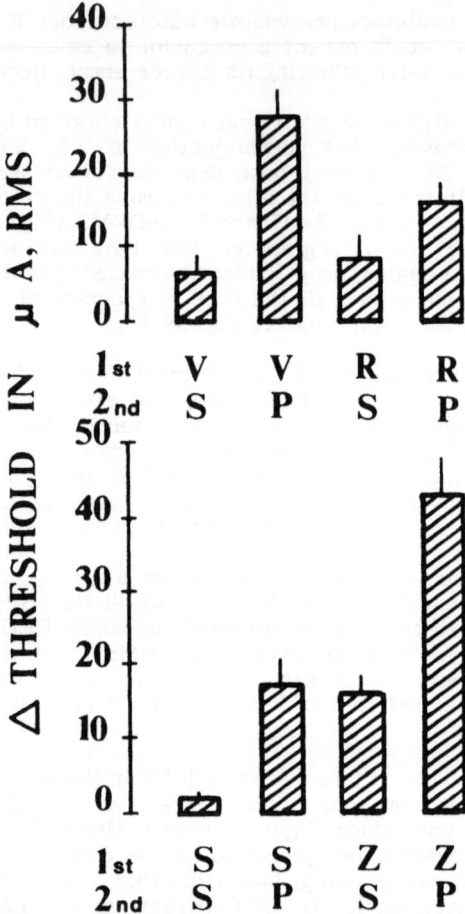

Fig. 38.4 Antagonism by ritanserin of the antiaversive effect of propranolol (P, upper panel), and summation of the effects or propranolol and zimelidine (lower panel). Other specifications in the legends of Figs. 2 and 3. A two-factor ANOVA showed a significant interaction between propranolol and ritanserin ($F_{1,32}=9.0$, $p=0.005$). The following group comparisons were significant: VP X VS ($t_{16}=5.3$, $p<0.001$) and VP X RP ($t_{16}=2.9$, $p=0.009$). Teh interaction between zimelidine and propranolol was nonsignificant (two-factor ANOVA, $F_{1,32}=2.4$, $p=0.129$), but there were significant effects of both zimelidine ($F_{1,32}=31.5$, $p<0.001$) and propranolol ($F_{1,32}=35.4$, $p<0.001$).

causes antipunishment effects. In addition, injection of 5-HT or the 5-HT$_{1A}$ receptor agonist 8-OH-DPAT into the lateral amygdala reduced punished responding below control level (Hodges et al., 1987). Since the BAS has been implicated in response suppression caused by punishment (Graeff & Rawlins, 1980), it may be concluded that 5-HT enhances aversive motivation in the amygdala, contrary to its suggested role in the MH and DCG.

Nevertheless, results with the social interaction test in rats are seemingly inconsistent with those obtained in punishment situations. Thus, File, James and MacLeod (1981)reported that a dose of the relatively specific neurotoxin 5,7-dihydrotryptamine (5,7-DHT), causing 80% depletion of 5-HT concentration in the amygdala (together with 29% noradrenaline depletion) following its microinjection into the amygdalar area, reduced social interaction as well as motor activity of pairs of rats placed in a lighted area, a result suggestive of enhanced fear. Accordingly, a lower dose of 5,7-DHT causing only 55% depletion of amygdalar 5-HT decreased dominance behaviours and increased submissiveness of home-cage rats toward intruders. Lesioned rats also became more submissive when they were intruders (File et al., 1981). As complementary evidence, injection of the direct 5-HT receptor agonist quipazine into the corticomedial amygdala of the rat decreased defensive aggression induced by electric foot-shock without affecting the pain threshold measured through the flinch and jump test. Predatory attack, however, was simultaneously reduced and both antiaggressive effects of intraamygdalar quipazine were attributed to increased rather than decreased fearfulness (Pucilowiski, Plaznick & Kostowiski, 1985). Clearly, further studies are needed to establish the role of 5-HT in amygdaloid areas related to aversion.

In summary, the afore described evidence indicates that 5-HT phasically inhibits the neuronal network of the DCG responsible for the elaboration and/or expression of aversive states. Although sparse, the results so far obtained in the MH support a similar conclusion. As regards the amygdala, however, existing evidence is inconclusive, the possibility remaining open for either a facilitatory or an inhibitory role of 5-HT in this component of the BAS.

Discussion
The implications of the experimental evidence reviewed in the preceding sections regarding anxiety disorders and their drug treatment has been discussed in recent publications (Graeff, 1987, 1988). As a consequence, only the main points will be considered here. Emphasis will be given to panic disorder because this affection seems to be the most directly related to the BAS. Indeed, electrical stimulation or microinjection of GABA antagonist drugs in the DCG of the rat causes initial behavioural immobility or freezing, followed by vigorous running and jumping (Brandao, De Aguiar & Graeff, 1982; Di Scala, Schmitt & Karli, 1984), a behavioural pattern that has been identified with panic (Panksepp, 1982). In humans, electrical stimulation of an area in or near the lateral edge of the midbrain CG induced feelings of intense fear or impending death, accompanied by autonomic changes such as sweating, piloerection, blushing of the face or neck and increases in heart and respiratory rates (Nashold et al., 1974), closely resembling a full-blown panic attack.

In spite of considerable evidence supporting the participation of noradrenergic mechanisms in panic (Charney & Redmond, 1983), the therapeutic efficacy of low doses (20 to 50 mg per day) of the potent and relatively selective 5-HT uptake inhibitor, chlorimipramine (Caetano, 1985; Glogers et al., 1981; Versiani et al., 1987) points to an important role of 5-HT as well.

Nevertheless, because several days of repeated drug administration are usually necessary for symptom reduction (Glogers et al., 1981; Versiani et al., 1987), allowing for the accumulation of desmethylchlorimipramine, which preferentially inhibits noradrenaline uptake, it is questionable to attribute the therapeutic effect of chlorimipramine to its action of serotonergic neurotransmission alone.

Two studies, however, found a relation between plasma levels of chlorimipramine and clinical improvement, but not with the metabolite (Stern, Mark & Mawson, 1980; Insel et al., 1983). In addition, zimelidine and fluvoxamine, both of which cause little change in noradrenergic mechanisms, have been shown to be highly effective in patients with panic disorder (Den Boer et al., 1987; Evans et al., 1986). Moreover, the immediate precursor of 5-HT synthesis, 5-hydroxytryptophan (5-HTP), also ameliorates the condition, though to a lesser extent than chlorimipramine (Kahn et al., 1987).

Assuming that 5-HT plays a role in the pathophysiology of panic attacks, the question nevertheless remains of where in the brain and how doses of the amine influence the outcome of panic. Based on their observations that patients vulnerable to lactate-induced panic show increased blood flow, blood volume and oxygen metabolism in the right parahippocampal area, assessed through positron emission tomography (PET scan), Reiman et al. (1986) hypothesized that increased noradrenergic input from the locus coeruleus to this region could be the cause of such abnormalities, making this area overresponsive to sensory stimuli. Given such a condition, a triggering event would cause the abnormal region to initiate an anxiety attack by the way of the septo-amygdala complex and its sequential projections to the hypothalamus and brain stem, in other words, through activation of the BAS. Because the experimental evidence reviewed in the preceding sections indicates that 5-HT exerts phasic inhibition on the neuronal network of the MH and DCG commanding the expression of defense reactions and elaborating aversive emotional-motivational states, these brain areas become potential sites for serotonergic modulation of panic attacks. Thus, it may be suggested that 5-HT uptake inhibitors and 5-HTP relieve panic attacks because they enhance 5-HT inhibition of the BAS.

Yet, the comparison between the drug effects so far obtained in experimental animals and clinical observations offers some difficulties for the above hypothesis.

As mentioned before, chronic administration of chlorimipramine is usually necessary before its beneficial effects on panic disorder appear. Actually, an increase of panic attacks often occurs during the initial phase of drug treatment (Caetano, 1985; Gloger et al., 1981; Versiani et al., 1987). Accordingly, we have recently observed that single administration of 25 mg of chlorimipramine enhanced the rise of anxiety scores induced by a simulated public speaking test in healthy volunteers, evaluated by a self-rating analog mood scale (Guimares, Zuardi & Graeff, 1979).

In contrast, the antiaversive effect of systematically administered chlorimipramine (Kiser et al., 1978) or intracerebrally injected zimelidine (Schuetz et al., 1985) occurred soon after the first injection.

Nevertheless, the above difficulty may be overcome when the mechanisms likely to be involved in these animal experiments and human studies are considered. As discussed before, the antiaversive effect of acutely administered 5-HT uptake inhibitors, in rats bearing electrodes implanted in the DCG, may be due to the potentiation of endogenous 5-HT directly released by the electrical stimulation from nerve terminals in the DCG. In the intact organism, however, these drugs are likely to exert other influences on serotonergic neurotransmission that may counteract this facilitatory action (Willner, 1985). Thus, the firing rate of raphe neurons is reduced by acute administration of chlorimipramine, decreasing 5-HT release from nerve endings, and postsynaptic 5-HT receptors may also be blocked to some extent. Therefore, the ultimate effect of acutely administered chlorimipramine and other 5-HT uptake inhibitors on overall 5-HT neurotransmission is difficult to assess, since it depends on the balance of opposing drug actions. Following chronic administration, however, these drugs as well other antidepressants appear to enhance 5-HT neurotransmission by both pre- and postsynaptic actions (Willner, 1985). Therefore, there is a possibility, compatible with the antiaversive role of 5-HT suggested above, that the initial aggravation of panic attacks following chlorimipramine administration is due to decreased serotonergic activity in the BAS, whereas symptom relief is determined by increased 5-HT inhibition of brain aversive areas. It should be kept in mind, however, that the role of 5-HT in anxiety is a matter of controversy and an opposite explanation for the same phenomena has been proposed (Den Boer et al., 1987; Gentil, 1986). Nevertheless, in agreement with the present view Gentil (1986) suggested a participation of the BAS in the pathophysiology of the panic syndrome.

Inhibitors of 5-HT uptake as well as less selective tricyclic antidepressants benefit several psychiatric conditions besides panic disorders. Among them stands obsessive-compulsive disorder, and arguments similar to those outlined above have led to a serotonergic hypothesis of this condition (Insel et al., 1985). In addition, phobias, generalized anxiety disorder, migraine, chronic psychogenic pain and other non-affective disorders have been reported to respond favourably to tricyclic antidepressants and related drugs (Murphy, Siever & Insel, 1985). As an explanation for such broad spectrum of therapeutic action, the last authors raise the possibility, among others, of these drugs to act upon a psychobiological substrate common to all the above conditions, which would be a mixture of anxiety and depression like Marks' (1983) 'dysphoria'. It may be added that such an affective state could be determined by an activation of the BAS and that its inhibition through 5-HT potentiation could be the basis of the polytherapeutic action of the so called antidepressant drugs. Yet, changes in catecholamines seem to be necessary for a full antidepressant drug effect to develop (Van Praag et al., 1987).

Also, some of the pathologies upon which potentiators of 5-HT are reported to act, such as bulimia, alcoholism and even compulsive disorder, may involve deficits of impulse control, another psychobiological function in which 5-HT is likely to be involved (Soubrie, 1986). In this sense, it is interesting to notice that low levels of the 5-HT metabolite, 5-hydroxy-indolacetic acid (5-HIAA) in the cerebrospinal fluid have been shown to be associated with both lack of impulse control and a propensity to "somatic" anxiety or panic attacks; in contrast, "psychic" anxiety was positively correlated with 5-HIAA levels.

The above dual concept of anxiety might reconcile the seemingly conflicting evidence on the role of 5-HT in anxiety, as discussed elsewhere (Graeff, 1988). Indeed, there is a considerable body of evidence, mainly from animal experiments with conflict procedures, indicating that 5-HT mediates behavioural inhibition induced by stressful situations, thought to be related to anxiety (Gray, 1982). The anxiolytic effect of benzodiazepines would be consequent to a reduction of 5-HT in serotonergic systems mediating response suppression, possibly those innervating the septo-hippocampal region (Wise, Berger & Stein, 1972; Gray, 1982). Clearly, this explanation does not apply to the antiaversive action of benzodiazepines (Graeff, 1984), since 5-HT seems to inhibit aversion in the MH and DCG. Even the unlikely possibility of 5-HT being released by benzodiazepines is not supported by the lack of ritanserin antagonism against the antiaversive effect midazolam observed in our laboratory. Therefore, the antiaversive action of benzodiazepines, which may be involved in their clinical anxiolytic effect (Graeff, 1984; Schenberg & Graeff, 1978), seems to be independent of 5-HT. Actually, there are several experimental results indicating that benzodiazepines substrate of aversion in the MH and DCG (Audi & Graeff, 1984, 1987; Brandao et al., 1982; Milani & Graeff, 1987). Nevertheless, it should be kept in mind that 5-HT may play an opposite role in the amygdala (Hodges et al., 1987) and thus, decrease in 5-HT activity caused by benzodiazepines could be a mechanism of their likely antiaversive action in this brain region (Scheel-Krueger & Petersen, 1982; Shibata et al., 1982).

Propranolol is sometimes used for treating abnormal anxiety, mainly of situational origin (Ananth & Lin, 1986). Because propranolol has been reported to selectively affect autonomic manifestations of anxiety (Tyrer & Lader, 1974), its anxiolytic effect has been largely attributed to beta-adrenoceptor blockade. Nevertheless, both animal experiments (Durel, Krants & Barret, 1986) and clinical observations (Greenblatt & Koch-Weser, 1974) are suggestive of central action. In this sense, the aforedescribed results showing an antiaversive effect of propranolol injected into the DCG of the rat, probably mediated through 5-HT may be an indication that the antianxiety effect of propranolol has a central component. If this suggestion is confirmed by further studies, propranolol would rank among anxioselective compounds, like buspirone or ritanserin, which seem to act primarily upon central serotonergic neurotransmission (Ceulemans et al., 1985; Eison & Temple, 1986).

Concerning the latter drugs, the experimental results to date do not support an involvement of 5-HT mechanisms of the BAS in their anxiolytic effect. In fact, we observed that the buspirone analog ipsapirone caused only mild antiaversive effects when injected into the DCG of the rat and ritanserin as well as ketanserin were not only devoid of such action, but instead antagonized the antiaversive effect of pro-serotonergic drugs. Nevertheless, a participation of the BAS in the anxiolytic action of 5-HT$_2$ receptor blockers should not be entirely dismissed, since it has been reported that microinjection of methysergide, ketanserin or cyproheptadine into the amygdala of the rat caused antipunishment effects similar to chlordiazepoxide and midazolam, while 5-HT enhanced the response-suppressant effect of punishment (Hodges et al., 1987; Kataoka et al., 1987; Petersen & Scheel-Krueger, 1984). Regarding buspirone, its clinical effect on anxiety becomes significant only after two weeks of repeated drug administration (Goa & Ward, 1986). Therefore, long-term adaptative changes in serotonergic neurotransmission may take place, the influence of which upon BAS functioning are so far unexplored. Also, buspirone-like drugs, given systemically, may act indirectly by stimulating 5-HT autoreceptors in the somadendrites of raphe neurones innervating the BAS (Eison & Temple, 1986).

In conclusion, the aforediscussed evidence seems to implicate serotonergic mechanisms modulating the neuronal network responsible for the elaboration and/or expression of aversive states in the pathophysiology of several types of anxiety disorders as well as in the mode of action of drugs used for their clinical treatment.

Acknowledgements: The author is indebted to Doctors Carlos B. Tomaz, Francisco, S. Guimaraes and Antonio W. Zuardi for their helpful comments on the manuscript, and to Mrs. Heloisa E.G. de Oliveira Greaff for advice on English writing and style.

References

Abrahams, V.C., Hilton, S.M. & Zbrozyna, A.W. (1960). Active vasodilatation produced by stimulation of the brain stem: its significance in the defence reaction. J. of Physiol., 154, 491-513.

Alexander, M. & Perachio, A.A. (1973). The influence of target sex and dominance on evoked attack in rhesus monkeys. Am. J. of Phys. Antropology, 38, 543-547.

Amano, K., Tanikawa, Y., Iseki, H., Kawabatabe, H., Notani, M., Kawamura, H. & Kitamura, K. (1978). Single neuron analysis of the midbrain tegmentum. Rostral mesencephalic reticulotomy for pain relief. Applied Neurophysiol., 41, 66-78.

Ananth, J. & Lin, K-M. (1986). Propranolol in psychiatry. Therapeutic uses and side effects. Neuropsychobiol., 15, 20-27.

Audi, E.A. & Graeff, F.G. (1984). Benzodiazepine receptors in the periaqueductal grey mediate anti-aversive drug action. Eur. J. of Pharmacol., 103, 279-285.

Audi, E.A. & Graeff, F.G. (1987). GABA-A receptors in the midbrain central grey mediate the antiaversive action of GABA. Eur. J. of Pharmacol., 135, 225-229.

Blanchard, D.C., Williams, G., Lee, E.M.C. & Blanchard, R.J. (1981). Taming wild rattus norvegicus by lesions of the mensencephalic central grey. Physiol. Psychol., 9, 157-163.

Brandao, M.L., De Aguiar, J.C. & Graeff, F.G. (1982). GABA mediation of the anti-aversive action of minor tranquilizers. Pharmacol. Biochem. and Behav., 16, 397–402.

Carobrez, A.B. (1987). Excitatory aminoacid mediation of the defense reaction. In: Neurosci and Behav. (Ed. Brandao, M.L.) pp 21–29, Vitoria, Brazil: UFES-Grafica.

Ceulemans, D.L.S., Hoppenbrowuers, M.L.J.A., Gelders, Y.G. & Reyntjens, A.J.M. (1985). The influence of ritanserin, a serotonin antagonist, in anxiety disorders: a double-blind placebo-controlled study versus lorazepam. Pharmacopsychiatry, 18, 303–305.

Chapman, W.P., Schroeder, H.R., Geyer, G., Brazier, M.A.B., Fanger, C., Poppen, J.S., Solomon, H.C. & Iakolev, P.I. (1954). Physiological evidence concerning importance of the amygdaloid nuclear regions in the circulatory function and emotion in man. Sci., 120, 949–950.

Charney, D.S. & Redmond, D.E.Jr. (1983). Neurobiological mechanisms in human anxiety. Evidence supporting central noradrenergic hyperactivity. Neuropharmacol., 22, 1531–1536.

Clarke, A. & File, S. (1982). Effects of ACTH, benzodiazepines and 5-HT antagonists on escape from periaqueductal grey stimulation in the rat. Progress in Neuro-Psychoharmacol. and Biol. Psych., 6, 27–35.

Delgado, J.M.R., Roberts, W.W. & Miller, N.E. (1954). Learning motivated by electrical stimulation of the brain. Am. J. of Physiol., 179, 587–593.

Den Boer, J.A., Westenberg, H.G.M., Kamerbeek, W.D.J., Verhoeven, W.M.A. & Kahn, R.S. (1987). Effect of serotonin uptake inhibitors in anxiety disorders; a double-blind comparison of clomipramine and fluvoxamine. Intern. Clin. Psychopharmacol., 2. 21–32.

Durel, L.A., Krantz, D.S. & Barrett, J.W. (1986). The antianxiety effect of beta-blockers on punished responding. Pharmacol. Biochem. and Behav., 25, 371–374.

Eison, A.S. & Temple, D.L. (1986). Buspirone: review of its pharmacology and current perspectives on its mechanism of action. The Am. J. of Med., 80, (suppl, 3B), 1–9.

Evans, L., Kenardy, J., Schneider, P. & Hoey, H. (1986). Effect of a selective serotonin uptake inhibitor in agoraphodia with panic attacks. A double-blind comparison of zimelidine, imipramine and placebo. Acta Psych. Scand., 73, 79–53

Fernandez de Molina, A. & Hunsperger, R.W. (1959). Central representation of affective reaction in the forebrain and brain stem: Electrical stimulation of amygdala, stria terminalis, and adjacent structures. J. of Physiol., 145, 251–265.

File, S., James, T.A. & Mac Leod, N.K. (1981). Depletion in amygdaloid 5-hydroxytryptamine concentration and changes in social and aggressive behaviour. J. of Neurotransm., 50, 1-12.

Gentil, V. (1986). Fisiopatologia da sindrome do panico. Revista da Associacao Medica Brasileira, 32, 101-107.

Glogers S., Grunhaus, L., Birmacher, B. & Troudart, T. (1981). Treatment of spontaneous panic attacks with chlorimipramine. Am. J. of Psychiatry, 138, 1215-1217.

Goa, K.L. & Ward, A. (1986). Buspirone: A preliminary review of its pharmacological properties and therapeutic efficacy as an anxiolytic. Drugs, 32, 114-129.

Graeff, F.G. (1974). Tryptamine antagonists and punished behaviour. J. of Pharmacol. and Exp. Therap., 189, 344-350.

Graeff, F.G. (1981). Minor tranquilizers and brain defense systems. Brazilian J. of Med. and Biolog. Res., 14, 239-265.

Graeff, F.G. (1984). The anti-aversive action of minor tranquilizers. Trends in Pharmacol. Sci., 5, 230-233.

Graeff, F.G. (1987). The anti-aversive action of drugs. In: Adv. in Behav. Pharmacol. (Eds. Thompson T., Dews P.B. & Barrett J.E.) Vol. 6, pp 129-156, Hillsdale, N.J.: Lawrence Erlbaum.

Graeff, F.G. (1988). Animal models of aversion. In: Animal models of Psychiatr. (Eds. Simon P., Soubrie P. & Widlocher D.) Vol. 9, Basel: Kargar, in press.

Graeff, F.G. & Rawlins, J.N.P. (1980). Dorsal periaqueductal gray punishment, septal lesions and the mode of action of minor tranquilizers. Pharmacol. Biochem. and Behav., 12, 41-45.

Graeff, F.G. & Schoenfeld, R.I. (1970). Tryptaminergic mechanisms in punished and nonpunished behaviour. J. of Pharmacol and Exp. Therap., 173, 277-283.

Graeff, F.G., Brandao, M.L., Audi, E.A. & Milani, H.L. (1986). Role of GABA in the anti-aversive action of anxiolytics. In: GABA and Neuroendocrine function. (Eds. Racagni G. & Donoso A.O.) pp. 79-86. New York: Raven Press.

Graeff, F.G., Brandao, M.L., Audi, E.A. & Schuetz, M.T. (1986). Modulation of the brain aversive system by GABAergic and serotonergic mechanisms. Behav. Brain Res., 21, 65-72.

Gray, J.A. (1982). The neuropsychology of anxiety: An enquire into the functions of the septo-hippocampal system. New York: Oxford University Press.

Greenblatt, D.J., Koch-Weser, J. (1974). Adverse reactions to beta-adrenergic receptor blocking drugs: a report from Boston collaborative drug surveillance program. Drugs, **7**, 118–129.

Guimaraes, F.S., Zuardi, A.W. & Graeff, F.G. (1988). Effect of chlorimipramine and maprotiline on expeirmental anxiety in humans. J. of Psychopharmacol., in press.

Hess, W.R. & Bruegger, M. (1943). Das subkortikale Zentrum der affectiven Abwehrreaction. Helvetica Physiologica Acta, **1**, 33–52.

Hodges, H., Green, S. & Glenn, B. (1987). Evidence that the amygdala is involved in benzodiazepine and serotonergic effects on punished responding but not on discrimination. Psychopharmacol., **92**, 491–504.

Insel, T.R., Mueller, E.A., Alterman, I., Linnoila, M. & Murphey, D.L. (1985). Obsessive–compulsive disorder and serotonin: Is there a connection? Biol. Psych., **20**, 1174–1188.

Insel, T.R., Murphy, D.L., Cohen, R.M., Alterman, I., Kilts, C. & Linnoila, M. (1983). Obsessive compulsive disorder: A double blind trail of clomipramine and clorgyline. Arch. of Gen. Psych., **40**, 605–612.

Iversen, S.D. (1984). 5–HT and anxiety. Neuropharmacol., **23**, SI, 1553–1560.

Johnston, A.L. & File, S.E. (1986). 5–HT and anxiety: promises and pitfalls. Pharmacol. Biochem. and Behav., **24**, 1467–1470.

Kahn, R.S., Westenberg, H.G.M., Verhoeven, W.M.A., Gispen-De Wied, C.C. & Kamerbeek, W.D.J. (1987). Effect of a serotonin precursor and uptake inhibitor in anxiety disorders; a double–blind comparison of 5–hydroxytryptophan, clomipramine and placebo. Int. Clin. Psychopharm., **2**, 33–45.

Kataoka, Y., Shibata, K., Yamashita, K. & Ueki, S. (1987). Differential mechanisms involved in the anticonflict action of benzodiazepines injected into the central amygdala and mammillary body. Brain Res., **416**, 243–247.

Kemble, E.D., Blanchard, C., Blanchard, R.J. & Takushi, R. (1984). Taming wild rats following medial amygdaloid lesions. Physiol. and Behav., **32**, 131–134.

Kim, Y.K. & Umbach, W. (1973). Combined stereotaxic lesions for treatment of behaviour disorders and severe pain. In: Surgical approaches in psychiatry. (Eds. Laitinen V. & Livingston K.E.) pp. 182–188. Baltimore: University Press.

Kiser, K.R. Jr. & Lebowitz, R.M. (1975). Monoaminergic mechanism in aversive brain stimulation. Physiol. and Behav., **15**, 47–53.

Kiser, R.S. Jr., German, D.C. & Lebowitz, R.M. (1978). Serotonergic reduction of dorsal central gray area stimulation–produced aversion. Pharmacol. Biochem. and Behav., **9**, 27–31.

Kiser, R.S., Brown, C.A., Sanghera, M.K. & German, D.C. (1980). Dorsal raphe nucleus stimulation reduces centrally-elicited fearlike behaviour. Brain Res.,

191, 265-272.

Leroux, A.G. & Meyers, R.D. (1975). Action of serotonin microinjected into hypothalamic sites at which electrical stimulation produced aversive responses in the rat. Physiol. and Behav., 14, 501-505.

Lipp, H.P. & Hunsperger, R.W. (1978). Threat, attack and flight elicited by electrical stimulation of the ventromedial hypothalamus of the marmoset monkey Cillitrix jacchus. Brain Behav. and Evolution, 15, 260-293.

Mancia, B. & Zanchetti, A. (1981). Hypothalamic control of autonomic function. In: Handbook of the hypothalamus. (Eds. Morgane P.J. & Panksepp J.) Vol. 3, pt. B, pp.47-202. New York, Marcel Dekker.

Marks, I. (1983). Are there anticompulsive of antiphobic drugs? Review of the evidence. Brit. J. of Psych., 143, 338-347.

Middlemiss, D.N. (1984). Stereoselective blockade at [^{3}H]5-HT binding sites and at the 5-HT autoreceptor by propranolol. Eur. J. of Pharmacol., 101, 289-293.

Milani, H. & Graeff, F.G. (1987). GABA-benzodiazepine modulation of aversion in the medial hypothalamus of the rat. Pharmacol. Biochem. and Behav., 28, 21-27.

Morato de Carvalho, S., De Aguiar, J.C. & Graeff, F.G. (1981). Effect of minor tranquillizers, tryptamine antagonists and amphetamine on behaviour punished by brain stimulation. Pharmacol. Biochem. and Behav., 15, 351-356.

Murphy, D.L., Siever, L.J. & Insel, T.R. (1985). Therapeutic responses to tricyclic antidepressants and related drugs in nonaffective disorder patient populations. Progress in Neurospychopharmacol. and Biol. Psych., 9, 3-13.

Nakao, H. (1958). Emotional behaviour produced by hypothalamic stimulation. Am. J. of Physiol., 194, 411-418.

Nashold, B.S. Jr., Wilson, N.P. & Slaughter, G.S. (1974). The midbrain and pain. In: Adv. in Neurol. (Eds. Bonica J.J.) Vol. 4: Int. symp. on pain. pp 191-196, New York, Raven Press.

Nieuwenhuys, R. (1985). Chemoarchitecture of the brain. Berlin: Springer-Verlag.

Olds, M.E. & Olds, J. (1962). Approach-avoidance interactions in rat brain. Am. J. of Physiol., 206, 515-520.

Panksepp, J. (1971). Aggression elicited by electrical stimulation of the hypothalamus in albino rats. Physiol. and Behav., 6, 321-329.

Panksepp, J. (1982). Toward a general psychobiological theory of emotions. The Behav. and Brain Sci., 5, 407-467.

Pazos, A., Engel, G. & Palacios, J.M. (1985). Beta-adrenoceptor blocking agents recognize a subpopulation of serotonin receptors in brain. Brain Research, 343, 403-408.

Petersen, E.N. & Scheel-Krueger, J. (1984). Anticonflict effects of 5-HT antagonists by intraamygdaloid injection. Collegium Internationale Neuro-Psychopharmacologicum, 14th CINP Congress Book of Abstract, p. 654.

Pucilowski, O., Plaznick, A. & Kostowski, W. (1985). Aggressive behaviour inhibition by serotonin and quipazine injected into the amygdala in the rat. Behav. and Neural Biol., 43, 58-68.

Reiman, E.M., Raichle, M.E., Robins, E., Butler, K., Herscovitch, P., Fox, P. & Perlmutter, J. (1986). The application of positron emission tomography to the study of panic disorder. Am. J. of Psych., 143, 469-477.

Robinson, B.W., Alexander, M. & Browne, G. (1969). Dominance reversal resulting from aggressive responses evoked by brain telestimulation. Physiol. and Behav., 4, 749-752.

Sepinwall, J. & Cook, L. (1978). Behavioural pharmacology of antianxiety drugs. In: Handbook of Psychopharmacol. (Eds. Iversen L.L., Iversen S.D. & Snyder S.) Vol. 13, pp.345-393. New York, Plenum Press.

Scheel-Krueger, J. & Petersen, E. (1982). Anticonflict effect of the benzodiazepines mediated by a GABAergic mechanism in the amygdala. Eur. J. of Pharmacol., 82, 115-116.

Schenberg, L.C. & Graeff, F.G. (1978). Role of the periaqueductal gray substance in the antianxiety action of benzodiazepines. Pharmacol. Biochem. and Behav., 9, 287-295.

Schmitt, P., Abou-Hames, H. & Karli, P. (1977). Effects aversifs et apetitifs induits par stimulations mesencephalique et hypothalamique. Brain Res., 130, 521-530.

Schmitt, P., Eclancher, F. & Karli, P. (1974). Etude des systemes de renforcement negatif et de renforcement positif au niveau de la substance grise centrale chez le rat. Physiology and Behav., 12, 271-279.

Schmitt, P., Di Scala, G., Jenck, F. & Sandner, G. (1984). Periventricular structures, elaboration of aversive effects and processing of sensory information. In: Modulation of sensorimotor activity during alterations in behavioural states. (Eds. Bandler R.) pp. 393-414. New York, Allan Liss.

Schmitt, P., Sandner, G., Colpaert, F.G. & De Witte, P. (1983). Effects of dorsal raphe stimulation on escape induced by medial hypothalamic or central gray stimulation. Behav. Brain Res., 8, 289-307.

Schmitt, P., Carrive, P., Di Scala, G., Jenck, F., Brandao, M.L., Bagri, A., Moreau, J.L. & Sandner, G. (1986). A neuropharmacological study of the periventricular neural substrate involved in flight. Behav. Brain Res., 22, 181-190.

Schuetz, M.T.B., De Aguiar, J.C. & Graeff, F.G. (1985). Anti-aversive role of serotonin in the periaqueductal grey matter. Psychopharmacol., 85, 340-345.

Schwarcz, J.R. (1977). Results of stimulation and destruction of the posterior hypothalamus: a long-term evaluation in neurological treatment. In: Psychiatry, pain and epilepsy. (Eds. Sweet W.H., Obrador S. & Martin-Rodriques J.G.) pp.429-439. Baltimore, University Park Press.

Shibata, K., Kataoka, Y., Gomita, Y. & Ueki, S. (1982). Localization of the site of the anticonflict action of benzodiazepines in the amygdaloid nucleus of rats. Brain Res., 234, 442-446.

Soubrie, P. (1986). Reconciling the role of central serotonin neurons in human and animal behaviour. The Behav. and Brain Sci., 9, 319-364.

Stern, R.S., Mark, I.M. & Mawson,D. (1980). Clomipramine and exposure for compulsive rituals: Plasma levels, side effects, and outcome. Brit. J. of Psych., 136, 161-166.

Tyre, P.J. & Lader, M.H. (1974). Response to propranolol and diazepam in somatic and psychic anxiety. Brit. Med. J., 2, 14-16.

Van Praag, H.M., Kahn, R., Asnis, G.M., Lemus, C.Z. & Brown, S.L. (1987). Therapeutic indications for serotonin-potentiating compounds: A hypothesis. Biol. Psych., 22, 205-212.

Versiani, M., Gentil, V., Guz, I., Paprocki, J., Ruschel, S., Pinheiro, T., Maciel, L., Nick, I. & Costa e Silva, J.A. (1986). Data about 508 cases of panic disorder and response to treatment with alprazolam, clomipramine, imipramine and tranylcypromine. In: Biol. Psych. 1985. (Eds. Shagass C., Josiassen R.C., Bridger W., Weiss K., Stoff D. & Simpson G.N.) pp 687-689. New York, Elsevier.

Willner, P. (1985). Antidepressants and serotonergic neurotransmission: an integrative review. Psychopharmacol., 85, 387-404.

Wise, C.D., Berger, B.D. & Stein, L. (1972). Benzodiazepines: anxiety-reducing activity by reduction of serotonin turnover in the brain. Scie., 17, 180-183.

Wise, R.A. & Bozarth, M.A. (1984). Brain reward circuitry: four circuit elements "wired" in apparent series. Brain Res. Bull., 12, 203-208.

Wolfle, T.L., Moyer, D.J., Carder, B. & Liebeskind, T.C. (1971). Motivational effects of electrical stimulation in the dorsal tegmentum of the rat. Physiol. and Behav., 7, 569-574.

STIMULUS PROPERTIES OF ARYLPIPERAZINE SECOND GENERATION ANXIOLYTICS

Richard A. Glennon, Richard Young, M.E. Pierson.
Department of Medicinal Chemistry, School of Pharmacy, Medical College of Virginia, Virginia Commonwealth University, Richmond, Virginia 23298, U.S.A.

The mechanism of action of diazepam and related benzodiazepine (BZ) anxiolytics seems to involve central BZ binding sites (Hommer et al., 1987). In animals, diazepam serves as a discriminative stimulus and the diazepam–stimulus generalizes to other BZs in a dose–dependant and stereoselective manner; the diazepam–stimulus can also be antagonized by the BZ antagonist flumazepil (Ro 15,1788). There exists, for a fairly large series of BZ derivatives, a significant correlation between their ED_{50} values for stimulus generalization in diazepam–trained animals and both their affinities for BZ binding sites and their human anxiolytic potencies (for a review see Young & Glennon, 1988).

Buspirone, ipsapirone and gepirone represent members of a new series of arylpiperazine "second generation anxiolytics" (SGAs). These agents are structurally unrelated to the benzodiazepines and display a low affinity for BZ binding sites. Early drug discrimination studies suggested that these agents might exert some of their effects via a serotonergic mechanism (e.g. Cunningham et al., 1985; Glennon, 1986b) and radioligand binding studies revealed that they possess significant affinity for $5-HT_1$ and, in particular, for $5-HT_{1A}$ sites (e.g. Peroutka, 1985). All three of these agents can conceivably give rise to a common arylpiperazine metabolite, 1-(2-pyramidinyl)piperazine (1-PP), although, to date, this has only been demonstrated for buspirone (Caccia et al., 1986). It has also been suggested that 1-PP may contribute to, or account for, the anxiolytic activity of buspirone (Garattini et al., 1982).

We have previously demonstrated that certain site–selective serotonergic agents can serve as discriminative stimuli in animals (for a review, see Glennon, 1987). For example, rats have been trained to discriminate the $5-HT_{1A}$ agonist 8-hydroxy-2-(di-n-propylamino) tetralin (8-OH-DPAT) and the $5-HT_{1B}$ agonist 1-(3-trifluoromethylphenyl)-piperazine (TFMPP) from saline, and stimulus generalization does not occur when animals trained to discriminate a site–selective serotonergic agent are administered doses of an agent selective for a different site. The purpose of the present study was to examine the stimulus effects of buspirone, ipsapirone, gepirone, and 1-PP in groups of animals trained to discriminate either 8-OH-DPAT (0.4 mg/kg) or TFMPP (0.5 mg/kg) from saline. 8-OH-DPAT trained animals were used because preliminary data suggested involvement of a $5-HT_{1A}$ mechanism; TFMPP-trained animals were also utilized because both the SGAs and 1-PP, like TFMPP, are arylpiperazines. The three SGAs were also examined in rats trained to discriminate diazepam (2 mg/kg) and ipsapirone (3 mg/kg) from saline.

Methods

Using standard operant chambers (Coulbourn model E10-10), groups of male Sprague–Dawley rats were trained to discriminate one of the above mentioned training drugs from saline using a two–lever operant choice task.

Procedures were similar to those previously reported (Glennon, 1986a; Young et al., 1986; Young & Glennon, 1987). Tests of stimulus generalization were conducted during extinction sessions using a 15-min presession injection interval. All injections were made via the intraperitoneal route and all solutions were made fresh daily in sterile 0.9% saline.

Results and discussion

The results of these studies are shown in Table 39.1. Where stimulus generalization occurred, ED_{50} values are presented. In the diazepam trained animals, buspirone, ipsapirone, and gepirone produced 40–50% diazepam-appropriate responding; higher doses resulted in disruption of behaviour (i.e., no responding). In all other cases where generalization did not occur, the animals made less than 25% of their responses on the drug-correct lever; higher doses resulted in disruption of behaviour.

Although both the benzodiazepines and the arylpiperazine SGAs produce anxiolytic profiles in certain animal models, and although members of both groups of agents are clinically effective antianxiety agents, their stimulus properties are dissimilar. The diazepam-stimulus did not generalize to any of the three SGAs examined; likewise, the ipsapirone stimulus failed to generalize to diazepam. The 8-OH-DPAT stimulus generalized to all three SGAs but not to diazepam. These results confirm earlier reports (Cunningham et al., 1985; Glennon, 1986b) that the stimulus properties of the SGAs may involve a 5-HT_{1A} mechanism and further suggest that a similar mechanism is not involved in the stimulus produced by diazepam. Supporting data are obtained with the ipsapirone-trained animals: the ipsapirone-stimulus generalized to buspirone, gepirone, the 5-HT_{1A} agonist 8-OH-DPAT, but not to diazepam. The TFMPP-stimulus failed to generalize to any of the agents studied.

It appears that the stimulus effects of the SGAs involve a 5-HT_{1A} mechanism. It is also unlikely that these stimulus effects can be attributed to the SGA metabolite 1-PP in that 1-PP failed to produce stimulus effects similar to those of diazepam, 8-OH-DPAT, or TFMPP.

Table 39.1 Results of stimulus generalization studies in animals trained to discriminate diazepam, 8-OH-DPAT, TFMPP or ipsapirone from saline.

		ED_{50} (mg/kg)			
Agent:	Group:	Diazepam	8-OH-DPAT	TFMPP	Ipsapirone
Diazepam		1.2	NSG	NSG	NSG
Buspirone		NSG	1.5	NSG	0.39
Ipsapirone		NSG	2.6	NSG	0.87
Gepirone		NSG	2.1	NSG	2.86
1-PP		NSG	NSG	NSG	–
8-OH-DPAT		NSG	0.17	NSG	0.09
TFMPP		–	NSG	0.2	NSG

NSG = No stimulus generalization.

These results are also in accord with the recent finding that in a conflict procedure 1-PP had no effect on punished or unpunished responding at intraperitoneal doses of up to 60 mg/kg (Young and Glennon, 1988). The diazepam-stimulus can be attenuated by pretreatment of the animals with the BZ antagonist flumazepil. Using rats trained to discriminate 0.2 mg/kg of 8-OH-DPAT from saline (Glennon, 1986a), we were unable to antagonize the stimulus effects of 8-OH-DPAT with flumazepil. In combination with the training dose of 8-OH-DPAT, flumazepil was inactive at intraperitoneal doses of up to 3 mg/kg and produced disruption of behaviour at higher doses.

Diazepam, buspirone, ipsapirone, gepirone, and 8-OH-DPAT, but not TFMPP or 1-PP, are active in conflict procedures (Young and Glennon, 1988). However, the stimulus properties of the active agents may be mediated via different mechanisms. The stimulus effects of diazepam appear to involve BZ receptors (Young and Glennon, 1987) whereas those of the three SGAs seem to involve a $5-HT_{1A}$ mechanism. Although serotonin has now been implicated as playing a role in the stimulus effects produced by the arylpiperazine SGAs, it remains to be determined whether the antianxiety effects of these SGAs also involve a 5-HT mechanism.

Acknowledgement: This work was supported in part by the Grants-In-Aid Program for Faculty of Virginia Commonwealth University.

References

Caccia, S., Conti, I., Vigano, G. & Garattini, S. (1986). 1-(2-Pyrimidinyl) piperazine as active metabolite of buspirone in man and rat. Pharmacology, 33, 46-51.

Cunningham, K.A., Callahan, P.M. & Appel, J.B. (1985). Similarities in the stimulus effects of 8-hydroxy-2-(di-n-propylamino) tetralin (8-OH-DPAT), buspirone, and TVX Q 7821: Implications for understanding the actions of novel anxiolytics. Soc. Neurosci. Abstr. 11.

Garattini, S., Caccia, S. & Mennini, T. (1982). Notes on buspirones mechanism of action. J. Clin. Psychiat., 43, 19-24.

Glennon, R.A. (1986a). Discriminative stimulus properties of the $5-HT_{1A}$ agonist 8-hydroxy-2-(di-n-propylamino)tetralin (8-OH-DPAT). Pharmacol. Biochem. Behav., 25, 135-139.

Glennon, R.A. (1986b). Site-selective serotonin agonists as discriminative stimuli. Psychopharmacology, 89, S42.

Glennon, R.A. (1987). Site-selective serotonin agonists as discriminative stimuli. In: Transduction Mechanisms of Drug Stimuli. (Eds. Colpaert, F.C., & Balster, R.L.) Berlin, Springer Verlag, in press.

Hommer, D.W., Skolnick, P. & Paul, S.M. (1987). The benzodiazepine/GABA receptor complex and anxiety. In: Psychopharmacology: The Third Generation of Progress. (Ed. H.Y.Meltzer) pp 977-983, New York, Raven Press.

Peroutka, S.J. (1985). Selective interaction of novel anxiolytics with the 5-HT$_{1A}$ receptors. Biol. Psychiat., 20, 971-979.

Young, R. & Glennon, R.A. (1987). Stimulus properties of benzodiazepines: Correlations with binding affinities, therapeutic potency, and structure activity relationships. Psychopharmacology, in press.

Young, R. & Glennon, R.A. (1988). Second generation anxiolytics and sero-tonin In: Serotonin Agonists as Psychoactive Drugs. (Eds. Gedulsky G. and Rech R.) NPP Press, in press.

Young, R.A., Glennon, R.A., Brase, D.A. & Dewey, W.L. (1986). Potencies of diazepam metabolites in rats trained to discriminate diazepam. Life Sci., 39, 17-20.

BEHAVIOURAL, NEUROCHEMICAL AND DISCRIMINATIVE STIMULUS EFFECTS OF SPIROXATRINE, A PUTATIVE 5-HT$_{1A}$ RECEPTOR ANTAGONIST

J.E.Barrett, S.N.Olmstead, L. Zhang C.Harrod, S.M.Hoffmann, K.Glover, M.A.Nader, B.A.Weissman[1].
Department of Psychiatry, Uniformed Services University of the Health Sciences. 4301 Jones Bridge Road. Bethesda, Maryland 20814-4799 USA, [1]Department of Pharmacology, Israel Institute for Biological Research, P.O. Box 19, Ness Ziona, Israel

Introduction
The discovery of clinically effective non-benzodiazepine anxiolytic compounds such as buspirone has propted a renewed interest in the role of serotonin (5-HT) in anxiety and in the development of suitable animal models for detecting such compounds. Several studies employing behavioural, neurochemical and neuropharmacological procedures have suggested that compounds such as buspirone act through a subtype of serotonin receptor designated as a 5-HT$_{1A}$ site (Cunningham, Callahan & Appel, 1987; Glennon, 1987; Mansbach & Barrett, 1987; Peroutka, 1986; Spencer & Traber, 1987). Compounds active at these sites have been described as 5-HT$_{1A}$ agonists (e.g. 8-OH-DPAT).

Although many conventional procedures, as well as those developed more recently, do not reflect the potential anxiolytic actions of this new class of compounds (Chopin & Briley, 1987) punished responding of pigeons appears to be very sensitive to and predictive of the effects of these novel anxiolytic drugs (Barrett & Witkin). The suggestion that spiroxatrine might function as a 5-HT$_{1A}$ antagonist (Nelson & Taylor, 1986) initiated the present series of experiments, conducted with pigeons, which were designed to evaluate the effects of spiroxatrine alone and in combination with agonists known to function at the 5-HT$_{1A}$ receptor.

Methods
Adult White Carneaux pigeons were used. In behavioural studies, pigeons were reduced to 85% of their free-feeding body weights and were trained to key peck in standard operant chambers using grain reinforcement. In one procedure (N=6), key pecking was maintained under a multiple fixed-interval (FI), fixed-ratio (FR) schedule; in the presence of red keylights the first response after 3 min produced 4-sec access to grain (FI), whereas in the presence of white keylights every 30th response produced gain (FR). Daily sessions consisted of alternating exposure to 10 components of each schedule. Under a second procedure (N=4), every 30th response in the presence of a red keylight produced food; in the presence of a white keylight, every 30th response produced both food and electric shock (delivered to electrodes implanted around the pubic bones). This punishment procedure, studied daily, consisted of 3 min exposure to each component of the schedule in an alternating manner with a session comprised of a total of 10 components. A third procedure (N=4) involved the training of 0.3 mg/kg spiroxatrine as a discriminative stimulus.

Pigeons were trained to respond on one key following the injection of spiroxatrine and on a different key following the injection of saline. Thirty consecutive responses on the correct key resulted in food delivery. Buspirone and 8-OH-DPAT were substituted on test days after spiroxatrine dose-effect curves were obtained. On test days, 30 consecutive responses on either key produced food.

In neurochemical studies (N=8) cerebrospinal fluid (CSF) was collected from pigeons with chronically implanted guide tubes located in the lateral ventricle. Neurochemical analysis of CFS from spiroxatrine-treated animals were performed using HPLC with electrochemical detection. Binding assays were performed in pigeons cerebrum membrane preparations using 1.2 nM [^{3}H] 8-OH-DPAT according to the methods described in Witkin et al. (1987).

In behavioural and neurochemical experiments, drugs were administered intramuscularly either immediately or 20 min (drug discrimination studies) before the session. When spiroxatrine was studied in combination with buspirone, spiroxatrine was administered 5 min before buspirone. Under the multiple FI, FR schedule and under the punishment procedure, drugs were administered twice per week, on Tuesday and Fridays, after performances had stabilized. Thursday's sessions or sessions when saline was administered instead of a drug served as control performances against which drug effects were evaluated. In drug discrimination studies, test doses of the training drug or drug substitutions occurred on Tuesday and Friday if on the preceding day performance was >90% and the first 30 responses occurred on the appropriate key. In the neurochemical experiments, CFS samples were collected once/week; drug effects were compared with samples collected after saline administration.

Results

Spiroxatrine produced dose-dependent decreases in responding under both FI and FR schedules. The FR was slightly more sensitive to the rate-decreasing effects with a dose of 0.3 mg/kg decreasing responding of all pigeons to 75% of control while not affecting FI performance. A dose of 3.0 mg/kg decreased rates under both components to zero. Buspirone (0.1 - 3.0 mg/kg) also showed dose-dependent decrased in performance under both components. When given in combination with spiroxatrine, there was little alteration of the spiroxatrine curve at lower buspirone doses (0.1 and 0.3 mg/kg); higher doses of buspirone potentiated the rate decreasing actions of spiroxatrine.

Spiroxatrine increased punished responding of all pigeons at doses between 0.01 - 0.3 mg/kg; these doses did not affect or decreased unpunished responding. Responding of all pigeons was increased to over 700 percent of control levels with 0.1 - 0.3 mg/kg doses. Increases in punished responding were comparable to those seen with ipsapirone and buspirone in this species.

In the drug discrimination procedure, 90-100% spiroxatrine-key responding occurred at doses between 0.03 - 0.1 mg/kg. All pigeons also responded 90-100% of the time on the key correlated with spiroxatrine when tested with buspirone doses between 0.03 - 1.0 mg/kg. Generalization to 8-OH-DPAT doses depended on whether this compound was studied before or after tests with buspirone.

If studied before buspirone, all pigeons responded 85-100% of the time on the drug-appropriate key to 8-OH-DPAT doses between 0.1 - 3.0 mg/kg. If studied after buspirone, however, pigeons did not respond or responded less than 50% of the time on the drug key across a wide range of 8-OH-DPAT doses (0.01 - 3.0 mg/kg).

Spiroxatrine (0.3 - 3.0 mg/kg) significantly elevated CFS levels of MHPG, DOPAG and HVA while producing decreases in 5-HIAA in a dose-dependent manner. Spiroxatrine inhibited [^{3}H] 8-OH-DPAT binding to pigeon cerebrum membranes with an IC_{50} value of 6.33 $\pm$ 1.01 x 10^{-9}M; the IC_{50} value for buspirone was 6.08 $\pm$ 0.54 x 10^{-8}M.

Discussion

Spiroxatrine produced many effects that are similar to buspirone in pigeons. Both compounds decrease performance maintained under FI and FR schedules, but produce large increases in punished responding (cf. Barrett et al., 1986; Witkin & Barrett, 1986). Similarly, buspirone substitutes for spiroxatrine in pigeons trained to discriminate spiroxatrine from saline and both compounds produce neurochemical changes in pigeons CSF and binding studies that are virtually identical (Mansbach et al., 1988). In view of these similarities in the behavioural, discriminative stimulus effects, and neurochemical characteristics, it would appear that spiroxatrine, like buspirone, is functioning behaviourally as a $5-HT_{1A}$ receptor agonist. It is also quite clear, however. that spiroxatrine, like buspirone, has strong dopamine (DA) antagonist properties. CSF levels of HVA and DOPAC were increased substantially in the present study by spiroxatrine and by buspirone in a separate experiment (Mansbach et al., 1988). Additionally, both buspirone and spiroxatrine block apomorphine- or piribedil-induced behaviour in pigeons (Witkin & Barrett, 1986; unpublished studies). Despite these strong DA antagonist effects for both buspirone and spiroxatrine, it appears that the anti-conflict and discriminative effects are mediated by or through the $5-HT_{1A}$ receptor system.

Effects with 8-OH-DPAT in pigeons trained to discriminate spiroxatrine from saline suggest an important role of pharmacological history in drug discrimination. Pigeons generalized from spiroxatrine to 8-OH-DPAT only if they had not previously been tested with buspirone. It may be that since buspirone and spiroxatrine show effects on both DA and 5-HT systems, and since 8-OH-DPAT is a relatively pure 5-HT compound, the order of test exposure may determine whether the DA, 5-HT or dual stimulus properties of spiroxatrine become dominant. If buspirone is tested first, after training, perhaps either the mixed DA/5-HT or DA actions of spiroxatrine are enhanced or become a stimulus complex which then impedes subsequent generalization to relatively selective $5-HT_{1A}$ compounds such as 8-OH-DPAT. If, however, 8-OH-DPAT is tested intitially, the $5-HT_{1A}$ properties serve as the discriminative stimulus.

Compounds active at the $5-HT_{1A}$ receptor site have recently appeared to be effective, potential anxiolytic drugs. Although many animal models have not been successful in reliably detecting activity of these substances, punished responding of the pigeon appears to be sensitive and selective for detecting both typical and atypical anti-anxiety compounds.

We thank Myra Zimmerman for excellent help in manuscript preparation. This work was supported by PHS Grant DA 02873.

References

Barrett, J.E., Witkin, J.M., Mansbach, R.S., Skolnick, P. & Weissman, B.A. (1986). Behavioural studies with anxiolytic drugs. III. Antipunishment actions of buspirone in the pigeon do not involve benzodiazepine receptor mechanisms. J. Pharmac. Exp. Ther., 238, 1009-1013.

Barrett, J.E. & Witkin, J.M (1988). Buspirone in animal models of anxiety. In Buspirone: mechanisms and clinical aspects. (Eds. Tunnicliff, G., Eison, A. & Taylor, D.) Orlando: Academic Press.

Chopin, P. & Briley, M. (1987). Animal models of anxiety: the effect of compounds that modify 5-HT neurotransmission. TIPS, 8, 383-388.

Cunningham, K.A., Callahan, P.M. & Appel, J.B. (1987). Discriminative stimulus properties of 8-hydroxy-2-(di-n-propylamino) tetralin (8-OH-DPAT): implications for understanding the actions of novel anxiolytics. Eur. J. Pharmac., 138, 29-36.

Glennon, R.A. (1987). Central serotonin receptors as targets for drug research. J.Med. Chem., 30, 1-12.

Mansbach, R.S. & Barrett, J.E. (1987). Discriminative stimulus properties of buspirone in the pigeon. J.Pharmac. Exp. Ther., 240, 364-369.

Mansbach, R.S., Harrod, H., Hoffman, S.M., Nader, M.A., Lei, Z., Witkin, J.M. & Barrett, J.E. (in press). Behavioural studies with anxiolytic drugs. V. Behavioural and IN VIVO neurochemical analysis in pigeons of drugs that increase punished responding. J. Pharmac. Exp. Ther.

Peroutka, S.J. (1986). Pharmacological differentiation and characterization of $5-HT_{1A}$, $5-HT_{1B}$ and $5-HT_{1C}$ binding sites in rat frontal cortex. J. Neurochem., 47, 529-540.

Spencer, D.G. Jr. & Traber, J. (1987). The interoceptive discriminative stimuli induced by the novel putative anxiolytic TVX Q 7821: behavioural evidence for the specific involvement of serotonin $5-HT_{1A}$ receptors. Psychopharmac., 91, 25-29.

Witkin, J.M. & Barrett, J.E. (1986). Interaction of buspirone and dopaminergic agents on punished behaviour of pigeons. Pharmacol. Biochem Behav., 24, 751-756.

RAPID TRAINING OF THE STIMULUS PROPERTIES OF SELECTIVE 5-HYDROXYTRYPTAMINE$_{1A}$ AGONISTS

Irwin Lucki, Judith A.South
Department of Psychiatry, University of Pennsylvania, Philadelphia, PA 19104-4283, U.S.A.

Drug discrimination has been used to study the behavioural effects produced by selecticve 5-hydroxytryptamine (5-HT) agonists because they produce distinct stimulus properties (see Glennon & Lucki, 1988). Rats trained to discriminate the stimulus properties of the 5-HT$_{1A}$ agonist 8-hydroxy-2-(di-n-propylamino) tetralin (8-OH-DPAT) respond similarly when they are administered other drugs that are selective for the 5-HT$_{1A}$ receptor, such as buspirone or ipsapirone (Glennon, 1986; Cunningham et al., 1987; Tricklebank et al., 1987). Drugs selective for other 5-HT receptor subtypes do not generalize to this cue.
Since drugs like buspirone are used to treat anxiety disorders in humans (Goldberg & Finnerty, 1979) and cause behavioural effects in animals typical of antianxiety agents (Eison et al., 1986), the stimulus properties of 5-HT$_{1A}$ agonists may provide a model for studying similar anxiolytic drugs.

The present study describes a method for rapidly training the stimulus properties of 8-OH-DPAT by employing conditioned taste aversion (CTA) as a behavioural baseline. The CTA is a rapidly-learned aversion for a novel flavor when consumption of a flavored solution is followed by sickness induced by a poison such as LiCl (Tuck & Riley, 1985). In the present study, rats rapidly learned to discriminate 8-OH-DPAT given prior to saccharin consumption as a cue that signalled the subsequent illness caused by administration of LiCl. Once the cue was established, other 5-HT$_{1A}$-selective compounds, such as buspirone or ipsapirone, mimicked the effects of the 5-HT$_{1A}$ agonist 8-OH-DPAT.

Methods
Adult male Spraque-Dawley rats (175-250 g) were placed on a restricted drinking schedule with tap water or 0.25% saccharin solution available for only 30 min during drinking sessions conducted each day in clear polycarbonate test cages located in the laboratory. Preweighed bottles of fluid were lowered into the cages through oppenings in the top and the amount of fluid consumed was measured by the difference in weight of the water bottle at the end of the session.

Rats were allowed to drink saccharin (0.25%) for 2 days prior to the start of conditioning. During conditioning, the drinking sessions were divided into drug trails or safe trials. During a drug trail (D), administration of 8-OH-DPAT (0.4 mg/kg IP) prior to drinking the saccharin solution was paired with injections of LiCl (1.8 mEq/kg) immediately after the session. On safe trials (S), saline (0.9% NaCl) was administered 15 min prior to saccharin drinking, and a second injection of saline instead of LiCl followed the session. Tap water (W) was given on days when conditioning trials could bot be conducted.
Acquisition of the discrimination was studied over the initial 9 drug or safe trails administered in the following order:
D-S-D-W-W-S-D-D-D-W-W-S-D-S-W-W-W-S-D-S-D-S-S.

A separate group of rats was used to examine the ability of repeated injections of 8-OH-DPAT (0.4 mg/kg IP) or saline administered 15 min prior to drinking sessions to affect saccharin consumption. This group served as unconditioned controls because they were never given LiCl.

Following acquisition of the discriminated CTA, the effects of altering the dose of 8-OH-DPAT or of substituting other drugs for 8-OH-DPAT were evaluated. For these studies, the following 4-day maintenance program was adopted: Day 1, drug trial; Day 2, presentation of tap water to rehydrate the animals; Day 3, safe trial, and Day 4, test trial. A test trial would only be conducted below 6 ml on the preceding drug trial. All drugs were administered 15 min prior to the drinking sessions. The presentation of drugs at test trial, other than the original dose of the training drug, was never followed by LiCl. Data for the test trials are expressed as percentage of drinking during the safe trials.

Results

Pairing the administration of the 5-HT_{1A}-selective agonist 8-OH-DPAT (0.4 mg/kg) prior to saccharin consumption with subsequent LiCl injections produced a conditioned taste aversion that discriminated the drug from saline after only 2-3 trials. During the last 3 acquisition sessions, rats drank 19.6 ± 1.7 ml saccharin (mean ± SEM) following saline, but only 4.8 ± 2.3 ml saccharin after administration of 8-OH-DPAT ($p<0.001$). 8-OH-DPAT produced a smaller, but significant, reduction of saccharin consumption in the unconditioned controls. During the last 3 drinking trials, mean saccharin consumption averaged 20.6 ± 2.2 ml following saline and 14.5 ± 3.5 ml following 8-OH-DPAT ($p <0.001$). This corresponded to a 29% reduction of drinking in unconditioned controls, compared with a 75% reduction of saccharin drinking in conditioned animals.

Test trials showed that 8-OH-DPAT produced a dose-dependent reduction of drinking behaviour in conditioned rats, significant at the training dose of 0.4 mg/kg, but did not significantly alter drinking by unconditioned controls (see Fig. 41.1). The attenuation of 8-OH-DPAT's mild effect on drinking in the unconditioned controls during this maintenance phase of the experiment may be due to their extended experience with saccharin, the less frequent testing schedule with 8-OH-DPAT, or involve the development of behavioural tolerance. The effects of ipsapirone were similar to those of 8-OH-DPAT. Ipsapirone reduced drinking in conditioned animals at doses (8-16 mg/kg) that failed to alter drinking behaviour in the control group. The effects of buspirone appeared similar to those of 8-OH-DPAT by producing a greater reduction of drinking behaviour in conditioned than in unconditioned animals (significant at 4.0 mg/kg).

In contrast, the selective reductions of saccharin drinking in 8-OH-DPAT-conditioned animals were not obtained with the 5-HT_{1B}-selective agonists 1-(m-trifluoromethylphenyl) piperazine (TFMPP) or 1-(m-chlorophenyl)piperazine (m-CPP). Each of these drugs failed to cause significant reductions of drinking behaviour in either the conditioned animals or the unconditioned controls when administered up to doses (0.8 mg/kg) that sustain independent stimulus effects (Lucki et al., 1987).

Discussion

This study demonstrates that the discriminated CTA procedure can be an

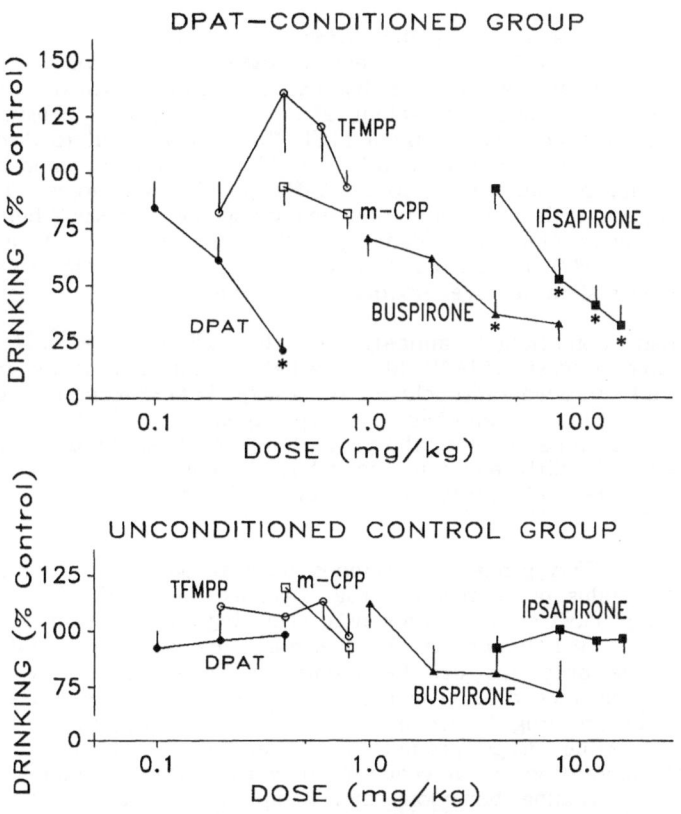

Figure 41.1 The effects of 5-HT agonists on the mean consumption of a 0.25% saccharin solution during test trial, expressed as a percentage of control (0.9% NaCl) ± 1 SEM. Values in the top panel were obtained in groups of rats (N=8) following acquisition of the discriminated CTA using 0.4 mg/kg H8-OH-DPAT as the training stimulus. Values in the bottom panel were obtained from unconditioned controls (N=8). Asterisks indicate that the value of the H8-OH-DPAT-conditioned group differs significantly from the corresponding value from the unconditioned controls, according to Neuman-Keuls test ($p<0.05$).

effective behavioural baseline to detect the stimulus properties of selective 5-HT agonists. Rats showed discrimination of the stimulus effects of the 5-HT$_{1A}$ agonist 8-OH-DPAT by decreasing their consumption of saccharin only in the presence of the drug cue that had been paired with the toxin LiCl. 8-OH-DPAT did not produce equivalent effects on saccharin consumption in control animals that were never given LiCl. Rats conditioned to discriminate 8-OH-DPAT generalized the cue to other 5-HT$_{1A}$-selective compounds, such as buspirone or ipsapirone, but not to the 5-HT$_{1B}$-selective agonists TFMPP or m-CPP. This agrees with previous studies employing lever pressing behaviour to measure drug discrimination (Glennon, 1986; Cunningham et al., 1987; Tricklebank et al., 1987) and supports the use of drug discrimination to study the behavioural effects of these novel anxiolytic compounds.

We have also conditioned animals to the stimulus effects of the 5-HT$_{1B}$-selective agonist TFMPP (0.8 mg/kg) within 2–3 trials using the discriminated CTA procedure (Lucki et al., 1987). TFMPP-conditioned animals generalized the cue to another 5-HT$_{1B}$ agonist m-CPP but not to 8-OH-DPAT. Taken together with the present results, these studies support the view that qualitative differences in the behavioural effects of 5-HT agonists may exist due to their selectivity for different 5-HT receptor subtypes (Glennon and Lucki, 1988).

The discriminated CTA procedure provides several potential advantages for studying the stimulus properties of drugs. Acquisition of the behaviour was extremely rapid by most previous standards, requiring only 2–3 pairings between 8-OH-DPAT and LiCl injections. Furthermore, by spacing training trials, acquisition of the drug cue can be studied without necessarily producing tolerance to the training drug. The use of a consummatory response (drinking behaviour) to measure drug discrimination may avoid the suppressive effects on responding that some drugs produce on lever pressing behaviour. Finally, expensive instrumentation is not required to conduct this procedure. Although drugs that alter drinking behaviour can complicate interpretation of these experiments, the comparison of drug effects on drinking behaviour between conditioned animals and unconditioned controls detected the learned aversion. It may also be possible to study a more sensitive measure of CTA, such as alterations of flavor preference, using this procedure.

Acknowledgements: The authors are grateful for the valuable assistance of Daniel Z. Press and Randy Berger in these experiments. This study was supported by USPHS grants MH 36262 and GM 34781.

References

Cunningham, K.A., Callahan, P.M. & Appel, J.B. (1987). Discriminative stimulus properties of 8-hydroxy-2-(di-n-propylamino) tetralin (8-OH-DPAT: implications for understanding the actions of novel anxiolytics. Eur. J. Pharmacol., **138**, 29–36.

Eison, A.S., Eison, M.S., Stanley, M. & Riblet, L.A. (1986). Serotonergic mechanisms in the behavioural effects of buspirone and gepirone. Pharmacol. Biochem. Behav., **24**, 701–707.

Glennon, R.A. (1986). Discriminative stimulus properties of the 5-HT$_{1A}$ agonist 8-hydroxy-2-(Di-n-propylamino tetralin (8-OH-DPAT). Pharmacol. Biochem. Behav., 25, 135-139.

Glennon, R.A. & Lucki, I. (1988). Behavioural models of serotonin receptor activation. In: The Serotonin Receptors. (Ed. Sanders-Bush, E.) New York, Humana Press, in press.

Goldberg, H.L. & Finnerty, R.J. (1979). The comparative efficiacy of buspirone and diazepam in the treatment of anxiety. Am J. Psychiat., 136, 1184-1187.

Lucki, I., South, J.A. & Berger, R. (1987). Rapid detection of the stimulus properties of 5-hydroxytryptamine (5-HT) agonists. Soc. Neurosci. Abstr., 13, 344.

Riley, A.L. & Tuck, D.L. (1985). Conditioned taste aversions: A behavioural index of toxicity. Ann. N.Y. Acad. Sci., 443, 272-292.

Tricklebank, M.D., Neill, J., Kidd, E.J. & Fozard, J.R. (1987). Mediation of the discriminative stimulus properties of 8-hydroxy-2-(di-n-propylamino) tetralin (8-OH-DPAT) by the putative 5-HT$_{1A}$ receptor. Eur. J. Pharmacol., 133, 47-56.

BEHAVIOURAL PHARMACOLOGY OF 5-HT: POSSIBLE PROGRESS IN THE LATERALITY OF CONCEPTS

P. Bevan, S. Lorens* and T. Archer°
Department of Pharmacology, Duphar B.V., Weesp, The Netherlands, * Department of Pharmacology, Stritch School of Medicine, Layola University of Chicago, Maywood, Illinois, U.S.A., ° Department of Psychology, University of Gothenburg, Gothenburg, Sweden

As indicated by Richard Green in his introductory chapter to this volume, there are at least two approaches to the study of the behavioural pharmacology of 5-HT. One essential purpose of this collection of papers would be then to review the basic state of research within the area. Obviously, such a wide endeavour is doomed to be incomplete, but there is reason to hope that all current issues are represented to a greater or lesser degree. An impressive array of evidence was presented suggesting that CNS 5-HT circuits play an important role in the modulation of several neuropsychological processes, including feeding, aggression and sexual behaviour, and the elaboration and expression of affective states (see chapters by Deakin, Willner, Soubrie, Broekkamp and Tyers). Inevitably, some topics concerning the behavioural pharmacology of 5-HT have not been touched upon or only in passing. Thus, this final chapter will include brief reviews of areas of 5-HT research that have previously shown or will show great promise. Another essential purpose must surely involve explicit suggestions as to potentially fruitful enterprises over the coming five years, perhaps even decade. This purpose is much more difficult (and dangerous) to fulfil because, amongst other evolutions, the rapid developments in technology, e.g. the synthesis of selective pharmacological tools or microdialysis-behaviour monitoring, will undoubtedly prove to facilitate certain trends, but not others. It is clear from the refinement of the procedures to observe and measure functional changes induced by 5-HT drugs that there is great awareness of the importance of methodological parameters, not least with regard to such considerations as species, strain, sex and housing conditions of the experimental animals employed. It is hoped that the different chapters will serve to demonstrate also the variety of behavioural effects that may or may not be influenced by drugs with interact with serotonergic systems.

The role of 5-HT in depression is reviewed briefly by both Willner (chapter 13) and Deakin (chapter 14) and the hypotheses offered by these authors concerning 5-HT dysfunction in depression do contribute towards a better understanding of the behavioural pharmacology. Willner's hypothesis is more formalized relating 5-HT turnover and social environment, impulsivity and social skills (or lack of them), and the involvement of antidepressants and 5-HT neurotransmission. It is probably an advantage to consider Willner's chapter in the context of Soubrie's chapter (chapter 29) relating depression and anxiety (but see also the chapter by Graeff in this context). Not least important are the attempts to correlate differential drug effects in animal models with their therapeutic efficacy in subgroups of anxiety and depressive disorders, a very necessary but thankless undertaking.

The organisation of brain 5-HT pathways, with topographically arranged 5-HT projections that are highly collateralized and extend to all levels of the neuraxis, would seem to allow some concomitant manifestation both of drug action and the dysfunctions arising from alterations to the system. It should be expected then that drugs affecting 5-HT neurotransmission or the behavioural disorders of varied origin will be observed as not one but an array of symptoms on a receptor site level, this situation becomes even more complex. One can then understand why affective disorders are associated not only with changes in mood but also alterations in food consumption, sexual activity, defensive and aggressive behaviour, nociception, hormonal responses and sleep/activity cycles. One challenge for the future will be to determine the circumstances under which 5-HT is released onto its distinct receptors ($5-HT_{1A-1D}$, $5-HT_2$, and $5-HT_3$) in response to different stimuli or activation of select afferent inputs. Is it possible that the terminals of different 5-HT neurons synapse onto distinct receptors located on neurons which utilize different neurotransmitters ? A "non-unitary" 5-HT influence may be achieved by these means, e.g. the 5-HT – NA brainstem – spinal interaction in nociception (see chapter by Berge, Post and Archer, this volume). Since none of the behavioural/physiological processes under study are under the exclusive control of 5-HT systems, the interactions with other neurotransmitters remain to be defined as a prerequisite to understanding.

The reviews provided by Broekkamp (chapter 28) and Tyers (chapter 30) pertain mainly to the evidence implicating 5-HT receptors, and $5-HT_3$ receptors particularly in the latter case, in animal models of anxiety. Thus, it may be fruitful to consider these contributions in the context of Colpert and Koek's chapter (chapter 37) and those of Barrett, Glennon, and Moser. Two points of the several made by Soubrie (chapter 29) ought to be attended to: (1) The possibility that compounds, like buspirone, may be of therapeutic value for treating anxiety in patients suffering from depressive states (Goa & Ward, 1986; Schweizer, Amsterdam, Rickels, Myron & Droba, 1986) and (2) The pharmacological significance of the presynaptic effects, as opposed to the postsynaptic effects, of the various $5-HT_{1A}$ receptor agonists. In recent years several excellent reviews have treated the role of 5-HT in drug screens for anxiety states and/or experimental animal models (e.g. File, 1984, 1985; Gardner, 1985; Iversen, 1984; Johnston & File, 1986; Thiebot, 1986; Witkin & Barrett, 1986). There appears to be a general consensus that 5-HT is involved in the pharmacological processes underlying the results obtained from the test models used to index some manner of anxiety reaction in animals (but see the Broekkamp, Soubrie and Tyers chapters, this volume). Graeff's chapter (chapter 38), in an analysis of the brain aversive systems and 5-HT, does indeed provide some very salient points regarding the preclinical and clinial aspects of 5-HT and anxiety.

With regard to the problems of pharmacological processes and the test models used to index behavioural reactions in animals, it should be noted that Fernandez-Guasti and Hong (this volume) showed that 8-OH-DPAT and RU 24969 were more potent than diazepam in the conditioned defensive burying procedure of Treit et al. (Treit, 1985; Treit, Pinel & Fibiger, 1981). The doses of 8-OH-DPAT that inhibited defensive burying in rats are comparable to those inhibiting marble burying in mice (Fredriksson & Archer, unpublished). However, diazepam (used by Fernandez-Guasti and Hong as a reference compound) appears to inhibit defensive burying only at doses higher than those needed to inhibit marble-burying (Archer, Fredriksson, Lewander & Söderberg, 1988).

It is interesting to note the growing number of investigations that implicate 5-HT_{1A} agonist involvement in fear and anxiety (e.g. Carli & Samanin, 1987; Engel et al., 1984; McMillen & Mattiace, 1983; Peroutka, 1985; but see also Chopin & Briley, 1987, Critchley & Handley, 1986, 1987; File, Johnston & Pellow, 1987; McCloskey, Paul & Commissaris, 1987; Rowan, Cullen, Moulton & Anwyl, 1987).

Throughout, it will be noted that certain basic considerations of behavioural pharmacology are stringently adhered to whereas others are applied only occasionally, whether as a result of design or by accident. It is quite apparent that dose-relationships have been carefully analysed, especially with regard to the relative affinities of 5-HT drugs for different 5-HT receptors and their subtypes. In this context it should be noted that quite meticulous attempts have been made to compare receptor-binding analyses with preclinical functional data and/or clinical symptomatology (see especially chapters by Willner, Olivier, Colpaert, Broekkamp, Mendelsohn and Gorzalka, and Van der Heyden et al.). Nevertheless, behavioural pharmacologists do need to perform more chronic or repeated administration drug studies (and not simply with regard to antidepressant research), and concomitantly further investigations of the development of <u>tolerance</u> to and <u>withdrawal</u> from the repeated administation of serotonergic agents. Here, one should consider more often the evaluation of drug effects in juvenile, adult or aged animals. In this regard, increasing evidence indicates that the actions of drugs affecting 5-HT neurotransmission can be very dependent on neurotoxin-induced denervations and/or supersensitivity of, for instance, catecholaminergic pathways in the CNS. Thus, information on drug action ought not to be confined solely to normal animals, but also to those with a CNS altered by the effects of selective pharmacological tools or aging processes.

Comprehensive reviews on the role of 5-HT in male and female sexual behaviour are presented in the chapters by Ahlenius and Larsson, for male sexual behaviour, and, Mendelson and Borzalka, for female sexual behaviour. As a general premise, it has been considered that treatments decreasing 5-HT neurotransmission in the brain facilitate male (e.g. intromissions, ejaculations) and female (lordosis) sexual behaviour whereas increases in 5-HT neurotransmission inhibit sexual behaviour. One potential problem for this seemingly straightforward situation was that 8-OH-DPAT facilitates male, but inhibits female, sexual behaviour (Ahlenius et al, 1986; Ahlenius et al., 1981). However, the paradoxical effect could be explained by recourse to either the diversity of 5-HT receptor classification or the partial agonist properties of the compound. Be that as it may, the 8-OH-DPAT paradox illustrated two major issues of general importance to the behavioural pharmacology of 5-HT, namely, the diverse nature of 5-HT receptors and their localization in the brain and spinal cord. Hillegaart et al. (1989) present an interesting study that may serve to illustrate a further aspect of this problem. Briefly, local application of 5-HT (10 µg) into the dorsal or median raphe nucleus facilitated male sexual behaviour whereas a 40 µg dose caused an inhibition. On the other hand, microinjections of 5-HT into the median raphe caused dose-dependent increases in motor activity while application into the dorsal raphe dose-dependently decreased motor activity. Thus, in addition to excitatory and inhibitory actions of 5-HT being dependent on the administered dose, the site of administration (i.e. action) may also profoundly influence the evoked behaviour. Too often, scant attention has been paid to such phenomena.

The behavioural pharmacology of serotonergic neurotransmission in pain, anxiety, depression and aggression may be inferred, in the preclinical laboratory, from experimental evidence derived from tests of nociception, drug screens presumed to reflect anxiolytic or antidepressant action, and test models of aggressive behaviour. Often, these tests are referred to as animal models, but such a classification is not strictly correct and should be avoided (McKinney & Bunney, 1969). In general, the investigation of pain and anxiety are related since some manner of unpleasant stimulation is presented and the response of the experimental animal is indexed. The difference between pain and anxiety for example, may be loosely explained on this basis: Pain involves the interpretation of nociceptive signals resulting from a more or less sharp sensation (for example strong heat or electric shock) whereas a state of anxiety requires the involvement of some form of fear, (be it as an hypothetical construct or as an intervening variable) whether conditioned or unconditioned. On the other hand, certain aspects of aggression are relevant to depressed states (see Willner's chapter). In sum, we must expect some overlap of the test models in the current state of preclinical research.

Certain drug screening procedures seem to offer an animal model status for anxiolytic action bearing in mind that the issues of 'validity' and 'reliability' have not generally been approached. Thus, for example, Young et al. (1987) combined brief electric shock and reward (sweetened milk) in a multiple variable-interval fixed-ratio (VI-FR) conflict schedule whereby the FR component scheduled shock presentation. They also trained rats to discriminate diazepam from saline in a two-lever operant choice procedure. Buspirone, gepirone and ipsapirone were all shown to increase responding suppressed by footshock in the conflict test but did not show any generalization to the diazepam stimulus control. This study (Young et al., 1987) dissociates an anxiolytic effect from any stimulus generalization to diazepam. The anxiolytic effects of these compounds may result from their abilities to interact with central 5-HT sites (Peroutka, 1985), and certainly the 5-HT implication in anxiety is gaining increasing prominence (e.g. Hjorth, Söderpalm & Engel, 1987). This evidence involving 5-HT$_{1A}$ receptor agonists in animal models of anxiety appears to receive some support from clinical investigations of the role of 5-HT in anxiety (e.g. Banki, 1977; Csanalosi et al., 1987; Goa & Ward, 1986; Goldberg & Finnerty, 1979; Jacobson et al., 1985; Rickels et al., 1982; Schweizer et al., 1986). However, the relationship of 5-HT and anxiety is not a simple one. For example, Charney et al. (1987) examined the effects of the 5-HT receptor agonist, m-chlorophenylpiperazine (mCPP) in agoraphobic and panic disorder patients and in healthy subjects. mCPP was shown to have an anxiogenic effect in both panic anxiety patients and healthy subjects as well as producing significant neuroendocrine changes in both groups.

Although Charney et al. (1987) question the 5-HT role in panic, there are other studies supporting 5-HT related drugs in the treatment of anxiety (e.g. Ceulemans et al., 1985; Eison & Eison, 1984; Gloger, Grunhaus & Birmacher, 1981; Kahn & Westenberg, 1985). Since most of the clinical investigations of 5-HT and anxiety relate to the compounds buspirone and gepirone, it may be worthwhile to consider the results of a recent study by Merlo Pich and Samanin

(1986) using experimental models of anxiety in rats. These authors compared the anxiolytic effects of diazepam, haloperidol, sulpiride and buspirone on punished and unpunished responding in conflict test (shocked drinking) as well as the number of crossings between the black and white compartments of a two-compartment apparatus. All four compounds appeared to exert an anxiolytic effect, i.e. increased punished responding and crossings between compartments, referred to as 'disinhibitory' effects by Merlo Pich and Samanin (1986). Despite some short-comings, this study represents an exemplary preclinical investigation for analysing the behavioural pharmacology of 5-HT with regard to anxiety, combining as it does conflict and environmentally induced neophobic procedures with a straightforward measurement of locomotor activity in photocell cages.

Wadenberg and Ahlenius (1988) provide evidence from an interesting little experiment showing that discriminative avoidance latency is increased (or perhaps 'disinhibited') by a low dose of 8-OH-DPAT. Such combinations of the available anxiety test procedures will prove to be an increasing requirement for valid and reliable drug screens (Treit, 1985; but see also Morinan & Parker, 1985; Parker & Morinan, 1986) if, indeed, the conclusive interpretations of the role of 5-HT in anxiety are to be resolved.

One notable omission in the analysis of the role of 5-HT neurotransmission in behaviour presented in this volume is a chapter on serotonergic drugs and the acoustic startle reflex in rats (e.g. Davis, 1980). As Davis et al. (1986) recently described, both inhibitory and excitatory effects of drugs acting on 5-HT receptors can be described since startle amplitude is dependent upon the characteristics of the eliciting stimulus. Although almost total depletion of 5-HT does not eliminate the startle reflex, much evidence suggests that 5-HT modulates startle. Table VIII.1 presents a summary of some of the modulatory effects of treatments altering 5-HT neurotransmission.

Certain trends from the available evidence may be summarised: (1) 5-HT depletion increased startle responses; (2) 5-HT itself injected into the forebrain depressed startle but injected into the spinal cord increased startle; (3) 5-HT$_{1A}$ agonists tend to increase startle (administered systemically) whereas 5-HT$_{1B}$ agonists depressed startle; (4) Intraventricular/intrathecal injections of the 5-HT$_{1A}$ and 5-HT$_{1B}$ agonists was also critical. The activation of 5-HT$_{1A}$ receptors-increase in startle amplitude is consistent with electrophysiological evidence showing that 5-HT enhanced the motor neuron response to excitatory input (e.g. McGall & Aghajanian, 1979). It is interesting to note that Geyer (this volume) has found that the 5-HT$_2$ antagonists cyproheptadine, ketanserin, ritanserin and cinanserin all significantly increased the rate of habituation to tactile startle since these compounds appear not to affect startle reactivity per se. Interestingly, the differential effects of 5-HT$_{1A}$ and 5-HT$_{1B}$ agonists in the forebrain may provide indication of receptor subtype selectivity for both agonist and antagonist action.

The role of 5-HT neurotransmission in the psychopharmacology of aggression is described at some length by both Olivier et al. (Chapter 9) and Miczek and Donat (Chapter 10). It would appear that almost all psychopharmacological agents are able to reduce aggressive behaviour apparently irrespective of this influence on 5-HT neurotransmission. The sophistication of the ethologist has, however, identified one group of drugs, serenics, which are quite specific suppressors of aggressive behaviours, in contrast to the other drugs whose effects are dismissed as non-specific. Serenic drugs share intrinsic activity at the 5-HT$_1$ autoreceptors, a receptor whose pharmacological characters seems to be species-dependent. Thus clinical data is crucial in helping us interpret the ethologists' findings in the context of man. It is of some interest to indicate that several of the 5-HT$_{1A}$ agonists (shown previously to lack selective action over a wide range of aggressive behaviours by Olivier et al., see chapter 9) have been applied by Blanchard et al., to investigate defensive threat and attack when

Table VIII.1 The effects of treatments altering 5-HT neurotransmission upon startle responses

Treatment	Duration (ac., chr)	Effect (↑, ↓, 0)	Reference
5-HT depletion			
Dorsal/median raphe lesion	chr	↑	6, 7, 11
p-chlorophenylalanine	chr	↑	1, 2
p-chloroamphetamine	ac/chr	↑	8
5,7-Dihydroxytryptamine	chr	↑	10
Tryptophan-free diet	chr	↑	15
5-HT agonists			
5-HT into lat. ventricle/HPC (ICV)	ac	↓	4, 12
5-HT intrathecally (IT)	ac	↑	4, 5
5-methoxy-N,N-dimethyltryptamine (5-MeO-DMT)	ac	↑	3
8-hydroxy-2-(di-n-propylamine) tetralin (8-OH-DPAT)	ac	↑	13, 14
1-(m-chlorophenyl) piperozine (mCPP)	ac	↓	5
8-OH-DPAT (ICV)	ac	0	5
8-OH-DPAT (IT)	ac	↑	5
mCPP (ICV)	ac	↓	5
mCPP (IT)	ac	0(↓)	5
Other treatments			
Tryptophan + MAOI	ac	↑	9
5-HTP + MAOI	ac	↑	9
Lithium + MAOI	ac	↑	9

Key: ac = acute, chr = chronic, ↑ = increase, ↓ = decrease, 0 = no effect

References:
1. Carlton & Advokat, 1973
2. Conner et al., 1970
3. Davis et al., 1980a
4. Davis et al., 1980b
5. Davis et al., 1986
6. Davis et al., 1982
7. Davis & Sheard, 1974
8. Davis & Sheard, 1976
9. Fechter, 1974
10. Geyer et al., 1980
11. Geyer et al., 1976
12. Geyer et al., 1975
13. Svensson, 1985
14. Svensson & Ahlenius, 1983
15. Walters et al., 1979

flight was either possible or blocked. It is important to note that neurotransmitter-neurohumoral interactions are considered by Miczek and Donat. These relations gain even greater importance when analysed together with environmental variables and the individual status of each experimental animal (Miczek, 1987; Winslow & Miczek, 1985).

It may be premature but we can speculate that the 5-HT$_{1A}$ agonists (so-called "second generation anxiolytic agents", e.g. buspirone, ipsapirone and gepirone, see chapter 39) may have an antidepressant action at least as interesting as their anxiolytic action. Note however that the most apparent obstacles, presently, to the eventual therapeutic usefulness of these compounds consist of the potential cardiovascular changes and poor bioavailability. These 5-HT$_{1A}$ agonists (including, for example, buspirone and 8-OH-DPAT) have been shown to produce and antidepressant action in at least relatively predictable animal models of depression, the differential-reinforcement- of-low-rates-72 sec schedule and the grouped-isolated housing model (Järbe and Archer, unpublished data).

Some mention ought to be made of the possible role of 5-HT in memory storage and retrieval processes as derived from avoidance learning studies (e.g. Essman, 1971, 1973; Joyce & Hurwitz, 1964; Ogren et al., 1977; Roffman & Lal, 1971; Woolley & Van der Hoeven, 1963). p-Chloroamphetamine (PCA) has been used, predominately in an acute administration in the dose range 1.0 to 5.0 mg/kg, to induce 5-HT as well as other neurotranmitter release and has been shown to cause clear performance deficits; 5-HT selectivity in the effects obtained has been demonstrated typically by the blockade of the deficits by pretreatment with various 5-HT uptake inhibitors. This acute effect of PCA (20-30 min after intraperitoneal administration) involves the release of endogenous 5-HT from presynaptic nerve endings in addition to an inhibition of the neuronal uptake of 5-HT (Fuller et al. 1974; Trulson & Jacobs, 1976). The effects of acute PCA treatment upon the retention of shock-induced fear reaction were studied in a series of experiments (Archer, 1982). This procedure was based on earlier work on freezing and immobility in rats (e.g. Blanchard & Blanchard, 1969; Bolles & Riley, 1973). Typically, rats were given fear conditioning trials and then tested for fear retention 24 hours later, and it was found that the fear reactions of PCA treated animals was considerably less than that of the control groups, i.e. rats receiving either saline, zimelidine, or pretreatment with zimelidine before acute PCA. Figure VIII.1 presents the results of a large experiment to demonstrate the selective effect of PCA treatment in disrupting fear retention by administering long-term PCA at neurotoxic doses (i.e. 2 x 10 mg/kg, 8 and 7 days prior to the acute PCA testing.

Long-term PCA (2 x 10 mg/kg) blocked the fear-retention deficit repeatedly observed following acute PCA (2.5 or 5 mg/kg) treatment. Long-term and selective depletion of noradrenergic terminals with the neurotoxin, DSP4, did not affect the acute effects of PCA upon fear retention. The neurochemical analyses indicated that the long-term PCA treatment caused a drastic depletion of 5-HT in forebrain regions (see Archer, 1982). It is of some interest to consider whether acute PCA treatment exerts an effect upon acquisition or on the retention of fear conditioning (typically 2 to 4 brief but inescapable shocks). Acquisition refers in this case to the short-term processes antedating the memory consolidation or storage processes assumed to occur during the retention interval. Some evidence (Archer et al. 1981) suggests that it is the consolidation rather than the acquisitory process that is affected. Other evidence (Archer, 1982, Experiments 5 and 6) suggest that the PCA-induced fear retention deficit may interfere with both storage and retrieval processes. There is some clinical relevance for linking 5-HT neurotransmission and memory processes. For instance, electroconvulsive shock (ECS), effective in the treatment of depression, both disrupts memory formation and increases 5-HT synthesis and receptor sensitivity in the rat brain as well as elevating the functional activity of catecholamines (Lebrecht & Nowak, 1980).

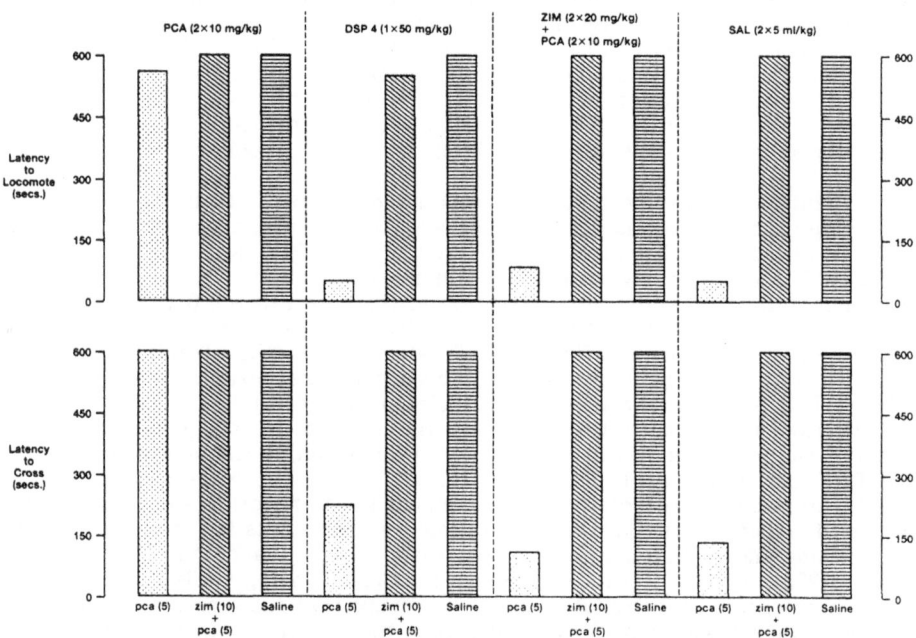

Fig. VIII.1   Long-term PCA (2 x 10 mg/kg) and acute PCA (5 mg/kg) interaction, a test for the specificity of the PCA-induced retention deficit.

ECS-induced retrograde amnesia can be antagonised by 5, 7-DHT lesions of the median raphe nucleus and by the 5-HT receptor antagonists bromolysergide and methysergide (Montanaro et al. 1979, 1981).

Further to the discussions of receptors site versus brain region in 5-HT action, an exhaustive series of experiments by Kennett and Curzon (1988) indicate that the inhibition of locomotion and feeding by mCPP and TFMPP may be the result of an activation of $5-HT_{1C}$ receptors which are widely distributed in rat brain regions including the hypothalamus (Blurton & Wood, 1986; Pazos & Palacios,

1985). Considerations of the functional roles of the $5-HT_{1C}$ receptor may be of importance in for instance the anorexic effects of certain 5-HT uptake inhibitors (Blundell, 1984) or in the sedative side-effects of trazadone (Feighner, 1980). At the same time it must be remembered that over 90 percent of the 5-HT in the body is localized in the periphery, in autonomic ganglia, in the gastrointestinal tract, and in blood platelets. One ought not to lose sight of the fact that some of the behavioural effects of orally or systemically administered drugs may be due to their peripheral actions either alone or in combination with their central effects, for example, changes in peristalsis, blood pressure and basal plasma horomonal levels. These diverse physiological processes, thus, ought to give us one important lesson, i.e. a more lateral approach to the psychopharmacological quest of effective drugs with 5-HT action must be developed.

The chapter by Samanin (chapter 22) provides a wide review of the role of 5-HT in feeding behaviour by focussing mainly on the anorectic effects of D-flenfuramine, 5-HT uptake inhibitors, 5-HT receptor agonists and antagonists. Samanin develops some salient concepts in this regard: test of selectivity using 5-HT neurotoxins, which aspects of feeding are regulated by brain 5-HT ? i.e. hunger vs satiety, the rewarding effects of food, nutrient selection, the 5-HT receptor subtypes modulating feeding. In the light of the earlier discussion concerning $5-HT_{1C}$ receptor agonist induced anorexia it is worth noting that Samanin suggests that 5-HT receptors other than $5-HT_2$ receptors mediate the anorectic activity of D-fenfluramine. Samanin further discusses the role of neuroanatomical sites (e.g. the paraventricular nucleus) sensitive to 5-HT activation, the role of peripheral 5-HT, and perhaps most relevantly the involvement of 5-HT and eating disorders.

The chapter by Goodall and Silverstone, reviewing as it does evidence for the involvement of serotonergic mechanisms in human feeding behaviour, provides an invaluable complement to Samanin's. Attention should be drawn to the discussion on the terms 'hunger', 'appetite', 'nutrient selection', and 'taste preferences' in considering the actions of drugs affecting 5-HT neurotransmission clinically. The clinical evidence is particularly intriguing although it would be of some interest to examine any data concerning the effects of a combination of tryptophan with a potential monoamine oxidase-A inhibitor, e.g. amiflamine, on feeding behavour. Like Samanin, Goodall and Silverstone suggest an important involvement for 5-HT in feeding behaviour but both reviews significantly side-step the considerably more difficult issue of mechanism. This tactic is a wise one since all the available evidence suggests that we have not as yet in behavioural and biochemical pharmacology derived the critical techniques for the eventual analysis of mechanism. Nor are the central neuroanatomical sites of interactions between 5-HT pathways and other neurotransmitter systems sufficiently well understood. One clear advantage of considering the functional role of 5-HT receptors in a wide range of behaviours may be the insights gained from both the parallels and the dissociations suggested by these analyses. Thus, for example the changes produced by drugs with 5-HT action on feeding behaviour and the likely concomitant changes in aggressive and sexual behaviour, albeit in possibly opposite directions would ensure a broader understanding of 5-HT function.

Finally, a worthwhile purpose for this volume may be in the pursuit of a wider degree of laterality in the formation of concepts to describe or explain the behavioural pharmacology of 5-HT. For instance, drugs affecting serotonergic systems seem to modulate male and female rat sexual behaviour in opposite ways and if one draws attention to the strain/species difference even more conflicting evidence may be expected, e.g. male mice are apparently insensitive to the 'triggering' properties of 8-OH-DPAT on sexual behaviour. Detailed studies are lacking in laboratory animal species other than rodents and it is hoped that encouraging these analyses particular considerations are made for the etho-experimental constraints affecting the behavioural procedures applied. The laterality concept, if seriously pursued, may provoke some reconsideration of current animal models of disease states, or, even at the simpler level of experimental analysis, the testing procedures upon which present pharmacological formulations of 5-HT receptor subtypes and their functional role are dependent.

This concept may also stimulate the pursuit of behavioural pharmacological analyses over neuroanatomically diversified 5-HT systems that encompass and extend onto further direct and indirect interactions of 5-HT with other neurotransmitters and, as important, the plethora of emerging neuropeptide pathways (see Nieuwenhuys, 1985). Perhaps the most useful consequence of the lateral approach would be the timely shift away from the "whole animal bio-assay" (which if not sufficient has at least proven a necessary tactic for analysing 5-HT receptor systems) procedures by considering behavioural pharmacology through a perspective defined by a broader biological emphasis. Thus, it may be that only through a more lateralised preclinical endeavour that sufficiently efficacious therapeutic drugs may be produced for the required clinical employment.

References

Ahlenius, S., Fernandez-Guasti, A., Hjorth, S. & Larsson, K. (1986). Suppression of lordosis behaviour by the putative 5-HT receptor agonist 8-OH-DPAT in the rat. Eur. J. Pharmacol., 124, 361-363.

Ahlenius, S., Larsson, K., Svensson, L., Hjorth, S., Carlsson, A., Lindberg, P., Wikström, H. & Sanchez, D. (1981). Effects of a new type of 5-HT receptor agonist on male rat sexual behaviour, Pharmacol., Biochem. Behav., 15, 785-792.

Archer, T. (1982). Serotonin and fear retention. J. Comp. Physiol. Psychol., 96, 491-516.

Archer, T., Ogren, S.O. & Johansson, C. (1981). The acute effect of p-chloroamphetamine on the retention of fear conditioning in the rat: Evidence for a role of serotonin in memory consolidation. Neurosci. Lett., 25, 75-81.

Archer, T., Frederiksson, A., Lewander, T. & Söderberg, U. (1988). Marble burying and spontaneous motor activity in mice: interactions over days and the effect of diazepam. Scand. J. Psychol., 28, 242-249.

Banki, C.M. (1977). Correlation of anxiety and related symptoms with cerebrospinal fluid 5-hydroxyindole acetic acid in depressed women. J. Neural Transm., 41, 135-143.

Blanchard, R.J. & Blanchard, D.C. (1969). Crouching as an index of fear. J. Comp. Physiol. Psychol., 67, 570-575.

Blundell, J.E. (1984). Serotonin and appetite. Neuropharmacol., 23, 1537-1551.

Blurton, P.A. & Wood, M.D. (1986). Identification of multiple binding sites of ^{3}H-5-hydroxytryptamine in the rat CNS. J. Neurochem., 46, 1362-1398.

Bolles, R.C. & Riley, A.L. (1973). Freezing as an operant response: Another look at the operant-respondent distinction. Learn. Motiv. 4, 268-275.

Carli, M. & Samanin, R. (1987). Potential anxiolytic properties of 8-hydroxy-2-(D-N-propylamino) tetralin, a selective serotonin 1A receptor agonist. Psychopharmacol., in press.

Carlton, P.L. & Advokat, C. (1973). Attenuation habituation due to parachlorophenylalamine. Pharmacol. Biochem. Behav., 1, 657-663.

Ceulemans, D.L.S., Hoppenbrouwers, H.J.A., Gelders, Y.G. & Reyntjens, A.J.M. (1985). The influence of ritanserin, a serotonin antagonist, in anxiety disorders: a double-blind placebo-controlled study vs lorazepam. Pharmacopsychiat., 18, 303-305.

Charney, D.S., Woods, S.W., Goodman, W.K. & Heninger, G.R. (1987). Serotonin function in anxiety II. Effects of the serotonin agonist MCPP in panic disorder patients and healthy subjects. Psychopharmacol., 92, 14-24.

Chopin, P. & Briley, M. (1987). Animal models of anxiety: the effect of compounds that modify 5-HT neurotransmission. Trends in Pharmacol. Sci., 8, 383-388.

Conner, R.J., Stolk, J.M., Barchas, J.D. & Levine, S. (1970). Parachlorophenylalamine and habituation to repetitive auditory startle stimuli in rats. Physiol. Behav., 5, 1215-1219.

Critchley, M.A.E. & Handley, S.L. (1986). 5-HT$_{1A}$ ligand effects in the X-maze anxiety test. Br. J. Pharmacol. (In press.)

Csanalosi, E., Schweizer, E., Case, W. & Rickels, K. (1987). Gepirone in anxiety: a pilot study. J. of Clin. Psychopharmacol., 7, 131-193.

Davis, M. (1980). Neurochemical modulation of sensory-motor reactivity: acoustic and tactile startle reflexes. Neurosci. Biobehav. Rev., 4, 241-263.

Davis, M., Astrachan, D.I., Gendelman, P.M. & Gendelman, D.S. (1980a). 5-methoxy-N,N-dimethyl- tryptamine: spinal cord and brainstem mediation of excitatory effects on acoustic startle. Psychopharmacol., 70, 123-130.

Davis, M., Astrachan, D.I. & Kass, E. (1980b). Excitatory and inhibitory effects of serotonin on sensorimotor reactivity measured with acoustic startle. Science, 209, 521–523.

Davis, M., Cassella, J.V., Wrean, W.H. & Kehne, J.H. (1986). Serotonin receptor subtype agonists: differential effects on sensorimotor reactivity measured with acoustic startle. Psychopharmacol. Bull., 22, 837–843.

Davis, M., Gendelman, D.S., Tischler, M.D. & Gendelman, P.M. (1982). A primary acoustic startle circuit: lesion and stimulation studies. J. Neurosci., 2, 791–805.

Davis, M. & Sheard, M.H. (1974). Habituation and sensitization of the rat startle response: effects of raphe lesions. Physiol. Behav., 12, 425–431.

Davis, M. & Sheard, M.H. (1976). p-chloramphetamine (PCA): acute and chronic effects on habituation and sensitization of the acoustic startle response in the rat. Eur. J. Pharmacol., 35, 261–273.

Eison, M.S. & Eison, A.S. (1984). Buspirone as a midbrain modulator: anxiolysis unrelated to traditional benzodiazepine mechanisms. Drug Dev. Res., 4, 109–119.

Engel, J.A., Hjorth, S., Svensson, K., Carlsson, A., & Liljequist, S. (1984). Anticonflict effect of the putative serotonin receptor agonist 8-hydroxy-2-(di-n-propylamino)tetralin (8-OH-DPAT). Eur. J. Pharmacol., 105, 365–368.

Essman, W.B. (1971). Drug effects and learning and memory processes. In: Advances in Pharmacology and Chemotherapy. (Eds. Garatini, S. & Shore, P.), pp 281–298. New York: Academic Press.

Essman, W.B. (1973). Age-dependent effects of 5-hydroxytryptamine upon memory consilidation and protein synthesis. Pharm. Biochem. Behav., 1, 7–14.

Fechter, L.D. (1974). Central serotonin involvement in the elaboration of the startle reaction in rats. Pharmacol. Biochem. Behav., 2, 161–171.

Feighner, J.P. (1980). Trazodone, a triazolopyridine derivative in primary depressive disorder. J.Clin. Psychiat., 41, 250–255.

File, S.E. (1984). The neurochemistry of anxiety. In: Antianxiety agents. (Eds. Burrows, G.D., Norman, T.R. & Davies, B.), pp. 13–32. Amsterdam: Elsevier.

File, S.E. (1985). Models of anxiety. J. Clin. Practice, 27, 15–19.

File, S.E., Johnston, A.L. & Pellow, S. (1987). Effects of compounds acting at CNS 5-hydroxy-tryptamine systems on anxiety in the rat. Br. J. Pharmacol., 90, 265P.

Fuller, R.W., Perry, K.W. & Molloy, B.B. (1974). Effect of an uptake inhibitor on serotonin metabolism in rat brain: Studies with 3-(p-trifluoromethylphenoxy)-N-methyl-3-phenylpropylamine (Lilly 1101040).

Life Sci., 15, 1161-1171.

Gardner, C.R. (1985). Pharmacological studies of the role of serotonin in animal models of anxiety. In: Neuropharmacology of Serotonin. (Ed. Green, A.R.), pp. 281-325. Oxford: Oxford University Press.

Gardner, C.R. (1986). Recent developments in 5-HT-related pharmacology of animal models of anxiety. Pharmacol. Biochem. Behav., 24, 1479-1485

Geyer, M.A., Petersen, L.R. & Rose, G.J. (1980). Effects of serotonergic lesions on investigatory responding by rats in a holeboard. Behav. Neural Biol., 30, 160-177.

Geyer, M.A., Puerto, A., Menkes, D.B, Segal, D.S. & Mandell, A.J. (1976). Behavioural studies following lesions of the mesolimbic and mesostriatal serotonergic pathways. Brain Res., 106, 257-270.

Geyer, M.A., Warbritton, J.D., Menkes, D.B., Zook, J.A., Mandell, A.J. (1975). Opposite effects of intraventricular serotonin and bufotenin on rat startle responses. Pharmacol. Biochem. Behav., 3, 687-691.

Gloger, S., Grunhaus, L. & Birmacher, B., Troudart, T. (1981). Treatment of spontaneous panic attacks with chlorimipramine. Am. J. Psychiat., 138, 1215-1217.

Goa, K.L. & Ward, A. (1987). Buspirone: a preliminary review of its pharmacological properties and therapeutic efficacy as an anxiolytic. Drugs, 32, 114-129.

Goldberg, H.L. & Finnerty, R.J. (1979). The comparative efficacy of buspirone and diazepam in the treatment of anxiety. Am. J. Psychiat., 136, 1184-1187.

Hillegaart, V., Ahlenius, S. & Larsson, K. (1989). Effects of local application of 5-HT into the median and dorsal raphe nuclei on male rat sexual and motor behavior. Behav. Brain Res., in press.

Hjorth, S., Söderpalm, B. & Engel, J.A. (1987). Biphasic effect of L-5-HTP in the Vogel conflict model. Psychopharmacol., 92, 96-99.

Iversen, S.D. (1984). 5-HT and anxiety. Neuropharmacol., 23, 1553-1563.

Jacobsen, A.F., Dominguez, R.A., Goldstein, B.J. & Steinbook, R.M. (1985). Comparison of buspirone and diazepam in generalized anxiety disorder. Pharmacotherapy, 5, 290-296.

Johnston, A.L. & File, S.E. (1986). 5-HT and anxiety: promises and pitfalls. Pharmacol. Biochem. Behav., 24, 1467-1470.

Joyce, D. & Hurwitz, H.M.B. (1964). Avoidance behaviour in the rat after 5-hydroxytryptophan (5-HTP) administration. Psychopharmacol., 5, 424-430.

Kahn, R.S. & Westenberg, H.G.M. (1985). L-5-hydroxy-tryptophan in the treatment of anxiety disorders. J. Affective Disorders, **8**, 197-200.

Kennett, G.A. & Curzon, G. (1988). Evidence that mCPP may have behavioural effects mediated by central 5-HT_{1C} receptors. Br. J. Pharmacol. (In press.)

Lebrecht, U. & Nowak, J.Z. (1980). Effect of single and repeated electroconvulsive shock on serotonergic system in rat brain: II Behavioural studies. Neuropharmacol., **11**, 1055-1061.

McCloskey, T.C., Paul, B.K. & Commissaris, R.L. (1987). Buspirone effects in an animal conflict procedure: comparison with diazepam and phenobarbital. Pharmacol. Biochem. & Behav., **27**, 171-175.

McKinney, W.T. & Bunney, W.E. (1969). Animal model of depression: review of evidence and implications for research. Arch. Gen. Psychiat., **21**, 240-248.

McMillan, B.A. & Mattiace, L.A. (1983). Comparative neuropharmacology of buspirone and MJ-13805, a potential anti-anxiety drug. J. Neural Transm., **57**, 255-265.

Merlo Pich, E. & Samanin, R. (1986). Disinhibitory effects of buspirone and low doses of sulpiride and haloperidol in two experimental anxiety models in rats: possible role of dopamine, Psychopharmacol., **89**, 125-130.

Miczek, K.A. (1987). The psychopharmacology of aggression. In: Handbook of Psychopharmacology. Vol. 19. Behavioural Pharmacology. (Eds. Iverssen, L.L., Iversen, S.D. & Snyder, S.H.), pp. 183-228. New York, Plenum Press.

Montanaro, N., Dall'Olio , R. & Gandolfi, O. (1979). Bromolysergide and methysergide protection against ECS-induced retrograde amnesia. Neuropsychobiol., **5**, 174-180.

Montanaro, N., Dall'Olio, R & Gandolfi, O. (1981). Reduction of ECS-induced retrograde amnesia of passive avoidance conditioning after 5,7-dihydroxytryptamine median raphe nucleus lesion in the rat. Neuropsychobiol., **7**, 57-66.

Morinan, A. & Parker, V. (1985). Are socially isolated rats anxious? Br. J. Pharmacol., **86**, 420P.

Nieuwenhuys, R. (1985). Chemoarchitecture of the brain. Springer-Verlag, Berlin.

Ogren, S.O., Ross, S.B., Hohn, A.C. & Bauman, L. (1977). 5-Hydroxytryptamine and avoidance performance in the rat. Antagonist of the acute effect of p-chloroamphetamine by zimelidine, an inhibitor of 5-hydroxytryptamine uptake. Neurosci. Lett., **7**, 331-336.

Parker, V. & Morinan, A. (1986). The socially-isolated rat as a model for anxiety. Neuropharmacol., **25**, 663-664.

Pazos, A. & Palacios, J.M. (1985). Quantitative autoradiographic mapping of serotonin receptors in the rat brain. I. Serotonin-1 receptors. Brain Res., **346**, 205-230.

Peroutka, S.J. (1985). Selective interaction of novel anxiolytics with 5-hydroxytryptamine 1A receptors. Biol. Psychiat., **20** 971-979.

Rickels, K., Waissman, K., Norstad, N., Singer, M., Stoltz, D., Brown, A. & Danton, I. (1982). Buspirone and diazepam in anxiety: a controlled study. J. Clin. Psychiat., **43**, (12) 81-86.

Roffman, M. & Lal, H. (1971). Facilatory effect of amphetamine on learning and recall of an avoidance response in rats. Arch. Intern. Pharmacol. Ther., **193**, 87-91.

Rowan, M.J., Cullen, W.K., Moulton, B. & Anwyl, R. (1987). Buspirone impairment of passive avoidance but not spatial learning in the rat. First Intern. Congress on the Behav. Pharmacol. of 5-HT (Amsterdam), P57.

Schweizer, E., Amsterdam, J., Rickels, K., Myron, K. & Droba, M. (1986). Open trial of buspirone in the treatment of major depressive disorder. Psychopharmacol. Bull., **22**, 183-185.

Svensson, L. (1985). Effects of 8-OH-DPA, lisuride and some ergot-related compounds on the acoustic startle response in rat. Psychopharmacol., **85**, 469-475.

Svensson, L. & Ahlenius, S. (1983). Enhancement by the putative 5-HT receptor agonist 8-OH-2-(di-n-propylamino) tetralin of the acoustic startle response in the rat. Psychopharmacol., **79**, 104-107.

Thiebot, M.H. (1986). Are serotonergic neurons involved in the control of anxiety and in the anxiolytic activitity of benzodiazepines? Pharmacol. Biochem. Behav., **24**, 1471-1477.

Treit, D. (1985). Animal models for the study of antianxiety agents: a review. Neurosci. & Biobehav. Rev., **9**, 203-222.

Treit, D., Pinel, P.J.P. & Fibiger, H.C. (1981). Conditioned defensive burying: a new paradigm for the study of anxiolytic agents. Pharmacol. Biochem. Behav., **15**, 619-626.

Trulson, M.E. & Jacobs, B.L. (1976). Behavioural evidence for the rapid release of CNS serotonin by PCA and fenfluramine. Eur. J. Pharmacol., **36**, 149-154.

Wadenberg, M.L. & Ahlenius, S. (1988). The putative 5-HT$_{1A}$ agonist 8-OH-DPAT affects reaction time as a function of stimulus intensity in a discriminative task in the rat. J. Neural Transm., in press.

Walters, J.K., Davis, M. & Sheard, M.H. (1979). Tryptophan free diet: effects on the acoustic startle reflex in rats. Psychopharmacol., **62**, 103-109.

Winslow, J.T. & Miczek, K.A. (1985). Social status as a determinant of alcohol effects on aggressive behaviour in Squirrel monkeys (Saimiri sciureus). Psychopharmacol., **85**, 953–958.

Witkin, J.M. & Barrett, J.E. (1986). Interaction of buspirone and dopaminergic agents on punished behaviour of pigeons. Pharmacol. Biochem. Behav., **24**, 751–756.

Woolley, D.W. & Van der Hoeven, T. (1963). Alteration of learning ability caused by changes in cerebral serotonin and catecholamines. Science, **139**, 610–611.

Young, R., Urbanic, A., Emrey, T.A., Hall, P.C. & Metcalf, G. (1987). Behavioural effects of several new anxiolytics and putative anxiolytics. Eur. J. Pharmacol., **143**, 361–371.

Author Index

A

Aaki, N., 206, 221
Abbot, D. H., 125, 126f, 144
Abbot, F. V., 311, 312
Abe, Y., 121f, 143
Abel, M. S., 185, 202
Abou-Hamed, H., 425, 442
Abrahams, V. C., 425, 437
Aceves, J., 378, 381
Ackerman, S., 262, 263, 270, 275
Adams, D. B., 124, 131
Adams, L., 245, 246
Adler, L. E., 206, 220, 212
Advokat, C., 464t, 469
Aghajanian, G. K., 40, 52, 55, 56, 59, 60, 61, 63, 64, 66, 67, 68, 70, 81, 82, 117, 131, 161, 165t, 177, 187, 200, 202, 338, 352, 463
Agmo, A., 377, 378, 380
Agnati, L., 43, 50, 57, 66
Agnati, L. F., 27, 28, 30, 31, 160, 170–171
Agren, 158t, 159t
Aguiar, J. C., 436
Ahlenius, S., 35, 37, 38, 38f, 39, 40, 41, 43, 44, 45f, 46, 47, 48, 50, 51, 53, 58, 62, 64, 66, 73, 76, 108, 461, 463, 464t, 468, 471, 473
Akesson, CH., 79, 82
Akil, H., 309, 312
Akiskal, 184
Alagna, S., 255
Albe-Fessard, D. G., 301, 317
Albert, D. J., 108, 161, 168
Aleotti, A., 254, 255, 260, 273
Alexander, M., 425, 437, 442
Algeri, S., 264, 274
Alkondon, M., 120f, 140
Allikmets, L. K., 119, 120f, 131
Alling, C., 158, 171
Alling, G., 184, 201
Almay, B. G. L., 301, 312
Alterman, I., 434, 435, 440
Altman, J. L., 120, 121f, 138
Al-Yassiri, M. M., 195, 199
Amano, K., 425, 437
Amin, M. M., 186, 199
Ammable, L., 206, 218
Amsterdam, J., 327, 334, 460, 462, 473
Amsterdam, J. D., 227, 228–229, 327, 330
Amsterdam, K., 346, 347, 351–352
Anand, M., 120f, 131
Ananth, J., 436, 437

Ananth, J. V., 186, 199
Ancill, R. J., 183, 199
Anden, N. E., 39, 43, 48
Anderson, C., 163, 173
Anderson, E. G., 305, 317
Anderson, G. H., 251, 256-257
Anderson, G. M., 158, 171
Anderson, H. A., 250, 256
Anderson, I. M., 191, 192, 193t, 199
Andersson, K., 160, 170-171
Andrews, M. J., 123, 139
Aneshensel, C. S., 164, 168
Ang, J., 161, 177
Angrist, B. M., 206, 218
Angst, J., 162, 168
Anisman, H., 163, 164, 168
Ankier, S. I., 195, 199
Antelman, S., 206, 223
Antelman, S. A., 269, 270, 271
Antin, J., 292, 293
Antkiewicz, L., 212, 220
Anton, Tay, F., 378, 381
Anwyl, R., 461, 473
Appel, J. B., 37, 49, 62, 65, 120, 121f,
  138, 157, 170, 399, 400, 403, 406, 407,
  409, 421, 410t, 413, 414t, 416t, 418t,
  419, 422, 445, 446, 447, 449, 452, 453,
  456
Appel, J. H. B., 399, 403-404
Apperley, E., 353, 356
Applegate, C. D., 89, 108, 121f, 131-132
Aprison, M. H., 157, 168
Araki, H., 121f, 143
Arasteh, K., 337, 338, 352
Archer, T., 162, 173, 299, 301, 306, 307,
  308, 310, 311t, 312-313, 313, 317, 397,
  459, 460, 465, 466, 468
Arendash, 61
Ariens, E. J., 415, 416, 421
Arieti, S., 166, 168
Arneric, S. P., 41, 48, 49
Arnt, J., 21, 23, 162, 171, 208, 213, 407,
  409, 410t, 414t, 416t, 418t, 421
Aron, C., 389, 394
Artaud, F., 27, 31
Arvidsson, L. E., 5, 7, 19, 35, 37, 38f,
  39, 40, 43, 44, 46, 47, 48, 51, 53, 90,
  91, 106, 108, 109, 206, 207, 220, 377,
  380, 381
Arwestrom, E., 311t, 312-313
Asberg, M., 125, 132, 158, 158t, 159,
  159t, 161, 162, 164, 165, 166, 168, 172,
  176, 184, 199
Ascroft, G. W., 158t, 164, 168-169
Ashcroft, G. W., 158t, 161, 176
Asnis, G. M., 349, 352, 435, 443
Assandri, A., 265t, 272
Astrachan, D. I., 464t, 469, 470
Atack, C. V., 91, 109
Atrens, D. M., 121f, 139
Atsmon, A., 206, 218
Audi, E. A., 105, 110, 403, 404, 426, 430,
  431, 436, 437, 439
Avery, G. S., 259, 278
Awouters, F., 21, 24, 236, 237
Azami, K., 309, 316
Azmitia, E., 76, 188, 203
Azmitia, E. C., 122, 132, 260, 271
Azrin, N. H., 103, 114
Azumi, K., 207, 223
Azzaro, A. J., 161, 165t, 169

B

Backer, B. J., 262, 263, 273
Baenninger, R., 106, 109
Bagri, A., 426, 442-443, 431
Bailey, H. E., 353, 356, 359
Bailey, W. H., 163, 177
Baldessarini, R. J., 206, 207, 218, 219
Ballenger, J. C., 125, 132-133, 166, 169,
  328, 333
Balster, R. L., 407, 421
Banerjee, U., 121f, 132
Banki, C. M., 158, 158t, 159t, 161, 169,
  462, 469
Barbaz, B. J., 40, 50
Barbeau, H., 12, 16
Barchas, 127t
Barchas, J., 119, 120f, 133
Barchas, J. D., 120f, 133, 464t, 469
Bardelay, C., 81, 82
Bareggi, S. R., 214, 222
Barfield, R. J., 101, 110
Barnes, N. M., 239
Barnett, A., 11, 19, 118f, 137, 206, 221
Barnett, S. A., 98, 109
Barone, D., 265t, 272
Barr, G. A., 121f, 134
Barrett, J. E., 321, 322, 330, 377, 380,
  449, 451, 452, 460, 474
Barrett, J. W., 436, 438
Barrett, R. J., 399, 400, 403, 404
Barry, H. III, 398, 403

AUTHOR INDEX

Barry, J. M., 385, 387
Bartfai, T., 28, 31, 301, 314
Bartness, T. J., 260, 280
Batta, S. K., 402, 405–406
Battaglia, G., 60, 70
Bauman, L., 465, 472
Beach, F. A., 36, 48
Bean, N. J., 259, 269, 274
Beck, A. T., 166, 169, 186, 203
Becker, J., 166, 169
Bedard, P., 11, 12, 16, 79, 81
Bednarcyk, B., 12, 16
Beecher, H., 302, 313
Beer, B., 354, 359
Behbehani, M. M., 309, 315
Beidel, D. C., 328, 334
Belcher, G., 306, 309, 313
Belenberg, A. J., 206, 219
Bell, J., 206, 212, 220
Bell, R., 89, 109, 120f, 132
Bellissimo, D. B., 264, 278
Bendotti, C., 254, 255, 260, 263, 264, 266, 268, 269, 270, 271, 272, 273, 275, 279, 284, 285, 338, 340, 343, 349
Benfenati, F., 28, 30, 160, 170–171
Benitez-King, G., 378, 381
Berendsen, 85
Berendsen, H. G., 83, 84, 86
Berendsen, H. H. G., 83
Beretta, C., 260, 270, 272, 275–276
Berge, O-G., 120f, 121f, 134, 260, 272, 301, 302, 303, 304, 305, 306, 307, 308, 309, 310, 311, 311t, 312, 313, 314, 315, 316, 317, 318
Berger, 158t
Berger, B. D., 165, 169, 179, 204, 321, 334, 354, 359, 436, 443
Berger, D., 165, 174
Berger, R., 454, 457
Berkelmans, B., 89, 92t, 97, 103, 104, 112, 215, 222, 390, 394, 395
Bernasconi, S., 89, 114, 119, 121f, 142
Bertalmio, A. J., 407, 408, 424
Bertillson, L., 158t, 159t, 165, 176
Bertilson, L., 158, 168
Bertilsson, L., 166, 172, 184, 199
Bertilsson, P., 164, 176
Berwish, N., 227, 327, 330
Besson, J. M., 179, 202, 304, 306, 314, 316
Bethea, C. L., 367, 368, 369
Bevan, P., 33, 87, 89, 92t, 97, 103, 104, 112, 215, 222, 390, 394, 395, 459
Bhargava, K. P., 120f, 131
Bhat, A. V., 186, 203
Bhatnagar, R. K., 41, 48
Bhattacharya, S. K., 306, 314
Bidzinski, A., 149, 152
Biegon, A., 76
Biggio, G., 214, 222
Bignami, G., 208, 218
Billey-Nichuck, J., 208, 219
Binns, J. K., 164, 168–169
Birley, J. L. T., 207, 221–222
Birmacher, B., 328, 332, 434, 439, 462, 471
Bischoff, S., 212, 222
Bitran, D., 35, 48
Bjorklund, H., 304, 315
Blackburn, T. P., 21, 22, 23, 24
Blackshear, M. A., 196, 199, 399, 400, 403
Blakeborough, L., 15, 20, 39–40, 52
Blanchard, 425, 440
Blanchard, C., 425, 440
Blanchard, D. C., 93, 98, 103, 109, 119, 132, 145, 147, 148, 425, 437, 465, 469
Blanchard, R. J., 93, 98, 103, 109, 119, 132, 145, 147, 148, 425, 437, 465, 469
Blas, C., 338, 352, 354, 359
Bleich, A., 349, 352
Blessed, G., 159, 160, 170, 185, 203
Blier, P., 61, 64, 161, 162, 165t, 169, 170, 227, 228, 338, 342, 349
Blum, I., 206, 218
Blum, K., 179, 200
Blundell, J. E., 249, 251, 252, 253, 255, 256, 259, 260, 261, 262, 263, 264, 268, 272, 276, 277, 287, 288, 295, 296, 298, 467, 469
Blurton, P. A., 466, 469
Boakes, R. J., 59, 64
Bockaert, J., 193t, 204
Bocknik, S. E., 121f, 132, 136
Boehrer, A., 89, 114
Boggan, W. O., 120, 121f, 138
Boissard, C. G., 119, 138
Boissier, J. R., 118f, 132, 389, 394
Bolles, R. C., 465, 469
Bonica, J. J., 301, 317
Booth, D. A., 262, 263, 273
Boradhurst, P. L., 377, 381
Borsini, E., 264, 279
Borsini, F., 254, 255, 260, 262, 263, 264,

270, 273, 278, 279, 284, 285
Bos, E. R. H., 162, 176
Bowden, C. L., 159, 172
Bowers, M. B., 158*t*, 159*t*, 161, 172
Bowery, N. G., 22, 24
Bowler, K., 328, 333
Boyar, W. C., 62, 69
Bozarth, M. A., 425, 443
Bradford, D., 186, 203
Bradford, L. D., 98, 99, 102, 107, 108, 109, 113, 206, 207, 208, 223
Bradley, P. B., 4, 16, 56, 59, 64, 65, 353, 356, 412, 421
Brady, L. S., 321, 330
Braff, D. L., 243, 245, 246
Brain, M., 207, 219
Brain, P. F., 93, 109
Brambilla, F., 225, 228
Brammer, G. L., 125, 129, 130, 140
Branchey, M. H., 206, 222
Brandao, M. L., 105, 110, 403, 404, 426, 431, 433, 436, 437–438, 439, 442–443
Brase, D. A., 446, 448
Braun, L. D., 30, 31
Brazier, M. A. B., 425, 438
Breese, G. R., 121*f*, 132
Breier, A., 328, 330
Bresler, D. E., 120*f*, 133
Bressa, G. M., 237
Bridger, W. H., 121*f*, 134
Bridges, K., 183, 201
Bridges, P. K., 195, 199
Bridges, P. V., 206, 219
Briggs, I., 59, 64
Brigham, P., 161, 170
Briley, M., 321, 330, 340, 345, 350, 371, 373, 374, 401, 404, 449, 452, 461, 469
Brittain, R. T., 183, 199, 214, 218, 353, 356
Broch, O. J., 306, 307, 315, 318
Brodie, B. B., 117, 132
Brodin, E., 27, 30, 301, 314
Broekkamp, C. L., 163, 169, 321, 326, 330, 361, 365
Broekkamp, C. L. E., 83, 84, 86
Brogden, R. N., 259, 278
Broussy, L., 118*f*, 136
Brown, A., 462, 473
Brown, C. A., 429, 440–441
Brown, G. C., 166, 169
Brown, G. L., 125, 132, 132–133, 159, 159*t*, 160, 161, 164, 165, 169

Brown, G. W., 164, 170
Brown, J., 79, 82
Brown, K., 89, 109, 120*f*, 132
Brown, S. L., 349, 352, 435, 443
Browne, G., 425, 442
Bruch, H., 269, 270, 273
Bruegger, M., 425, 440
Brydon, J., 252, 257
Bryner, C. A., 262, 277
Buczko, W., 259, 275
Budhram, P., 322, 330, 361, 364, 365
Buech, O., 212, 222
Bullard, W. P., 259, 275
Bullock, T. H., 123, 133
Bunney, W. E., 125, 132, 462, 472
Burg, W. van den., 158, 172
Burke, J. C., 121*f*, 134
Burke, J. D., 328, 334
Burt, J., 121*f*, 139
Burton, M. J., 260, 262, 269, 273, 274, 275, 277
Butcher, L. L., 39, 48, 118*f*, 120*f*, 133, 134
Butcher, S. G., 39, 43, 48
Butler, A., 183, 199, 214, 218, 241, 242, 353, 355, 356, 358
Butler, K., 434, 442
Buus Lassen, J., 321, 334, 322
Buxton, D. A., 377, 381
Bymaster, F. P., 43, 48–49

C

Caccia, S., 260, 264, 266, 275, 279, 445, 447
Caetano, 434
Caggiula, A. R., 269, 270, 271
Cairncross, K. D., 162, 163, 170
Calas, A., 27, 30–31
Callahan, P. M., 37, 49, 410*t*, 414*t*, 422, 445, 446, 447, 449, 452, 453, 456
Camba, R., 121*f*, 133
Campeas, R. B., 328, 332
Camps, F. E., 159, 175
Candalaresi, G., 264, 279
Candolfi, F., 225, 228
Candy, J. M., 159, 160, 170
Cannon, J. G., 41, 48
Carder, B., 425, 443
Carey, R. J., 259, 273
Carli, M., 340, 342, 347, 350, 461, 469
Carlsson, A., 5, 7, 19, 35, 37, 38*f*, 39, 40,

AUTHOR INDEX

41, 43, 44, 46, 47, 48, 49, 50, 51, 53, 91, 108, 109, 160, 170, 206, 207, 218, 220, 322, 331, 338, 340, 343, 350, 377, 380, 381, 461, 468, 470
Carlsson, M., 46
Carlton, J., 260, 263, 279
Carlton, P. L., 464t, 469
Carlton, S. M., 310, 316
Carmon, A., 301, 317
Carney, J. M., 401, 405
Carney, M. W. P., 183, 199
Carobrez, A. B., 426, 438
Carrer, H., 59, 68
Carrive, P., 426, 431, 442–443
Carruba, M. O., 260, 273
Carside, R. F., 183, 199
Carstens, E., 309, 314
Carter, C. J., 212, 218
Carter, R. B., 157, 170, 399, 403–404
Carwley, J., 389, 395
Case, W., 345, 350, 462, 469
Caspar, R. C., 158, 175–176
Cassella, J. V., 463, 464t, 470
Casu, M., 214, 222
Cavagnini, F., 225, 228
Cavanaugh, J., 206, 220
Cavicchioli, L., 160, 170–171
Cervo, L., 270, 273, 307, 317
Ceuleman, D. L. S., 180, 180f, 199, 436, 438, 462, 469
Chadwick, D., 161, 176, 354, 358
Challoner, T., 356, 357
Chambers, D. R., 159, 160, 170, 184, 185, 200
Chamove, A. S., 129, 133
Chance, W. T., 399, 406
Chan-Palay, V., 27, 30
Chaouch, A., 304, 314
Chaouloff, F., 163, 172, 225, 228
Chapman, W. P., 425, 438
Chaput, Y., 161, 162, 165t, 170
Charanjit, S., 196, 199
Charig, E. M., 191, 192, 193t, 199, 200
Charney, D. S., 159t,, 161, 170, 172, 189, 191, 199–200, 202, 328, 330, 434, 438, 462, 469
Charry, J. M., 163, 177
Chase, S., 398, 404
Chase, T. N., 164, 171
Checkley, S. A., 253, 257, 271, 278
Chiocchio, S. R., 76, 77
Chojnmacka-Wojcik, E., 417, 423

Chopin, P., 321, 330, 340, 345, 350, 371, 373, 374, 401, 404, 449, 452, 461, 469
Chorev, M., 27, 32
Chouinard, G., 206, 218
Ciarenello, R. D., 41, 43, 49
Claassen, V., 186, 200
Claassen, V. C., 97, 109
Clark, D., 41, 46, 49
Clark, J. T., 41, 43, 49, 53
Clarke, A., 427, 438
Clement, L. G., 56, 59, 65
Clineschmidt, B. V., 62, 259, 273, 328, 330
Clinical Research Centre, 183, 200
Clody, D. E., 354, 359
Coates, I. H., 183, 199, 214, 218, 353, 356
Coe, 127t
Cohen, A., 120f, 135
Cohen, D. J., 158, 171
Cohen, M. L., 58, 61, 65
Cohen, M. S., 63, 65
Cohen, R. M., 196, 199, 434, 440
Cole, B. T., 120f, 121f, 137
Cole, J. O., 161, 170
Cole, W., 251, 255
Coleman, A., 186, 204
Coleman, B. S., 186, 199
Collinge, J., 354, 356
Collins, D., 353, 356
Collins, P. J., 254, 256
Colpaert, F. C., 11, 16, 198, 200, 235, 236, 237, 238, 321, 322, 325, 331, 333, 398, 400, 404, 407, 408, 409, 410, 411, 415, 416, 417, 420, 421, 421–422, 422, 429
Commissaris, R. L., 321, 322, 333, 340, 341, 351, 354, 356, 461, 472
Conn, P. J., 55, 70, 93, 94t, 109
Connell, P. H., 206, 218
Conner, R. J., 464t, 469
Conner, R. L., 120f, 133
Connor, J. D., 79, 82
Consolo, S., 260, 266, 275
Conte, H., 161, 176
Conti, I., 445, 447
Conti, L., 327, 331
Cook, L., 120f, 141, 208, 219, 321, 331, 334, 427, 442
Cools, A. R., 1, 153, 247
Coombes, R. C., 356, 357
Cooper, B. R., 121f, 132

Cooper, S. J., 159, 160, 170, 184, 185, 200, 262, 273, 288, 289
Cooper, S. M., 353, 356
Coppen, 158*t*
Cordon, J. J., 321, 322, 333
Corne, S. J., 11, 12, 16
Cornfeldt, M. L., 208, 219
Cornford, E. M., 30, 31
Corrigan, S., 163, 177
Corrodi, H., 39, 43, 48, 160, 170
Corsellis, J. A. N., 160, 170
Cortes, R., 21, 24, 89, 108, 113
Coryell, W., 225, 228
Coscina, D. V., 121*f*, 134
Cossery, J-M., 266, 268, 275
Costa, E., 5, 16
Costa e Silva, J. A., 434, 443
Costall, B., 58, 65, 183, 202, 206, 212, 219, 239, 242, 270, 273, 353, 354, 355, 356, 357, 358, 359, 365, 383, 385, 387, 389, 391, 394, 395
Cote, T. E. D., 39, 43, 53
Cotecchia, S., 264, 276
Coughlan, J., 268, 273, 293
Cousty, D., 81
Covi, L., 183, 202
Cowen, P. J., 190, 191, 192, 193*t*, 199, 200, 201, 202, 203, 251, 256
Cox, B., 21, 22, 24, 162, 163, 170, 353, 357
Craddock, D., 253, 256
Crane, P. D., 30, 31
Crawford, T. B. B., 164, 168-169
Crawley, J., 391
Crawley, J. N., 347, 350
Creese, I., 212, 219
Criborn, C. O., 43, 44, 47, 49
Critchley, M. A. E., 236, 237, 325, 331, 371, 374, 461, 469
Cross, A. J., 159, 160, 170, 185, 185*t*, 200, 203
Crow, T. J., 159, 160, 170, 184, 185, 185*t*, 200, 203
Crowley, W. R., 35, 49, 56, 57, 71, 76
Croy, D. J., 321, 332
Csanalosi, E., 462, 469
Csanalosi, I., 345, 350
Cullen, W. K., 461, 473
Cummings, S. W., 161, 176
Cunningham, D., 356, 357
Cunningham, K. A., 37, 49, 62, 65, 400, 403, 407, 409, 410*t*, 413, 414*t*, 416*t*,
418*t*, 419, 421, 422, 445, 446, 447, 449, 452, 453, 456
Cunningham Owens, D. G., 183, 202
Curtis-Prior, P. B., 263, 273-274
Curzon, G., 5, 19, 44, 46, 47, 50, 62, 63, 65, 67, 79, 81, 83, 86, 108, 110, 125, 140, 151, 152, 163, 172, 197, 202, 206, 219, 225, 226, 227, 228, 268, 269, 274, 283, 284, 285, 287, 288, 289, 291, 293, 294, 306, 316, 318, 338, 340, 343, 346, 347, 350, 351, 364, 366, 466, 472
Cytawa, J., 262, 274
Czlonkowski, A., 121*f*, 135

D

da Prada, M., 41, 52
Dahle, H., 308, 314
Dahlof, L. G., 44, 45*f*, 49, 50
Dakoski, 193*t*
Dal Tose, G., 264, 274
Dalessandro, M., 121*f*, 142
Dall'Olio, R., 466, 472
Dallob, A., 11, 16
Danish University Antidepressant Group, 186, 200
Danton, I., 462, 473
Danysz, W., 310, 313
Darcourt, G., 186, 199
Daruna, J. G., 89, 109
Darwin, C., 145, 148
Daugherty, M., 118*f*, 133
Daval, G., 8, 20, 35, 51, 266, 280, 337, 338, 352
Davey, A., 183, 199
Davidson, A. B., 208, 219
Davidson, J., 206, 219
Davidson, J. M., 41, 43, 49, 53
Davies, J., 212, 219, 302, 314
Davies, J. E., 97, 109
Davies, M., 306, 307, 310, 311*t*, 317
Davies, M. A., 377, 378, 381
Davies, R. F., 259, 269, 274
Davies, S. O., 328, 332
Davis, D., 160, 174
Davis, G. A., 57, 61, 65, 70
Davis, J. D., 262, 274
Davis, J. M., 158, 159, 172, 175-176
Davis, J. N., 91, 109
Davis, K. H., 61, 62, 67
Davis, M., 89, 113-114, 120*f*, 141, 463, 464*t*, 469, 470, 473

AUTHOR INDEX

Dawsey, W. J., 259, 275
De Aguiar, J. C., 105, 113, 321, 333, 427, 428f, 429, 433, 435, 437–438, 441, 443
De Blasi, A., 264, 276
de Haan, S., 164, 176
De Hey, A. M. D. M., 101, 111
De Jonge, A., 83, 86
de la Riva, C., 125, 126f, 144
De la Vega, E., 191, 203
de las Nieves Parisi, M., 76, 77
De Montigny, C., 61, 64, 161, 162, 165t, 169, 170, 187, 200, 227, 228, 338, 342, 349
De Simoni, M. G., 264, 274
De Souza, R. J., 44, 51, 83, 86
De Vos-Frerichs, T. P., 100, 111
De Wied, D., 206, 223
de Wit, H., 270, 281
De Witte, P., 429
De Witte, P. H., 398, 404
Deacon, R., 321, 322, 330, 331, 340, 350
Deakin, J. F. W., 5, 16, 160, 170, 179, 200
Dealy, R., 328, 333
Dedek, J., 158, 173
DeFeo, J. J., 120f, 136
Deisz, R., 119, 139–140
Dekirmenjian, H., 129, 140
del Fiacco, M., 39, 53
Del Ponte, A., 253, 257
del Rio, J., 164, 171
Delecluse, F., 237
Delgado, J. M. R., 425, 438
Delini-Stula, A., 118f, 119, 120f, 121f, 133, 138, 212, 222
Delitala, G., 255
Delizio, R. D., 365, 366
Della Loggia, F., 253, 257
Delwaide, P. J., 237, 238
Demassey, Y., 22, 24–25, 81, 82, 284, 286
Dement, W., 207, 223
Dement, W. C., 120f, 133
Den Boer, J. A., 434, 435, 438
Dencker, S. J., 158t, 164, 170
Denenberg, V. H., 101, 110
Dennis, S. G., 303, 314
DePaulis, A., 89, 114
Desa, A., 288, 289
Deshmukh, P. P., 21, 24, 35, 50
DeSouza, R. J., 7, 8, 9f, 13, 15f, 16, 17, 196, 200, 201
Deutch, A. Y., 368, 369

Devanzo, J. P., 118f, 133
Devilla, L., 255
Dewey, W. L., 446, 448
Di Chiara, G., 121f, 133
Di Scala, G., 426, 433, 442, 442–443, 431
Dickenson, A. H., 304, 306, 314, 316
Dickenson, A. M., 179, 202
Dickinson, M. H., 122, 137
Dickinson, S., 197, 202
Dickinson, S. L., 79, 81, 225, 228
Dietrich, A. P., 120f, 133
Dixon, A. K., 103, 109
Dlabac, A., 40, 52
D'Mello, G. D., 401, 404
Dlohozkova, N., 118f, 136
Dobson, C., 208, 219
Dols, L. C. W., 162, 176
Domeney, A. M., 183, 202, 206, 219, 239, 242, 353, 354, 355, 357, 358, 359, 365, 383, 387, 389, 391, 394, 395
Dominguez, R. A., 345, 351, 462, 471
Dompert, W. U., 35, 50–51, 377, 378, 380, 381
Dompert, W. V., 377, 378, 379, 380
Donald, P., 166, 175
Donaldson, S. R., 206, 219
Donat, P., 117
Donatsch, P., 59, 69–70, 239, 242, 353, 359
Donohoe, T. P., 225, 228, 283, 285, 288, 289, 364, 366
Doods, H. N., 83, 86
Doran, A., 164, 174
Dorovini-Zis, K., 158, 175
Dosen, E., 11, 19
Dourish, C. J., 291, 293, 294
Dourish, C. T., 27, 28, 31, 44, 46, 47, 50, 62, 63, 65, 67, 83, 86, 108, 110, 151, 152, 226, 227, 228, 268, 269, 273, 274, 283, 284, 285, 287, 288, 289, 293, 294, 338, 340, 343, 346, 347, 350, 351
Dourish, D. T., 293
Dowie, C., 183, 202
Downing, K., 327, 332
Downing, R., 183, 202
Dray, A., 59, 64, 212, 219, 302, 314
Dren, A. T., 206, 219
Dresse, A., 162, 174
Drewnowski, A., 262, 275
Droba, M., 227, 228–229, 327, 334, 346, 347, 351–352, 460, 462, 473
Drugan, R. C., 347, 350

Dube, L., 122, 139
Dubinsky, B., 121f, 133
Dubner, R., 301, 317
Dubuisson, D., 303, 314
DuCret, R. P., 259, 275
Duggan, A. W., 302, 314
Dugovic, C., 236, 237
Dugovic, G., 238
Duncan-Jones, P., 183, 201
Dunnmer, D. L., 184, 201
Dupont, A. F., 232, 233
Durel, L. A., 436, 438
Dust, E. G., 79, 82
Dykstra, L. A., 157, 170, 399, 403–404
Dysken, M. W., 206, 221

E

Ebert, M. H., 125, 132, 166, 172
Eccleston, D., 164, 168–169
Eccleston, E. G., 159, 175
Eclancher, F., 429, 442
Eichelman, B., 89, 110, 111, 117, 118f, 119, 120f, 121f, 133, 135
Eichelman, B. S. Jr., 120f, 121f, 133
Eichler, A. J., 270, 271
Eide, P. K., 305, 306, 314, 318
Eisemann, M., 301, 318
Eison, A. S., 62, 65, 436, 437, 438, 453, 456, 462, 470
Eison, M. S., 62, 65, 196, 203, 453, 456, 462, 470
El Mestikawy, S., 7, 8, 18–19, 20, 35, 51, 212, 220, 231, 266, 280, 337, 338, 347, 352
Elfvin, L. G., 304, 315
Elghozi, J-L., 158, 173
Eliasson, M., 57, 59, 65, 68, 69
Ellis, W. G., 123, 137
Ellison, G. D., 120f, 133
Emerit, M. B., 7, 8, 18–19, 35, 51
Emrey, T. A., 322, 335, 462, 474
Emson, P. C., 27, 30–31, 31
Eneroth, P., 76, 77, 160, 170–171
Engel, G., 4, 16, 55, 56, 57, 59, 61, 62, 65, 66, 67, 68, 69–70, 83, 86, 90, 91, 94t, 106, 110, 113, 239, 242, 266, 274, 353, 356, 357, 359, 379, 381, 412, 421, 430, 442
Engel, J., 38, 41, 47, 361, 366
Engel, J. A., 40, 50, 293, 294, 322, 331, 340, 343, 350, 377, 380, 461, 462, 470, 471
Ennis, C., 353, 357
Enns, M., 262, 275
Eriksson, B., 162, 168
Eriksson, E., 41, 46, 49
Eriksson, U., 301, 318
Ermisch, A., 30
Erskine, M. S., 101, 110
Ervin, F. R., 129, 138, 161, 175, 177
Escalante, A., 377, 378, 380
Eskay, R. L., 39, 43, 53
Espino, C., 62, 66
Esposito, E., 311, 312, 318
Essman, W. B., 465, 470
Essman, W. D., 407, 408, 424
Evans, K. R., 270, 274
Evans, L., 434, 438
Evans, W. O., 303, 315
Everitt, B., 43, 50, 304, 315
Everitt, B. J., 55, 57, 59, 60, 61, 66, 69, 179, 204

F

Fairbairn, C. G., 190, 201
Fairbanks, L., 158, 171
Fairburn, C. G., 251, 256
Fake, C. S., 239, 242, 353, 357, 383, 387
Fallon, S. L., 62, 69
Fang, V. S., 162, 173
Fanger, C., 425, 438
Fanselow, M. S., 304, 315
Farabegoli, C., 160, 170–171
Farattini, S., 260, 275–276
Farkas, T., 186, 199
Farnebo, L. O., 259, 260, 275
Fasmer, O. B., 27, 30, 260, 272, 302, 303, 305, 306, 308, 309, 311t, 313, 315, 316
Faull, 127t
Fechter, L. D., 464t, 470
Feer, H., 149, 151, 152
Feighner, J. P., 467, 470
Feinber, I., 207, 219
Feldman, R. S., 232, 233, 400, 405
Fellows, E. J., 120f, 141
Feniuk, W., 353, 356, 357, 358, 412, 421
Fenuik, W., 4, 16, 56, 65
Ferguson, J. M., 254, 255
Fernandez de Molina, A., 425, 438
Fernandez-Guasti, A., 47, 50, 58, 62, 64, 66, 73, 76, 377, 378, 380, 461, 468
Fernstrom, J. D., 250, 255, 255–256, 263,

AUTHOR INDEX

264, 274
Ferraris, A., 264, 279, 284, 285
Ferrier, I. N., 159, 160, 170
Ferrini, R., 43, 50
Ferron, A., 338, 352, 354, 359
Feuerstein, T. J., 328, 331
Fibiger, H. C., 270, 280, 460, 473
Fiedling, S., 208, 219
File, S., 427, 433, 438, 439
File, S. E., 321, 323, 325, 331, 332, 334, 337, 340, 350, 354, 357, 371, 373, 374, 377, 381, 383, 387, 425, 440, 460, 461, 470, 471
Fincham, J., 252, 257
Finer, N., 253, 256
Finiger, H. C., 378, 381
Finnerty, R. J., 196, 201, 377, 380, 453, 457, 462, 471
Fisher, S., 183, 202, 327, 332
Fitzgerald, R. G., 158t, 164, 173
Flannelly, K., 98, 111
Flannely, K. J., 145, 147, 148
Flannery, J. W., 125, 129, 140
Flatmark, T., 260, 272, 311t, 313
Fleetwood-Walker, S., 27, 31
Fleminger, J. J., 183, 204
Fletcher, P., 293, 294
Fletcher, P. J., 260, 266, 269, 274, 275, 277, 288, 289
Flynn, J., 41, 48
Fodritto, F., 264, 274
Foldes, A., 5, 16
Foote, W. E., 60, 64, 117, 131
Forchetti, C. M., 27, 30
Ford, H. T., 356, 357
Fordon, E. K., 184, 201
Fordyce, D. J., 89, 111
Foreman, M. M., 43, 50, 56, 57, 59, 66
Forler, C., 5, 7, 20, 40, 53, 79, 81, 82, 225, 229, 284, 286, 373, 375
Forster, C., 162, 163, 170
Fortune, D. H., 214, 218, 353, 356, 358
Foster, G. A., 21, 22, 23
Fotherby, 158t
Fowler, P. J., 102f, 141
Fox, P., 434, 442
Fozard, J. R., 4, 5, 7, 15, 16, 19, 20, 35, 40, 52, 53, 56, 58, 65, 66, 79, 81, 82, 97, 112, 214, 220, 225, 229, 266, 277, 284, 286, 337, 351, 353, 356, 357, 371, 373, 374, 375, 410t, 412, 414t, 421, 424, 453, 457

Fram, D. H., 207, 222
Franck, J. E., 56, 57, 66
Frank, H. J. L., 30, 31
Frankenthaler, L. M., 183, 202, 327, 328, 332, 334
Frankfurt, M., 76
Franklin, K. B. J., 312, 316
Fratta, W., 39, 53
Fraunhoffer, M., 309, 314
Frazer, A., 158t, 164, 173, 212, 222
Frazer, G. A., 328, 331
Frederiksson, A., 460, 468
Fredholm, B., 57, 66, 160, 170-171
Freedman, D. X., 121f, 134
Freedman, J., 304, 315
Freedman, R., 206, 212, 220
Frenken, M., 410, 422
Frey, E. A., 39, 43, 53
Friedman, E., 11, 16, 163, 175
Freidman, R., 400, 404
Frith, C. D., 183, 202
Fuenmayor, L. D., 46, 50
Fukushima, H., 207, 221
Fuller, R. W., 58, 61, 65, 162, 175, 195, 200, 465, 470-471
Furset, K., 304, 313
Fuxe, K., 27, 28, 30, 31, 39, 43, 48, 50, 57, 59, 60, 61, 66, 160, 162, 170, 170-171, 173, 259, 260, 275
Fyer, A. J., 328, 332
Fyer, M. R., 328, 332
Fyro, B., 164, 174, 225, 229

G

Gaddum, J. H., 4, 16, 55, 66, 205, 220
Galantar, M., 206, 223
Gallberg, B., 225, 229
Galliani, G., 265t, 272
Gandolfi, O., 466, 472
Gannon, P. J., 122, 132
Garattini, S., 89, 114, 118f, 121f, 142, 164, 171, 249, 254, 255, 256, 259, 260, 262, 264, 265t, 266, 270, 271, 272, 273, 275, 279, 280, 284, 285, 445, 447
Garcia, S., 46, 50
Garcia-Cabrera, I., 302, 304, 313
Gardner, C. R., 5, 13, 18, 193t, 200, 321, 322, 323, 330, 331, 340, 341, 350, 361, 364, 365, 371, 374, 377, 378, 380, 381, 460, 471
Garrigou, D., 163, 169

Garzon, J., 164, 171
Gastpar, M., 161, 174
Gauen, R., 129, 137
Gazet, J. C., 356, 357
Gebhart, G. F., 302, 311, 317
Gebhart, G. L., 41, 48
Gelders, Y., 235, 237, 238
Gelders, Y. G., 180, 180f, 199, 235, 237, 238, 365, 366, 436, 438, 462, 469
Gelenberg, A. J., 161, 171
Geller, E., 125, 129, 140
Geller, I., 321, 322, 331–332, 332
Geller, N. E., 179, 200
Gendelman, D. S., 464t, 469, 470
Gendelman, P. M., 464t, 469, 470
Gentil, V., 434, 435, 439, 443
Gentsch, C., 149, 151, 152
Gerall, A. A., 57, 67
Gerber, G. J., 270, 281
Gerber, R., 40, 50
German, D. C., 325, 333, 428, 429, 435, 440, 440–441
Gerner, R. H., 158, 171
Gerrard, P. A., 353, 354, 357, 365, 383, 391, 394
Gershon, S., 163, 175, 185, 206, 218, 327, 333
Gessa, G. L., 39, 52, 53, 214, 222
Geyer, G., 425, 438
Geyer, M. A., 243, 245, 246, 259, 275, 464t, 471
Ghezzi, D., 259, 279
Giacalone, E., 118f, 142, 164, 171
Gibbons, J. L., 121f, 134
Gibbs, J., 262, 263, 270, 275, 292, 293
Gibson, C. J., 161, 171
Gibson, E. L., 262, 263, 273
Giglio, J. S., 329, 332
Gilbert, F., 268, 273, 283, 285, 293, 294
Gilbert, W., 250, 257
Gillin, J. C., 207, 220
Gillman, P. K., 206, 219
Gilon, C., 27, 32
Giral, P., 341, 346, 350
Girault, M. M. T., 44, 51
Gispen-De Wied, C. C., 434, 440
Gjerris, A., 158, 171, 184, 201
Glaeser, B. S., 62, 69
Glaser, T., 35, 50–51, 179, 201, 377, 378, 379, 380, 381, 408t, 410t, 423
Glasser, A., 43, 50
Glenn, B., 338, 340, 343, 351, 431, 433, 436, 437, 440
Glennard, R., 235, 236, 238
Glennon, R. A., 4, 13–14, 17, 37, 51, 60, 61, 62, 66, 70, 90, 91, 106, 109, 110, 207, 220, 399, 400, 404, 406, 407, 409, 410t, 412t, 413, 414, 414t, 415, 416t, 418t, 420, 422, 423, 424, 445, 446, 447, 448, 449, 452, 453, 457
Glogers, S., 328, 332, 434, 439, 462, 471
Glover, K., 449
Glowinski, J., 27, 31, 338, 352, 354, 359
Glowinsky, J., 212, 220
Glusman, S., 122, 136
Goa, K. L., 401, 404, 437, 439, 460, 462, 471
Goetz, D., 328, 332
Gold, A., 183, 202
Goldberg, D. P., 183, 201
Goldberg, H. L., 196, 201, 377, 380, 453, 457, 462, 471
Goldberg, M. E., 120, 121f, 133, 134, 141, 322, 332
Goldman, B. D., 101, 110
Goldman, B. S., 101, 110
Goldstein, B., 186, 199
Goldstein, B. J., 345, 351, 462, 471
Goldstein, M., 27, 30, 163, 175, 304, 315
Gomita, Y., 436, 443
Gommeren, W., 188, 202, 235, 236, 238, 245, 246, 265, 276–277
Gonzalez, J. P., 193t, 201
Goodale, D. B., 41, 48
Goodall, E., 249, 252, 253, 255, 256, 257, 260, 262, 279
Goodlet, I., 259, 280
Goodman, P. A., 163, 177
Goodman, W. K., 328, 330, 462, 469
Goodwin, 85, 158t
Goodwin, E. K., 164, 171
Goodwin, F. K., 125, 126, 132, 132–133, 137, 159, 159t, 160, 161, 164, 165, 166, 169, 171, 172–173, 184, 201, 389, 391, 395
Goodwin, G. M., 4, 5, 7, 8, 9f, 10, 11, 12f, 13, 15f, 16, 17, 44, 51, 83, 86, 160, 171, 190, 193t, 196, 200, 201, 251, 256
Gorman, J. M., 328, 329, 332
Gorman, L. K., 329, 332
Gorsalka, 193t
Gorski, 61
Gorzalka, B. B., 55, 56, 57, 58, 61, 62, 63, 68, 69, 73, 76, 77

AUTHOR INDEX

Gothert, M., 55, 59, 62, 66, 83, 86, 91, 94*t*, 110, 266, 274, 353, 357
Gottfries, C. G., 159, 171
Gottlieb, F., 207, 219
Goudie, A. J., 249, 256
Govini, S., 214, 222
Goyer, P. F., 125, 132, 132-133, 164, 169
Gozlan, E., 266, 280
Gozlan, H., 7, 8, 18-19, 20, 21, 24, 35, 51, 212, 220, 266, 268, 275, 337, 338, 352
Grabowsky, M., 212, 220
Graeff, F. G., 105, 110, 113, 179, 201, 321, 325, 329, 332, 333, 334, 377, 379, 380-381, 381, 403, 404, 425, 426, 426*f*, 427, 428*f*, 429, 430, 431, 433, 434, 435, 436, 437, 437-438, 439, 440, 441, 442, 443
Graham, C. W., 129, 140
Grahame-Smith, D. G., 5, 10*f*, 11, 15, 17, 17-18, 18, 39, 51, 191, 202, 353, 358
Grant, L. D., 121*f*, 132, 134
Grasset, G., 118*f*, 132
Graw, P., 161, 174
Gray, G., 184, 201
Gray, J. A., 436, 439
Grayson, D., 183, 201
Green, 127*t*
Green, A. R., 3, 4, 5, 6*f*, 7, 8, 9*f*, 10, 10*f*, 11, 12*f*, 13, 15, 15*f*, 16, 17, 18, 39, 44, 51, 79, 81, 108, 110, 160, 171, 193*t*, 196, 200, 201, 353, 358, 377, 381
Green, R., 206, 223
Green, S., 338, 340, 343, 351, 431, 433, 436, 437, 440
Greenblatt, D. J., 436, 439-440
Gregori, S., 237
Grewe, C. W., 39, 43, 53
Griersmith, B. T., 302, 314
Griffith, J. D., 206, 220
Grilly, D. M., 208, 220
Grimm, V. E., 262, 280
Grinker, J. A., 262, 275
Grinspoon, L., 207, 222
Grossman, S. P., 121*f*, 134
Growdon, J., 250, 257
Growdon, J. H., 263, 281
Gruenberg, E., 328, 334
Grunhaus, L., 328, 332, 434, 439, 462, 471
Guagnano, M. T., 253, 257
Gudelsky, G. A., 193*t*, 201, 322, 332

Guimaraes, F. S., 434, 440
Gulevich, G., 207, 223
Gullberg, B., 164, 174
Gungras, M., 329, 332
Gunning, S. J., 353, 356, 357, 359
Gupta, G. P., 120*f*, 131
Guy, A. P., 5, 13, 18, 193*t*, 200, 377, 378, 380, 381
Guz, I., 434, 443
Gwirtsman, H. E., 166, 172

H

Hacksell, U., 5, 7, 19, 35, 37, 38*f*, 39, 40, 43, 44, 46, 48, 51, 90, 91, 106, 108, 109, 206, 207, 220, 377, 380, 381
Haefely, W., 41, 52
Hafstad, K., 161, 170
Hafstead, E. M., 189, 199-200
Hagan, R., 214, 218, 353, 356
Hagan, R. M., 241, 242, 355, 358
Hagg, T., 101, 111
Haigler, H. J., 55, 56, 59, 60, 63, 66, 67
Hall, H., 79, 82
Hall, J. E., 6*f*, 79, 81
Hall, J. L., 43, 50
Hall, M. D., 7, 8, 18-19
Hall, P. C., 322, 335, 462, 474
Halmi, K. A., 262, 263, 270, 275
Hamberger, B., 259, 260, 275
Hammon, M., 353, 358
Hammond, M., 10*f*, 11, 18
Hamon, K. A., 268
Hamon, M., 7, 8, 18-19, 20, 21, 24, 35, 51, 212, 220, 231, 266, 275, 280, 337, 338, 347, 352
Handley, S. L., 79, 82, 236, 237, 325, 331, 371, 374, 461, 469
Hanin, I., 159, 172
Hansen, S., 40, 53
Hard, E., 361, 366
Hare, T. A., 158, 171
Harper, A. E., 263, 264, 278
Harris, G. D., 309, 315
Harris, T., 164, 170
Harrison-Read, P. E., 193*t*, 201
Harris-Warrick, R. M., 122, 136, 137
Harrod, C., 449
Harrod, H., 451, 452
Hart, Y., 27, 32
Hartig, P. R., 93, 94*t*, 109
Hartmann, E., 161, 170

Hartmann, R. J., 321, 322, 331-332, 332
Hartog, J., 107, 107-108, 113
Harvey, J. A., 305, 306, 315
Hauck, A. E., 410t, 418t, 422
Hawkins, R., 245, 246
Hawthorn, J., 353, 356, 357, 359
Headley, P. M., 302, 314
Heal, D. J., 4, 7, 8, 10f, 11, 13, 15f, 16, 17, 18, 19, 83, 86, 160, 171, 188, 196, 200, 202-202
Heapy, C. G., 21, 22, 23
Hegstrand, L. R., 89, 111, 118f, 120f, 121f, 135
Heinemann, E. G., 398, 404
Held, J., 206, 220
Helzer, Y. E., 328, 334
Henderson, G., 353, 358-359
Hendrie, C. A., 145, 383, 387
Hendrik, E., 57, 67
Heninger, G. R., 159, 161, 170, 172, 189, 191, 199-200, 202, 328, 330, 462, 469
Henriksson, Ch., 43, 44, 47
Henry, P., 263, 281
Hepper, P. G., 89, 109
Herbert, J., 125, 126f, 144
Herbut, M., 103, 113, 120f, 140
Herscovitch, P., 434, 442
Hertting, G., 97, 109, 328, 331
Herz, A., 208, 220
Hess, J., 30
Hess, W. R., 117, 134, 425, 440
Heuring, R. E., 4, 19
Hibert, M., 371, 373, 374
Hietala, O., 206, 221
High, J. P., 121f, 134
Hildebrand, J., 237
Hill, 263
Hill, A. J., 251, 252, 253, 255, 256
Hill, D. F., 55, 71
Hill, J. L., 196, 199, 364, 366
Hill, J. M., 214, 218, 241, 242, 353, 355, 356, 358
Hillegaart, V., 39, 47, 461, 471
Hillenbrand, K., 55, 59, 62, 66, 83, 86, 91, 94t, 110, 266, 274
Hilton, S. M., 425, 437
Himmelhoch, J., 328, 334
Hinde, R. A., 365, 366
Hintgen, J. M., 157, 168
Hjorth, A., 108
Hjorth, S., 5, 7, 19, 35, 37, 38f, 39, 40, 41, 43, 44, 46, 47, 48, 49, 50, 51, 58, 62, 64, 66, 73, 76, 81, 82, 206, 207, 220, 322, 331, 338, 340, 343, 350, 364, 366, 377, 380, 381, 461, 462, 468, 470, 471
Hlinak, Z., 47, 51, 62, 67
Ho, B. T., 398, 399, 406
Hobbelen, P., 321, 335
Hodge, G. K., 118f, 134
Hodges, H., 338, 340, 343, 351, 431, 433, 436, 437, 440
Hoebel, B. G., 259, 275
Hoey, H., 434, 438
Hofer, M. A., 361, 366, 391, 395
Hoffman, B. J., 93, 94t, 109
Hoffmann, S. M., 449, 451, 452
Hofmann, H. P., 121f, 136
Hohn, A. C., 465, 472
Hokfelt, T., 27, 28, 30, 39, 48, 57, 59, 60, 66, 160, 170, 304, 315
Hole, D. R., 57, 58, 61, 67, 73, 74, 75, 76
Hole, K., 27, 30, 120f, 121f, 134, 260, 272, 302, 303, 304, 305, 306, 307, 308, 309, 310, 311, 311t, 312, 313, 314, 315, 316, 318
Holets, V. R., 304, 315
Holfelt, T., 27, 30-31
Hollister, L. E., 206, 220, 327, 332
Holloway, F. A., 401, 405
Holm, A. C., 118f, 138, 162, 173, 305, 307, 317
Holt, J., 292, 293
Holtzman, S. G., 397, 404-405, 407, 415, 416, 423
Hommer, D. W., 445, 447
Hong, E., 377, 378, 378-379, 381
Honma, T., 207, 221
Hoofdakker, R. H. van den., 158, 172
Hoppenbrouwers, H. J. A., 462, 469
Hoppenbrouwers, M. L., 235, 237, 238
Hoppenbrouwers, M. L. J. A., 180, 180f, 199, 235, 237, 238, 365, 366, 436, 438
Hori, K., 145, 147
Hornykiewicz, O., 206, 221
Horovitz, Z. P., 121f, 134
Horowitz, M., 254, 256
Horwitt, D., 245, 246
Houser, V. P., 206, 208, 221
Howard, J. L., 122, 134
Hoyer, D., 4, 20, 21, 24, 35, 52, 55, 57, 59, 61, 62, 66, 67, 68, 69, 83, 86, 90, 91, 94t, 110, 110-111, 264, 266, 268, 274, 275, 278, 353, 357, 379, 381, 408, 423

AUTHOR INDEX

Hrboticky, N., 250, 251, 256, 256–257
Hruska, R. E., 40, 53
Hudson, A. L., 22, 24
Huges, J. P., 5, 18
Hull, E. M., 35, 48
Humber, D. C., 214, 218, 353, 356
Humphrey, P. P. A., 4, 16, 56, 65, 214, 218, 353, 356, 357, 358, 412, 421
Hunkskaar, S., 302, 303, 305, 307, 314, 315, 316, 318
Hunsperger, R. W., 425, 438, 441
Hunt, P., 81
Hunt, P. J., 79, 82
Hunter, A. J., 56, 57, 58, 59, 61, 62, 63, 67, 71, 73, 74, 75, 76, 77, 193t, 204
Hunter, D. C., 214, 218
Huntingford, F. A., 119, 134
Huot, S., 158, 174
Hurwitz, H. M. B., 465, 471
Hutson, P. H., 44, 46, 47, 50, 62, 65, 83, 86, 108, 110, 125, 140, 226, 228, 268, 269, 274, 283, 284, 285, 287, 288, 289, 291, 293, 294, 306, 316, 318, 338, 340, 343, 350, 351, 364, 366
Hyde, J. R. G., 354, 357
Hylden, J. L. K., 27, 30, 309, 316
Hyslop, D. K., 417, 424
Hytell, J., 162, 171
Hyttel, J., 21, 23

I

Iakolev, P. I., 425, 438
Idzikowski, C., 235, 236, 238
Ieni, J. R., 128, 134
Insel, T. R., 166, 171, 328, 335, 364, 366, 434, 435, 440, 441
Invernizzi, R., 260, 264, 270, 272, 275, 275–276, 276, 279
Invitti, C., 225, 228
Ioro, L. C., 206, 221
Ireland, S. J., 214, 218, 241, 242, 353, 355, 356, 358
Irwin, S., 120f, 134
Iseki, H., 425, 537
Itoms, S., 206, 221
Iversen, L. L., 27, 28, 31
Iversen, S. D., 27, 28, 31, 179, 204, 268, 273, 293, 337, 338, 351, 425, 440, 460, 471

J

Jaarsma, I., 98, 99, 107, 113
Jack, D., 214, 218, 353, 356
Jackisch, R., 328, 331
Jackson, A., 125, 140
Jackson, D. M., 121f, 139
Jacob, J. J., 44, 51
Jacobs, B. L., 5, 19, 20, 120f, 135, 160, 171, 245, 246, 465, 473
Jacobs, R. G., 328, 334
Jacobsen, A. F., 462, 471
Jacobson, A. F., 345, 351
Jacoby, D., 208, 220
Jagenau, A. H. M., 93, 111
Jalowiec, J. E., 120f, 138
James, M. D., 73
James, T. A., 433, 439
James, V., 193t, 200
Janowsky, A., 162, 173
Janssen, P. A., 11, 16, 245, 246
Janssen, P. A. J., 21, 24, 56, 58, 59, 67, 93, 111, 188, 198, 200, 202, 206, 222, 235, 236, 237, 238, 238, 321, 325, 331, 408, 409, 410, 415, 416, 417, 420, 421–422, 422
Janssen, P. F. M., 235, 236, 238, 245, 246, 265, 276–277
Jarbe, T. U. C., 397, 398, 399f, 405
Javaid, J., 159, 172
Jean, A., 262, 276
Jeffcoats, S. L., 27, 30–31
Jenck, F., 321, 324f, 361, 365, 426, 431, 442, 442–443
Jenner, P., 354, 358
Jesberger, J. A., 162, 163, 171–172
Jessell, T. M., 27, 31
Jimerson, D. C., 125, 132
Jisheng, H., 302, 317
Jnston, J. A., 159, 170
Johannessen, J. N., 310, 316
Johansson, C., 306, 307, 317, 466, 468
Johansson, F., 301, 316
Johansson, G., 162, 173
Johansson, O., 27, 30–31
Johnson, G. E., 120f, 121f, 134
Johnson, J. A., 160, 170, 184, 185, 200
Johnson, M. D., 76
Johnson, P., 4, 10f, 11, 13, 17, 18
Johnson, S. K., 208, 220
Johnson, S. M., 353, 358–359
Johnston, A. L., 321, 325, 332, 334, 371,

374, 377, 381, 425, 440, 460, 461, 470, 471
Johnston, J., 159, 160, 170
Johnstone, E. C., 183, 202
Joly, D., 163, 172
Joly-Gelouin, D., 326, 330
Jones, B. J., 4, 19, 58, 67, 91, 111, 183, 202, 346, 351, 353, 354, 355, 357, 358, 359, 365, 383, 385, 387, 391, 394
Jonsson, G., 27, 30, 57, 59, 60, 66, 307, 308, 310, 311t, 312–313, 313, 317
Jordan, C. C., 241, 242, 355, 358
Jordan, G. C., 353, 356
Jordau, C. C., 214, 218
Jorgensen, H. A., 308, 313
Jori, A., 158t, 259, 275
Joseph, M. H., 160, 170
Joyce, D., 465, 471

K

Kaesermann, H. P., 103, 109
Kahn, R., 435, 443
Kahn, R. J., 183, 202, 327, 328, 332, 334
Kahn, R. S., 349, 352, 434, 435, 438, 440, 462, 472
Kalish, H. T., 397, 405
Kalkman, H. O., 57, 59, 61, 67, 83, 86, 90, 110
Kallman, M. J., 399, 406
Kalman, H. O., 379, 381
Kamerbeek, W. D. J., 434, 435, 438, 440
Kametani, H., 231, 233
Kanarek, R. B., 263, 278
Kane, N. L., 89, 111
Kang, L., 118f, 133
Kantak, K. M., 89, 98, 111, 118f, 120f, 121f, 135
Kantamaneni, B. D., 206, 219
Kantamaneni, D. B., 125, 140
Kaplan, J., 206, 207, 220, 223
Kaplan, M., 227, 228–229, 327, 334
Karli, P., 121f, 122, 135, 425, 429, 433, 442
Karniol, I. G., 329, 332
Karpowicz, J. K., 121f, 133
Kass, E., 464t, 470
Kataoka, Y., 436, 437, 443
Katayama, Y., 353, 358–359
Katoaka, Y., 431, 440
Katz, J. L., 270, 276
Katz, R. J., 121f, 135

Kaumann, A. J., 410, 422
Kawabatabe, H., 425, 437
Kawamura, H., 425, 437
Kaye, W. H., 166, 172
Kebabian, J. W., 39, 43, 53
Keen, H., 253, 256
Kehne, J. H., 463, 464t, 470
Kehr, W., 39, 40, 41, 51, 53, 91, 109
Keller, R. W. Jr., 271, 276
Kelley, M. E., 206, 219
Kelly, M. E., 58, 65, 183, 202, 239, 353, 354, 357, 358, 359, 365, 383, 385, 387, 391, 394
Kelly, P. H., 5, 18
Kelly, S. J., 312, 316
Kemble, E. D., 425, 440
Kemp, D. E., 193t, 204
Kemp, J. D., 21, 22, 23, 24
Kempf, E., 89, 114, 115, 121, 121f, 139, 142, 143
Kenardy, J., 434, 438
Kennerson, A., 183, 199
Kennett, G., 163, 172
Kennett, G. A., 44, 50, 63, 67, 151, 152, 197, 202, 225, 226, 227, 228, 268, 269, 274, 283, 284, 285, 343, 346, 347, 351, 466, 472
Kennis, L., 21, 24
Kerr, F. W. L., 301, 317
Keshary, P. R., 306, 314
Kessler, J. P., 262, 276
Keverne, E. B., 125, 126f, 144
Kidd, E. J., 371, 373, 374, 410t, 414t, 424, 453, 457
Kilpatrick, G. J., 4, 19, 58, 67, 91, 111
Kilpatrick, G. K., 355, 358
Kilts, C. D., 321, 322, 333
Kim, Y. K., 425, 440
Kind, F. D., 383, 387
King, F. D., 239, 242, 353, 357
Kinohi, R., 120f, 134
Kirch, R., 206, 212, 220
Kirkham, T. C., 262, 276
Kisara, K., 121f, 138
Kiser, J. R. R. S., 325, 333
Kiser, K. R. Jr., 427, 440
Kiser, R. S., 429, 440–441
Kiser, R. S. Jr., 428, 435, 440
Kissileff, H., 262, 275
Kitamura, K., 425, 537
Klein, 166
Klein, D. F., 328, 332

AUTHOR INDEX

Klein, W. J., 125, 132
Klemfuss, H., 5, 19, 160, 171
Klimek, V., 270, 277
Kline, W. J., 164, 169
Kling, A., 129, 140
Klippel, R. A., 21, 24
Klodzinska, A., 417, 423
Knapp, S., 164, 175, 259, 275
Knudson, K., 262, 268, 280
Knutson, J. F., 89, 111
Koch-Weser, J., 436, 439-440
Kocsis, J. D., 342, 351
Koe, B. K., 117, 135
Koek, W., 407, 416, 417, 423
Koenig, J. I., 322, 332
Kohl, R. L., 57, 65
Konig, J. F. R., 21, 24
Konig, J. I., 193*t*, 201
Koob, G. F., 269, 278
Korduba, C. A., 206, 221
Koresko, R. L., 207, 219
Korf, J., 158, 158*t*, 172
Korn, M. L., 349, 352
Kornfeld, E. C., 43, 48-49
Kosar, E., 118*f*, 136
Koslow, S. H., 158, 159, 172, 175-176
Kostowski, W., 117, 118*f*, 120*f*, 121*f*, 123, 135, 136, 139, 149, 152, 212, 221, 433, 442
Kow, L-M., 61, 67
Koyama, T., 160, 172
Kraemer, G. E., 125, 136
Krantz, D. S., 436, 438
Kravitz, E. A., 122, 136, 137
Kreiskott, H., 121*f*, 136
Kreusi, M. J. P., 164, 169
Krsiak, K., 118*f*, 139
Krsiak, M., 93, 94, 95, 98, 103, 111, 111-112, 112, 118*f*, 119, 136, 138
Kruk, M. R., 100, 101, 111, 112, 114, 126, 138
Kruse, C. G., 217, 222
Kruse, H., 208, 219
Kruszewska, A., 270, 271, 276
Kruszewska, A. Z., 311, 312, 318
Krynock, G. M., 399, 406
Kuczenski, R., 164, 175
Kuhn, D. M., 399, 403
Kulkani, A. S., 121*f*, 132
Kulkarni, A. S., 121*f*, 136
Kupferberg, A., 270, 276
Kuribara, H., 208, 213, 221

Kurz, K. D., 58, 61, 65
Kyriakides, M., 262, 276

L

Lacoste, V., 161, 174
Lader, M. H., 186, 204, 365, 366, 436, 443
Ladinsky, H., 260, 266, 275
Laduron, P. M., 4, 19, 21, 24, 235, 236, 238, 245, 246, 265, 276-277
Lakosi, J. M., 41, 48
Lakoski, J. M., 187, 202
Lal, H., 120*f*, 136, 402, 406, 465, 473
Lammers, J. H. C. M., 100, 101, 111, 112
Lamon, S., 122, 137
Lampertilo, M., 225, 228
Landgraf, R., 30
Lane, S. M., 73
Langer, S. Z., 157, 172
Lapierre, Y. D., 186, 199, 251, 255, 328, 331
Lapin, I. P., 119, 120*f*, 131, 136
Laporte, A. M., 231
Lara, P. P., 206, 223
LaRiccia, J., 208, 220
Larousse, C., 389, 394
Lars, J., 377, 380
Larsen, J. J., 21, 23
Larsson, K., 35, 36, 37, 38, 38*f*, 39, 40, 41, 43, 44, 45*f*, 46, 47, 48, 50, 51, 52, 53, 58, 62, 64, 66, 73, 76, 108, 461, 468, 471
Lassen, J. B., 354, 359
Latham, C. J., 249, 255, 261, 272
Laude, D., 158, 173
Laufer, R., 27, 32
Laurence, B. E., 4, 10*f*, 11, 18
Lavielle, R., 253, 256
Lawton, C. L., 295
Le Bars, D., 179, 202, 304, 305, 306, 311, 314, 316
Le Biahn, C., 345, 352
Le Douarec, J. C., 118*f*, 136, 137
Le Magnen, K., 262, 263, 276
Le Moal, M., 271, 277
Leaf, R. C., 121*f*, 122, 134, 137
Leander, J. D., 157, 170, 172, 263, 276, 287, 288, 289, 399, 403-404
Leather, S. R., 15, 20, 39-40, 52
Leathwood, P., 264, 276
Lebovitz, R. M., 4, 11, 20, 56, 59, 61, 63,

69, 325, 333
Lebowitz, R. M., 427, 428, 435, 440
Lebrecht, U., 183, 204, 466, 472
Leckman, J. F., 159t, 161, 172
Lee, E. M. C., 425, 437
Leger, L., 304, 318–319
Lehman, E., 162, 174
Lei, Z., 449, 451, 452
Leibowitz, S. F., 121f, 134, 262, 268, 276, 280
Leiter, L. A., 250, 251, 256, 256–257
Leitz, F. H., 206, 221
Lemberger, 254, 257
Lemus, C. Z., 435, 443
Lent, C. M., 122, 137
Leonard, B. E., 163, 172
Leone, C. M. L., 321, 333
Leroux, A. G., 428, 441
Lesham, M. B., 261, 272
Levin, M. W., 262, 274
Levine, 127t
Levine, A. S., 269, 277
Levine, S., 120f, 133
Levine, S., 464t, 469
Levy, G. P., 353, 356
Lewander, T., 79, 460, 468
Lewinsohn, P. M., 164, 172
Lewis, J. K., 365, 366
Leysen, J. E., 4, 19, 21, 24, 57, 59, 60, 67, 188, 202, 235, 236, 238, 245, 246, 265, 276–277
Lichsteiner, M., 149, 151, 152
Lidberg, L., 166, 172
Lieberman, D. A., 163, 173
Liebeskind, T. C., 425, 443
Liebman, J. M., 40, 50
Liebowitz, M. R., 328, 332
Light, R., 245, 246
Liljequist, S., 40, 50, 322, 331, 340, 343, 350, 377, 380, 461, 470
Lima Filho, E. C., 329, 332
Lin, K-M., 436, 437
Lin, M. T., 305, 318
Lindberg, P., 5, 7, 19, 35, 37, 38f, 39, 40, 43, 44, 46, 48, 51, 108, 206, 207, 220, 377, 380, 381, 461, 468
Lindblom, U., 301, 317
Lindgren, J. A., 304, 315
Lindgren, T., 98, 111
Lindh, B., 304, 315
Lindquist, M., 91, 109, 206, 218
Linnoila, M., 126, 137, 158, 160, 161, 164, 165, 166, 171, 172–173, 174, 184, 201, 206, 221, 222, 434, 435, 440
Lipinski, J. F., 207, 218
Lipman, R. S., 327, 332
Lipman, S. R., 183, 202
Lipp, H. P., 425, 441
Lipper, S., 191, 203
Lipscomb, A., 263, 281
Livingstone, M. S., 122, 136, 137
Ljungberg, T., 207, 221
Ljungdahl, A., 27, 30
Llewelyn, M. B., 309, 316
Lloyd, C., 164, 173
Lloyd, K. G., 163, 169
Lloyd, K. L., 326, 330
Lobeck, W. G., 196, 203
Lofthouse, R., 160, 170
Long, J. P., 41, 48, 49
Lopachin, R. M., 303, 316–317
Lore, R., 98, 111
Lorens, S., 459
Lorens, S. A., 347, 352, 367, 368, 369
Losito, B. G., 163, 177
Louilot, A., 271, 277
Lowy, M., 160, 173, 194, 202
Lowy, M. T., 159, 160, 172, 173
Lucik, R. R., 270, 271
Lucki, I., 346, 347, 352, 453, 454, 457
Lucot, 62
Luine, V. N., 76
Lundberg, J., 304, 315
Luttinger, D., 321, 322, 333
Luzzani, F., 265t, 272
Lydiard, R. B., 328, 333
Lynes, W. H., 354, 356
Lyon, R. A., 61, 62, 67

M

Maas, J. W., 129, 137, 140, 159, 172, 175, 175–176
Maayani, S., 417, 424
MacDonals, A. D., 302, 319
MacElroy, J. F., 232, 233
Mach, E., 22, 24–25, 284, 286
Maciel, L., 434, 443
Mack, G., 121, 121f, 139, 142, 143
MacKenzie, R. G., 259, 275
Mackintosh, N. J., 398, 405
Macko, E., 120f, 141
MacLeod, N. K., 433, 439
Madge, N., 164, 174

AUTHOR INDEX

Maertens de Noordhout, A., 237, 238
Maffi, G., 208, 221
Maggioni, M., 225, 228
Magoul, R., 27, 31
Magre, J., 284, 286
Magson, L. D., 252, 256
Maher, J. B., 302, 314
Maier, S. F., 163, 173, 347, 350
Maj, J., 270, 277, 417, 423
Major, L. F., 125, 132-133, 166, 169
Makman, M. H., 40, 52-53
Maleck, J. B., 322, 332
Maler, L., 123, 137
Malick, J. B., 11, 19, 118f, 121f, 137
Malm, V., 158t, 164, 170
Malmnas, C. O., 35, 52, 55, 69
Mancia, B., 425, 441
Mandell, A. J., 164, 175, 245, 246, 259, 275, 464t, 471
Mann, J. J., 160, 173, 175, 184, 203
Mann, M. A., 128, 141
Mansbach, R. S., 322, 330, 449, 451, 452
Mantegazza, P., 260, 273
Marcinkiewicz, M., 21, 24
Marco, E. J., 27, 30
Marcou, M., 163, 172, 225, 227, 228, 343, 346, 347, 351
Margules, D. L., 56, 57, 71
Marini, S., 237, 266, 269, 277
Mark, S., 250, 257
Markiewicz, L., 121f, 135
Markowska, L., 121f, 123, 135, 136
Marks, I., 435, 441
Marks, I. M., 434, 443
Marks, P. C., 121, 121f, 137
Markstein, R., 55, 68
Marquet, A., 35, 51
Marsden, C. A., 5, 19
Marsden, C. D., 161, 176, 270, 273, 354, 358
Marshall, E. F., 185, 203
Martensson, B., 158, 162, 168
Martin, D. A., 21, 22, 24
Martin, L. L., 40, 50, 61, 62, 68, 69, 196, 199
Martin, P., 231, 233, 341, 347, 350, 352
Masala, A., 255
Masek, K., 118f, 136, 139
Mason, N. R., 195, 200
Mason, P., 321, 322, 333
Massi, M., 266, 269, 277
Matis, P. A., 208, 219

Matthews, W. D., 11, 19
Matthysse, A., 206, 221
Mattiace, L. A., 461, 472
Mattis, P. A., 120f, 141
Mauk, M. D., 342, 351
Mausback, R. S., 377, 380
Mavissakalian, M., 328, 333
Mawson, D., 434, 443
May, J., 212, 220
Mayer, D. J., 309, 310, 312, 316
Maynert, E. W., 305, 319
McArthur, R. A., 261, 263, 272, 277
McBride, A., 160, 173
McCall, R. B., 60, 61, 68
McCance, S. L., 191, 202
McClellan, C. M., 353, 356
McCloskey, T. C., 322, 333, 340, 341, 351, 461, 472
McClure, 158t
McDougall, E. J., 164, 168-169
McElroy, J. F., 400, 405
McEwen, B. S., 160, 273
McGall, 463
McGarty, R. C., 122, 137
McGrath, M. A., 353, 358
McGuffin, 62
McGuire, M. T., 125, 129, 130, 140
McHugh, P. R., 254, 257, 262, 269, 277, 278
McIndewar, I., 259, 280
McKeith, I. G., 159, 160, 170
McKenney, J. D., 62, 66, 106, 110, 207, 220, 410t, 416t, 422, 423
McKinney, W. T., 365, 366, 462, 472
McLain, W. C., 120f, 121f, 137
McLeod, 158t
McMillan, B. A., 35, 51-52, 97, 111, 461, 472
McMillan, D. E., 399f, 405
McNair, D. M., 327, 332
McNair, M., 183, 202
McNaur, D. M., 328, 334
McPherson, K., 183, 202
McQueen, J. K., 161, 176
McRitchie, B., 353, 356
Meek, J. L., 27, 30
Meelis, W., 100, 101, 114
Meert, T. F., 198, 200, 235, 236, 237, 238, 321, 322, 325, 331, 333, 408, 410, 420, 422
Megens, A. A. H. P., 236, 237
Meierl, G., 122, 141

Meisenberg, G., 30, 31
Meister, B., 304, 315
Melander, T., 304, 315
Meltzer, H. Y., 157, 159, 160, 162, 172, 173, 175, 193t, 194, 201, 202, 322, 332
Melzack, R., 311, 312
Memo, M., 260, 273
Mendels, J., 158t, 164, 173
Mendelson, D. S., 73, 76, 77
Mendelson, S. D., 55, 56, 57, 58, 61, 62, 63, 68
Mendlewicz, J., 162, 173
Meneses, A., 377, 378, 380
Menkes, D. B., 464t,, 471
Mennini, T., 260, 264, 265t, 266, 272, 275, 276, 279, 284, 285, 445, 447
Menon, M. K., 212, 221
Mereu, G., 39, 52
Merlo, M., 340, 351
Merlo Pich, E., 322, 333, 462–463, 463, 472
Mersky, H., 301, 317
Metcalf, G., 322, 335, 462, 474
Metz, A., 11, 19, 188, 201–202
Meyers, R. D., 428, 441
Meyerson, B. J., 35, 52, 55, 57, 59, 60, 62, 65, 68, 69, 70
Meyerson, L. R., 185, 202
Michaluk, J., 212, 220
Michanek, A., 59, 65
Miczek, K. A., 89, 93, 94, 95, 98, 103, 111, 111–112, 112, 117, 119, 120, 121f, 126, 127, 129, 130f, 137–138, 138, 143, 394, 395, 465, 472, 474
Middlemiss, D. N., 4, 5, 15, 16, 19, 20, 35, 39–40, 52, 56, 59, 63, 65, 69, 70, 79, 81, 82, 91, 97, 112, 266, 277, 284, 337, 351, 353, 356, 371, 373, 374, 377, 378, 382, 412, 421, 430, 441
Mignot, E., 158, 173
Miksic, S., 402, 406
Milani, H., 436, 441
Milani, H. L., 426, 431, 439
Miller, E. B., 120f, 141
Miller, N. E., 425, 438
Miller, R., 206, 219
Millman, N., 118f, 144
Mills, F. J., 235, 236, 238
Minardo, R., 208, 220
Miner, W. D., 353, 356, 358, 390, 395
Minfeng, R., 302, 317
Minichiello, M. D., 164, 169

Minor, B. G., 306, 307, 308, 310, 311t, 312–313, 313, 317
Minor, T.H., 347, 350
Mir, A. K., 371, 373, 374
Miranda, F., 264, 279
Mishra, R., 162, 173
Missale, C., 260, 273
Mitchell, R., 27, 31
Mithani, S., 371, 374
Mitsushio, H., 367
Mittleman, G., 269, 277
Mobarok Ali, A. T. M., 353, 357
Modigh, K., 119, 138
Modrow, H. E., 401, 405
Mogilnicka, E., 119, 120f, 138
Mohammed, A. K., 310, 317
Mohrland, J. S., 311, 317
Molloy, B. B., 465, 470–471
Molnar, G., 158, 161, 169
Molyneux, S. G., 11, 19, 188, 201–202
Moniz, E., 261, 272
Montanaro, N., 466, 472
Montgomery, A. M. J., 159t, 269, 277, 291, 293, 294
Mook, D. G., 262, 277
Morali, G., 44, 52
Moran, T. H., 254, 257, 262, 269, 277, 278
Morand, C., 129, 138
Morato de Carvalho, S., 427, 441
Moreau, J. L., 426, 431, 442–443
Moret, C., 7, 19
Morgane, P. J., 120f, 138
Morinan, A., 463, 472
Morley, J. E., 269, 271, 277
Moroji, T., 206, 221
Morris, J. B., 186, 203
Morris, R. W., 208, 219
Mos, J., 89, 92t, 95, 97, 98, 100, 101, 102, 103, 104, 108, 112, 113, 114, 215, 222, 361, 365, 366, 389, 390, 394, 395
Moser, P., 371
Moss, R. L., 56, 57, 59, 66
Mott, J., 41, 48
Moulton, B., 461, 473
Moyer, D. J., 425, 443
Mucha, A., 120f, 141
Mueller, E. A., 435, 440
Mueser, K. T., 216, 221
Muhlbauer, 89, 112
Muller, E. E., 225, 228
Muller-Schweinitzer, E., 40, 52

AUTHOR INDEX 493

Mumford, J. M., 301, 317
Munro, J. F., 166, 175
Murphy, D. L., 191, 196, 199, 203, 434, 435, 440, 441
Murphy, R. M., 309, 315
Murray, M. R., 207, 221-222
Muscat, R., 291
Musi, B., 361, 366
Mylecharane, E. J., 4, 16, 56, 65, 353, 356, 412, 421
Myron, K., 346, 347, 351-352, 460, 462, 473

N

Nader, M. A., 449, 451, 452
Nagayama, H., 157, 168
Nakao, H., 425, 441
Nashold, B. S. Jr., 425, 433, 441
Nathan, P. W., 301, 317
Naylor, P. L. R., 353, 356, 359
Naylor, R. J., 58, 65, 183, 202, 206, 212, 219, 239, 242, 270, 273, 353, 354, 355, 356, 357, 358, 359, 383, 385, 387, 389, 394, 395
Neale, R., 40, 50
Neale, R. F., 62, 69
Neekers, L. M., 27, 31
Neill, J., 63, 70, 79, 81, 82, 377, 378, 382, 410t, 414t, 424, 453, 457
Nelson, 449
Nelson, D. L., 4, 20, 21, 24, 35, 50, 55, 69, 264, 278, 353, 359, 408, 409, 423
Nelson, J. C., 159t, 161, 172
Nelson, L. D., 21, 24
Ness, T. J., 302, 317
New, J. S., 196, 203
Newgrosh, G., 353, 357
Newlon, P. G., 399, 406
Neyt, H. C., 91, 110-111
Nichols, C. L., 43, 48-49
Nick, I., 434, 443
Nickolson, V. J., 321, 335
Nielsen, E. B., 399, 405
Niemegeers, C. J. E., 4, 19, 188, 198, 200, 202, 206, 222, 235, 236, 237, 321, 325, 331, 408, 409, 410, 417, 420, 422
Nieuwenhuys, R., 304, 318, 429, 441, 468, 472
Nikulina, E. M., 121f, 138
Nilsson, G., 27, 30, 377, 380
Nilsson, J. L. G., 5, 7, 19, 35, 37, 38f, 39, 40, 43, 44, 46, 48, 51, 79, 82, 108, 206, 207, 220, 377, 381
Nimgoankar, V. L., 4, 10f, 11, 18
Nomuras, 231, 233
Noordenbos, W., 301, 317
Nordstrom, P., 159, 159t, 161, 168
Norelli, C., 259, 275
Norstad, N., 462, 473
North, R. A., 353, 358-359
Norton, K. R. W., 186, 203
Notani, M., 425, 437
Nowak, J. Z., 466, 472
Nutt, D. J., 192, 193t, 199
Nuutila, A., 126, 137, 161, 165, 166, 172-173
Nuveu, C., 118f, 137
Nyback, G., 164, 174, 225, 229

O

Oakley, N. R., 183, 202, 212, 219, 346, 351, 353, 354, 357, 358, 359, 365, 383, 387, 391, 394
Oates, J. A., 206, 220
O'Shaughnessy, K., 10f, 11, 18, 353, 358
Oberlander, C., 22, 24-25, 79, 81, 82
Oblin, A., 27, 30-31
Ogren, S-O., 27, 31, 118f, 138, 160, 162, 170-171, 173, 259, 260, 275, 301, 304, 305, 306, 307, 308, 310, 311, 312, 314, 317, 465, 466, 468, 472
Ogura, Y., 121f, 138
O'Brien, M., 121, 121f, 137
Oldendorf, W. H., 30, 31
Olds, J., 425, 441
Olds, M. E., 425, 441
Olivier, 117, 128
Olivier, B., 89, 92t, 93, 94, 95, 97, 98, 99, 100, 101, 102, 103, 104, 107, 107-108, 108, 109, 112, 113, 114, 126, 138, 215, 222, 361, 365, 366, 389, 390, 394, 395
Olivier, J., 164, 171
O'Callaghan, M., 262, 280
Olmstead, S. N., 449
Olpe, H. R., 161, 174
Olson, L., 304, 315
O'Donnell, V., 93, 109
Onodera, K., 121f, 138
Onteneinte, B., 27, 31
Oon, M. C. H., 207, 221-222
Oostvegel, S., 101, 111
Opitz, K., 118f, 143

Oreland, L., 301, 312
Orthen-Gambill, N., 263, 278, 296, 298
Ortiz, A., 327, 333
Ortmann, R., 212, 222
Orvaschel, H., 328, 334
Osborne, N. N., 108, 113
Oscos, A., 377, 378, 380
Overo, K. F., 162, 171
Overton, D. A., 398, 402, 405, 405–406
Owen, F., 159, 160, 170, 184, 185, 185t, 200, 203
Owen, J. E., 195, 200
Oxford, A., 214, 218, 353, 356

P

Pachtman, E., 206, 212, 220
Paglietti, E., 39, 52
Pagni, C. A., 301, 317
Palacios, J. M., 4, 20, 21, 23, 24, 35, 52, 55, 69, 89, 108, 113, 264, 268, 275, 278, 337, 342, 351, 408, 423, 430, 442, 466–467, 473
Palannapian, 159t
Palay, S. L., 27, 30
Palfreyman, M. G., 158, 174
Panerai, A. E., 225, 228
Panksepp, J., 120f, 138, 259, 269, 274, 425, 433, 441
Papadakos, P., 262, 268, 280
Papeschi, 158t
Paprocki, J., 434, 443
Pardridge, W. M., 30, 31
Parent, A., 122, 139
Paris, J. M., 367, 369
Parker, V., 463, 472
Partlett, J. R., 206, 219
Pataccini, R., 264, 276, 279
Patel, J. B., 322, 332
Patey, A., 8, 20, 266, 280, 337, 338, 352
Paul, B. K., 322, 333, 340, 341, 351, 461, 472
Paul, S. M., 164, 174, 347, 350, 445, 447
Pavinich, G., 161, 177
Paxinos, G., 121, 121f, 137, 139
Paykel, E., 186, 199
Paykel, E. S., 186, 203
Payne, A. P., 123, 139
Pazos, A., 4, 20, 21, 23, 24, 35, 52, 55, 69, 89, 108, 113, 264, 268, 275, 278, 337, 342, 351, 408, 423, 430, 442, 466–467, 473

Pecevich, M., 206, 212, 220
Pecknold, J. C., 328, 334
Pedigo, N. W., 4, 20, 21, 24, 55, 69, 264, 278, 353, 359, 408, 423
Pellegrini-Quarantotti, B., 39, 52
Pellow, S., 325, 334, 371, 373, 374, 461, 470
Penington, N. J., 60, 69
Penot, C., 121, 121f, 139, 143
Perachio, A. A., 425, 437
Perel, J., 328, 333
Perez de los Cobos, 159t
Perline, R., 160, 173, 194, 202
Perlmutter, J., 434, 442
Pernow, B., 27, 30, 30–31, 304, 315
Peroutka, 371
Peroutka, S. I., 55, 56, 59, 61, 63, 69
Peroutka, S. J., 4, 11, 13, 19, 20, 21, 24, 41, 43, 44, 49, 53, 90, 113, 160, 162, 174, 183, 184, 203, 243, 246, 264, 278, 342, 351, 353, 359, 368, 369, 400, 406, 407, 408, 408t, 423, 445, 448, 449, 452, 461, 462, 473
Perris, C., 301, 318
Perris, H., 301, 318
Perry, E. K., 159, 160, 170, 185, 203
Perry, K. W., 465, 470–471
Perry, R. H., 159, 160, 170
Persson, M., 304, 318–319
Persson, M-L., 311t, 312–313
Peters, J. C., 263, 264, 278
Petersen, E., 436, 442
Petersen, E. N., 321, 322, 334, 354, 359, 431, 437, 442
Petersen, L. R., 464t, 471
Peterson, L., 245, 246
Peterson, L. L., 301, 314
Pettibone, D. J., 328, 330
Petty, F., 197, 203, 231, 233
Pfaff, D. W., 61, 63, 65, 67
Pflueger, 62
Philpot, J., 11, 19
Philpot, K., 188, 201–202
Piala, J. J., 121f, 134
Picarelli, Z. P., 4, 16, 55, 66
Pich, E. M., 311, 312, 318, 340, 351
Pichat, L., 7, 8, 18–19, 21, 24, 212, 220
Pickar, D., 164, 174
Pickering, R. W., 11, 12, 16
Pieri, L., 41, 52
Pieri, M., 41, 52
Pierson, M. E., 445

AUTHOR INDEX

Pihl, R. O., 161, 175, 177
Pilc, A., 183, 204
Pinder, R. M., 259, 278, 321, 327, 331, 335
Pinel, P. J. P., 378, 381, 460, 473
Pinheiro, T., 434, 443
Pinnock, R. D., 41, 52
Pivik, T., 207, 223
Pizzi, M., 260, 273
Placheta, P., 97, 109
Plaznik, A., 433, 442
Plaznik, A., 120f, 121f, 135, 139
Plewako, M., 149, 152
Plutchik, R., 161, 176
Podvalova, I., 40, 52
Pohl, R., 327, 333
Poitras, D., 122, 139
Pollack, C. P., 270, 276
Pollard, G. T., 122, 134
Pollock, J. D., 269, 278
Pople, A., 356, 357
Popova, N. K., 121f, 138
Poppen, J. S., 425, 438
Popplewell, D. A., 262, 273
Poschlova, N., 118f, 139
Post, 158t
Post, C., 27, 30, 301, 302, 304, 306, 307, 308, 309, 310, 311t, 312-313, 313, 315, 317
Post, R. M., 160, 164, 171, 174, 184, 201
Poterzio, F., 225, 228
Potkin, S. G., 206, 222
Potter, L., 227, 327, 330
Poulter, M., 159, 160, 170, 184, 185, 200
Powell, D. A., 120f, 121f, 137
Poyser, J., 183, 199
Prange, A. J. Jr., 206, 223
Prasad, V., 119, 139
Pretel, S., 27, 31
Price, G. W., 22, 24, 328
Probst, A., 268, 275
Prosen, 166
Proudfit, H. K., 305, 310, 317, 318
Prouteau, M., 263, 273-274
Przewlocka, B., 119, 120f, 138, 270, 273
Psychotropics, 213t, 222
Pucilowski, O., 117, 120f, 121f, 135, 139, 433, 442
Puech, A. J., 341, 346, 350
Puerto, A., 259, 275, 464t, 471
Puhringe, W., 161, 174
Pujol, J. F., 22, 24-25, 81

Pujol, J-F., 284, 286
Pycock, C. J., 11, 16, 79, 81, 212, 218, 270, 273, 354, 356

Q

Quadbeck, H., 162, 174
Quik, M., 188, 203
Quineaux, N. B., 162, 174

R

Raab, A., 119, 123, 124f, 139-140, 143
Racagni, G., 186, 203
Rachlin, H., 165, 174
Radeke, E., 212, 222
Rafaelsen, O. J., 158, 171, 184, 201
Ragozzino, P. W., 121f, 134
Rahwan, R. G., 121f, 136
Raible, L. H., 57, 69
Raichle, M. E., 434, 442
Rainbow, T. C., 76
Rainey, L. D., 262, 277
Raisman, R., 157, 172
Raleigh, M. J., 125, 129, 130, 140
Ramsey, T. A., 158t, 164, 173
Rankin, H., 161, 176
Rao, B., 186, 203
Raptis, L., 261, 280-281
Raskin, A., 164, 174
Rasmussen, K., 64
Rasmussen, S. A., 328, 334
Rawlins, J. N. P., 433, 439
Ray, A., 120f, 140
Rech, R. H., 262, 278, 310, 318, 321, 322, 333, 354, 356
Redmond, D. E., 129, 140
Redmond, D. E. Jr., 129, 137, 434, 438
Rees, A. R., 6f, 79, 81
Regier, D. A., 328, 334
Reid, L. R., 43, 48-49, 151, 152
Reiffenstein, R. J., 60, 69
Reiman, E. M., 434, 442
Reinhard, J. F., 161, 170, 189, 199-200
Reisine, T., 27, 31
Reiss, D. R., 328, 330
Renaer, M. J., 301, 317
Renver, P., 206, 220, 212
Renyi, A. L., 118f, 138
Renyi, L., 37, 43, 52, 79, 80, 81, 82
Reubi, J. C., 27, 31
Reynolds, E. H., 161, 176

Reynolds, E. J., 354, 358
Reyntjens, A., 235, 237, 238, 365, 366
Reyntjens, A. J. M., 180, 180f, 199, 436, 438, 462, 469
Riblet, L. A., 62, 65, 453, 456
Riblett, C. A., 196, 203
Richards, R., 251, 257
Richardson, B. D., 353, 356, 359
Richardson, B. P., 4, 16, 56, 59, 65, 69–70, 90, 106, 113, 239, 242, 353, 356, 359, 412, 421
Richardson, J. S., 162, 163, 171–172
Richardson, K. D., 347, 352, 367, 368, 369
Rickels, K., 183, 202, 227, 228–229, 327, 330, 332, 334, 345, 346, 347, 350, 351–352, 352, 460, 462, 469, 473
Rijk, H. W., 326, 330
Riley, A. L., 453, 457, 465, 469
Riley, G., 183, 202
Rimon, R., 126, 137, 161, 165, 166, 172–173
Ringberger, V., 184, 199
Rion, R., 378, 378–379, 381
Rivot, J. P., 304, 314
Rizley, R., 328, 334
Robbins, T. W., 269, 270, 278, 280
Roberts, M. H. T., 55, 59, 70, 309, 316
Roberts, S., 55, 71
Roberts, W. W., 425, 438
Robertson, A., 160, 173, 194, 202
Robins, E., 434, 442
Robins, L. N., 165, 174, 328, 334
Robinson, B. W., 425, 442
Robinson, J. L., 328, 330
Robinson, J. M., 192, 193t, 199
Robinson, P. H., 253, 254, 257, 269, 271, 278
Rochat, C., 307, 317
Rodgers, R. J., 103, 113, 145
Rodin, J., 271, 278
Rodnight, R., 206, 207, 221–222, 222
Rodriguez-Sierra, J. F., 57, 61, 70
Roffman, M., 465, 473
Rogawski, M. A., 40, 52
Rogers, P. J., 261, 272
Rolinski, Z., 103, 113, 120f, 121f, 140
Rom, M., 165, 174
Romandini, S., 270, 271, 276, 307, 311, 312, 317, 318
Roos, B-E., 158t, 164, 170, 175
Rose, G., 245, 246

Rose, G. J., 464t, 471
Rosecrans, J. A., 61, 66, 106, 110, 398, 399, 400, 404, 406, 409, 410t, 412t, 418t, 423, 424
Rosenfeld, M., 40, 52–53
Rosland, J. H., 302, 318
Ross, C. A., 5, 20
Ross, S. B., 118f, 138, 465, 472
Rossi, A. C., 119, 140
Rossi, J. III, 259, 269, 274
Roth, M., 183, 199
Roth, R. H., 368, 369
Rothenberg, A., 270, 279
Rovasio, P. P., 255
Rowan, M. J., 461, 473
Rowland, N., 269, 278
Rowland, N. E., 260, 263, 269, 277, 279
Roy, A., 164, 174
Roy-Byrne, P., 159t, 160, 174
Rubenson, A., 39, 48
Rubinow, D. R., 160, 174
Ruckart, R., 118f, 133
Ruda, M. A., 27, 31
Rudy, T. A., 303, 318
Ruhle, H. J., 30
Ruschel, S., 434, 443
Russell, G. F. M., 253, 257, 271, 278
Ruter, M., 164, 174
Ryall, R. W., 306, 309, 313
Ryan, S. M., 347, 350
Rydi, T. A., 303, 316–317

S

Sachs, B. D., 36, 53
Sagen, J., 310, 318
Sahakian, B. J., 125, 140
Sala, A., 89, 114
Salama, A. I., 120, 141, 322, 332
Sali, L., 225, 228
Samanin, R., 249, 254, 255, 256, 259, 260, 261, 262, 263, 264, 266, 268, 269, 270, 271, 272, 273, 275, 275–276, 276, 278, 279, 280, 284, 285, 307, 311, 312, 317, 318, 322, 333, 338, 340, 342, 343, 347, 349, 350, 351, 461, 462–463, 463, 469, 472
Samuel, C., 311, 312
Samuelsson, B., 304, 315
Sanchez, D., 5, 7, 19, 35, 37, 38f, 39, 40, 43, 44, 46, 48, 51, 206, 207, 220, 377, 380, 381, 461, 468

AUTHOR INDEX

Sandchen, D., 108
Sander, A., 41, 52
Sanders-Bush, E., 55, 61, 62, 68, 70, 93, 94*t*, 109, 196, 199, 399, 400, 403, 404
Sandner, G., 426, 442, 442–443, 429, 431
Sanger, D. J., 163, 172, 208, 222
Sanger, G. J., 239, 242, 353, 356, 357, 358, 383, 387, 390, 395
Sanghera, M. K., 35, 51–52, 97, 111, 429, 440–441
Sano, M., 62, 66
Sanyal, A. K., 306, 314
Sarna, G. S., 125, 140
Sauer, G., 41, 53
Savard, R., 160, 174
Sawyer, P. R., 259, 278
Saxena, P. R., 4, 16, 56, 65, 353, 356, 412, 421
Scalia-Tomba, G-P., 158, 166, 168, 172
Scatton, B., 158, 173
Schacter, M., 10*f*, 11, 18, 353, 358
Schaffner, R., 306, 309, 313
Schalling, D., 125, 132, 161, 164, 166, 176
Schalling, M., 304, 315
Schechter, N. D., 402, 406
Scheel-Krueger, J., 399, 405, 436, 437, 442
Scheinin, M., 126, 137, 158, 161, 165, 166, 171, 172–173
Schellekens, K. H. J., 93, 111
Schellenberg, A., 161, 174
Schell-Krueger, J., 431, 442
Schenberg, L. C., 325, 334, 379, 381, 426*f*, 427, 436, 442
Schipper, J., 89, 92*t*, 95, 97, 98, 101, 102, 103, 104, 108, 109, 112, 206, 207, 215, 217, 222, 223, 390, 394, 395
Schlicker, E., 55, 59, 62, 66, 83, 86, 91, 94*t*, 110, 266, 274, 353, 357
Schlosberg, A. J., 89, 111
Schmidt, W. J., 122, 141
Schmitt, P., 425, 426, 429, 431, 433, 442, 442–443
Schneider, P., 434, 438
Schneider, R., 212, 219
Schnellmann, R. G., 21, 24
Schoenfeld, R. I., 377, 380–381, 427, 439
Schoepf, H., 162, 168
Schrieber, R., 120*f*, 121*f*, 137
Schroeder, H. R., 425, 438
Schuckit, M. A., 371, 375
Schuetz, M. T., 426, 431, 439

Schuetz, M. T. B., 403, 404, 428*f*, 429, 435, 443
Schulterbrandt, R. G., 164, 174
Schuman, T., 377, 378, 381
Schuster, R., 165, 169
Schuster, R. H., 165, 174
Schutz, M. T. B., 105, 110, 113
Schutzberg, M., 304, 315
Schuurman, T., 35, 50–51, 322, 323, 334, 377, 380
Schwarcz, J. R., 425, 443
Schwartz, J. P., 27, 31
Schwartz-Giblin, S., 63, 65
Schweitzer, E. E., 227, 228–229
Schweizer, E., 345, 346, 347, 350, 351–352, 352, 460, 462, 469, 473
Schweizer, E. E., 327, 334
Sclafani, A., 295, 298
Scott, J. P., 122, 141
Scott, M. A., 97, 111
Scott, S. M., 35, 51–52
Scuvee-Moreau, J., 162, 174
Seaman, S. F., 365, 366
Sebben, M., 193*t*, 204
Sedvall, F., 164, 174
Sedvall, G., 225, 229
Segal, D. S., 164, 175, 464*t*, 471
Seidel, P. R., 377, 378, 381
Selinger, Z., 27, 32
Sen, P., 120*f*, 140
Sensi, S., 253, 257
Sepinwall, J., 321, 331, 427, 442
Sepinwall, J. L., 321, 334
Serrano, A., 158, 173
Shair, H., 361, 366, 391, 395
Shannon, M., 60, 70, 84
Sharma, K. K., 120*f*, 140
Sharman, D. F., 158*t*, 164, 168–169
Sharpley, A. L., 193*t*, 203
Shaw, D. M., 159, 175
Shaw, E., 205, 223
Shaywitz, B. A., 158, 171
Sheard, M. H., 60, 64, 89, 103, 113, 113–114, 117, 119, 120*f*, 121*f*, 127, 131, 139, 141, 464*t*, 470, 473
Shearman, D. J. C., 254, 256
Shearman, G., 402, 406
Shephard, R. A., 377, 381
Sherman, A. D., 197, 203, 231, 233
Shibata, K., 431, 436, 437, 440, 443
Shimizu, J., 231, 233
Shopsin, B., 163, 175

Shore, P. A., 117, 132
Shor-Posner, 284
Shrivastava, R. M., 206, 222
Shulgin, A. T., 206, 222
Shyu, K. W., 305, 318
Sietnieks, A., 57, 58, 59, 60, 62, 70
Siever, L. J., 191, 203, 435, 441
Silbergeld, E. K., 40, 53
Sills, M. A., 212, 222
Silverman, P. B., 398, 399, 406
Silverstone, T., 249, 251, 252, 253, 255, 256, 257, 260, 262, 276, 279
Simansky, K. J., 89, 111, 305, 306, 315
Simmons, W. H., 30, 31
Simon, H., 271, 277
Simon, P., 118f, 132, 231, 233, 341, 345, 350, 352, 389, 394
Simonovic, M., 162, 173
Simpson, G. M., 206, 222
Singer, M., 462, 473
Sinton, C. M., 62, 69
Sireling, L. I., 186, 203
Sitonen, L., 353, 357
Sjostrand, L., 158t, 159t, 165, 176
Sjostrom, R., 158t, 164, 175
Skinner, J., 321, 322, 333
Skolnick, P., 322, 330, 347, 350, 377, 380, 445, 447, 451, 452
Slangen, J. L., 398, 404, 407, 422
Slater, P., 21, 24
Slater, S., 191, 203
Slaughter, G. S., 425, 433, 441
Sloviter, R. S., 79, 82
Smets, R. J. M., 83
Smith, A., 207, 221-222
Smith, C. D., 11, 19
Smith, D. W., 417, 424
Smith, E. R., 41, 43, 49, 53
Smith, G., 251, 253, 257, 262, 263, 270, 275
Smith, G. P., 292, 293
Smith, L. M., 44, 53
Smith, R. D., 122f, 132
Smith, S. E., 161, 175, 177
Snoddy, H. D., 195, 200
Snyder, S. H., 4, 11, 13, 20, 21, 24, 55, 56, 59, 61, 63, 69, 160, 162, 174, 183, 184, 203, 212, 219, 243, 246, 264, 278, 353, 359, 408, 423
Soderberg, U., 460, 468
Soderpalm, B., 462, 471
Sodersten, P., 76, 77

Sofia, R. D., 57, 70, 121f, 141, 164, 175
Sokola, A., 264, 274
Solomon, H. C., 425, 438
Solomon, R. A., 193t, 203
Solti, M., 28, 31
Sorensen, A. S., 158, 171, 184, 201
Soubrie, P., 27, 31, 108, 114, 179, 203, 231, 233, 337, 338, 340, 341, 345, 346, 347, 350, 352, 354, 359, 361, 366, 425, 435, 443
South, J. A., 453, 454, 457
Spampinato, U., 266, 268, 275
Spano, P. F., 121f, 133, 214, 222, 260, 273
Speciale, S. G., 27, 31
Speight, T. M., 259, 278
Spencer, 323, 334
Spencer, D. G., 35, 50-51, 408t, 410t, 423, 424
Spencer, D. G. Jr., 377, 380, 449, 452
Spindel, E., 27, 30-31
Spindler, J., 270, 281
Sprouse, J. S., 55, 64, 70, 81, 82, 338, 352
Spyraki, C., 270, 280
Stables, R., 353, 356, 359
Stadler, P. A., 59, 69-70, 239, 242, 353, 357, 359
Stahl, S. M., 157, 175
Staines, W., 304, 315
Stanger, R. L., 118f, 144
Stanley, 185
Stanley, B. G., 260, 280
Stanley, M., 62, 65, 158, 160, 173, 175, 184, 203, 453, 456
Stanton, J. B., 164, 168-169
Stark, P., 162, 175
Stefanick, M. L., 41, 43, 49, 53
Stein, L., 179, 204, 321, 334, 354, 359, 436, 443
Steinbook, R. M., 345, 351, 462, 471
Steinbusch, H., 27, 30
Steinbusch, H. W. M., 27, 30-31, 89, 114, 304, 318
Stern, M., 207, 222
Stern, R. S., 434, 443
Stern, W. C., 120f, 138
Sternbach, R. A., 301, 317
Sternberg, D., 161, 170
Sternberg, D. E., 189, 191, 199-200, 202
Stewart, I. C., 166, 175
Stillman, R., 207, 220

AUTHOR INDEX

Stoessl, A. J., 27, 28, 29, 31
Stokes, J. W., 158t, 164, 173
Stokes, P. E., 158, 175–176
Stolerman, I. P., 401, 404
Stolk, J. M., 120f, 133, 464t, 469
Stoll, P. M., 158, 175–176
Stoll, W. A., 206, 223
Stoltz, D., 462, 473
Stolz, J. F., 193t, 201
Stone, G. A., 62, 69
Stone, J. D., 164, 168
Stramentinolo, G., 207, 218
Straughan, D. W., 55, 59, 70, 353, 356, 358
Stricker, E. M., 271, 276
Stunkard, A. J., 269, 280
Sturgeon, R. D., 162, 173
Subramanyam, 158t,
Sugarman, J., 206, 221
Sugrue, M. F., 259, 280, 283, 285, 305, 318
Sulcova, A., 118f, 136
Sulser, F., 162, 173
Sunderland, S., 301, 317
Sundstrom, E., 311t, 312–313
Suomi, S. J., 365, 366
Svare, B. B., 128, 141
Svensson, K., 40, 46, 47, 49, 50, 53, 79, 82, 322, 331, 340, 343, 350, 377, 380, 461, 470
Svensson, L., 35, 37, 38, 38f, 39, 40, 41, 43, 44, 46, 47, 48, 53, 108, 461, 464t, 468, 473
Svensson, U., 206, 207, 220
Swedberg, M. D. B., 397, 405
Sze, P. Y., 164, 177

T

Tadokoro, S., 208, 213, 221
Tagliamonte, A., 39, 53
Takahashi, L. K., 103, 109, 119, 132
Takushi, R., 425, 440
Tan, C. C. W., 58, 65
Tanikawa, Y., 425, 437
Tarchalska, B., 123, 135
Tarchalska-Krynska, B., 123, 136
Tarsy, D., 207, 223
Tatarczynska, E., 417, 423
Tattersall, F. D., 58, 65, 353, 357
Taylor, 83, 449
Taylor, D. A., 11, 20, 79, 82

Taylor, E. W., 409, 423
Taylor, J. R., 270, 280
Taylor, P., 183, 204
Tedeschi, D. H., 120f, 141
Tedeschi, R. E., 120f, 141
Tegeler, J., 162, 174
Temple, D. L., 196, 203, 436, 437, 438
Terenius, L., 304, 315
Thiebot, M. H., 337, 338, 341, 345, 352, 460, 473
Thisvol, 193t
Thoa, N. B., 120f, 121f, 133
Thompson, J., 161, 176
Thonen, P., 184, 199
Thoolen, M. J. M. C., 83, 86
Thorberg, S-O., 79, 82
Thoren, P., 125, 132, 158, 159, 159t, 164, 168, 176
Thornton, E. W., 249, 256
Thurlby, 264, 280
Thurly, P. L., 261, 262, 280
Thurmond, J. B., 128, 134
Thut, P., 120f, 136
Tilson, H. A., 310, 318
Timmermans, P. B. M. W. M., 83, 86
Tischler, M. D., 464t, 470
Titeler, M., 60, 61, 62, 67, 70, 207, 220, 407, 409, 414, 423
Tjolsen, A., 306, 318
Tollenaere, J. P., 4, 19, 57, 59, 60, 67
Tomasikova, Z., 118f, 136
Tomlinson, B. E., 159, 160, 170
Tomosky, T. K., 122, 137
Tongroach, P., 212, 219
Tordorff, A. F. C., 5, 18
Traber, J., 35, 50–51, 179, 201, 377, 378, 379, 380, 381, 408t, 410t, 423, 424, 449, 452
Trabucchi, M., 214, 222
Tramezzani, J. H., 76, 77
Traskman, L., 125, 132, 158t, 159, 159t, 165, 176, 184, 199
Traskman-Bendz, L., 125, 132, 158, 159, 159t, 161, 162, 164, 166, 168, 175
Treit, D., 377, 378, 381, 382, 460, 463, 473
Trenchard, E., 249, 256
Tricklebank, M. D., 5, 7, 20, 40, 53, 63, 70, 79, 81, 82, 83, 85, 97, 114, 225, 229, 284, 286, 306, 316, 318, 371, 373, 374, 375, 377, 378, 382, 410t, 414t, 417, 424, 453, 457

Tricou, B. J., 160, 173, 194, 202
Trimble, M. R., 161, 176
Trojniar, W., 262, 274
Troudart, T., 328, 332, 434, 439, 462, 471
Trulson, M. E., 5, 20, 259, 275, 337, 338, 352, 465, 473
Tsay, R., 250, 257
Tsuruta, K., 39, 43, 53
Tuck, D. L., 453, 457
Tuck, J. R., 166, 172
Tuckwell, V., 254, 256
Tulp, M., 89, 215, 222, 390, 394, 395
Tulp, M. T. M., 89, 92t, 97, 103, 104, 112, 206, 207, 217, 222, 223
Turner, D. H., 353, 356
Turner, S. N., 328, 334
Turnsky, B., 207, 222
Tye, N. C., 179, 204
Tyers, M. B., 4, 19, 58, 67, 91, 111, 183, 202, 206, 214, 218, 219, 239, 241, 242, 346, 351, 353, 354, 355, 356, 357, 358, 359, 365, 383, 387, 389, 391, 394, 395
Tyre, P. J., 436, 443

U

Uberkoman-Wiita, B., 160, 173
Ueki, S., 121f, 143, 431, 436, 437, 440, 443
Ulibarri, C., 57, 71
Ulrich, R. E., 103, 114
Umbach, W., 425, 440
Ungerstedt, U., 39, 43, 48, 207, 221
Urban, J. H., 347, 352, 367, 368, 369
Urbancic, A., 322, 335, 462, 474

V

Vaccarino, F. J., 270, 274
Vaillant, 165
Valenstein, E. S., 269, 277
Valzelli, L., 89, 93, 98, 103, 114, 117, 118f, 119, 121f, 122, 136, 141–142, 142, 161, 164, 171, 176, 259, 279
Van Aken, H., 98, 99, 107, 113
Van Aken, J. H. M., 100, 101, 114
Van Dalen, D., 93, 95, 102, 107, 107–108, 108, 109, 113
Van de Kar, L. D., 367, 368, 369
Van de Velde, D., 81, 82
Van den Burg, M., 162, 176
Van den Bussche, G., 235, 237, 238, 365, 366
Van der Heyden, 158t
Van der Heyden, J., 215, 222, 389
Van der Heyden, J. A. M., 89, 92t, 97, 103, 104, 112, 205, 206, 207, 208, 217, 222, 223, 389, 390, 394, 395
Van der Hoeven, T., 465, 474
Van der Kar, L. D., 347, 352
Van der Kolk, 329, 334–335
Van der Poel, A. M., 100, 101, 111, 112, 114, 394, 395
Van der Veen, F., 321, 335
Van Gompel, P., 188, 202, 235, 236, 238, 245, 246, 265, 276–277
Van Kammen, D. P., 206, 223
Van Oorschot, R., 98, 99, 101, 103, 107, 112, 113
Van Praag, H. M., 158, 158t, 159t, 161, 162, 164, 165, 176, 213t, 223, 349, 352, 435, 443
Van Ree, J. M., 206, 223
Van Riezen, H., 321, 335
Van Sloten, M., 120f, 134
Van Voxel, P., 206, 219
Van Zwietan, P. A., 83, 86
Vandenburg, J., 21, 24
Vandermaelen, 338
Vassar, H. B., 57, 70
Vassout, A., 118f, 119, 120f, 121f, 133
Vaughan, T., 206, 223
Vellucci, S., 212, 219
Verdu, A., 81, 82
Verge, D., 8, 20, 21, 24, 35, 51, 266, 280, 337, 338, 352
Vergnes, M., 89, 114, 115, 121, 121f, 139, 142, 143
Verhoeven, W. M. A., 434, 435, 438, 440
Verhofstad, A., 27, 30
Verhofstad, A. A. J., 27, 30–31
Vernon-Roberts, J., 254, 256
Versiani, M., 434, 443
Verweimp, M., 188, 202
Vestergaard, 159t
Vetulani, J., 12, 16, 183, 204
Vidrio, H., 378–379, 381
Vigano, G., 445, 447
Virgilio, 185
Virkkunen, M., 126, 137, 161, 165, 166, 172–173
Vitale, M. L., 76, 77
Viukari, M., 206, 221
Vivovia, C. A., 212, 221

AUTHOR INDEX

Vogel, J. R., 354, 359
Vojnik, M., 161, 169
Von Hungen, K., 55, 71
Von Knorring, L., 301, 312, 316, 318

W

Wachtel, H., 41, 53
Wade, G. N., 62, 66
Wadenberg, M. L., 463, 473
Wagner, A., 125, 132, 162, 168
Wagner, J., 158, 174
Waissman, K., 462, 473
Wakelin, J. S., 186, 199, 365, 366
Waldbillig, R. J., 121f, 143, 260, 262, 280
Waldmeier, P. C., 119, 138
Walker, R. F., 73, 77
Wall, C. L., 262, 277
Waller, H., 191, 202
Walletschek, H., 123, 124f, 143
Walsh, B. T., 270, 276
Walsh, L. L., 106, 115
Walsh, M. L., 108, 161, 168
Walters, J. K., 464t, 473
Walther, B., 306, 315
Wang, R. Y., 161, 165t, 177
Warbritton, J. D., 464t, 471
Warburton, D. M., 89, 109
Ward, A., 401, 404, 437, 439, 460, 462, 471
Ward, I. L., 56, 57, 66, 71
Ward, N. G., 161, 177
Warner, B. T., 11, 12, 16
Warner, L. H., 207, 223
Wasley, J. W. F., 62, 69
Watamabe, N., 206, 221
Waters, S. J., 21, 24
Watkins, L. R., 310, 316
Watts, A. D., 353, 357, 358
Wauquier, A., 236, 237, 238
Wederlin, L., 184, 201
Wedzony, K., 270, 277
Weidley, E., 208, 219
Weidley, E. F., 208, 219
Weidman, H., 40, 52
Weil-Malherbe, H., 208, 223
Weinberger, D. R., 206, 222
Weiner, H., 270, 276
Weinstock, M., 118f, 143
Weischer, M-L., 118f, 143
Weismann-Nanopoulos, D., 284, 286
Weiss, C., 118f, 143

Weiss, G. F., 268, 280
Weiss, J. M., 163, 177
Weiss, S., 193t, 204
Weiss, S. R., 166, 172
Weissman, A., 117, 135
Weissman, B. A., 322, 330, 377, 380, 449, 451, 452
Weissman, D., 22, 24–25
Weissman, M. H., 328, 334
Welch, A. S., 119, 143
Welch, B. L., 118f, 119, 143
Wennogle, L. P., 185, 202
Werdelin, L., 158, 171
Werdinius, B., 158t, 164, 170
Westenberg, H. G. M., 434, 435, 438, 440, 462, 472
Wetzler, S., 349, 352
Wheatley, D., 327, 335
Wheeler, T. J., 249, 256
White, F. J., 399, 400, 403, 406
White, N., 27, 30–31
Whitesides, G. H., 122, 137
Whitman, J., 118f, 121f, 135
Wiesel, F-A., 164, 174, 225, 229
Wijkstrom, A., 39, 47
Wiklund, L., 304, 318–319
Wikstrom, A., 5, 7, 19
Wikstrom, H., 35, 37, 38f, 39, 40, 43, 44, 46, 48, 51, 108, 206, 207, 220, 377, 380, 381, 461, 468
Wikstrom, J., 35, 39, 48
Wilcox, G. L., 27, 30, 309, 316
Wilffert, B., 83, 86
Wilker, J., 208, 219
Willard, A. L., 122, 143
Williams, B. J., 27, 28, 31
Williams, G., 425, 437
Williams, H. L., 35, 51–52, 97, 111
Williams, M., 40, 41, 48, 50, 62, 69
Willner, P., 157, 158t, 160, 161, 162, 163, 164, 177, 291, 293, 294, 349, 352, 425, 435, 443
Wilson, C. A., 56, 57, 58, 59, 61, 62, 63, 67, 71, 73, 74, 75, 76, 77, 123, 139, 193t, 204
Wilson, I. C., 206, 223
Wilson, N. P., 425, 433, 441
Wilson, P. R., 308, 319
Winger, G., 407, 408, 424
Wingfield, M., 206, 219
Winker, M. A., 310, 318
Winslow, J. T., 129, 130f, 143, 465, 474

Winter, J. C., 321, 335, 398, 406, 413, 424
Wirz-Justice, A., 161, 174
Wise, C. D., 179, 204, 321, 334, 354, 359, 436, 443
Wise, R. A., 261, 270, 280, 280–281, 281, 425, 443
Witkin, J. M., 322, 330, 377, 380, 449, 450, 451, 452, 460, 474
Wnek, D. J., 122, 137
Wode-Helgodt, B., 164, 174
Wodehelgodt, B., 225, 229
Woestenborghs, R., 188, 202
Woggon, B., 162, 168
Wojcik, J. D., 161, 171
Wolf, M. E., 368, 369
Wolfe, B. B., 212, 222
Wolfle, T. L., 425, 443
Wong, D. T., 43, 48–49, 151, 152, 162, 175
Wood, A. J., 7, 17
Wood, M. D., 466, 469
Woods, J. H., 407, 408, 416, 417, 423, 424
Woods, L., 21, 24, 35, 50
Woods, S. W., 328, 330, 462, 469
Woolfe, G., 302, 319
Woolley, D. W., 205, 223, 465, 474
Workman, M. P., 120f, 134
Wormser, U., 27, 32
Wrean, W. H., 463, 464t, 470
Wren, A. F., 162, 163, 170
Wu, T. C., 305, 318
Wurtman, J. J., 250, 257, 263, 281, 298
Wurtman, R. J., 250, 255, 255–256, 257, 263, 274, 281, 295, 298
Wyatt, R., 207, 222
Wyatt, R. J., 27, 31, 206, 207, 220, 222, 223
Wynants, J., 235, 236, 238, 245, 246, 265, 276–277

Y

Yahr, P., 57, 71
Yaksh, T. L., 301, 303, 308, 318, 319
Yamamoto, T., 121f, 143
Yamamura, H. I., 4, 20, 21, 24, 35, 50, 55, 69, 264, 278, 353, 359, 408, 423
Yamamura, H. J., 21, 24
Yamashita, K., 431, 437, 440
Yanai, J., 164, 177
Yap, C. Y., 11, 20, 79, 82, 83
Yates, C. M., 161, 176
Yen, C. Y., 118f, 144
Yevich, J. P., 196, 203
Yocca, F. D., 417, 424
Yodyingyuad, U., 125, 126f, 144
York, J. L., 305, 319
Youdim, M. B. H., 5, 18, 162, 173
Young, A. M., 407, 408, 424
Young, J. G., 158, 171
Young, J. P. R., 186, 204
Young, R., 61, 62, 66, 106, 110, 322, 335, 399, 400, 404, 406, 407, 409, 410t, 412t, 414, 418t, 422, 423, 424, 445, 447, 448, 462, 474
Young, R. A., 446, 448
Young, R. C., 292, 293
Young, S. C., 27, 28, 31
Young, S. N., 129, 138, 161, 175, 177
Yuwiler, A., 125, 129, 130, 140

Z

Zacharko, R. M., 163, 164, 168
Zanchetti, A., 425, 441
Zarcone, V., 207, 223
Zayed, I., 321, 335
Zbrozyna, A. W., 425, 437
Zeisel, S. H., 263, 281
Zemlan, F. P., 35, 49, 56, 57, 61, 67, 71, 259, 275, 309, 315
Zethof, T., 98, 99, 100, 107, 112, 113
Zigmond, M. J., 271, 276
Zimmermann, M., 225, 228, 309, 314
Zini, I., 28, 30
Zohar, J., 166, 171, 196, 199, 328, 335
Zolovick, A. J., 120f, 138, 259, 269, 274
Zook, J. A., 464t, 471
Zook, L., 245, 246
Zuardi, A. W., 329, 332, 434, 440
Zumoff, B., 270, 276
Zung, W., 206, 219

Subject Index

Acetylcholines, in CAR procedure, 211t
Aggression. *See also* Aggressive behaviour
  animal, 117–119
  defensive, 103–105
  discussion on, 106–108
  ethophramacological study of, 126–130
    schematic representation of, 128f
  5-HT and, 87–88, 91, 93
  5-HT$_3$ antagonists and, 389–391, 392–393t, 394
  isolation-induced, 118f, 391, 393t
  maternal, 104f, 394
  offensive
    hypothalamically-induced, 100–101
    isolation induced, 93–98
    maternal, 101–102
    resident-intruder, 98–99
    serontonergic drugs and, 93–102
  predatory, 105–106, 121f
  shock-induced, 120f
Aggressive behaviour. *See also* Aggression
  brain 5-HT system and, 117, 130–131
  isolation and, 118f
  serotonin and, 89–108
Agonistic behaviour, 122–126
Agoraphobia, 327–328
(-)Alprenolol, in antagonism tests, 412t

Amiflamino, effect on postsynaptic responses at 5-HT synapses, 165t
Amitriptyline
  effect of
    on burying and grooming, 326t
    in CAR procedure, 211t
Amphetamine(s)
  effects of
    on attack bites, 130f
    on sideways threats, 130f
    on walking duration in confrontations, 130f
  ejactulatory response and, 79–81, 80t
  as neurotransmitter psychotogen in schizophrenia, 205t
Animal aggression, 5-HT and, 117–119. *See also* Aggression
Anorexia, 5-HT and, 271
Antiaggressive activity
  isolation-induced aggression and, 391, 393t
  maternal aggression and, 393t, 394
  test methods, 390
  results, 391, 393t, 394
Anticonvulsants, in CAR procedure, 211t
Antidepressant(s)
  effect of

503

anxiolytic, 183
  on depression, 160-161
  on head twitch behavior, 11*t*
  on 5-HT$_2$ receptor characteristics, 11*t*
5-HT function and, 186-197
  enhancement of, 187-188
5-HT precursors as, 161-162
5-HT reuptake blockers as, 186-187, 187*t*
5-HT uptake inhibitors as, 162
mechanism of action of, 195-197, 195*t*
reversal of therapeutic actions of, 162-163
and systems involving 5-HT receptor subtypes, 195*t*
Antipsychotic test
  apomorphine-induced climbing, 390-391, 392*t*
  conditioned avoidance response, 390, 392*t*
  methods of, 389
  results of, 390-391, 392*t*
Ants, serotonin influences in, 123
Anxiety
  antidepressants and, 183
  experimentally induced, 338, 339*f*, 340-342, 341*f*, 342*f*
  5-HT and, 299-300
  5-HT$_{1A}$ agonists and, 371-374, 372*t*
  5-HT$_2$ receptors and, 181-183, 180*f*, 181*f*, 182*f*
  5-HT$_3$ and, 353-356
  5-HT$_3$ antagonists and, 389-391, 392-393*t*, 394
  models of
    agoraphobia, 327-328
    anxiety disorders, 326-327
    aversive brain stimulation, 324*f*, 325
    citalopram in conflict-test, 323*f*
    conflict tests, 321-324
    defensive burying, 325-326, 326*t*, 327*t*
    elevated plus maze, 325
    5-HT and, 321-330
    5-H$_3$ antagonists in, 383-385, 386*f*, 387
    generalized anxiety disorder, 327
    obsessive-compulsive disorders, 328
    panic disorders, 327-328
    posttraumatic stress disorder, 329
    social interaction, 323-324
    social phobia, 329

neurobiology of, 183-184
reduced, 337-338
serotonin and
  as bridge between depression and, 337-349
  implications for future research, 345-346
Anxiety disorders, 326-327. *See also specific disorders*
Anxiolytic tests
  4-plat test, 391, 392*t*
  light-dark test, 391, 393*t*
  methods of, 389-390
  results of, 391
Apomorphine-induced climbing, 390-391, 392*t*
Arylpiperazine second generation anxiolytics, 445-447, 446*t*
Atropine, in CAR procedure, 211*t*
Autoradiographs, showing [^{3}H]8-OH-DPAT binding, 23*f*
Aversion
  drug discrimination and, 397-403, 399*f*, 402*f*
  serotonin and, 425-437
    discussion on, 433-437
    intracerebral drug injection, 428-433, 428*f*, 430*f*, 432*f*
    systemic drug injection, 426*f*, 427-428
Aversive brain stimulation, 324*f*, 325

Baclofen, schizophrenia and, 205*t*
Barbiturates, in CAR procedure, 211*t*
Befiperide
  activity of, 216*t*
  effect of
    on aggression, 107*t*
    on intermale aggression, 97, 97*f*
    on muricide in rats, 105*t*
    on resident-intruder aggression, 99, 99*f*
    in isolation-induced aggression, 95*t*
  schizophrenia and, 205*t*
  structure of, 218*f*
Behaviour, serotonin-mediated, 27-30. *See also specific behaviour;* Sexual behaviour
Behavioural pharmacology, 3
Benzodiazepines
  effect of
    in CAR procedure, 211*t*

SUBJECT INDEX

differential, on defensive patterns, 145–147
on ultrasonic cries, 363t
schizophrenia and, 205t
Brain 5-HT system, 117, 130–131
BRL 43694, effect of
  in anxiety models, 386f
  on hyperactivity, 240f
Bromo-lisuride, sexual behaviour and, 41, 42f
Burying. *See* Defensive burying
Buspirone
  activity of, 216t
  affinities of, 408t, 414t
  drug-appropriate response by, 419t
  effect of
    on aggression, 107t
    on behaviour in elevated plus-maze, 372t
    on defensive behaviour, 146–147, 146t
    on burying and grooming, 327t
    on intermale aggression, 97, 97f, 97–98
    in isolation-induced aggression, 95t
    on muricide in rats, 105t, 106
    on resident-intruder aggression, 98–99, 99f
    on ultrasonic cries, 363t
    on in vitro 5-HT release, 94t
    on in vitro PI hydrolysis, 94t
    on in vivo 5-HTP formation, 94t
    on water-lick conflict test, 339f, 340, 341f
  stimulus generalization and, 446t
Butaclamol, activity of, 213t
Butobarbital, in CAR procedure, 211t

CAR. *See* Condition avoidance response
Carbamazepine, in CAR procedure, 211t
Carbidopa, 10f
Carbohydrates, 295–298, 297f. *See also* Feeding behaviour
Catecholaminergic antagonists, effect on 8-OH-DPAT effects, 39–40
CCK-analogue, schizophrenia and, 205t
Central nervous system (CNS), senktide and, 30
Cercopithecus aethiops, serotonin influences on, 125
Chlordiazepoxide
  effect of

    in CAR procedure, 211t
    on ultrasonic cries, 363t
Chlorpromazine
  activity of, 213t
  shock intensity and effect of, 210, 210t
Chlorprothixene, activity of, 213t
Cinanserin, 8t
  lordosis behaviour and, 57
  startle habituation and, 244t, 245
Circling behaviour, 22
Citalopram
  effect of
    on burying and grooming, 326t
    on conflict behaviour, 323f
    on depression, 186–187, 187t
    on postsynaptic responses at 5-HT synapses, 165t
Climbing, apomorphine-induced, 390–391, 392t
Clomipramine, effect on burying and grooming, 326t
Clonidine
  in CAR procedure, 211t
  schizophrenia and, 205t
Clopipazan, activity of, 213t
Clorgyline, effect on postsynaptic responses at 5-HT synapses, 165t
Clovoxamine, in CAR procedure, 211t
Clozapine, activity of, 213t
CNS. *See* Central nervous system
Condition avoidance response (CAR)
  antipsychotic test results, 390, 392t
  antipsychotics and reduction of, 208
  centrally acting drugs and CAR procedure, 211t
  inhibition of, 208
Conditioned taste aversion, 453–456, 455f
Conflict test(s)
  in anxiety disorders, 321–324
  citalopram in, 323f
  water-lick
    buspirone in, 339f, 340, 341f
    diazepam in, 339f, 340
    proadifen in, 341, 342f
CS. *See* Conditioned stimuli
CSF 5-HIAA, in depressed suicide attempters, 159t
Cyproheptadine, 8t
  effect of
    on escape latency, 426f
    on startle habituation, 243, 244t, 245
  lordosis behaviour and, 57–58

Defense Test Battery, 145–146
Defensive aggression, 103–105
Defensive burying
  5-HT$_1$ agonists and, 377–380, 379f
  in anxiety disorders, 325–326, 326t, 327t
Depression
  central 5-HT turnover in, 157–160
    studies of, 158t
  citalopram and, 186–187, 187t
  5-HT receptor subtypes in, 179–180, 194–195, 197–198
    conflicting views on, 188t, 189f
    geometric mean spontaneous fluctuations, 183f
    S.C.R. spontaneous fluctuations, 181f
  hormonal responses to tryptophan in, 189–192, 190f, 191f, 192f
  mechanism of action of 8-OH-DPAT on, 225–227, 226t
  neurobiology of, 183–184
  neurochemical changes in, 157–160
  neuroendocrinology in, 193–194, 194f
  psychosis and, 153–155
  research on, 153
  serotonin and, 153
    as bridge between anxiety and, 337–349
  social isolation and, 164–165
  stress and, 163–164
  stress-induced, 346–347, 348f, 349
  symptomatology of, 161
  theory of serotonergic dysfunction in, 157–168, 167t
    acute 5-HT manipulations and mood, 161
    ambiguous data, 157–160, 158t, 159t
    antidepressants and, 160–161
    central 5-HT turnover, 158t
    characterization, 157
    chronic antidepressant treatment, 165t
    construction of, 166, 167t, 168
    CSF 5-HIAA in suicide attempters, 159t
    decreased transmission, 160–161
    5-HT precursors as antidepressants, 161–162
    5-HT uptake inhibitors as antidepressants, 162
    functional significance, 165–166, 165t
  neurochemical changes, 157–160
  origin, 163
  reversal of antidepressant action, 162–163
  social isolation, 164–165
  stress, 163–164
  symptomatology, 161
Desipramine, 9f, 11t
Diazepam
  activity of, 386f
  effect of
    on behaviour in elevated plus-maze, 372t
    in CAR procedure, 211t
    on burying behaviour, 379f
    on defensive behaviour, 146–147, 146t
    shock intensity and, 210, 210t
    on ultrasonic cries, 363t
    on water-lick conflict test, 339f, 340
  stimulus generalization and, 446t
  test results for
    in antagonism of apomorphine-induced climbing, 390–319, 392t
    in conditioned avoidance response, 390, 392t
    and ED$_{50}$ to inhibit isolation-induced aggression in mice, 391, 393t
    and four-plate test in mice, 391, 392t
    and light-dark cage exploration, 391, 393t
    and rat pup ultrasonic vocalizations, 391, 393t
2,5-Dimethoxy-4methylamphetamine (DOM), lordosis behaviour and, 60. *See also* DOM
2,5-Dimethoxy-4-methyl-phenyl-ethylamine (DOMPE), lordosis behaviour and, 60. *See also* DOMPE
N,N,Dimethyltryptamine (DMT), lordosis and, 60–61. *See also* DMT
DMI, effect on postsynaptic responses at 5-HT synapses, 165t
DMT, lordosis and, 60–61
DOI
  ED$_{50}$-values of, in isolation-induced aggression, 95t
  effect of
    on aggression, 107t
    on competition-rate in male rats, 150f
    on maternal aggression, 102, 102f

SUBJECT INDEX

on muricide in rats, 105t
on ultrasonic cries, 363t
on in vitro 5-HT release, 94t
on in vitro PI hydrolysis, 94t
on in vivo 5-HTP formation, 94t
head shakes and, 84, 85, 86t
DOM
  in antagonism tests, 412t
  in drug discrimination studies, 410t
  stimulus effects of, 411t
  drug-appropriate response by, 419t
  5-HT_2 binding site affinity of, 418t
  lordosis behaviour and, 60
DOMPE, lordosis behaviour and, 60
Dopamine
  in CAR procedure, 211t
  in schizophrenia, 205t
Dorsal raphe nucleus (DRN), rotational behaviour and lesioning of, 21–22
DRN. *See* Dorsal raphe nucleus
Drug discrimination
  aversion and, 397–403, 399f, 402f
  characteristics of 5-HT binding sites, 408–409, 408t
  5-HT and, 407–421
  5-HT_1 antagonists and, 445–447, 446t, 449–452
  mediation through one or more receptors, 411–412, 411t, 412t
  partial generalization, 414–415, 419t
  analysis of, 415–420, 420t
  recent studies of, 409, 410t
  receptor mediation, 409–410
  relationship
    to 5-HT binding sites, 413–414, 414t, 416t, 418t
    to functional 5-HT receptor subtypes, 412–413
Drugs. *See specific agents*
DU 28853
  ED_{50}-values of, in isolation-induced aggression, 95t
  effect of, on intermale aggression, 96f

Eating disorders, 269–271, 270t
Eating habits. *See* Feeding behaviour
ECS, 11t
Ejaculatory response
  5-HT_{1A} and, 79–81, 80t
  raclopride and, 80t
Electric fish, serotonin influences in, 123
Electrolytic lesions

of ascending pathways, 305
of descending pathways, 305–306
Elevated plus maze, in anxiety disorders, 325
Eltoprazine
  activity of, 216t
  ED_{50}-values of, in isolation-induced aggression, 95t
  effect of
    on aggression, summary of, 107t
    on hypothalamically-induced aggression, 100f, 101
    on intermale aggression, 96f, 96–97
    on maternal aggression, 102, 102f, 104f
    on muricide in rats, 105t, 106
    on resident-intruder aggression, 98, 99f
    on ultrasonic cries, 363t
    on in vitro 5-HT release, 94t
    on in vitro PI hydrolysis, 94t
    on in vivo 5-HTP formation, 94t
  test results for
    in antagonism of apomorphine-induced climbing, 390–391, 392t
    in conditioned avoidance response, 390, 392t
    and ED_{50} to inhibit isolation-induced aggression, 391, 393t
    and four-plate test in mice, 391, 392t
    and light-dark cage exploration, 391, 393t
    and rat pup ultrasonic vocalizations, 391, 393t
γ-Endorphine, schizophrenia and, 205t
Ergot drugs, effect of, on male rat sexual behaviour, 38–39

Fear
  5-HT_{1A} agonists and, 367–369
  5-HT_2 receptors and, 181–183, 180f, 181f, 182f
Feeding behaviour
  carbohydrate suppression, 5-HT and, 295–298, 297f
  central manipulations of serotonergic system, 268–269
  eating disorders, 269–271, 270t
  8-OH-DPAT and, 291–293, 292t
  5-HT receptor subtypes effects on, 264–266, 265t, 267f, 268
  5-HT_1 agonists and

effects of, 283-284
  proposed roles of, 284-285
  5-HT$_{1B}$ agonists and, 284
  5-HT$_{1C}$ agonists and, 284
food intake, 283-285
hunger control, 259-260, 261f
hunger vs satiety, 261-262
liquid diets, 291-293, 292t
nutrient selection, 263-264
  serotonergic control of, 249-250
palatability-induced ingestion, 287-288
peripheral 5-HT and, 269
rewarding effect of food, 262-263
serotonin and, 247-248
  control of nutrient selection, 249-250
  drugs enhancing neurotransmission, 250-254
  drugs reducing neurotransmission, 254-255
solid diets, 291-293, 292t
Fenfluramine
  activity of, 216t
  ED$_{50}$-values of, in isolation-induced aggression, 95t
  effect of
    on aggression, summary of, 107t
    on carbohydrate suppression, 295-298, 297f
    on feeding behaviour, 251-254
    on maternal aggression, 102, 102f
    on muricide in rats, 105t, 106
  feeding control and, 259-260, 261f
  overeating and, 271
Flesinoxan
  ED$_{50}$-values of, in isolation-induced aggression, 95t
  effect of
    on aggression, summary of, 107t
    on intermale aggression, 97, 97f
    on maternal aggression in rats, 102
    on ultrasonic cries, 363t
    on in vitro 5-HT release, 94t
    on in vitro PI hydrolysis, 94t
    on in vivo 5-HTP formation, 94t
Fluoxetine (FXT), effect of
  on burying and grooming, 326t
  on feeding behaviour, 254
  on ultrasonic cries, 363t
Fluphenazine
  activity of, 213t
  effect of on hyperactivity, 240f
Fluprazine

ED$_{50}$-values of, in isolation-induced aggression, 95t
effect of
  on aggression, summary of, 107t
  on hypothalamically-induced aggression, 101
  on intermale aggression, 96f, 97
  on muricide in rats, 105t, 106
  on in vitro 5-HT release, 94t
  on in vitro PI hydrolysis, 94t
  on in vivo 5-HTP formation, 94t
Fluvoxamine
  activity of, 216t
  ED$_{50}$-values of in isolation-induced aggression, 95t
  effect of
    on aggression, summary of, 107t
    on burying and grooming, 326t
    in CAR procedure, 211t
    on hypothalamically-induced aggression, 101
    on intermale aggression, 96f, 97
    on maternal aggression, 102, 102f
    on muricide in rats, 105t, 106
    on resident-intruder aggression, 98-99, 99f
    on ultrasonic cries, 363t
Food, rewarding effect of, 262-263
Food intake. See Feeding behaviour
Foot-shock-induced defensive aggression, 103
Forepaw treading
  following ics senktide, 28
  5-HT and, 8t
  5-MeODMT and, 8t
  8-OH-DPAT and, 9f
  serotonin receptor subtypes and, 84, 85t
Formica rufa, serotonin influences on, 123
Four-plate test, 391, 392t
FXT. See Fluoxetine

GABA, as neurotransmitter psychotogen in schizophrenia, 205t
Generalized anxiety disorder, 327
Gepirone
  effect on defensive behaviour, 146-147, 146t
  stimulus generalization and, 446t
GR 38032F
  activity of, 215t
  in anxiety models, 386f

SUBJECT INDEX 509

ED$_{50}$-values of, in isolation-induced aggression, 95t
  effect of
    on aggression, summary of, 107t
    on burying and grooming, 327t
    on hyperactivity, 240f
    on intermale aggression, 97, 97f
    on maternal aggression, 102, 102f
    on muricide in rats, 105t, 106
    on ultrasonic cries, 363t
    on in vitro 5-HT release, 94t
    on in vitro PI hydrolysis, 94t
    on in vivo 5-HTP formation, 94t
  schizophrenia and, 205t
  test results for
    in antagonism of apomorphine-induced climbing, 390–391, 392t
    in conditioned avoidance response, 390, 392t
    and ED$_{50}$ to inhibit isolation-induced aggression in mice, 391, 393t
    and four-plate test in mice, 391, 392t
    and light-dark cage exploration, 391, 393t
    and rat pup ultrasonic vocalizations, 391, 393t
Growth hormone, depression and, 190, 190f
Guanabenz, in CAR procedure, 211t
Guanfacine, in CAR procedure, 211t
Gymnotidea, serotonin influences on, 123

^{3}H-5-HT, in frontal cortex of suicides, 185, 185t
Haloperidol, 8t
  activity of, 213t
  shock intensity and effects of, 210, 210t
  test results for
    in antagonism of apomorphine-induced climbing, 390–391, 392t
    in conditioned avoidance response, 390, 392t
    and ED$_{50}$ to inhibit isolation-induced aggression, 391, 393t
    and four-plate test in mice, 391, 392t
    and light-dark cage exploration, 391, 393t
    and rat pup ultrasonic vocalizations, 391, 393t
Hampsters, serotonin influences in, 123
Head shakes, 84, 85t
Head twitch(es)
  antidepressant treatments and, 11t
  8-OH-DPAT and, 15f
  5-HTP and, 10f
  5-MeODMT and, 15f
  following ics senktide, 27–28
  ipsapirone and, 15f
  propranolol and, 12f
  ritanserin and, 12f
Headweaving
  5-HT and, 8t
  5-MeODMT and, 8t
  8-OH-DPAT and, 9f
Helplessness, reversal of, 231–233, 232ff
5-HIAA
  effect of steroid treatment on, 74, 75t
  in frontal cortex of suicides, 185, 185t
HInd limb abduction
  8-OH-DPAT and, 9f
  5-HT and, 8t
  5-MeODMT and, 8t
5-HT. *See also* Serotonin
  aggression and, 87–88, as assessed by traditional laboratory tests, 117–119
  aggressive behaviour and, 89–93, 106–108
  agonistic behaviour and, 122–126
  anxiety and, 299–300
    models of, 321–330
    agoraphobia, 327–328
    anxiety disorders, 326–327
    aversive brain stimulation, 324f, 325
    citalopram in conflict-test, 323f
    conflict tests, 321–322, 323f
    defensive burying, 325–326, 326t, 327t
    elevated plus maze, 325
    generalized anxiety disorder, 327
    obsessive-compulsive disorders, 328
    panic disorders, 327–328
    posttraumatic stress disorder, 329
    social interaction, 323–324
    social phobia, 329
  behavioural pharmacology of, 459–468, 464t, 466f
    in aggression, 462
    aggression and, 463–465
    in anxiety, 460, 462
    in depression, 459, 462
    fear-retention deficit and, 465–466, 466f
    feeding behaviour and, 467
    memory storage and, 465

in pain, 462
panic and, 462-463
problems in, 460
retrieval processes and, 465
in sexual behaviour, 461-462
startle responses and, 463, 464*t*
binding sites, in drug discrimination
characteristics of, 408-409, 408*t*
relationship to, 413-414, 414*t*,
carbohydrate suppression and, 295-298, 297*f*
conditioned taste aversion and, 453-456, 455*f*
control of hunger and, 259
deficiency vs excess, 188-189, 188*t*, 189*f*
depression and
central turnover in, 157-160
conflicting views on, 188*t*
drug discrimination and, 407-421
analysis of partial generalization, 415-420, 420*t*
mediation through one or more receptors, 411-412, 411*t*, 412*t*
partial generalization, 414-415, 419*t*
recent studies on, 409, 410*t*
receptor mediation, 409-410
eating disorders and, 269-271, 270*t*
effects of
on aversive threshold, 428*f*
on mood, 161
on in vitro 5-HT release, 94*t*
on in vitro PI hydrolysis, 94*t*
on in vivo 5-HTP formation, 94*t*
excess vs deficiency, 188-189, 188*t*, 189*f*
feeding control and, 259-260, 261*f*
female rat sexual behaviour and, 73-76
function
antidepressants and, 186-197
in depressed patients, 184-185, 185*t*
enhancement of, 187-188
human feeding and, 249-255
drugs enhancing neurotransmission, 250-254
drugs reducing neurotransmission, 254-255
killing and, 119-122
neuropharmacological manipulations of,
on aggressive behaviour, 118, 118*f*
neurotransmission
fenfluramine and, 251-254
tryptophan and, 250-251

pain and, 299-300, 301-312
anatomical considerations, 304
behavioural tests of nociception, 301-304
depletion of serotonin by p-Cl-phenylalanine, 304
lesions of ascending pathways, 305
lesions of descending pathways, 305-306
noradrenergic involvement in serotonergic antinociception, 310
precursors effects, 306-310
receptor agonists effects, 306-310
receptor antagonists effects, 304-305
releasing agents effects, 306-310
opiate analgesia, 310-312
palatability-induced ingestion and, 287*t*-288
peripheral, in feeding behaviour, 269
precursor(s)
as antidepressants, 161-162
effects of, on pain, 306-310
releasing agents, pain and, 306-310
reuptake blockers
as antidepressants, 186-187, 187*t*
effect on ultrasonic cries, 363*t*
reuptake inhibitors, feeding behaviour and, 254
schizophrenia and, 206
sexual behaviour and, 33-34
steroid treatment and hypothalamic activity of, 74, 75*t*
system, modulation of, 205-217
ultrasonic vocalisation and, 361-365, 363*t*
uptake inhibitors
as antidepressants, 162
effects of, on postsynaptic responses at 5-HT synapses, 165*t*
5-HT agonist(s)
activity of, 216*t*
correlation coefficients for, 217*t*
effect of
on competition-rate in male rats, 149-151, 150*f*
on mean consumption, 455*f*
on sexual behaviour, 75
effects on, 8*t*
5-HT antagonist(s)
activity of, 215*t*
effect of
on aversive brain stimulation, 324*f*

SUBJECT INDEX

on sexual behaviour, 75
effectiveness of, 56
8-OH-DPAT as, 40-41
lordosis behaviour and, 56-59
putative, effects of, 8*t*
5-HT receptor(s)
agonists, pain and, 306-310, 311*t*
antagonists, 56-59
feeding behaviour and, 254-255
pain and, 304-305
in frontal cortex of suicides, 185, 185*t*
neuroendocrinology in depression and, 193-194, 194*f*
subtypes, 3-4
antidepressant drugs and, 195-197, 195*t*
depression and, 179-180, 194-195, 197-198
differential roles of, 55-63
feeding behaviour and, 264-266, 265*t*, 267*f*, 268
functional, 412-413
functional interplay of, 83-86, 85*t*
functional tests for, 13-14, 14*t*
synthesis of, 83
systems involving, 195*t*
5-HT$_1$ agonist(s)
defensive burying and, 377-380, 379*f*
food intake and, 283-285
liquid diets and, 291-293, 292*t*
solid diets and, 291-293, 292*t*
5-HT$_1$ antagonist(s)
drug discrimination and, 445-447, 446*t*, 449-452
ultrasonic cries and, 363*t*
5-HT$_1$ receptor(s)
agonists, cumulative burying behaviour and, 379*f*
5-HT$_2$ opposed function of, 193, 193*t*
behavioural functions of, 197
subtypes, effects of drugs on feeding and, 284-285
systems involving, 198*t*
5-HT$_{1A}$
lordosis and, 62
rotational behaviour and, 21-23
5-HT$_{1A}$ agonist(s)
anxiety and, 371-374, 372*t*
effect of
on defensive patterns in wild rattus, 145-147
on food intake, 283-284

modulation of, in anxiety, 342-344
on ultrasonic cries, 363*t*
ejaculatory response and, 79-81, 80*t*
experimentally induced anxiety and, 338, 339*f*, 340-342, 341*f*, 342*f*
fear and, 367-369
reversal by, 226*t*
stress-induced depression and, 346-347, 348*f*, 349
5-HT$_{1A}$ receptor(s), functional models to, 5, 6*f*, 7-8, 8*t*, 9*f*, 10, 10*f*, 11*t*
5-HT$_{1B}$, rotational behaviour in rats and, 21-23
5-HT$_{1B}$ agonists
effect of
on food intake, 284
on ultrasonic cries, 363*t*
ejaculatory response and, 79-81, 80*t*
5-HT$_{1C}$ agonist(s), food intake and, 284
5-HT$_{1C}$-phosphoinositol, production of, in choroid plexus of pigs, 93
5-HT$_2$, lordosis behaviour and, 58
5-HT$_2$ agonist(s), ultrasonic cries and, 363*t*
5-HT$_2$ antagonist(s)
effect of
on habituation deficit in schizophrenia, 243, 244*t*, 245
on ultrasonic cries, 353*t*
startle habituation and, 243, 244*t*, 245
as thymosthenics, 235-237
5-HT$_2$ receptor(s)
antidepressants and, 183
anxiety and, 181-183, 180*f*, 181*f*, 182*f*
characteristics of, 11*t*
fear and, 181-183, 180*f*, 181*f*, 182*f*
5-HT$_1$ opposed function of, 193, 193*t*
models of function, 11-13, 12*f*
systems involving, 198*t*
5-HT$_3$
anxiety and, 353-356
lordosis behaviour and, 58-59
5-HT$_3$ antagonist(s)
aggression and, 390, 391, 392*t*
anxiety and
anxiolytic activity, 391
anxiolytic tests, 389-390
models of, 383-385, 386*f*, 387
effects of
anxiolytic, 355*f*
on ultrasonic cries, 363*t*
psychosis and, 239-241, 240*f*

antipsychotic activity, 390–391
antipsychotic test, 389
5-HTP
  activity of, 216t
  effects of, 10f
Human feeding. See Feeding behaviour
Hunger
  control of, 259
  vs. satiety, 261–262
5-Hydroxytryptamine$_{1A}$ agonists, stimulus properties of, 453–456, 455f
5-Hydroxytryptophan, effects of, 10f
Hyperactivity, 5-HT$_3$ inhibition of, 239–241, 240f
Hypoactivity, serotonin receptor subtypes and, 84, 85t
Hypothalamically-induced aggression, 100–101
Hypothermia, serotonin receptor subtypes and, 84, 85t

ICS 205-930
  activity of, in anxiety models, 386f
  effect of
    on aversive brain stimulation, 324f
    on burying and grooming, 327t
    on hyperactivity, 240f
  lordosis behaviour and, 58
Imipramine
  effect of
    on burying and grooming, 326t
    on postsynaptic responses at 5-HT synapses, 165t
^{3}H-Imipramine, in frontal cortex of suicides, 185, 185t
Indalpine
  effect of, on burying and grooming, 326t
  reversal of helpless behaviour by, 347, 348f
Indorenate, effect of, on cumulative burying behaviour, 379t
Ingestion, palatability-induced, 287–288. See also Feeding behaviour
Invertebrate(s), serotonin influences in, 122–123
Ipsapirone
  activity of, 216t
  affinities of, 408t, 414t, 418t
  chronic administration effects of, 14t
  discriminative stimulus effects of, 411t
  in drug discrimination studies, 410t

drug-appropriate responding by, 419t, 420t
ED$_{50}$-values of, in isolation-induced aggression, 95t
effect of
  on aggression, summary of, 107t
  on behaviour in elevated plus-maze, 372t
  on competition-rate in male rats, 150f
  on cumulative burying behaviour, 379f
  on intermale aggression, 97f, 97
  on muricide in rats, 105t, 106
  on ultrasonic cries, 363t
  on in vitro 5-HT release, 94t
  on in vitro PI hydrolysis, 94t
  on in vivo 5-HTP formation, 94t
  on regional forebrain dopac concentrations, 232f
head twitch and, 15f
stimulus generalization and, 446t
Isolation. See Social isolation
Isolation-induced aggression, 93–98, 118f

Ketanserin
  activity of, 215t
  in antagonism tests, 412t
  effect of
    on aversive brain stimulation, 324f
    on startle habituation, 244t, 245
  rotational behaviour in rats and, 21
^{3}H-Ketanserin, in frontal cortex of suicides, 185, 185t
Killing, 5-HT and, 119–122

Leeches, serotonin influences in, 122
Light-dark test, 391, 393t
Liquid diet(s), 291–293, 292t. See also Feeding behaviour
Lisuride
  effect of
    catecholaminergic antagonists and, 39–40
    on male rat sexual behaviour, 38–39
    serotonergic antagonists and, 39–40
  pharmacological profile of, 46
Lobster(s), serotonin influences in, 122–123
Locomotor activity, serotonin receptor subtypes and, 84, 85t
Lordosis
  cinanserin and, 57

SUBJECT INDEX

cyproheptadine and, 57–58
DMT and, 60–61
DOM and, 60
DOMPE and, 60
5-HT receptor antagonists and, 56–59
5-HT receptor subtypes and, 56–63
5-HT$_{1A}$ and, 62
5-HT$_2$ antagonists and, 58
5-HT$_3$ and, 58–59
5-MeODMT and, 60–61
ICS 205-930 and, 58
LSD and, 59–60
LY53857 and, 58
mescaline and, 60
metergoline and, 57
methiothepin and, 57–58
methysergide and, 56–57
mianserin and, 57–58
MK 212 and, 62
phenylalkylamines and, 60
pizotefin and, 57–58
psilocybin and, 60–61
quipazine and, 61–62
ritanserin and, 58
serotonin receptor agonists and, 59–63
TFMPP and, 62–63
TMA and, 60
tryptamine derivatives and, 60–61
Lower lip retraction, serotonin receptor subtypes and, 84, 85t
Loxapine, activity of, 213t
LSD
  activity of, 216t
  affinities of, 408t, 414t, 416t, 418t
  in antagonism tests, 412t
  in drug discrimination studies, 410t, 411t
  drug-appropriate responding by, 419t, 420t
  lordosis behaviour and, 59–60
  as neurotransmitter psychotogen in schizophrenia, 205t
LTP. *See* L-Tryptophan
LY 141865. *See* Quinpirole
LY53857, lordosis behaviour and, 58

Macaca arctoides, serotonin influences on, 125
Macaca mulatta, serotonin influences on, 125
Male rats. *See* Rats, male
MAO-A inhibitors, effects of, on postsynaptic responses at 5-HT synapses, 165t
Maprotyline, effect of, on burying and grooming in mice, 326t
Maternal aggression
  effects of serotonergic drugs on, 104
  5-HT$_3$ antagonists and, 394
  in rats, 101–102
mCPP
  effect of
    on burying and grooming in mice, 326t
    on competition-rate in male rats, 150f
  penile erections and, 84, 85, 85t
MDL 72222
  activity of, 215t
  ED$_{50}$-values of, in isolation-induced aggression, 95t
  effect of
    on aggression, summary of, 107t
    on aversive brain stimulation, 324t
    on maternal aggression in rats, 102, 102f
    on in vitro 5-HT release, 94t
    on in vitro PI hydrolysis, 94t
    on in vivo 5-HTP formation, 94t
  test results for
    in antagonism of apomorphine-induced climbing, 390–391, 392t
    in conditioned avoidance response, 390, 392t
    in isolation-induced aggression in mice, 391, 393t
5-MeODMT, 8t
  activity of, 216t
  affinities of, 408t, 414t, 416t, 418t
  in drug discrimination studies, 410t
  drug-appropriate responding by, 419t, 420t
  ED$_{50}$-values of, in isolation-induced aggression, 95t
  effect of
    on aggression, summary of, 107t
    on aversive threshold, 428f
    on burying and grooming in mice, 327t
    on competition-rate in male rats, 150f
    on muricide in rats, 105t, 106
    on resident-intruder aggression, 98, 99f
    on in vitro 5-HT release, 94t
    on in vitro PI hydrolysis, 94t

on in vivo 5-HTP formation, 94t
head twitch and, 15f
lordosis and, 60–61
Mescaline
lordosis behaviour and, 60
as neurotransmitter psychotogen in schizophrenia, 205t
Metergoline, 8t
activity of, 215t
effect of
on aversive brain stimulation, 324f
on DF anorexia, 265t, 266
on food intake, 261f
lordosis behaviour and, 57
responses to LTP, 191–192, 192f
Methiothepin, 8t
activity of, 215t
effects of, 6f
lordosis behaviour and, 57–58
5-Methoxy-dimethyl-tryptamine (5-MeODMT). See 5-MeODMT
Methysergide, 8t
activity of, 215t
ED_{50}-values of
insolation-induced aggression, 95t
effect of, 6f
on aggression, summary of, 107t
on attack bites, 130f
on lactating female mouse aggression, 129f
on maternal aggression in rats, 102, 102f
on muricide in rats, 105t, 106
on sideways threats, 130f
on ultrasonic cries, 363t
on in vitro 5-HT release, 94t
on in vitro PI hydrolysis, 94t
on in vivo 5-HTP formation, 94t
on walking duration in confrontations, 130f
lordosis behaviour and, 56–57
Mianserine, 8t
activity of, 215t
ED_{50}-values of
in isolation-induced aggression, 95t
effect of
on aggression, summary of, 107t
on aversive brain stimulation, 324f
on burying and grooming in mice, 326t
on maternal aggression in rats, 102, 102f

on muricide in rats, 105t, 106
on in vitro 5-HT release, 94t
on in vitro PI hydrolysis, 94t
on in vivo 5-HTP formation, 94t
lordosis behaviour and, 57–58
Mice
apomorphine-induced climbing in, 390–391, 392t
foot-shock-induced defensive aggression in, 103
isolation-induced aggression in, 93–98
lactating female, aggression toward male in, 128, 129f
serotonin influences in, 118f, 123
Miopithecus talapoin, serotonin influences on, 125, 126f
MK 212, lordosis and, 62
Molindone, activity of, 213t
Monkey(s), serotonin influences in, 125
Monoamine inhibitors, in CAR procedure, 211t
Mood, effects of 5-HT on, 161
Morphine
in CAR procedure, 211t
shock intensity and, 210, 210t as neurotransmitter psychotogen in schizophrenia, 205t
Mouse-killing by rats. See Muricide
Muricide, 105–106, 105t

Naloxone
in CAR procedure, 211t
schizophrenia and, 205t
Neuroleptics
clinically effective
activity of, 213t
correlation coefficients for, 214t
schizophrenia and, 205t
Ninaprine, effect of, on burying and grooming in mice, 326t
Nociception, behavioural tests of, 301–304
Noradrenaline
in CAR procedure, 211t
in schizophrenia, 205t
Nortriptyline, effect of, on burying and grooming in mice, 326t
Nutrient selection, 263–264

Obesity. See Eating disorders
Obsessive-compulsive disorders (OCD), 328
OCD. See Obsessive-compulsive disorders

SUBJECT INDEX

Offensive aggression, serotonergic drugs and, 93–102
8-OH-DPAT
  activity of, 216t
  affinities of, 408t, 414t, 416t, 418t
  in antagonism tests, 412t
  antidepressant-like action of, 227
  behavioural response to, 9f
  bromo-lisuride and, 41, 42f
  chronic administration of, effect of, 14t
  circling behaviour and, 22
  discriminative stimulus effects of, 411t
  in drug discrimination studies, 410t
  drug-appropriate responding by, 419t, 420t
  ED$_{50}$-values of, in isolation-induced aggression, 95t
  effect of
    on aggression, summary of, 107t
    attenuation of, 227
    on behaviour of rats in elevated plus-maze, 372t
    on burying and grooming in mice, 327t
    catecholaminergic antagonists and, 39–40
    on consumption of various diets, 291–293, 292t
    on cumulative burying behaviour, 379t
    on food-deprived rats, 267f
    on free-feeding rats, 267f
    on hypothalamically-induced aggression, 100f, 101
    on intermale aggression, 96f, 97
    on male rat sexual behaviour, 36–39, 38f, 43–44, 45f
    on maternal aggression in rats, 101–102, 102f, 104f
    on muricide in rats, 105t, 106
    serotonergic antagonists and, 39–40
    on ultrasonic cries, 363t
    on in vitro 5-HT release, 94t
    on in vitro PI hydrolysis, 94t
    on in vivo 5-HTP formation, 94t
    on waiting capacity for food reward, 344f, 345
  as 5-HT antagonist, 40–41
  forepaw treading and, 84, 85, 85t
  head shakes and, 84, 85, 85t
  head twitch and, 15f
  hyperphagic response to, 227
  hypoactivity and 84, 85t
  hypothermia and, 84, 85t
  lack of effect of pretreatment with, 227
  lower lip retraction and, 84, 85t
  mechanism of action of, 225–227, 226t
  penile erections and, 84, 85, 85t
  pharmacological profile of, 46
    comparison of, 41, 43
  reversal of helpless behaviour by, 347, 348f
  selective actions of, in limbic brain areas, 44, 46
  stimulus generalization and, 446t
[^{3}H]8-OH-DPAT binding, 22, 23f
Opiate(s)
  in CAR procedure, 211t
  in schizophrenia, 205t
Opiate analgesia, serotonergic involvement in, 310–312
Org 6997, effect of, on burying and grooming in mice, 326t
Overeating. See Eating disorders
Oxazepam, effects of, on ultrasonic cries, 363t

Pain, 5-HT and, 299–300, 301–312
  anatomical considerations, 304
  behavioural tests of nociception, 301–304
  depletion of serotonin by p-Cl-phenylalanine, 304
  lesions of ascending pathways, 305
  lesions of descending pathways, 305–306
  noradrenergic involvement in serotonergic antinociception, 310
  precursors effects, 306–310
  receptor agonists effects, 306–310
  receptor antagonists effects, 304–305
  releasing agents effects, 306–310
  serotonergic involvement in opiate analgesia, 310–312
Pain tests, effects of serotonin receptor agonists in, 311t
Palatability-induced ingestion, 287–288. See also Feeding behaviour
Panic disorders, 327–328
Paroxetine, effect of, on burying and grooming in mice, 326t
PCP, as neurotransmitter psychotogen in schizophrenia, 205t
PCPS, effects of, on lactating female

mouse aggression, 129f
Penile erections, serotonin receptor
    subtypes and, 84, 85t
Peptides, as neurotransmitter psychotogen
    in schizophrenia, 205t
Pergolide. See Quinpirole
Phenobarbital, in CAR procedure, 211t
p-Cl-Phenylalanine, depletion of serotonin
    by, 304
Pirenperone
    aversive brain stimulation and, 324f
    rotational behaviour in rats and, 21
Pizotefin, lordosis behaviour and, 57–58
Plasma LH
    effect of steroid treatment on, 74, 75t
Posttraumatic stress disorder, 329
1-PP, stimulus generalization and, 446t
3-PPP
    pharmacological profile of, 46
    schizophrenia and, 205t
Prazosine, schizophrenia and, 205t
Predatory aggression, 105–106, 105t, 121t
Primates, non-human, serotonin influences in, 125
Proadifen, effects of, on water-lick conflict test, 341, 342f
Prochloroperazine, activity of, 213t
Propranolol, 8t
    activity of, 215t
    isolation-induced aggression and, 95t
    effects of, 6f
        on aggression, summary of, 107t
        on aversive threshold, 432f
        in CAR procedure, 211t
        on head twitch behaviour, 12f
        on maternal aggression in rats, 102, 102f
        on muricide in rats, 105t, 106
    schizophrenia and, 205t
Psilocybin, lordosis and, 60–61
Psychosis
    5-HT system and, 205–217
    5-HT$_3$ antagonists and, 239–241, 240f, 398–391, 392–393t, 394
    depression and, 153–155

Quinpirole, effects of, 39–40
Quipazine, 8t
    activity of, 216t
    affinities of, 408t, 414t, 416t, 418t
    drug-appropriate responding by, 419t
    ED$_{50}$-values of, in isolation-induced aggression, 95t
    effect of
        on aggression, summary of, 107t
        on competition-rate in male rats, 150f
        on hypothalamically-induced aggression, 100f, 101
        on maternal aggression in rats, 102, 102f, 104f
        on muricide in rats, 105t, 106
        on in vitro 5-HT release, 94t
        on in vitro PI hydrolysis, 94t
        on in vivo 5-HTP formation, 94t
    lordosis and, 61–62

R 55 667. See Ritanserin
Raclopride, ejaculatory response and, 80t
Rats
    food-deprived, 267f
    foot-shock-induced defensive aggression in, 103
    free-feeding, 267f
    hypothalamically-induced aggression in, 100f, 100–101
    maternal aggression in, 92t, 101–102, 102f, 104f
        effects of serotonergic drugs on, 104f
    resident-intruder aggression in, 98–99, 99f
    rotational behaviour in, 21–23
Rats, female
    lordosis behaviour in
        cinanserin and, 57
        cyproheptadine and, 57–58
        5-HT receptor antagonists and, 56–59
        5-HT receptor subtypes and, 56–63
        metergoline and, 57
        methiothepin and, 57–58
        methysergide and, 56–57
        mianserin and, 57–58
        pizotefin and, 57–58
    sexual behaviour in, steroids and, 73–74, 74, 75t
Rats, male
    sexual behaviour in, 35–36
        bromo-lisuride and, 41, 42f
        drug-induced changes in, 37f
        8-OH-DPAT and, 36–38, 38f, 43–44, 45f
        ergot drugs and, 38–39
        lisuride and, 38–39
        quinpirole and, 39
        serotonergic antagonists and, 39–40

SUBJECT INDEX

serotonergic modulation of, 35–47
sexual behaviour in, 35–36
  normal, 36f
Reactivity, 8t
Resident-intruder aggression in rats, 98–99
Rhesus macaques, serotonin influences in, 125
Rimcazole, schizophrenia and, 205t
Ritanserin
  activity of, 215t
  chemical structure of, 235f
  clinical findings with, 236–237
  description of, 235, 237
  dose-related anxiolytic effect of, 180, 180f
    geometric mean spontaneous fluctuations, 182, 182f
    S.C.R. spontaneous fluctuations, 181–182, 181f
  ED_{50}-values of, in isolation-induced aggression, 95t
  effects of, 94t
    on aggression, summary of, 107t
    anxiolytic, dose-related, 180, 180f
    on aversive threshold, 430f, 432f
    on DF anorexia, 265t, 266
    on head twitch behaviour, 12f
    on maternal aggression in rats, 102, 102f
    on muricide in rats, 105t, 106
    on startle habituation, 244t, 245
    on ultrasonic cries, 363t
    on in vitro PI hydrolysis, 94t
    on in vitro 5-HT release, 94t
    on in vivo 5-HTP formation, 94t
    on waiting capacity for food reward, 344f, 345
  lordosis behaviour and, 58
  neurophysiological effects of, 236
  pharmacology of, 235–236
  psychopharmacology effects of, 236
  responses to LTP, 192, 192f
  tridimensional structure of, 235f
Ro-15-1788, effects of, in CAR procedure, 211t
Rotational behaviour, in rats, 21–23
RSD-127, pharmacological profile of, 41, 43
RU 24969
  activity of, 216t
  affinities of, 408t, 414t, 418t
  drug-appropriate responding by, 419t

ED_{50}-values of, in isolation-induced aggression, 95t
effect of
  on aggression, summary of, 107t
  on burying and grooming in mice, 327t
  on competition-rate in male rats, 150f
  on cumulative burying behaviour, 379f
  on intermale aggression, 96f, 97
  on maternal aggression in rats, 102, 102f, 104f
  on muricide in rats, 105t, 106
  on ultrasonic cries, 363t
  on in vitro 5-HT release, 94t
  on in vitro PI hydrolysis, 94t
  on in vivo 5-HTP formation, 94t

Saimiri sciureus, serotonin influences on, 125
SCH23390, schizophrenia and, 205t
Schizophrenia
  5-HT and, 206
  5-HT hypothesis of, 206–207
  5-HT system and, 205–217
  altered noradrenergic transmission and, 206
  dopamine theory of, 206
  habituation deficit in
    $5-HT_2$ antagonists and, 243, 245
    effects of serotonin-2 antagonists on, 244t
  neurotransmitters involved in, 205–206, 205t
  treatment of
    development of new drugs for, 207–208
Senktide
  behavioural effects of, 27
  central nervous system and, 30
  ratio of 5-HIAA to 5-HT and, 28–29, 29f
Serotonergic antagonists, effects of, on effects produced by 8-OH-DPAT, 39–40
Serotonergic anticociception,
  noradrenergic involvement in, 310
Serotonergic compounds, effect of, 94t
Serotonergic drugs
  ED_{50}-values of, in isolated-induced aggression, 95t
  effects of

on maternal aggression in rats, 101–102, 102f, 104f
on predatory aggression, 105t
on sexual behaviour, 33–34
Serotonergic dysfunction, in depression, 157–168, 167t
Serotonergic system, central manipulations and, 268–269
Serotonin. See also 5-HT
  affinity of, 92t
  aggressive behaviour and, 106–108
  anxiety and, 345–346
    aversion and, 397–403, 399f, 402f
    discussion on, 433–437
    intracerebral drug injection, 428–433, 428f, 430f, 432f
    systemic drug injection, 426f, 427–428
  background of, 1–2
  behaviour mediated by, 27–30
  as bridge between anxiety and depression, 337–349
  defensive aggression and, 103–105
  depression and, 153
  effects of
    on aversive threshold, 428f
    in CAR procedure, 211t
    in different pain tests, 311t
    on startle habituation, 244t
  feeding behaviour and, 247–248
    control of nutrietn selection, 249–250
    drugs enhancing neurotransmission, 250–254
    drugs reducing neurotransmission, 254–255
    in humans, 249–255
  influence of, 118f, 122–126
  modulatory action of, 89–108
  as neurotransmitter psychotogen in schizophrenia, 205t
  offensive aggression and, 93–102
  pain and, 301–312
  predatory aggression and, 105–106
  receptor agonists, lordosis behaviour and, 59–63
  receptor subtypes
    functional interplay of, 83–86, 85t
    synthesis of, 83
  rotational behaviour in rats and, 21–23
  tolerance to reward delay and, 344f, 345
  uptake inhibitors, reversal of helpless behaviour by, 231–233, 232f
Sertraline, effect of, on burying and grooming in mice, 326t
Sexual behaviour, in rats
  bromo-lisuride and, 41, 42f
  drug-induced changes in, 37f
  8-OH-DPAT and, 36–38, 38f
    nature of effects of, 43–44, 45f
  ergot drugs and, 38–39
  5-HT and, 33–34
  lisuride and, 38–39
  normal, 36f
  quinpirole and, 38–39
  serotonergic antagonists and, 39–40
  serotonergic modulation of, 35–47
  steroid treatment and, 74, 75t
Shock-induced aggression, 120f
Shuttle box sessions, escape deficits, 347, 348f
Social interaction, anxiety disorders and, 323–324
Social isolation, depression and, 164–165
Social phobia, 329
Solid diet(s), 291–293, 292t. See also Feeding behaviour
Spiroxatrine
  in antagonism tests, 412t
  behavioural effects of, 449–452
  discriminative stimulus effects of, 449–452
  neurochemical effects of, 449–452
Squirrel monkeys
  CSF monoamine metabolite concentrations in, 127t
  serotonin influences in, 125
Startle habituation, 5-HT$_2$ antagonists and, 243, 244t, 245
Steroids, effect of, on female rat sexual behaviour, 73–74, 74, 75t
Stress, depression and, 163–164
Stumptail macaques, serotonin influences in, 125
Suicide(s)
  depressed attempters of, CSF 5-HIAA in, 159t
  substances in frontal cortex of, 185, 185t
Sulpiride, activity of, 213t

NK-3 Tachykinin receptors, activation of, 28
Talapoin monkeys, serotonin influences

in, 125, 126f
TFMPP
  5-HT$_{1A}$ binding site affinity of, 414t
  5-HT$_{1B}$ binding site affinity of, 416t
  5-HT$_2$ binding site affinity of, 418t
  activity of, 216t
  affinities of
    at 5-HT binding subtypes, 408t
    in antagonism tests, 412t
  discriminative stimulus effects of, 411t
  in drug discrimination studies, 410t
    drug-appropriate responding by, 419t
  ED$_{50}$-values of
    in isolation-induced aggression, 95t
  effects of
    on aggression, summary of, 107t
    on competition-rate in male rats, 150f
    on intermale aggression, 96f, 97
    on maternal aggression in rats, 102, 102f, 104f
    on muricide in rats, 105t, 106
    on resident-intruder aggression, 98, 99f
    on ultrasonic cries, 363t
    on in vitro 5-HT release, 94t
    on in vitro PI hydrolysis, 94t
    on in vivo 5-HTP formation, 94t
  lordosis and, 62–63
  stimulus generalizations and, 446t
Thioridazine, activity of, 213t
Thymosthenics, 5-HT$_2$ antagonists as, 235–237
TMA, lordosis behaviour and, 60
Trans-dihydro-lisuride, pharmacological profile of, 46
Tranylcypromine, 6f
Trazodone, effect of, on burying and grooming in mice, 326t
Tree-shrews, serotonin influences in, 123–124, 124f
TRH-analogue, schizophrenia and, 205t
Tricyclics, effects of, on postsynaptic responses at 5-HT synapses, 165t

1-(3-Trifluoromethylphenyl)piperazine (TFMPP). See TFMPP
Trifluperazine, activity of, 213t
2,4,5-Trimethoxyamphetamine (TMA). See TMA
Tryptophan, 6f
  effects of, on feeding behaviour, 250–251
  hormonal responses to, in depression, 189–192, 190f, 191f, 192f
Tupaia belangeri, serotonin influences on, 123–124, 124t

UCS. See Unconditioned stimulus
Ultrasonic calling
  5-HT and, 361–365, 363t
  test results, 391, 393t
Unconditioned stimulus (UCS), effect of shock strength variation, 209f

Vervet monkeys, serotonin influences in, 125
Vocalisation, ultrasonic, 361–365, 363t

WB-4101, competition-rate in male rats and, 150f
Wild rattus,
  defensive patterns in, 145–147

Xylamidine, effects of, on food intake, 261f

Yohimbine
  in CAR procedure, 211t
  in schizophrenia, 205t

Zimeldine, 9f, 11t
  effects of
    on aversive threshold, 430f, 432f
    on burying and grooming in mice, 326t
    on postsynaptic responses at 5-HT synapses, 165t
    on ultrasonic cries, 363t